Maxwell on Molecules and Gases

Maxwell on Molecules and Gases

Edited by
Elizabeth Garber,
Stephen G. Brush,
and
C. W. F. Everitt

The MIT Press
Cambridge, Massachusetts
London, England

Publication of this volume was supported by a grant from the General Research Board of the University of Maryland.

This book was set in Times Roman by Asco Trade Typesetting Ltd., Hong Kong, and printed and bound by Halliday Lithograph in the United States of America.

Library of Congress Cataloging-in-Publication Data

Maxwell, James Clerk, 1831–1879.
  Maxwell on molecules and gases.

  Includes index.
  1. Maxwell, James Clerk, 1831–1879—Correspondence. 2. Maxwell, James Clerk, 1831–1879—Manuscripts. 3. Gases, Kinetic theory of. 4. Molecules. 5. Atoms. 6. Statistical physics. I. Garber, Elizabeth. II. Brush, Stephen G. III. Everitt, C. W. F. (C. W. Francis), 1934–   . IV. Title.
QC16.M4A4   1986   533′.2   85-18838
ISBN 0-262-07094-4

Contents

QC 16
M4
A41
1986
PHYS

List of Serial Abbreviations   *xiii*

Preface   *xvii*

I Kinetic Theory and the Properties of Gases: Maxwell's Work in Its Nineteenth-Century Context   *1*

II Documents on Atomic and Statistical Physics   *65*

1

"On the Properties of Matter"   *65*

2

Lecture at King's College London, 1860   *68*

3

Letter from John Herapath to Maxwell, February 23, 1864   *80*

4

Letter from Maxwell to Unknown Correspondent, February 7, 1866   *82*

5

[Discussion Remarks on Brodie's Theory at Chemical Society Meeting, 1867]   *84*

6

Letter from Maxwell to Peter Guthrie Tait, November 13, 1867   *86*

7

Letter from Maxwell to Peter Guthrie Tait, 1867?   *87*

8

[Notes for British Association Address, 1870]   88

9

"Address to the Mathematical and Physical Sections of the British Association," September 15, 1870   89

10

Letter from Maxwell to William Thomson, 1871   105

11

"Introductory Lecture on Experimental Physics," Cambridge, October 1871   110

12

Challis's "Mathematical Principles of Physics"   126

13

A. J. Mott, "Atoms and Ether"   131

14

"Atoms and Ether" [reply to Mott]   131

15

[Notes for "Molecules" Article]   133

16

"Molecules"   137

17

Letter from Herbert Spencer to Maxwell, December 4, 1873   156

18

Letter from Maxwell to Herbert Spencer, December 5, 1873   158

19

Letter from Maxwell to Herbert Spencer, December 17, 1873   159

20

Letter from Herbert Spencer to Maxwell, December 30, 1873   160

21

"Plateau on Soap Bubbles"   162

22

[Draft of "Atom" Article for the Encyclopedia Britannica]   *170*

23

"Atom"   *175*

24

"On the Dynamical Evidence of the Molecular Constitution of Bodies"   *216*

25

Postcard from Maxwell to Peter Guthrie Tait, March 19, 1875   *238*

26

Letter from Maxwell to George Gabriel Stokes, September 25, 1875   *239*

27

Letter from the Right Rev. C. J. Ellicott, D. D., Lord Bishop of Gloucester and Bristol, to Maxwell, November 21, 1876   *241*

28

Letter from Maxwell to the Right Rev. C. J. Ellicott, D. D., Lord Bishop of Gloucester and Bristol, November 22, 1876   *242*

29

Letter from Samuel Tolver Preston to Maxwell, December 5, 1876   *243*

30

"Constitution of Bodies"   *245*

31

Letter from Maxwell to Peter Guthrie Tait, August 7, 1878   *255*

32

[Notes for a Lecture on Kinetic Theory]   *257*

33

[Fragment on Statistical Regularity]   *260*

34

"Gases, Kinetic Theory of" [Draft of Editorial Note for Papers of Henry Cavendish]   *262*

35
"On the Molecular Constitution of Air"   265

36
"Physical Sciences"   267

37
Letter from Maxwell to Simon Newcomb, May 31 and June 2, 1879   272

III   Documents on the Kinetic Theory of Gases   277

1
Letter from Maxwell to George Gabriel Stokes, May 30, 1859   277

2
"On the Dynamical Theory of Gases"   281

3
Letter from Maxwell to George Gabriel Stokes, October 8, 1859   282

4
[Notes for "Illustrations of the Dynamical Theory of Gases"]   283

5
[Notes on Graham's Diffusion Data]   284

6
"Illustrations of the Dynamical Theory of Gases"   285

7
Letter from Maxwell to Lewis Campbell, January 5, 1860   319

8
"On the Results of Bernoulli's Theory of Gases as Applied to Their Internal Friction, Their Diffusion, and Their Conductivity for Heat"   320

9
[Miscellaneous Problems]   321

10
Letter from Maxwell to H. R. Droop, January 28, 1862   336

11

Letter from Maxwell to Lewis Campbell, April 21, 1862   337

12

"On the Conduction of Heat in Gases"   339

13

Letter from Maxwell to C. H. Cay, January 5, 1865   349

14

Letter from Maxwell to Thomas Graham, May 1, 1865   351

15

Letter from Maxwell to Peter Guthrie Tait, June 17, 1865   352

16

Letter from Maxwell to H. R. Droop, July 19, 1865   353

17

*"On the Viscosity or Internal Friction of Air and other Gases"
[Abstract]   354*

18

*"On the Viscosity or Internal Friction of Air and other Gases," Bakerian
Lecture, February 8, 1866   359*

19

*[Notes for the Section "On the Mutual Action of Two Molecules" in "On the
Dynamical Theory of Gases" (1867 Memoir)]   387*

20

"To Find the Equations of Motion . . ." [Notes for 1867 Memoir]   391

21

*"Law of Volumes" [Notes for Section "The Equilibrium of Temperature
Between Two Gases" (1867 Memoir)]   392*

22

*[Draft of Sections "Specific Heat of Unit of Mass at Constant Volume" and
"Specific Heat of Unit of Mass at Constant Pressure" (1867 Memoir)]
395*

23

[Notes for Section "Determination of the Inequality of Pressure . . ." and Part of Following Section on Viscosity for 1867 Memoir]   396

24

"Encounter of Two Molecules"   398

25

[Heat Conduction for the Case of Varying Temperature and Pressure]   404

26

[Miscellaneous Notes and Calculations]   405

27

Letter from Maxwell to Peter Guthrie Tait, April 4, 1866   413

28

Letter from Peter Guthrie Tait to Maxwell, April 6, 1866   414

29

"On the Dynamical Theory of Gases" [Abstract]   415

✓ # 30

"On the Dynamical Theory of Gases"   419

31

Letter from Maxwell to Peter Guthrie Tait, March 12, 1868   473

32

Letter from Maxwell to Lewis Campbell, October 19, 1872   475

33

[Notes on Boltzmann's H Theorem]   476

34

[Draft of "On Loschmidt's Experiments on Diffusion in Relation to the Kinetic Theory of Gases"]   477

35

"N August 1, 1873" [Calculations]   483

36
[Calculations for the Number of Molecules of Hydrogen in Unit Volume]   486

37
[Comparison of Data on Diffusion]   488

38
[Notes on Diffusion]   491

39
[Notes on Viscosity]   492

40
Postcard from Maxwell to Peter Guthrie Tait, Summer 1873   493

41
Postcard from Maxwell to Peter Guthrie Tait, Summer 1873 [?]   495

42
"On Loschmidt's Experiments on Diffusion in Relation to the Kinetic Theory of Gases"   496

43
Letter from Maxwell to John William Strutt, Third Baron Rayleigh, August 28, 1873   505

44
Letter from Alexander Crum Brown to Maxwell, September 4, 1873   506

45
Letter from Maxwell to Peter Guthrie Tait, February 14, 1876   508

46
"Diffusion of Gases through Absorbing Substances"   510

47
Letter from Maxwell to William Garnett, June 30, 1877   515

48
[Notes on Diffusion]   516

49
"Quiet Diffusion July 3, 1877"   518

50
"On Steady Diffusion"   518

51
[Notes on the Diffusion of Gases]   522

52
"Diffusion"   524

53
"Liquid Vapour Mixture"   547

54
[Notes on the Evaporation of Liquids]   547

55
[Notes on Diffusion and Transpiration]   549

Index   551

List of Serial Abbreviations

Abh. Boehm. Ges. Wiss.	Abhandlungen der Königlich Boehmischen Gesellschaft der Wissenschaften, Prague
Amer. J. Phys.	American Journal of Physics
Amer. J. Sci.	American Journal of Science
Amer. Sci.	American Scientist
Ann. Chim.	Annales de Chimie
Ann. Chim. Phys.	Annales de Chimie et de Physique
Ann. Phil.	Annals of Philosophy
Ann. Phys.	Annalen der Physik
Ann. Sci.	Annals of Science
Arch. Hist. Exact Sci.	Archive for History of Exact Sciences
Arch. Int. Hist. Sci.	Archives Internationales d'Histoire des Sciences
Berlin Akad. Monats.	Monatsberichte der Königlich Preussischen Akademie der Wissenschaften zu Berlin
British J. Hist. Sci.	British Journal for the History of Science
Cambridge Dublin Math. J.	Cambridge and Dublin Mathematical Journal
Chem. Rev.	Chemical Reviews
Edinburgh J. Sci.	Edinburgh Journal of Science
Edinburgh Rev.	Edinburgh Review
Hist. Sci.	History of Science
Hist. Studies Phys. Sci.	Historical Studies in the Physical Sciences
J. Chem. Educ.	Journal of Chemical Education
J. Chem. Soc.	Journal of the Chemical Society (London)

J. Chem. Soc. London	Journal of the Chemical Society, London
J. Hist. Ideas	Journal of the History of Ideas
J. Math. Pures Appl.	Journal de Mathématiques Pures et Appliquées
J. Rational Mech. Anal.	Journal of Rational Mechanics and Analysis
J. Reine Angew. Math.	Journal für die reine und angewandte Mathematik
J. Res. Nat. Bur. Stand.	Journal of Research of the National Bureau of Standards
Mem. Acad. Sci. Paris	Mémoires de l'Académie des Sciences, Paris
Mem. Manchester Lit. Phil. Soc.	Memoirs of the Manchester Literary and Philosophical Society
Monats. Preuss. Akad., Berlin	Monatsberichte der k. Preussischen Akademie, Berlin
Notes Records R. Soc. London	Notes and Records of the Royal Society of London
Phil. J.	Philosophical Journal
Phil. Mag.	Philosophical Magazine
Phil. Trans. R. Soc. London	Philosophical Transactions of the Royal Society of London
Phys. Blaetter	Physikalische Blaetter
Phys. Rev.	Physical Review
Phys. Z.	Physikalische Zeitschrift
Proc. Manchester Lit. Phil. Soc.	Proceedings of the Manchester Literary and Philosophical Society
Proc. Oklahoma Acad. Sci.	Proceedings of the Oklahoma Academy of Science
Proc. R. Soc. Edinburgh	Proceedings of the Royal Society of Edinburgh
Proc. R. Soc. London	Proceedings of the Royal Society of London
Proc. Sec. Sci., K. Akad. Wet. Amsterdam	Proceedings of the Section of Sciences, K. Akademie van Wetenschappen te Amsterdam
Quart . J. Math.	Quarterly Journal of Mathematics
Quart. J. Sci.	Quarterly Journal of Science
Railway Mag.	Railway Magazine
Rep. (no.) Meeting BAAS	Report of the (no.) Meeting of the British Association for the Advancement of Science
Sitz. Math.-Naturwiss. Cl. Akad.	Sitzungsberichte der

Wiss., Wien	Mathematische-Naturwissenschaftliche Classe der kaiserlichen Akademie der Wissenschaften, Wien
Sov. Phys. Tech. Phys.	Soviet Physics, Technical Physics
Studies in Hist. Phil. Sci.	Studies in History and Philosophy of Science
Trans. Cambridge Phil. Soc.	Transactions of the Cambridge Philosophical Society
Trans. R. Soc. Edinburgh	Transactions of the Royal Society of Edinburgh
Z. Phys.	Zeitschrift für Physik

Preface

The major achievements of James Clerk Maxwell's scientific career are generally acknowledged to be his theory of electromagnetic fields and waves and his kinetic theory of gases (with its generalization, statistical mechanics). Although the range of Maxwell's work in kinetic theory is somewhat narrower, its long-term influence on scientific thought is in many ways just as profound, and some of the problems Maxwell wrestled with remained matters for debate—occasionally fierce debate—over the succeeding fifty or more years. Thus the letters and manuscripts published in this volume (and in our next) reveal a great deal about nineteenth century science. But to understand them, one has to grasp the general outline of the development of kinetic theory and see the relation of Maxwell's work to that of Clausius and Boltzmann in Germany, Gibbs in the United States, and Maxwell's British friends and colleagues, Tait, William and James Thomson, Stokes, Andrews, Reynolds, and Watson.

On the surface Maxwell's kinetic theory papers represent a far less cohesive development of their subject than do his writings on electromagnetism. It was H. W. Watson, not Maxwell, who wrote the Clarendon Press's *Treatise on the Kinetic Theory of Gases* corresponding to Maxwell's *Treatise on Electricity and Magnetism*. In reality there are notable parallels, which we shall discuss shortly, between the evolution of Maxwell's thought in the two fields, but first it is convenient to list Maxwell's chief papers in kinetic theory and to give some indication of the scientific ideas they contained and the questions they raised.

Maxwell published four important papers on gas theory over twenty years:

1. "Illustrations of the Dynamical Theory of Gases" (1860)

2. "On the Dynamical Theory of Gases" (1867)

3. "On Boltzmann's Theorem on the Average Distribution of Energy in a System of Material Points" (1879)

4. "On Stresses in Rarefied Gases Arising from Inequalities in Temperature" (1880)

Besides these Maxwell published the important experimental paper "On the Viscosity or Internal Friction of Air and other Gases" (1865); the textbook *Theory of Heat*, which went through four editions in Maxwell's lifetime and six more thereafter; and a series of short but often significant articles and lectures published mostly in *Nature* during the 1870s on questions such as the size of molecules, the constitution of bodies, the continuity of the liquid and gaseous states, and the equilibrium of temperature in a column of gas.

Because of the large quantity of material to be presented, we have divided the documents into two parts. The present volume covers the kinetic theory of transport phenomena in gases as Maxwell developed it up to about 1875, based mainly on papers 1 and 2, and Maxwell's ideas about the nature of atoms and molecules. Our next volume will include documents related to Maxwell's later papers on statistical mechanics (3) and rarefied gas dynamics (4) and a number of items dealing with thermodynamics, the equation of state of gases, and irreversibility. The documents are arranged topically (and then chronologically within each topic) rather than strictly chronologically in order to help the reader understand the structure and development of Maxwell's theories. In so doing, we hope that our book will be more than just a reference work. In chapter I a general account of these theories in the context of nineteenth-century physical science is provided.

To appreciate fully the personal remarks in Maxwell's correspondence one must keep in mind the salient facts about his life and career. The reader seeking more detailed biographical information should consult the book by Everitt[1] and other works cited therein.

James Clerk Maxwell was born in Edinburgh on June 13, 1831. His father, John Clerk, who added the name Maxwell to his own when he married Frances Cay in 1826, had been trained as a lawyer but spent most of his time managing the Maxwell family estate, Glenlair, near Dalbeattie in southwest Scotland. The family fortune, though somewhat dissipated over the centuries, was still adequate to permit Maxwell the option of retiring to Glenlair to devote himself full time to research when he did not have a suitable academic post.

John Clerk Maxwell and Frances had one other child, a daughter born in 1828 who died at the age of two. Frances herself died when James was eight; after two unhappy years in which his education was entrusted to a tutor, James was sent to Edinburgh Academy, where he acquired a strong background in classics. Two other students there who became Maxwell's lifelong friends were Lewis Campbell and Peter Guthrie Tait.

In 1847 Maxwell was enrolled at the University of Edinburgh, where he stayed for three years before going to Cambridge University. He was strongly influenced in Edinburgh by the physicist J. D. Forbes and the philosopher Sir William Hamilton, the Professor of Logic and Metaphysics (not to be confused with the Irish mathematician William Rowan Hamilton). At Cambridge, Maxwell started in St. Peter's College but quickly transferred to Trinity College, whose Master was the formidable scientist-philosopher William Whewell. Maxwell prepared for the Mathematical Tripos by studying with the famous tutor William Hopkins; but another pupil of Hopkins, E. J. Routh, beat him to the position of Senior Wrangler in the Tripos examination of 1854. (They were tied for the Smith's Prize in Mathematics, the other major test for aspiring scientists at Cambridge.)

The leading physicist in Cambridge at that time was George Gabriel Stokes (1819–1903), the Lucasian Professor of Mathematics. Stokes's fame survives through his theoretical and experimental work on hydrodynamics, elasticity, and optics. His critical papers on optics provided a background for the optical side of Maxwell's electromagnetic theory and nicely complemented William Thomson's contributions on the electrical side. Maxwell had the opportunity to meet and talk with Stokes and corresponded with him later about gas viscosity (documents III-1 and 3) and other matters. After Maxwell returned to Cambridge in 1871, the two men and their wives became close friends; Stokes was one of the executors of Maxwell's will.

The starting point of Maxwell's work on kinetic theory was his reading in April 1859 of a translation in the *Philosophical Magazine* of a paper on this subject by Rudolf Clausius.[2] The idea of attributing pressure in gases to the random impacts of molecules against the walls of the containing vessel had been suggested many times before; Clausius himself had written one paper on kinetic theory eighteen months earlier,[3] but until 1859 the theory had never really caught on. Much of the prevailing opinion favored Newton's hypothesis that gas pressure is due to static repulsion between particles—the theory that Maxwell himself had been taught while a student at Edinburgh and had alluded to in his early essay "On the Properties of Matter" (document II-1) in which he stated that when the force between particles "is repulsive and inversly as the distance, the body is called gaseous." Another rival to the random impact theory was the molecular vortex hypothesis promoted by Davy and worked out in detail by Rankine.[4] The novelty of Clausius's paper lay not so much in its explanatory power as in its providing a highly original new concept—the "mean free path" of a gas molecule—which opened the way to determining statistically the motions of large numbers of colliding bodies. This was a problem Maxwell had already considered but had regarded as hopelessly intractable in his essay "On the Stability of the Motion of Saturn's Rings," for which he was awarded the Adams Prize for the year 1856. (See our previous volume, *Maxwell on Saturn's Rings.*) He read Clausius just as he was preparing the final revised manuscript of the essay for publication and enthusiastically took up the fresh challenge of gas theory.

Maxwell's first kinetic theory paper was based on the provisional hypothesis that the molecules of a gas are elastic bodies of definite dimensions: the so-called "billiard-ball" model of the gas—though Maxwell's billiard balls, like Pooh-bah's in *The Mikado*, could in some circumstances be elliptical. The technical details of this, and of Clausius's two papers, are discussed in chapter I. Three theoretical novelties should be mentioned here, with some of the implications Maxwell drew from them, since these determined the shape of much that was to come.

1. *Velocity distribution law*   Maxwell derived formulas for the statistical distribution of velocities among large numbers of gas molecules.

2. *Concept of transport phenomena*   Maxwell introduced the notion that the phenomena of diffusion, viscosity, and heat conduction in a gas are parallel dynamical processes, depending on the transfer of mass, momentum, and energy, respectively, from one part of a gas to another or between two gases.

3. *Equipartition theorem*   Clausius (and also Waterston) had deduced that in a mixture of two gases the mean translational energies of the two systems of molecules are equal in all three axes. Maxwell gave a more rigorous proof of this result, allowing for the velocity distribution law, and then reached the much more surprising conclusion that in any number of systems of colliding particles the mean *rotational* energies of each system about each of the three principal axes of every particle are equal to each other and equal to the mean translational energies of each system along each of the three axes.

Among the new detailed results were formulas for the coefficients of viscosity, diffusion, and heat conduction of a gas in terms of the mean velocity and mean free path of the molecules and numerical estimates for the mean free path. An unexpected conclusion, right though apparently contrary to common sense, was that over a wide range of pressures the viscosity should be *independent* of the pressure of the gas. This and other results of Maxwell's first kinetic theory are discussed further in chapter I and in the documents in chapter III. Some conclusions were to be strikingly verified; others turned out to be wrong and called for modifications of the theory.

An important test of good science lies not only in what it does but in what it leads to. Maxwell's kinetic theory paper of 1860 (document III-6) contributed directly or indirectly to research in the following areas of physics:

1. *Mean free path and the size of molecules*   Although Maxwell did not in 1860 make estimates of the size of molecules, his estimates of mean free path contributed at two levels to the progress of molecular physics. First, in general, merely by providing a measure of a one-dimensional quantity related to the molecular structure of a gas, they encouraged many physicists who would not otherwise have done so to take molecules seriously, as something in their domain of interest, not just a theoretical construct of the chemists. Second, more specifically, they supplied one line of evidence from which Josef Lo-

schmidt, G. J. Stoney, and William Thomson did actually make the first estimates of molecular diameters.[5]

2. *Mistakes in the 1860 paper and the revised transport theory*   Maxwell's first theory of heat conduction contained mistakes, partly the result of carelessness and partly symptomatic of deeper problems in the elementary kinetic theory. Consideration of these mistakes led Maxwell in 1866 to formulate a radically different theory. His attempt circa 1863 to patch up the old theory of transport phenomena formed the subject of his second major paper.

3. *The viscosity law and interatomic forces*   One result of Maxwell's 1860 paper which was mathematically correct but inconsistent with experiment was that the coefficient of viscosity μ of a "billiard ball" gas is proportional to the square root of the absolute temperature T. The experiments on the viscosity of air and other gases done by Maxwell and his wife, Katherine, between 1863 and 1865 demonstrated that this result is untrue of real gases: They found in fact that μ is more nearly proportional to T than to $T^{1/2}$. In 1866 Maxwell proved that the law proportional to T would follow from the assumption of an inverse fifth-power repulsive force between the gas molecules. Although ultimately neither the $T^{1/2}$ nor T law was found to hold true, this divergence between experiment and the billiard-ball model of a gas was one of a number of experimental data that prompted physicists in the 1860s and 1870s to investigate seriously interatomic forces.

4. *The velocity distribution law, thermodynamics, and Maxwell's demon*   Once a definite statistical law had been proposed to describe the distribution of velocities among gas molecules, only time was needed before questions would be asked about the relation of molecular statistics and the laws of thermodynamics. These were studies by Boltzmann, Maxwell, William Thomson, and others; they led among other more solemn items to one of the most charming personages in the history of physics: Maxwell's demon. Maxwell's contributions to this field and to thermodynamics and statistical mechanics in general are presented and discussed in our next volume.

5. *The specific heat anomaly and difficulties in the equipartition theorem*   The equipartition theorem, a remarkable theoretical result, had an equally remarkable, indeed alarming, experimental implication. When combined with one of Clausius's formulas, it gave a rigorous expression for the ratio γ of the specific heats of a gas at constant pressure and constant volume that was in sharp disagreement with experiment. This result, which Maxwell in one excited moment said "overturns the whole hypothesis" (document III-8), was to become the subject of interminable debate throughout the nineteenth century, the difficulties only increasing as proofs of the equipartition theorem were widened and deepened by Boltzmann, Maxwell, and their followers. Equipartition became one of the grand mysteries of classical physics and was not to be resolved until the emergence of quantum mechanics. Maxwell's

contributions to the discussion are reserved for our next volume, but his final pronouncement on the problem can be found in document II-19.

Only one major area of Maxwell's work on kinetic theory can be said definitely not to find its germ in the 1860 paper. That was one of the greatest, however: the creation of the science of rarefied gas dynamics, which followed partly from the revised transport theory of 1866 and partly from the experiments of William Crookes and others during the 1870s on the radiometer. This topic is also included in the volume on thermodynamics and statistical mechanics.

We return now to the comparison of Maxwell's work in kinetic theory and electromagnetism. Each displays a pattern of advancing abstraction and steadily increasing sophistication in mathematical technique. In electromagnetism Maxwell began with an avowedly incomplete theory, based on scientific analogies, then developed a specific model of the ether, his "molecular vortex hypothesis" (partly suggested, strangely enough, by Rankine's theory of gases), and finally proceeded in two stages to his general Lagrangian formulation of the field equations. In gas theory similarly he began with the billiard-ball model, recognized as inadequate, and then proceeded through two stages of refinement to the paper of 1879 which should, in Larmor's words, be regarded as marking "the emergence of Statistical Dynamics into the rank of a special exact science."[6]

Maxwell's molecular vortex ether yielded the brilliant insight that electromagnetic forces can propagate as waves with the velocity of light. Having grasped that truth Maxwell immediately jettisoned his model as arbitrary and clumsy and sought a foundation that would "clear the electromagnetic theory of light from all unwarrantable assumption, so that we may safely determine the velocity of light by measuring the attraction between bodies kept at a given difference of potential, the value of which is known in electromagnetic measure."[7] The billiard-ball model of a gas similarly helped Maxwell take kinetic theory beyond the derivation of the simple gas laws to the calculation of transport coefficients, but detailed empirical data—much of it obtained because of the qualitative success of the billiard-ball model—finally pointed to the need to take into account the long-range interatomic forces, and this together with Clausius's criticism and Maxwell's own opinion that his first derivation of the velocity distribution function "may appear precarious"[8] forced him to develop a more sophisticated mathematical technique for computing the observable properties of gas models.

In the theory of transport phenomena developed by Maxwell in the 1867 paper it was still necessary to assume a specific molecular model in order to compute transport coefficients. Experimental data and mathematical convenience combined to suggest the $1/r^5$ law of molecular force. But the value of the theory transcended this special case, the inadequacy of which Maxwell was soon to recognize: It brought a surer grasp of the physical problems and the capability for greater abstraction and generality of solution, though these were not fully realized until well into the twentieth century.

Maxwell's final gas theory, written in 1878, the year before his death, completed the process of abstraction. By extending earlier investigations by Ludwig Boltzmann and H. W. Watson, Maxwell applied Hamilton's dynamical methods to determine the statistical behavior of large numbers of material points, represented as an ensemble of mechanical systems in "phase space," without reference to any specific molecular model or force law and indeed without restriction to binary encounters between molecules. This, with some of Boltzmann's work, supplied the foundation for the later research of J. Willard Gibbs on statistical mechanics. Thus Maxwell carried through in kinetic theory the same process of liberation from restricted hypotheses that he had used in developing the Lagrangian formulation of electromagnetic field theory. Behind all lay his dissatisfaction, both philosophical and empirical, with the existing molecular models. Maxwell's many reviews and short papers of the 1870s on molecules can be seen as attempts to clarify the foundations of kinetic theory and as part of a search for a "nonspecific" basis for gas theory.[9]

Just as the Lagrangian formulation of electromagnetic field theory had been much more than an attempt to generalize the laws of electromagnetism, so Maxwell's last paper, "On Boltzmann's Theorem," was much more than an attempt at a generalized gas theory. In fact, as Maxwell emphasized in conversations with some of his younger colleagues at the Cavendish Laboratory, Boltzmann and he had proved too much because there was no reason on the surface of things why the results should not apply to solids and liquids as well as gases.[10] Yet to reaffirm the parallel, gas theory and electromagnetic theory underwent in Maxwell's hands closely similar developments from the use of a specific model to the successive reformulation of the original ideas in more and more abstract terms. The attempt with electromagnetic fields was more successful because all known phenomena could be brought within the formulation, whereas in kinetic theory there was on one side the embarrassment of "proving too much" and on the other the difficulty that only equilibrium properties could be included: Transport theory was still dependent on special models and assumptions. Overproof was removed with the appearance of energy quantization; the restrictions of transport theory have been removed only in the past two decades.

Maxwell's application of the Lagrangian and Hamiltonian methods coincided with a general movement among British and European mathematicians toward wider use of the methods of analytical dynamics in physical problems. The course of that movement in Great Britain can be followed through Cayley's two British Association reports on advanced dynamics, Routh's *Treatise on the Dynamics of a System of Rigid Bodies* and Thomson and Tait's

Treatise on Natural Philosophy.[11] In kinetic theory, Boltzmann's papers of 1868, 1872, and 1877 provided the background for Maxwell's later work.[12] Maxwell pursued matters in his own way, emphasizing especially the need to express the ideas of the mathematicians in physical language. He explained his mature preference in physical theorizing as follows:

The aim of Lagrange was, as he tells us himself, to bring dynamics under the power of the calculus, and therefore he had to express dynamical relations in terms of the corresponding relations of numerical quantities.

In the present day it is necessary for physical inquirers to obtain clear ideas in dynamics that they may be able to study dynamical theories in the physical sciences. We must therefore avail ourselves of the labours of the mathematician, and selecting from symbols those which correspond to conceivable physical quantities, we must retranslate them into the language of dynamics.

In this way our words will call up the mental images, not of certain operations of the calculus, but of certain characteristics of the motion of bodies.

The nomenclature of dynamics has been greatly developed by those who in recent times have expounded the doctrine of the Conservation of Energy, and ... most of the following statement is suggested by the investigation in Thomson and Tait's *Natural Philosophy*, especially the method of beginning with the case of impulsive forces.

I have applied this method in such a way as to get rid of the explicit consideration of the motion of any part of the system except the coordinates or variables on which the motion of the whole depends. It is important to the student to be able to trace the way in which the motion of each part is determined by that of the variables, but I think it is desirable that the final equations should be obtained independently of this process....

In following dynamical theories of the physical sciences, it has been too frequent a practice to invent a particular dynamical hypothesis and then by means of the equations of motion to deduce certain results. The agreement of these results with real phenomena has been supposed to furnish a certain amount of evidence in favour of the hypothesis.

The true method of physical reasoning is to begin with the phenomena and to deduce the forces from them by a direct application of the equations of motion. The difficulty of doing so has hitherto been that we arrive, at least during the first stages of the investigation, at results which are so indefinite that we have no terms sufficiently general to express them without introducing some notion not strictly deducible from our premises.

It is therefore very desirable that men of science should invent some method of statement by which ideas, precise so far as they go, may be conveyed to the mind, and yet sufficiently general to avoid the introduction of unwarrantable details.[13]

The Lagrangian formulation of electromagnetic theory and the Hamiltonian formulation of statistical mechanics furnished Maxwell with methods of statement that came close to fulfilling these ideals.

Such philosophical reflections were very much a part of Maxwell's physics. As an undergraduate at Edinburgh in 1849 he had, like many of his contemporaries, had his interest in philosophical questions fired by the teachings of Sir William Hamilton, and he became no mean philosopher of science himself. Maxwell's commitment to the mechanical world view was tempered with an acute sense of the limitations of mechanics and of the scientific method in general. When, following William Thomson, he began to exploit analogies between different branches of mathematical physics—heat, electricity, hydrodynamics, electromagnetism, and gas theory—he brought a rare philosophical insight to bear on them. The phrases Maxwell used to describe the method of scientific analogy reappear in later apologies for mechanism (when that world view was under siege at the end of the nineteenth century) by Oliver Lodge, Joseph Larmor, and J. J. Thomson, though they were seldom expressed with Maxwell's subtlety of argument.

Beautiful and satisfying as the processes of abstraction and generalization are, we should never forget, as too many historians of ideas do forget, that scientific theory becomes fruitful only insofar as it derives specific results and compares them with experiment. It is a measure of Maxwell's immense stature as a physicist that like Newton he took pleasure while reaching out to great generalizations in also making detailed practical calculations. Thus in his 1865 paper "A Dynamical Theory of the Electromagnetic Field," in which he introduced the Lagrangian formulation of the field equations, we find a new method, more accurate than any previously given, for calculating the self-inductance of a circular coil of many turns on a channel of rectangular cross section, with a numerical evaluation of the self-inductance of the coil used in the British Association experiment to determine the ohm in absolute units. Similarly the paper "On Boltzmann's Theorem on the Average Distribution of Energy in a System of Material Points," the most abstract of all Maxwell's scientific papers, ends with a suggestion for an experiment based on considering the rotational degrees of freedom. Maxwell proved that the densities of the constituent components in a rotating mixture of gas would be the same if each gas were present by itself. Hence gaseous mixtures could be separated by means of a centrifuge. The method also promised much more accurate diffusion data than was hitherto available. Maxwell's correspondence (document III-47) discloses a plan to set up experiments at Cambridge. Many years later diffusion became a standard technique for separating gases commercially.

The reader who peruses the documents presented here, many of them printed for the first time, will find throughout a wonderful range of concrete practical detail and wide-ranging philosophical abstraction.

In the documents ⟨ ⟩ indicates a word that was deleted in the original, ? a scribble that is indecipherable to all three editors, and [ ] our own insertions (usually a suggested title for some unpublished notes). [space] means that there is a gap equivalent to several lines of text in the manuscript, perhaps indicating that Maxwell wanted to leave room to insert material. Editorial notes to a document, indicated by superscript letters, appear at the end of the document. These notes identify the correspondent and his or her relation to Maxwell, contemporaries of historical interest, and references to any of Maxwell's own published papers or works in progress. For papers reproduced from *The Scientific Papers of James Clerk Maxwell*, we have retained the original page numbers in that book to facilitate cross references.

Maxwell was sometimes in the habit of making notes, or composing a draft on one side of a sheet of paper; then, at a later time, using the reverse side for notes or a draft on an entirely different subject. The initial sheets could get transposed in the process. Ordering the sheets of paper so that one side reads consecutively disorders the notes or draft on the other and they become "fragments." This has happened to document III-34 which is on the reverse of a draft of sections from the *Treatise on Electricity and Magnetism* as well as other documents from this period.

We would like to thank Brigadier Wedderburn-Maxwell, N. H. Robinson of the Royal Society of London, and A. E. B. Owen of Cambridge University for giving us permission to publish this material and for helping us obtain copies of it.

The research of Stephen G. Brush involved in the preparation of this book was supported by a grant from the History and Philosophy of Science Program of the National Science Foundation.

Notes

1. C. W. F. Everitt, *James Clerk Maxwell: Physicist and Natural Philosopher* (New York: Scribner, 1975). See also the detailed analysis of Maxwell's family and personal background in Everitt's essay, "Maxwell's Scientific Creativity," in *Springs of Scientific Creativity*, R. Aris, H. T. Davis, and R. H. Stuewer, eds. (Minneapolis: University of Minnesota Press, 1983), 71–141. Another recent biography is Ivan Tolstoy, *James Clerk Maxwell: A Biography* (Chicago: University of Chicago Press, 1981); see Everitt's review in *The Physics Teacher* 22 (1984), 264–266.

2. Cited in note 19, chapter I.

3. Cited in note 20, chapter I.

4. W. J. M. Rankine, "On the Hypothesis of Molecular Vortices, and Its Application to the Mechanical Theory of Heat," *Proc. R. Soc. Edinburgh* 2 (1850), 275–288; "On the Centrifugal Theory of Elasticity, as Applied to Gases and Vapours," *Phil. Mag.* [4] 2 (1851), 509–542; "On the Centrifugal Theory of Elasticity, and Its Connection with the Theory of Heat," *Trans. R. Soc. Edinburgh* 20 (1853), 425–440; "On the Hypothesis of Molecular Vortices, or Centrifugal Theory of Elasticity, and its Connexion with the Theory of Heat," *Phil. Mag.* [4] 10 (1855), 354–363, 411–420.

5. See chapter I, notes 139, 140, and 142.

6. J. Larmor, *Mathematical and Physical Papers*, vol. 2 (Cambridge: Cambridge University Press, 1929), appendix 3, p. 743.

7. Letter from Maxwell to C. Hockin, September 7, 1864, quoted in Lewis Campbell and William Garnett, *The Life of James Clerk Maxwell* (New York: Johnson Reprint, 1969), 340.

8. Document III-30; *The Scientific Papers of James Clerk Maxwell*, W. D. Niven, ed. (New York: Dover, 1965), vol. 2, 43.

9. See J. Dorling, "Maxwell's Attempt to Arrive at Non-Speculative Foundations for the Kinetic Theory," *Studies in Hist. Phil. Sci.* 1 (1970), 229–248, for an interpretation of this search.

Maxwell did not accept the view (adopted already in 1857 by Clausius) that a molecule of oxygen, for example, consists of two atoms of oxygen. His atoms were the ultimate indivisible constituents of matter, all identical and not necessarily related to molecules in the way imagined by chemists (see pp. 138, 140, and 211). In fact, he argued from spectroscopic evidence that molecules *cannot* be systems of two or three point atoms (p. 232).

10. See also document III-47, and Arthur Schuster, "The Clerk-Maxwell Period," in *A History of the Cavendish Laboratory, 1871–1910* (London: Longmans, Green, 1910), 14–39 (esp. p. 31); Schuster, *The Progress of Physics during 33 Years (1875–1908)* (Cambridge: Cambridge University Press, 1911), 29.

11. Arthur Cayley, "Report on the Recent Progress of Theoretical Dynamics," *Rep. 27th Meeting BAAS* (1857), 1–42; "Report on the Progress of the Solution of Certain Special Problems of Dynamics," *Rep. 32nd Meeting BAAS* (1862), 184–252. E. J. Routh, *An Elementary Treatise on the Dynamics of a System of Rigid Bodies* (Cambridge, 1860, 1868). William Thomson and P. G. Tait, *A Treatise on Natural Philosophy* (Oxford: Clarendon Press, 1867; new edition, Cambridge: Cambridge University Press, 1879).

12. L. Boltzmann, "Studien über das Gleichgewicht der lebendigen Kraft zwischen bewegten materiellen Punkten," *Sitz. Math.-Naturwiss. Cl. Akad. Wiss., Wien*, Teil II, 58 (1868), 517–560; "Weitere Studien über das Wärmegleichgewicht unter Gasmolekülen," *Sitz. Math.-Naturwiss. Cl. Akad. Wiss., Wien* 66 (1872), 275–370; "Ueber die Beziehung eines allgemeinen mechanischen Satzes zum zweiten Hauptsatze der Wärmetheorie," *Sitz. Math.-Naturwiss. Cl. Akad. Wiss., Wien* 75 (1877), 67–73. Translations of the last two papers are in Brush, *Kinetic Theory* (New York: Pergamon Press, 1966), vol. 2, 88–175, 188–193.

13. *Scientific Papers*, vol. 2, 308–309.

Maxwell on Molecules and Gases

I Kinetic Theory and the Properties of Gases: Maxwell's Work in Its Nineteenth-Century Context

When James Clerk Maxwell died in 1879, Peter Guthrie Tait eulogized him as "the leading molecular scientist" of his day. His work in electricity and magnetism then commanded far less attention and esteem. From the perspective of the twentieth century the value of Maxwell's original work in electromagnetism and statistical mechanics seems more important than his papers in kinetic theory or molecular theory. Yet we should take these nineteenth-century assessments seriously if we are to judge the role of kinetic theory and Maxwell in the development of modern physics.

Since the middle of the eighteenth century, theories of gases were closely bound to theories of heat and of the constitution of matter. The kinetic theory of gases was no exception. Kinetic theory emerged in the same decade as the "mechanical theory of heat" and helped to establish an empirically based and thus scientifically plausible mechanical, molecular theory of matter.[1] Thus the development of the kinetic theory of gases is intimately tied to the emergence of thermodynamics and mechanical theories of matter. Although all the intellectual changes involved in the passage from the caloric to the mechanical theory of heat (and its generalization, the energy principle) are still unknown, even less known is the history of theories of matter, particularly in the early nineteenth century. In both heat and matter theory chemists' concerns were as important as physicists', and in the nineteenth century engineers materially aided the intellectual development of the theory of heat. No one consistent overall picture of the emergence of thermodynamics[2] or mechanical molecular theory yet exists, although there is a growing body of specialist studies.[3] What is certain is that there were two main theories of heat—"substance" and "mode of motion"—developed during the first half of the nineteenth century. The substance theory—based on the assumption that heat is a material fluid, usually called "caloric"—was really a complex set of different theories about the behavior of heat and gases. The problem of the nature of heat, and the caloric theory in particular, were part of chemistry, and Lavoisier's commitment to it secured caloric a position in the new rationalized chemistry.[4] A formidable array of experimental studies of heat and gases had accumulated by the 1820s.[5] This was accompanied by a growing, mathematical theoretical explanation of the nature of gases, and the operation of caloric.[6]

The idea of heat as simply a mode of motion of particles of matter goes back to the seventeenth century, but it seemed inadequate to explain most phenomena of interest to chemists. Yet the caloric theory was never satisfactory, witness its myriad varieties, and as an explanatory scheme lost its cogency in the middle of the nineteenth century.[7] It was replaced by the first successful invasion of the domain of chemistry by physicists, the wave theory of heat. Motion was ascribed primarily to the ether and only secondarily to atoms of ordinary matter. The theory was developed by physicists interested in radiation phenomena and the analogies that could be drawn between light and heat.[8] The justification for such a view goes back to the experiments on radiant heat conducted by William Herschel, Macedonio Melloni, and James Forbes. These experiments attracted considerable attention around 1800 and again around 1830[9] and convinced most physicists that radiant heat has all the qualitative properties of light—reflection, refraction, polarization, and interference—and differs from light only in quantitative degree with respect to some parameter. Moreover, no clear distinction was made between the nature of radiant heat and the nature of heat in general; indeed, experiments on radiant heat seemed at the time to offer the best hope for determining the nature of heat.

So long as the particle ("emission") theory of light was generally accepted, a particle (caloric) theory of heat was plausible. The fact that heat can go through "empty" space was cited as an argument against the idea that heat was molecular motion. But when the wave theory of light was accepted after 1820 as a result of the work of Thomas Young in England and especially of Augustin Fresnel in France, the balance shifted, and physicists tended to accept an analogous wave theory for heat. Put into a fairly definite quantitative form by Ampère in 1835, the theory described heat as the kinetic energy of ether vibrations; these are generally in equilibrium with the vibrations of material molecules even if there is no direct transfer of energy between molecules other than by means of the intervening ether.

The wave theory of heat provided a transition from the caloric theory to thermodynamics by initially identifying caloric with ether and then by admitting that heat could be either the *quantity* of ether or the *quantity of motion* of the ether. The final steps toward the mechanical theory were taken by identifying heat as molecular motion, with a plausible mechanism for transferring energy from one molecule to another across empty space and, finally, by returning to a mechanical theory of heat, forgetting about the role of the ether except when radiant heat is involved.

There is some evidence that this line of reasoning did play some part in the discovery of energy conservation, though it was insufficient to secure that discovery until there was independent evidence for the mechanical equivalent of heat.[10]

At the time Maxwell composed the first document presented here (1847), there was no one established theory of heat. Some physicists had already accepted a purely mechanical theory of heat, and, significantly, there is no reference to caloric theory in Maxwell's essay "On the Properties of Matter," although Newton's image of gases remains intact. All Maxwell states is that "the properties of bodies relative to heat and light are—Transmission, Reflection, and Destruction" and, a little further on, that heat can be reduced to force. This reads like neither the wave nor the caloric theory of heat. Maxwell never took either one seriously; in his later papers, he always clearly distinguished between radiant heat and ordinary heat, as we do today. He also was spared the agony that William Thomson, for example, went through in first embracing the caloric theory (up to 1850) and then explicitly rejecting it.

Neither the wave nor the caloric theory, however, was adequate to explain the range of phenomena known by the 1840s, especially those familiar to engineers dealing with the operations of machines, such as the steam engine. The study of gases and of mechanics, based on the operations of machines, thus had a special urgency for physicists, chemists, and engineers in the first half of the nineteenth century. In this period there had been a striking succession of discoveries, mostly as by-products of experiments in chemistry, concerning the properties of gases: Dalton's law of partial pressures; Gay-Lussac's law of equivalent volumes; the establishment of the law of thermal expansion of gases by Charles, Dalton, and especially Gay-Lussac; Gay-Lussac's experiments on free expansion of gases into a vacuum; and the relation between specific heat and sound velocity uncovered by Laplace and Poisson. More complicated phenomena, such as thermal conduction and convection, diffusion, and transpiration, were also being investigated.[11]

Even though the mechanical theory of heat superseded the wave theory in the 1850s, it never gave a complete account of all thermal phenomena either. Joule and others simply chose to exclude radiation phenomena from the domain of their theory. And indeed it was just in the region of the interaction of radiation (or "ether") with matter that the classical theory eventually broke down. Later, the goal of Lorentz, van der Waals, Jr., and Jeans, in their efforts to explain blackbody radiation, was to provide an account of the mechanism for exchange of energy between matter and the ether and thus fill this gap in the mechanical theory of heat.[12] At no time in the nineteenth century was there a single self-consistent theory that could claim to account satisfactorily for all properties of heat.

Any kinetic theory of gases, even after the acceptance of the wave theory of heat, faced a formidable array of data to explain and the equally formidable barrier of the traditional coupling of the nature of heat and matter to explicate or explain away. And it was not until after the general acceptance of a mechanical theory of heat that a kinetic theory of gases was admitted into the mainstream of physical theory. Before about 1857 kinetic theories of gases had been proposed, but by theorists whose works stand in isolation against a background of indifference or outright hostility. Kinetic theorists, especially after the development of caloric theory in the context of a gas theory, also had to explain both specific heats, adiabatic change, and the velocity of sound in gases and provide an alternative theory of matter to replace the Newtonian static, corpuscular theory of gases.[13] Explaining gaseous pressure and Boyle's law was only a small beginning. We can therefore begin to understand the neglect Daniel Bernoulli's theory suffered even in 1738[14] and to appreciate the theoretical range necessarily explored by the later kinetic theorists John Herapath and J. J. Waterston.[15] Alternative static theories of gases provided some of the same results[16] and were entrenched by the nineteenth century.[17] The confluence of the mechanical theory of heat and kinetic theory was shown further by J. R. Joule's adoption of the mechanical theory of heat and his adoption of Herapath's kinetic theory and methods to calculate the velocity of a molecule in hydrogen gas.[18] Similarly, all three were concerned with developing theories of molecules; Waterston even estimated their size.

Despite this theoretical activity in kinetic theory in Britain, Maxwell's attention was first drawn to the theory by a paper by Rudolf Clausius, translated into English in early 1859.[19] This was Clausius's second paper on kinetic theory; his first had been published in 1857[20] after the appearance of a brief and somewhat carelessly drawn sketch by August Krönig in 1856.[21] In his first paper Clausius mentioned that, before his papers on the mechanical theory, he had conceived of heat as the motions of molecules but had severed this connection to present his conclusions about heat independently of any particular hypothesis about molecular motion. Maxwell himself delineates (in document II-10) what he believed Clausius had introduced into kinetic theory in his two papers. Clausius allowed his molecules to move freely through the volume of the gas; their impacts were random, and their combined volume was much smaller than that occupied by the gas. The duration of the impacts was small compared with the interval between them so that intermolecular forces could be ignored. Finally, from a definition of temperature as the mean vis viva (mv^2) and from specific heat measurements involving the total energy of the molecules, Clausius inferred that molecules have internal structure. He could qualitatively account for changes of state and electrolytic action, and he could also encompass the diffusion of gases. He had no way, however, of estimating the number of molecules in a given volume of gas or their size. And, despite Maxwell's suggestion to the contrary, Clausius used only average velocities, though he explicitly stated that there is a wide range of velocities among the gas molecules. Clausius was nevertheless the first to recognize that the molecular velocities he deduced from the gas law were averages.[22]

All expositions of the kinetic theory of gases start by deriving the *ideal gas law*, which relates changes in pressure, volume, and temperature; this is a combination of the well-known laws of Boyle and Gay-Lussac, usually written in the form

$$PV = nRT, \tag{1}$$

where P is pressure, V is volume, n represents the amount of gas present (measured in gram-moles in modern terminology), T is the absolute temperature, and R is the gas constant. According to the kinetic theory, the corresponding relation for a gas of N molecules of mass m, each moving at velocity v, is

$$PV = (1/3)Nmv^2. \tag{2}$$

Note that if we consider the velocity to be a vector with components v_x, v_y, v_z, then the quantity v^2 on the right-hand side of Eq. (2) will be the sum of the squares of these components. If the velocities of the molecules in the gas are not equal, then v^2 is replaced by the average value of v^2, usually called the "mean square velocity" $\overline{mv^2}$. Since the total kinetic energy of the molecules is $(1/2)Nmv^2$, the right-hand side of Eq. (2) is equal to two thirds of the total kinetic energy of the motions of the gas molecules. In the nineteenth century the term "vis viva" (Latin for "living force"), defined as mv^2, was still commonly used; according to the kinetic theory, the product PV is equal to one third of the vis viva. Thus the kinetic theory was sometimes called the vis viva theory.

The derivation of Eq. (2) was set as a problem in an examination at St. Peter's College (Peterhouse), Cambridge, in June 1852:

(a) If an infinite number of perfectly elastic material points equally distributed through a hollow sphere, be set in motion each with any velocity, show that the resulting continuous pressure (referred to a unit of area on the internal surface) is equal to one-third of the *vis viva* of the particles divided by the volume of the sphere.

(b) Prove the same proposition for a hollow sphere of any form.

The problem and two solutions were published in 1853 in the *Cambridge and Dublin Mathematical Journal*. C. Truesdell[23] credits it to William Thomson, who was one of the editors of the *Journal*. But the writing style is unlike Thomson's, and the solutions are dated from Chamonix, Switzerland, August 18, 1853. At that time Thomson had just cut short a Mediterranean cruise and had returned to Britain on account of his wife's ill health. A more plausible author is W. J. Steele (1831–1855), then a bachelor-scholar of Peterhouse and a successful private tutor and co-author with P. G. Tait of the elementary *Treatise on the Dynamics of a Particle* (Cambridge, 1856).

Be that as it may, Maxwell most likely read this item because he too contributed questions to the same volume of the *Cambridge and Dublin Mathematical Journal*. It is impressive that the second solution provided, unlike many others (including even Clausius's investigation four years later), explicitly shows that the result is independent of the velocity distribution: The pressure is related to the total vis viva of the particles "whatever be their velocities." [24]

Clausius also insisted that rotational motions of molecules must be included in the kinetic theory. [25] Since collisions between rough extended bodies cause exchange of energy between their translational and rotational motions, Clausius guessed that the energy associated with the two types of motion among large numbers of colliding molecules would settle down to a constant ratio σ. From thermodynamical arguments he showed that should be related to γ, the ratio of the two specific heats of a gas, by the equation

$$\sigma = \tfrac{3}{2}(\gamma - 1), \tag{3}$$

and from the experimental value $\gamma = 1.421$ he concluded that σ was 0.6315.

Clausius's second paper (1858) was written to counter criticism by the Dutch meteorologist C. H. D. Buys-Ballot, who objected that the molecular velocity in a gas could not be as high as kinetic theory indicated because the odor of a pungent gas takes several minutes to permeate a room. [26] Clausius [27] pointed out that molecules of finite size must continually collide and rebound in different directions and deduced from a statistical argument that the probability W of a molecule traveling a distance L without collision is

$$W = e^{-L/l} \tag{4}$$

Though Clausius had an analytical expression for l/d, the ratio of mean free path to molecular diameter, he had no way of obtaining it numerically and thus could not completely answer Buys-Ballot's original objection. He demonstrated the plausibility of his theory by claiming that if $l/d = 1000$, the ideal gas law would be nearly satisfied, at least within the limits of deviation from this law found experimentally for real gases by Regnault. On the other hand, for distances L more than a few times the mean free path, the probability W given by Eq. (4) becomes extremely small. Although this argument showed that for a certain range of values of d it was possible for the theory to account for both the ideal gas law and the absence of diffusion over macroscopic distances, there was no assurance that the diameters of real molecules would in fact lie within that range. Having arrived at the concept of mean free path, Clausius could not take full advantage of it because he had no way of estimating the molecular diameter.

Clearly, in remembering the title of Clausius's paper of 1857, "On the Kind of Motion that We Call Heat," and in reading both his papers, Maxwell's attention was still within that complex of problems traditionally tied to the theory of gases. His theory was also dependent on empirical parameters for its development, and many of his conclusions were qualitative and suggestive rather than exact and mathematically derived.

Maxwell refocused and transformed the theory into a mathematical, conceptually self-contained, predictive one. From his first paper on kinetic theory Maxwell ignored the problem of the nature of heat. Kinetic theory was the key to understanding the nature of molecules, and in his papers molecular science was transformed from a qualitative and speculative theory to a mathematically precise one. The boldness of this step is seen in the focus of the 1860 papers: to use the transport properties of gases to find the mean free path and hence the size of the molecule. Thus, in this first paper on kinetic theory, Maxwell gave two explicit estimates of the mean free path for air molecules. Applying his formula for viscosity he calculated from Stokes's data a mean free path of 1/44,700 of an inch and stated that "each particle makes 8,077,200,000 collisions per second" (document III-6, p. 391). Applying his formula for the diffusion coefficient he obtained an independent estimate of 1/389,000 of an inch from Graham's experiments on the diffusion of olefiant gas (ethylene) into the atmosphere "quoted by Herapath" (document III-6, p. 403) and noted that this value is "not very different from that deduced from experiments on friction" (i.e., from viscosity) (document III-6, p. 409). Maxwell's continuing interest in estimates of molecular size shows that he was well aware of the importance of his own research in that area.

The key to developing the connections between the motions of microscopic molecules and the macroscopic, measurable transport properties was Maxwell's velocity distribution function. The first paper on gas theory presents a series of propositions, the first five of which lead to a statistical formula expressing the distribution of velocities in a gas at uniform pressure, the molecules being treated as elastic spheres.[28] When a moving sphere collides with another fixed sphere randomly placed in its path, the direction in which it rebounds is distributed with equal probability over each element of solid angle. Hence, Maxwell argued, the distribution law has an explicit functional form, which he proceeded to derive as follows. Let the components of velocity in three axes be x, y, z. Then the number dn of molecules whose velocities lie between x and $x + dx$, y and $y + dy$, z and $z + dz$ is $Nf(x)f(y)f(z)\,dx\,dy\,dz$. But since the three coordinates are arbitrary, dn depends only on the molecular velocity v, where $v^2 = x^2 + y^2 + z^2$, and therefore the distribution must satisfy the functional relation

$$f(x)f(y)f(z) = \phi(x^2 + y^2 + z^2), \tag{5}$$

the solution of which is an exponential. By applying the fact that the total number of particles is finite, one finds that the velocity components in a given direction obey the relation

$$dN_x = N \frac{1}{\alpha \sqrt{\pi}} e^{-x^2/\alpha^2} \, dx, \tag{6}$$

where α is a quantity with the dimensions of velocity; the number of particles with velocities lying between v and $v + dv$ is

$$dN_v = N \frac{4}{\alpha^3 \sqrt{\pi}} v^2 e^{-v^2/\alpha^2} \, dv. \tag{7}$$

From these formulas mean values of various products and powers of velocities used in calculating the viscosity and diffusion coefficients could be expressed in terms of α. Maxwell also obtained related formulas for the distribution of velocities among systems of two or more different kinds of molecules.

The derivation of Eqs. (6) and (7) marks the beginning of a new epoch in physical science. The application of statistics to physical and social observations was already widely known, and Maxwell's distribution law was, as he remarked, identical in form with the "normal distribution" introduced by Laplace and Gauss in the theory of errors; but the idea that physical *processes* should be described by a statistical function was a new step. Both its origin and validity deserve careful study. Intuitively, Eq. (7) is plausible enough, for dN_v approaches zero as v approaches zero or infinity, and reaches a maximum at the intermediate speed $v = \alpha$, consistent with the natural physical expectation that only a few molecules will have very high or very low velocities. This was verified empirically years later in experiments with molecular beams. Yet Maxwell's assumption that the three resolved components of velocity could be treated as independent variables is, as he afterward conceded, "precarious," and the whole derivation gives a strange appearance of having nothing to do with molecules or their collisions.

In his second paper on kinetic theory, published in 1867, Maxwell offered another derivation based on an analysis of molecular encounters. To maintain equilibrium between the number of molecules in a certain state of motion before and after collision, the distribution function must satisfy the relation

$$f(v_1)f(v_2) = f(v_1')f(v_2'), \tag{8}$$

where v_1 and v_1' are the velocities of the first molecules and v_2 and v_2' those of the second before and after an encounter. Equations (6) and (7) are obtained by combining Eq. (8) with the equation for energy conservation during encounters. The new derivation established the equilibrium of the exponential distribution but not its uniqueness. From considerations of cyclic collision processes Maxwell sketched an argument (which was afterward established rigorously by Boltzmann) that all other distributions converge to the same form. Other approaches were initiated by Boltzmann from 1868 onward and eventually transcended gas theory and led to the separate science of statistical mechanics.[29]

Maxwell's kinetic theory is a link between nineteenth-century traditions in statistical analysis and in atomic speculation, and as such it holds a wider interest for historians of science. Maxwell's acquaintance with statistical ideas began while he was a student at Edinburgh. In part it was awakened by the professor of Natural Philosophy, J. D. Forbes. In 1848 Forbes sparked in the *Philosophical Magazine* a discussion of statistical problems by reviving the Rev. John Michell's old argument for the existence of binary stars, namely, that the proportion of stars paired in close proximity is much higher than would be expected if they were isolated objects scattered at random over the sky.[30] Not long afterward Maxwell read Laplace's classic treatise on the calculus of probabilities, followed by George Boole's writings on the laws of thought.[31]

The work of greatest influence on Maxwell's development of gas theory may well be a review in the July 1850 *Edinburgh Review* of the magnificently titled collection of essays by Adolphe Quetelet, *Letters Addressed to H.R.H. the Grand Duke of Saxe-Coburg and Gotha on the Theory of Probabilities as Applied to the Moral and Political Sciences*. The author of the review was Sir John Herschel.[32] It ranged over many statistical questions, social and otherwise; a contemporary letter from Maxwell to his friend and future biographer, Lewis Campbell, strongly suggested that he had read it. The letter was undated and Campbell from memory put it as "June ? 1850," but there can be little doubt that it was written just after the publication of Herschel's review in July 1850.[33] Maxwell discoursed on probability theory with remarks such as the following:

[T]he true Logic for this world is the Calculus of Probabilities, which takes account of the magnitude of the probability (which is, or which ought to be in a reasonable man's mind). This branch of Math., which is generally thought to favour gambling, dicing, and wagering, and therefore highly immoral, is the only "Mathematics for Practical Men," as we ought to be.

This can be compared with a passage in the Herschel review:[34]

The calculus of Probabilities, under the less creditable name of the doctrine of Chances, originated at the gambling table; and was for a long time confined to estimating the chances of success and failure in throws of dice, combinations of cards, and drawings of lotteries. It has since effectually obliterated the stain of its cradle.

Whether, indeed, Maxwell read the review in 1850, it was reprinted in Herschel's *Essays* in 1857, and we know that Maxwell read and admired these essays.[35]

One passage in Herschel's essay embodied a new derivation of the law of least squares that was intended to make the subject more popularly comprehensible. The proof is reproduced here so that the reader can compare it directly with Maxwell's proof (document III-6). Note that Herschel deals with a distribution in two dimensions rather than in three.

We set out from three postulates. 1st, that the probability of a compound event, or of the concurrence of two or more independent simple events, is the product of the probabilities of its constituents considered single; 2ndly, that there exists a relation or numerical law of connexion (at present unknown) between the amount of error committed in any numerical determination and the probability of committing it, such that the greater the error the less its probability, according to some regular LAW of progression, *which must necessarily be general and apply alike to all cases, since the causes of error are supposed alike unknown* in all; and it is on this ignorance, and not upon any peculiarity in cases, that the idea of probability in the abstract is founded; 3rdly, that the errors are equally probable if equal in numerical amount, whether in excess, or in defect of, or in any way beside the truth. This latter postulate necessitates our assuming the function of probability to be what is called in mathematical language *an even function*, or a function of the square of the error, so as to be alike for positive and negative values; and the postulate itself is nothing more than the expression of our state of *complete* ignorance of the causes of error, and their mode of action. To determine the form of this function, we will consider a case in which the relations of space are concerned.

Suppose a ball dropped from a given height, with the intention that it shall fall on a given mark. Fall as it may, its deviation from the mark is *error*, and the probability of that error is the unknown function of its square, *i.e.*, of the sum of the squares of its deviations in any two rectangular directions. Now, the probability of any deviation depending solely on its magnitude, and not on its direction, it follows that the probability of each of these rectangular deviations must be the same function of *its* square. And since the observed oblique deviation is equivalent to the two rectangular ones, supposed concurrent, and which are essentially independent of one another,[36] and is, therefore, a compound event of which they are the simple independent constituents, therefore its probability will be the product of their separate probabilities. Thus the form of our unknown function comes to be determined from this condition, viz., that the product of such functions of two independent elements is equal to the same function of their sum. But it is shown in every work on algebra that this property is the peculiar characteristic of, and belongs only to, the exponential or logarithmic function. This, then, is the function of the square of the error, which expresses the probability of committing that error. That probability decreases, therefore, in geometrical progression, as the square of the error increases in arithmetical. And hence it further follows, that the probability of successively committing any given system of errors on repetition of the trial, being, by postulate I, the product of their separate probabilities, must be expressed by the same exponential function of the sum of their squares however numerous and is, therefore, a maximum when that sum is a minimum.[37]

The resemblance between Herschel's proof and Maxwell's derivation of the velocity distribution law is obvious; it also helps to explain why Maxwell's derivation was puzzling to the scientists who read it in the context of gas theory rather than statistics. In view of the debate over the velocity distribution law, it is worth adding that Herschel's proof also was immediately attacked by Boole and others.[38] Whether Maxwell knew of the criticisms is uncertain. In the reprint of the review in 1857, Herschel added an important footnote in defense of the derivation that Maxwell does not seem to have taken into account in his application to kinetic theory: "That is, *the increase or diminution in one or which may take place without increasing or diminishing the other.* On this, the whole force of the proof turns."[39]

Theodore M. Porter has argued recently that since the inspiration for Maxwell's introduction of statistical methods into physics came from a review of Quetelet's works, one should associate Maxwell's early thinking about molecular velocity distributions with the kind of statistical reasoning used in the social sciences by Quetelet and others.[40] In particular, "statistical" did not have the connotation of "disorderly" or "uncertain" that it later acquired in physics, but rather meant the emergence of regularity out of the apparently chaotic behavior of large numbers of individuals. Hence "Maxwell's discovery that the distribution of molecular velocities will conform to the error function, which introduced advanced probability techniques into physics, had little or nothing to do with his later recognition that the macroscopic gas laws are only probabilistic."[41] Porter attributes Maxwell's later emphasis on uncertainty to his dislike for the antireligious writings of Huxley and Tyndall, who claimed that scientific laws could provide deterministic explanations of all natural phenomena.

But none of this is pertinent to Maxwell's kinetic theory of 1860. And herein lies the rub. Tracing cultural influences in science is an extremely slippery business, and even when the evidence seems explicit, as in the case of Thomas Malthus's influence on Charles Darwin, key changes in the meanings of words undermine any simplistic coupling of culture and particular scientific results.[42] Also, in his review Herschel clearly ranged over much of statistics in general before even discussing the *content* of Quetelet's book. Herschel was sensitive to his audience, who could not be expected to know much about statistical reasoning. Maxwell's remarks on this review, which he clearly read, are general. And, finally, unless Maxwell states explicitly what he actually got from Quetelet, all is conjecture. What Maxwell received from statistics he received from Herschel; this is suggestive but not definitive. "Creative synthesis" gets us no further in understanding what precisely Maxwell took from social statistics in 1859. Creation and synthesis are psychological processes about which we know nothing, and describing what happens is not explaining the process; it is merely detailing the data. And in this case all we have are generalities described by Maxwell, a sketch provided by Herschel on the one hand and precisely derived mathematical results on the other. This leaves a

chasm that can be filled only by speculation. Detailing Maxwell's later contemplations on the matter does not tell us his sources, or state of mind, in
1859. The issue for historians is not that "the introduction of statistics to the
kinetic theory was ... not the result of the internal development of a single,
well-demarcated discipline, but a creative synthesis of a great range of knowledge,"[43] but what the nature of that synthesis is. The process remains labeled
but unexplored.[44]

Whatever the influences and sources of Maxwell's distribution function and
the flaws in his derivation, the focus of his attention in 1860 was on the
transport properties of gases. Although his analysis has some serious faults, it
was the first time anyone had proposed such a unified treatment displaying the
similarity of these three "transport processes," and the first time anyone had
made a direct quantitative estimate of the coefficients from kinetic theory. One
does not have to use a velocity distribution function in order to estimate
transport coefficients. In fact, the sophisticated mathematical apparatus of
Maxwell's treatment sometimes obscured the physical significance of *qualitative* results that could be derived from a cruder theory. One example is the
prediction that viscosity and heat conduction coefficients are independent of
density (except at densities several times larger or smaller than those normally
encountered). Maxwell's letter to Stokes (document III-1) gives a more comprehensible explanation of this result than that which can be found in his
published papers, and it was left to Clausius to point out explicitly that the
result applies to heat conduction as well as to viscosity.[45]

One study of special importance for Maxwell in 1860 was that on the
viscosity of gases by G. G. Stokes for a paper on the damping of pendulums.[46]
Maxwell had used Stokes's data in treating the hypothesis of gaseous rings for
Saturn.[47] He naturally made viscosity one of his first subjects for calculation
and found to his astonishment that viscosity should, according to kinetic
theory, be independent of the pressure of the gas. The experimental confirmation of this seemingly paradoxical prediction by Maxwell and his wife,
Katherine, in 1865 ultimately became the most telling piece of evidence in
favor of the theory.

Maxwell attributed the viscosity of gases to transfer of momentum between
moving layers of molecules. He proceeded by calculating the transverse momentum carried by a single molecule from a layer dz to another layer dz' moving past it with known velocity, and then applied Clausius's formula (Eq. (4)) to
find the probability of a molecule's starting in dz and ending in dz'. Integration
over all space gives the total friction drag and hence the viscosity coefficient μ:

$$\mu = \tfrac{1}{3}\rho\bar{l}\bar{v},\tag{9}$$

where ρ is the density, $\bar{l}$ the mean free path, and $\bar{v}$ the mean velocity of
agitation. Since $\bar{l}$ is inversely proportional to ρ, μ is independent of the pressure

of the gas. Although the number of molecules carrying momentum from one layer to the next increases with pressure, the average distances over which the momentum is carried decrease in the same ratio. The result is actually independent of the special assumption about molecular interactions and applies to any short-range force. On the other hand, the temperature dependence predicted by Eq. (9) is peculiar to the model; μ is proportional to $\bar{v}$ and thus to the square root of the absolute temperature. For other force laws the viscosity coefficient is found to increase as T^n, where n is generally some fraction between 1/2 and 1. Yet the fact that viscosity *increases* with temperature for gases goes against commonsense notions based on experience with liquids.

The conclusion that gas viscosity is independent of pressure was known as "Maxwell's law" among some nineteenth-century physicists, but this usage is no longer common. Maxwell himself could hardly believe it at first, and when he wrote to Stokes asking for experimental evidence to test the prediction (document III-1), he learned that the only data known at the time (an observation Sabine made in 1829) suggested that the viscosity of a gas *does* vary with pressure. Stokes later admitted that the analysis leading to that conclusion implicitly involved the assumption that the viscosity vanishes at low pressures. New experimental work by Maxwell and by other physicists inspired by Maxwell's theory established his "law" within a few years, and this success seems to have been a major factor in persuading scientists to adopt the kinetic theory of gases.[48]

The diffusion and heat conduction calculations proceeded along similar lines but were less successful in accounting for experimental results. The phenomena themselves are more complicated, and their eventual understanding was due in part to experiments stimulated by Maxwell's gas theory of 1867.

The general problem of heat transfer in gases can be traced back to Newton's law of cooling (1701), which seems to be the first attempt at a quantitative treatment. Newton observed the rate of cooling of a piece of hot iron under conditions that involved primarily convection with some radiation and conduction. Later experiments were motivated by the desire to separate these three modes of heat transfer.[49] This line of investigation culminated in the work of Dulong and Petit (1817), who measured the rate of cooling of various substances when surrounded by air at pressures down to 2 or 3 mm Hg. They assumed that the effect of heat transfer through air could be estimated by extrapolating their results to zero pressure, where the only heat transfer would be that due to radiation. The extrapolated zero-pressure heat transfer could then be subtracted from the heat transfer at a finite pressure, giving the rate of heat transfer at that pressure resulting from the effects of air (they did not distinguish between conduction and convection).[50]

According to Dulong and Petit, the total rate of cooling of a body at temperature $t + \theta$ surrounded by air pressure p and temperature t is the sum of two parts: the cooling due to radiation,

$$v_r = ma^\theta(a^t - 1),$$ (10)

and the cooling due to heat transfer through air,

$$v_a = mp^c t^b,$$ (11)

where a, b, c, and m are numerical constants (e.g., $b = 1.233$; $c = 0.45$). This result was known as the *Dulong-Petit law* throughout most of the nineteenth century. It should not be confused with what is now called the Dulong-Petit law, which relates specific heats and molecular weights, though in fact the measurement of specific heats was based on measurements of cooling times and thus depends indirectly on the heat transfer law.

When Maxwell began working on the kinetic theory in the late 1850s, physicists already distinguished between conduction and convection. Conduction, however, was thought of as a process of internal radiation of heat from one molecule to another in all states of matter. Any heat transfer resulting from the actual motion of the molecules was regarded as convection. The prevailing opinion was that it was extremely difficult to disentangle the effects of radiation, conduction, and convection in fluids, and Maxwell seems to have shared this opinion.[51] Yet his own theoretical ideas about heat conduction in gases played a major role in clarifying the experimental situation and indirectly led to the replacement of the Dulong-Petit radiation law (Eq. (10)) by the Stefan-Boltzmann law.[52]

In Maxwell's theory fast molecules move from hot regions of a gas to cooler ones and collide with slower moving molecules, giving up some of their kinetic energy. This was the process of heat conduction. It is analogous to viscosity, also attributed to the motion of fast molecules from regions where the average velocity is higher to regions where the average speed is lower, with a resulting transfer of excess momentum. In viscosity, however, there is an overall motion of the gas in a particular direction, and this overall motion varies from one region to another; in heat conduction the gas is at rest in every macroscopic region, whereas the intensity of the random thermal motion (which is the same, on the average, in every direction) varies from one region to another. Heat conduction must also be distinguished from convection, in particular, "free" or "natural" convection, in which density differences between hot and cold regions set up macroscopic currents in the gas. Heat conduction is due not to such coordinated motions of many molecules but rather to the random wanderings of individual molecules from one region to another. (It is sometimes called "molecular convection.")

Because Maxwell used his equilibrium distribution function, he assumed implicitly that the molecules moved equally in all directions from any infinitesimal region of the gas. The number of molecules that reach another region at a certain distance depends on the molecules' mean free paths. These paths in turn depend on the number of molecules in a unit of volume and the molecular diameter. Integrating these quantities over all emitting regions, with an exponential factor that decreases with distance from the plane (taking the mean free path into account), gives the rate of transfer of matter across an imaginary plane in the gas. Maxwell expressed this as

$$q = \frac{1}{3}\frac{d}{dx}(mn\bar{v}\bar{l}), \tag{12}$$

where m is the mass of the molecule, n the number of molecules, $\bar{v}$ the "mean velocity of agitation," and $\bar{l}$ the mean free path.

Maxwell assumed that he could simply replace the mass by the total kinetic energy of the molecule, $E = \frac{1}{2}\beta mv^2$, where β represents the ratio of the total energy to the internal energy of the molecule taken from Clausius, and obtain the rate of heat transfer across the plane.[53] The velocity gradient is replaced by the temperature gradient dT/dx, and J is the mechanical equivalent of heat:

$$Jq = -\frac{3}{4}\beta p l \bar{v} \frac{1}{T}\frac{dT}{dx}. \tag{13}$$

By putting in numerical values for pressure (p) and other parameters, Maxwell concluded that "the resistance of a stratum of air to the conduction of heat is about 10,000,000 times greater than that of a stratum of copper of equal thickness."[54]

From Eq. (13) we can see that ($3\beta p l \bar{v}/4T$), the thermal conductivity, is independent of density and proportional to the square root of the temperature, following the viscosity. Lighter gases, such as hydrogen, should have the greatest conductivity at a given temperature and pressure. This last result was (or should have been) especially significant because it was the unusually high thermal conductivity of hydrogen, compared with air and other gases, that led Gustav Magnus and other experimenters to the conclusion that gases must be able to transfer heat by conduction in addition to convection (convection would presumably be more effective in gases with greater specific gravity because it depends on density differences).[55] Maxwell did not state this conclusion explicitly or refer to the experimental results available.

Although Maxwell's treatment of heat conduction broke new ground in kinetic theory, there were errors of principle and of arithmetic in some of his calculations; Clausius exposed these errors (not without a certain scholarly relish) in an 1862 paper. The chief error lay in the use of an isotropic distribution function even when density and pressure gradients existed in the gas. This procedure led to incorrect numerical constants in various formulas and introduced an erroneous term in the heat conduction equation that implied that a steady flow of heat is accompanied by bodily motions of the gas.

Clausius offered a corrected theory, but because he persisted in assuming that all molecules have equal velocities, it too contained a wrong numerical constant. Maxwell wrote out his own revised theory of heat conduction after 1862 (document III-12) but never published the details as he had meanwhile become dissatisfied with the whole mean free path method.[56] The true value of Clausius's criticisms was in showing the need for a formulation of kinetic theory consistent with established macroscopic equations. Maxwell produced this in 1867.

For the third transport property, diffusion, there was actually a small body of accepted quantitative data available when Maxwell began his research. The German chemist J. W. Döbereiner had noticed in 1823 that hydrogen gas escapes rapidly through small cracks in glass containers. Thomas Graham, in Glasgow, followed up this observation by launching an extensive series of studies of diffusion phenomena.[57] He found that the rates of diffusion of different gases through a small opening are inversely proportional to the square roots of their densities. Graham did not initially attribute this behavior to differences in size or velocity of the atoms of the different gases; in fact he explicitly rejected Döbereiner's suggestion that hydrogen atoms go through holes more easily because they are smaller, saying that "we have no reason to suppose that the particles of hydrogen are smaller than those of the other gases."[58] Instead, he proposed the empirical concept of "equivalent volumes of diffusion" of gases, these volumes being inversely proportional to the square roots of the densities of the gases. Thus, taking the equivalent volume of air as 1, he suggested that the equivalent volume of hydrogen would be about 4. Diffusion was conceived as a process of *exchange* of equivalent volumes of one gas for another; when one volume of hydrogen leaks out, 1/4 volume of air comes in to replace it.

In later experiments Graham found that, although his law (that the rate of flow is inversely proportional to the square root of density) holds for effusion through thin plates and for diffusion of one gas into another, it does not apply to the flow of gases through longer tubes. As the tube is made longer, the velocity diverges from its value for diffusion or effusion, and eventually the time of flow becomes proportional to the length of the tube. The relative velocities of this kind of flow, which he called "transpiration," have constant ratios for different gases flowing through tubes of the same length, but these ratios are not the ones applicable to diffusion. Graham suggested that his results were comparable with those of Poiseuille for the flow of liquids through tubes; and that they "prove the incompleteness of the received view of gaseous constitution."[59]

In his experiments in the 1840s Graham also investigated the variation of the velocity of transpiration with density and temperature. He reported that "for equal volumes of air at different densities, the times of transpiration are inversely as the densities." [60] Although he did not state explicitly that the flow of a given *mass* of gas takes place at the same rate regardless of density, this conclusion could be extracted from his work with the advantage of hindsight and thus interpreted as a confirmation of Maxwell's prediction that gas viscosity is independent of density. The same is true of Graham's other conclusion[61] that the effect of temperature is to increase the time of flow for a given mass.

Graham's experiments were discussed at some length by John Herapath,[62] who pointed out that they provide excellent confirmation for the kinetic theory. Herapath, however, did not accept Graham's distinction between the phenomena of effusion and transpiration and failed to realize that the latter involved *internal* friction (viscosity) of the gas. Nevertheless, Herapath did receive Graham's praise for reviving the kinetic theory and "first applying it to the facts of gaseous diffusion." [63]

Maxwell's first theory of diffusion, in his 1860 paper, was based on two distinct physical effects: the random motion of molecules, governed by his velocity-distribution law, and an overall flow of the two gases through each other. The theory of the first effect, like that of heat conduction, was based on Eq. (12); in fact, Maxwell treated heat conduction as a special case of a general theory of diffusion. For diffusion through a porous plug, the effect gives a contribution to the diffusion rate that does not follow Graham's square root law, but "the closer the material of the plug, the less will this term affect the result." [64] Maxwell quickly abandoned this case because he did not think his assumptions were valid for it, and, moreover, he could not find a way of fixing the values of the four numerical constants that come into the theory.

Maxwell then turned to diffusion through a tube, and after making a number of assumptions and approximations, including the assumption that both gases have the same density, he derived the following equation:

$$\text{rate of diffusion} = \frac{dy}{dt} = -\frac{4}{3}\sqrt{\frac{2k}{\pi}}\frac{sl}{P}\frac{dp}{dx}, \tag{14}$$

where y is the volume of gas in a container, s is the cross section of the tube leading out of this container, P is the total pressure, p is the partial pressure of the diffusing species, and k is proportional to the average molecular speed. Although Maxwell used this equation to estimate a value for the mean free path, he did not discuss whether the equation itself was in harmony with experiment.

In his 1867 memoir Maxwell admitted that his earlier theory of gas diffusion was erroneous, and later he gave a revised formula for the diffusion coefficient of gases composed of elastic spheres.[65]

All the areas of triumph and difficulty for the kinetic theory of gases are rehearsed in these initial papers of 1860. Maxwell even touches on the specific heats problem and the distribution of energy between translational and rotational motions for rough spherical bodies. He found that after sufficient time these energies become equal; Maxwell was to return many times to this "equipartition theorem." As the theory stood in 1860, it could not explain the experimental result that the specific heats ratio, and hence the ratio of total to internal energy, is 1.408 for several gases. Theoretically, for point particles, γ should be 1.667 and for rough spheres, 1.33. This discrepancy "overturned the whole hypothesis" according to Maxwell's 1860 report to the British Association (document III-8). His closing sentence for the "Illustrations of the Dynamical Theory of Gases" was calmer but equally decisive:

Finally by establishing a necessary relation between the motions of translation and rotation of all particles not spherical, we proved that a system of such particles could not possibly satisfy the known relation between the two specific heats of all gases.[66]

On the plus side, however, Maxwell rederived a result previously deduced by both Waterston and Clausius, namely, the equality of the average translational kinetic energies of two sets of colliding particles:

$$m_1 \overline{v_1^2} = m_2 \overline{v_2^2}, \tag{15}$$

where m_1 and m_2 are the respective molecular weights.[67] On combining this result with the pressure equation (Eq. (2)), Maxwell obtained a kinetic theory justification for Avogadro's hypothesis before it was discussed by Cannizzaro in his famous "Sketch of a Course of Chemical Philosophy" distributed at the Karlsruhe Congress in 1860.[68] In 1872, Cannizzaro declared that chemical atomic theory was strongly supported by kinetic theory of gases, in part because Avogadro's hypothesis was common to both.[69]

Before the publication of Maxwell's first paper on kinetic theory, the transport properties of gases were investigated only rarely. The experiments themselves are inherently difficult, involving the measurement of very small quantities that can be obscured by competing transport processes unless extraordinary precautions are taken. (This is especially true of heat conduction, in which radiation and convection are often much more important.) In addition, the relation of transport properties to current physical theory before 1860 was tenuous, their investigation esoteric. Gases were considered as molecular, but the molecules were centers of force and static, the intermolecular forces keeping them apart. Transport coefficients were useless in the development of such a static theory of gases. While Graham was publishing on diffusion and Stokes on viscosity and while Magnus and Oskar Meyer were working on thermal conduction, the interest of the scientific community in most of this research was minimal.

All this changed with Maxwell's kinetic theory. The measurement of transport coefficients could now give quantitative values for the size and mean free path of a molecule, as Maxwell suggested. Moreover, by checking the prediction that the viscosity and heat conduction coefficients are independent of pressure, experimentalists could provide direct confirmation of the kinetic theory; and by determining the temperature dependence of the coefficients, they could obtain detailed information about the intermolecular forces. Thus molecular science became, instead of a collection of disjointed speculations, a respectable theory based firmly on empirical data.

It was from the group of experimentalists already measuring the transport coefficients that Maxwell gained the earliest and most positive support. Oskar Emil Meyer (1834–1909) had begun experiments on the viscosity of gases before the appearance of Maxwell's earliest papers on kinetic theory. Even so, Meyer found that his value for viscosity $\mu = 0.000360$ for air at $18°$ C (cgs units) was hardly compatible with the usual static molecular model for gases.[70] In the static model viscosity was attributed to forces between neighboring molecules, and one had to assume stronger forces between air molecules than those between water molecules. But he noted that such a result was compatible with the kinetic theories of Bernoulli and Clausius in which viscosity was a measure of the equalization of the velocities of molecules in different layers of the gas.

Thus, even before considering Maxwell's theory, Meyer was sensitive to its implications and therefore focused on the "striking result" of Maxwell's theory that viscosity should be independent of density or pressure and that it should increase with temperature, contrary to the behavior of liquids. After reading Maxwell, Meyer, a true experimentalist, set about establishing "*through measurement how the viscosity coefficient of air varies with pressure and temperature*" as a means of testing the "admissibility of the hypothesis of molecular collisions."[71] In analyzing earlier viscosity measurements, Meyer discovered the hidden assumption in Stokes's analysis. Yet for all his detailed care Meyer's result was not the unequivocal support for kinetic theory that one senses he had hoped for. The viscosity varied slowly with pressure and increased only slightly with temperature.[72] Theoretically, Meyer chose to follow Clausius and recognized that, at low pressures when the mean free path becomes large, the theory is not valid.

Despite his experimentalist's caution Meyer's work on viscosity helped to establish the legitimacy of kinetic theory.[73] Indeed, Meyer's career centered on viscosity measurements of gases, partially in an effort to understand molecular interactions. He closely analyzed earlier as well as contemporary experimental estimates, and in places links up with his brother Julius Lothar's concerns with molecular volume and molecular weight.[74]

Other early support for Maxwell's theory came from Josef Stefan, who, like Meyer had begun experimental work on gases before Maxwell's theory was published.[75] Thus, by the time Maxwell had completed his own measurements on the viscosity of air in collaboration with his wife, he had support for his theory and the results of other's experiments to consider. Maxwell observed the damping of a torsional pendulum consisting of a stack of horizontal disks 8 inches in diameter suspended from a long fiber and interleaved between a second set of stationary disks inside a sealed box. The pendulum was set oscillating by a magnet, and its amplitude and period were measured with the aid of a mirror attached to the suspension. By varying the number and spacing of the disks, losses from extraneous sources were eliminated, and the viscosity of gas in the vessel was determined over a range of pressures between 0.5 and 30 inches Hg. The temperature could also be varied between $0°$ and $100°$ C by surrounding the vessel with a steam, oven or ice. Over the ranges studied, the viscosity was found to be independent of pressure as predicted and nearly a linear function of the absolute temperature T.[76]

Maxwell's results were significantly different from those of Meyer, and he defended his method in a postscript. Meyer's disks were in an airtight vessel, whereas Maxwell's were merely protected from drafts. Maxwell varied the distance between the disks, and the motion of the air around the disk edges was material to his results; he appealed to William Thomson for a mathematical theory of this effect (document III-13). Finally, Meyer measured $\sqrt{\mu\rho}$, whereas Maxwell measured μ directly.[77]

Maxwell's paper was enthusiastically supported by William Thomson, and Thomson's call for more experimental work of that caliber was typical of early responses to Maxwell's kinetic theory:

My dear Stokes

I return Maxwell's paper on the Viscosity of air and other gases by book post. I am very glad to have had an opportunity of reading it before its publication in the Transactions. It is most interesting and valuable. The evidence it contains as to the accuracy of the results in absolute measure is very satisfactory.

I hope the author will be induced to continue the investigation for other gases and liquids.

The plan he adopts is quite the same as one I had long contemplated, and intended applying myself for liquids (except that I intended to have only one vibrating disc). It seems to me ⟨best⟩ better than any other plan adapted both for determining the viscosity, and the coefficient of slip between the fluid & the disc. I hope Maxwell will soon be able to settle the very important question of slip for liquids. Investigations of this kind are of national importance, and anything that money can do to promote them, whether by supplying a convenient experimental laboratory, and a sufficient number of thoroughly qualified operators to carry out the work, or in any other way, ought to be done by the government. If the government knew its own interest even on a strictly and simply economical grounds, it would do everything that money can do to promote the *execution* of good experiments. You of all others know how much there is between the planning & the execution of experiments for most important objects. Could not the Royal Society move government for the establishment of laboratories for *investigation* in which teaching would be thoroughly subordinate to the search for new knowledge of properties of matter.

Suppose such a laboratory was established in London with you or Maxwell, or both, in charge, the results would be worth hundreds of fold the annual cost, even in material economies wh [which] would arise from the knowledge gained. Besides, the directors of such institution or institutions would naturally form an advisory council for the govt (admiralty, army, customs etc) which would save hundreds of thousands wasted in useless experiments on a large scale for every thousand spent in keeping up the proposed laboratory. I do think the Royal Society might move in this matter with effect. Your always truly,

W. Thomson[78]

In these early reactions to Maxwell the problems that now seem so obvious in his first derivation of the distribution function remained undetected in the enthusiastic support for a theory that predicted quite unexpected yet confirmed results. The only negative voice was Clausius's, and he referred to Maxwell's use of the symmetrical distribution function to derive the transport properties, not the derivation of the distribution function itself.[79] One might wonder why Clausius criticized Maxwell on such grounds. Clausius showed little appreciation of the value of a velocity distribution function for any system, and he got an incorrect numerical factor in his formula for the mean free path because he did not make appropriate use of such a function.[80] Nevertheless, Clausius was a good critic and saw that any gradient of concentration, momentum, or energy in the gas must be reflected in the distribution function, otherwise there could be no transport. In heat conduction the molecules have an additional component of momentum in the direction in which temperature is changing, and this asymmetry has to be included in the velocity distribution.

Clausius then developed his own theory of heat conduction (see note 45), which became important in Maxwell's reassessment of his own theory. Clausius considered the rate of change of some (initially unspecified) property of the gas resulting from collisions of gas molecules and the corresponding change in a quantity defined on the molecular level (every transport coefficient involves the relation between a particular pair of (macroscopic) gas and (microscopic) molecular quantities). In thermal conduction the vis viva carried across a unit of a plan perpendicular to the x axis in unit time becomes

$$G = (1/2)mN \int_{-1}^{+1} I(\overline{c^3})\mu \, d\mu,$$

where m is the mass of a molecule, N the number of molecules per unit of volume, and I the proportion of the molecules whose velocity direction lies at an angle between μ and $\mu + d\mu$ with the x axis; I is not a symmetrical function; $(\overline{c^3})$ is the average of the cube of the velocity of the molecules. Corresponding expressions exist for viscosity and diffusion.

This was a promising start, but Clausius could not carry it through; his functions became so complicated that he had to make frequent power-series expansions and neglect higher-order terms. He never really improved on Maxwell's calculation of the velocity distribution function. His final formula for the thermal conductivity coefficient was almost identical to Maxwell's: When Maxwell's result was translated into the same variables, there was a factor of $1/2$ in place of the $5/12$ found by Clausius. Clausius took some pains to point out that this near-agreement was only apparent, that Maxwell had arrived at his result by erroneous reasoning and a fortuitous cancellation of two errors, for his formulas refer to the situation in which there is a progressive motion of mass as well as energy in one direction.

In commenting on his final result for heat conduction, Clausius noted that the coefficient increases as the square root of the absolute temperature but is independent of pressure. The former is similar to that of the velocity of sound and therefore requires no particular explanation. The latter

is explained by the circumstance that, although the number of molecules which can convey the heat is greater in a gas which is rendered more dense by increased pressure, the distances traversed by the individual molecules are smaller. This latter conclusion might lead to absurdity, if it were assumed to be applicable to the gas under every conceivable condition of compression or expansion. It must, however, be borne in mind that there are obvious limits to the application of it to conditions of the gas which depart very much from the mean condition: on the one hand, the gas must not be so much compressed as to produce a too great departure from the laws of permanent gases which have been taken as the foundation for the whole course of reasoning; and on the other hand, it must not be so much expanded that the mean length of excursion of the molecules becomes so great that its higher powers cannot be disregarded.[81]

The same remarks apply equally well to Maxwell's earlier results for viscosity, diffusion, and heat conduction, but Maxwell himself did not make them. All he said about the pressure independence of viscosity was that "such a consequence of a mathematical theory is very startling, and the only experiment I have met with on the subject does not seem to confirm it."[82] In the early 1860s, Clausius seemed much more committed to the kinetic theory and felt a need to justify its results and to point out the limits of its validity. He also had to deal with a few skeptics—Buys-Ballot, Hoppe, Jochmann, and others. So Clausius, not Maxwell, published the first qualitative explanation of the density independence of gas transport coefficients even though Maxwell discovered it (see document III-1).

The impact of Clausius's criticism can be gauged in Maxwell's initial attempts to correct his 1860 paper (the manuscript published here as document III-12).[83] Maxwell refers to Clausius in his first remarks:

M. Clausius has recently published an investigation of the particular case of the conduction of heat through a gas which was very imperfectly treated by me in the paper referred to. He has pointed out several oversights in my calculation, I have reexamined it and found some other[s] the influence of which extends to other parts of my investigation.

The additional oversight to which Maxwell referred was

the question which I neglected to consider in my former paper. When the density and temperature of a gas or the composition of a system of mixed gases vary from one place to another, what is the proportion of particles, which, starting from one given place, arrive at another given place without a collision.

Yet this attempt at a revision remained unfinished and unpublished. Although Clausius provided Maxwell with the skeleton of a new approach, his theory was still dependent on the mean free path concept. Maxwell tried to work with a variable path, depending on its position in the gas, but found this approach too cumbersome: it led to nonintegrable expressions; in addition when nonuniformities of pressure and temperature were present, the mean free path seemed to vary from one place to another in the gas. When the particular molecular model, using intermolecular forces, was considered, the mean free path was ill-defined: molecules traveled in complicated orbits with deflections and distances that were variable functions of the velocities and initial path. Yet some quantitative description of the heterogeneous structure of a gas was needed. Maxwell proceeded by using not a characteristic distance but a characteristic *time*, the "modulus of the time of relaxation" of stresses in a gas. This was a concept he afterward applied to many other areas of physics besides gas theory. Its origins and meaning depend on Stokes's theories of viscosity and elasticity.

Elasticity can be defined as a stress developed in a body in reaction to change of form. Both solids and fluids exhibit elasticity of volume; only solids are elastic against changes of shape. A fluid *resists* changes in shape (through its viscosity), but the resistance is evanescent: motion generates stresses proportional to velocity rather than to displacement. Nevertheless there are useful mathematical similarities between the theories of elastic solids and viscous fluids. In 1845 Stokes wrote a powerful paper giving a new treatment of the equations of motion of a viscous fluid.[84] His fundamental postulate was that the stresses on an element of fluid depend only on the displacement in its immediate vicinity. With this assumption and some general analytical arguments, he obtained equations of motion for a viscous medium, now generally called the *Navier-Stokes equations*. His concept of a fluid was not that embodied in the kinetic theory of gases:[85]

If we suppose a fluid to be made up of ultimate molecules, it is easy to see that these molecules must, in general, move among one another in an irregular manner, through spaces comparable with the distances between them, when the fluid is in motion. But since there is no doubt that the distance between two adjacent molecules is quite insensible, we may neglect the irregular part of the velocity, compared with the common velocity with which all the molecules in the neighborhood of the one considered are moving.

Nevertheless, Stokes's equations do provide a fairly good description of real gases, and two of his results impinge on Maxwell's revised kinetic theory. One concerns the relations between shear and compression in solids and fluids, the other the role of time.

Poisson had analyzed the elastic properties of solids, assuming that they are due solely to central forces between the constituent particles.[86] He found that the coefficients of shear and compression are in a fixed ratio and can be reduced to a single constant. Real solids, however, do not obey Poisson's relation. Stokes, therefore, rejected the molecular theory of elasticity and assumed ad hoc two independent elastic constants. A similar question arose in the theory of fluids: the dissipation of energy in shear and compression may depend on one or two coefficients of viscosity. Stokes gave a physical argument for retaining one coefficient only. (His assumption is now known to be false for nearly all real fluids, although it survived in the literature of kinetic theory well into the twentieth century.)[87]

Stokes had also observed in the 1845 paper that if the time derivatives in the equations of motion of a viscous fluid are replaced by spatial derivatives, they become the equations of stress for an elastic solid. Poisson had also noticed this transformation, but Stokes went further. He remarked that some substances commonly regarded as solids, such as pitch or glass, flow viscously when left for long periods; he applied that observation in a new explanation of the aberration of starlight, supposing that the luminiferous ether acts like a solid for high-frequency optical vibrations and like a viscous fluid in the relatively slow motions of the earth about the sun.[88] Other physicists also became interested in viscoelastic processes about the same time, among them William Thomson (Lord Kelvin) and J. D. Forbes. Forbes, one of the early British Alpinists, applied the idea to the motions of glaciers.[89]

In his first paper on elastic solids Maxwell assumed, as had Stokes, two independent constants and made several contributions to the theory, of which the most important was a conjectural explanation of Sir David Brewster's discovery of induced double refraction in elastically strained glass.[90] Maxwell again investigated hydrodynamical questions in his paper on Faraday's lines of force and only returned to elasticity when he had to correct for viscous effects in the torsion wires from which his viscosity apparatus was suspended.[91] Reflection on the problem led him to a new method of specifying viscosity that extended Stokes's treatment. In an ideal solid body free from viscosity, a distortion or strain S of any kind gives rise to a constant stress F according to the equation $F = ES$, where E is the coefficient of elasticity for that particular kind of strain. In a viscous body, F is not constant but decreases to zero gradually, and if the rate of relaxation of stress is proportional to F, the process can be described empirically by the differential equation[92]

$$\frac{dF}{dt} = E\frac{ds}{dt} - \frac{E}{\tau},\tag{16}$$

which gives an exponential decay of stress governed by a relaxation time τ.

Equation (16) gives the simplest possible representation of the phenomena discussed by Stokes. Maxwell may have formulated it by comparison with William Thomson's theory of inductive electrical circuits, inverting an analogy between mechanical and electrical systems, which Maxwell himself utilized in his "Dynamical Theory of the Electromagnetic Field."[93] The relation equates the viscosity μ of any substance with the product $(E_s \tau)$, where E_s is the instantaneous rigidity against shearing stresses. A given substance can depart from perfect solidity by having either small rigidity or short relaxation time or both. For an ideal solid, τ is infinite; for a glacier τ is of the order of 10^{-10} sec. A test that occurred to Maxwell was to examine a moving liquid for the effects of elastic distortion on polarized light, such as those discovered by Brewster and analyzed by Maxwell in his early paper on elasticity. After some difficulty Maxwell demonstrated in 1873 that a solution of Canada balsam in water exhibits temporary double refraction with a relaxation time of about 10^{-2} sec.[94]

One might suppose that if a fluid ever acts like a solid, it would only do so at high densities. But, as Maxwell discovered, things are not so simple. Consider a group of molecules moving about in a box. Their impact on the walls exert pressure. If the volume is changed from V to $V + dV$, the pressure changes by an amount dp proportional to $-p(dV/V)$. But in the theory of elasticity the stress dp in a substance arising from an isotropic change of volume is $-E(dV/V)$, where E is the coefficient of elasticity. The elasticity of a gas is proportional to its pressure. Suppose now that the pressure is reduced until the free paths of the molecules are much greater than the dimension of the box, and let the walls be flexible and rough so that the molecules rebound in random directions. In addition to the pressure caused by the normal components of molecular impacts, there is a continual transfer of the transverse components of momentum from wall to wall, and even though the box is flexible, it exhibits a resistance to shearing stresses. A rarefied gas behaves like an elastic solid! Moreover, the ratio of rigidity to compressibility is identical with that found by Poisson for a solid of stationary molecules governed by central forces. Such coincidences, Maxwell dryly remarked, "deserve notice from those who speculate on theories of physics."[95] Let us call this property of a rarefied gas *quasisolidity*. Following the ideas expressed in Eq. (16), the relaxation of quasisolidity by mutual collisions between molecules can represent the viscosity μ of the gas. Since $\mu = E_s \tau$ and E_s is proportional to the pressure p, μ must be proportional to p. But Maxwell's experiments indicate that μ is independent of p and directly proportional to the temperature T. With p proportional to T from Eq. (1), the time τ is independent of T, that is, of the mean square velocity of the molecules. This implies that the number of encounters between pairs of molecules producing a given angular deflection is independent of the molecules' initial relative velocities. Such a result holds only when the force between the molecules varies inversely as the fifth power of the distance.

Two more observations on gases as quasisolids are of interest. As the pressure increases, a given group of molecules passes from being a quasisolid to a gas, from a gas to a liquid, and then to a true solid. The relaxation times for both a true solid and a quasisolid approach infinity; there is a minimum value for τ in some intermediate condition. The viscosities of liquids and gases may be expected on quite general grounds to have reciprocal aspects, and this is borne out by experiment. The viscosity of a gas increases with temperature, that of a liquid decreases. The second observation concerns the relation between relaxation time and mean free path. Although the concept of mean free path becomes elusive when there are forces between molecules, some link must exist between the two. Maxwell worked out this link in his 1879 paper in a footnote added in response to a query by William Thomson.[96] For a gas composed of elastic spheres the product of τ and the mean velocity $\bar{v}$ of the molecules is a characteristic distance λ, whose ratio to the mean free path $\bar{l}$ is $8/3\pi$. The concept of free path is a special formulation of the relaxation concept and is applicable only to freely colliding particles of finite diameter.

Maxwell's interest in elasticity and his understanding of the limitations of the mean free path idea came together in his second gas theory in which he used the center-of-force molecular model. To calculate the motions of a pair of molecules subject to an inverse nth-power repulsion was a straightforward exercise in orbital dynamics. Maxwell expressed the result in terms of four variables: b, the distance between parallel asymptotes before and after an encounter; V_{12}, the initial relative velocity; θ, the angle between the line of apses; ϕ, the angle determining the plane in which V_{12} and b lie. One difficulty was the possible transfer of energy between translational and internal motions of the molecules: Maxwell took the amount transferred during any one encounter to be negligibly small. The problem then was the statistical specification of encounters between large numbers of molecules obeying the same force law. As in his first paper, Maxwell studied a mixture of two gases with molecular weights M_1, M_2 and wrote the number dN_1 of molecules of type 1 with velocities ranging between ξ_1 and $\xi_1 + d\xi_1$, etc., as $f(\xi_1, \eta_1, \zeta_1) d\xi_1 d\eta_1 d\zeta_1$, with a similar expression for molecules of type 2. The velocities being defined, the relative velocity V_{12} of molecules in the two groups is also a definite quantity, and the number of encounters between them occurring in time δt is determinate. It is $V_{12} b \, db \, d\phi \, dN_1 \, dN_2 \, \delta t$.

Consider now some quantity Q describing the motion of molecules in group 1; Q may be any power, or products of powers, of the velocity or its components. Let Q be changed to Q' by the encounter. The net rate of change in the quantity for the group is $(Q' - Q)$ times the number of encounters per second, or

$$\frac{\delta}{\delta \tau} Q \, dN_1 = (Q' - Q) V b \, db \, d\phi \, dN_1 \, dN_2. \tag{17}$$

Equation (17) is the fundamental equation of Maxwell's revised transfer theory. It replaces the earlier equations based on Clausius's probability (Eq. (4)); those equations were restricted to elastic molecules and failed in accuracy because the mean free path is a variable function of position and direction. By integrating with respect to b, ϕ, N_1, and N_2, the total rate of change $(\delta/\delta\tau)(QN_1)$ is found for different Q, and equations are obtained that have terms identifiable with the coefficients of diffusion, viscosity, thermal conductivity, and other properties of gases.

The integrations with respect to ϕ are straightforward: they result in formulas containing M_1, M_2, ζ, b, θ, V_{12}, etc., in which b and θ enter in only two combinations, $B_1 = \int 4\pi b \, db \sin^2 \theta$ and $B_2 = \int \pi b \, db \sin^2 2\theta$. Now θ, the angle of deflection, is a function of b and V that depends on intermolecular forces. With a force proportional to r^{-n}, the integrals B_1, B_2 reduce to definite numerical quantities for each n, and the resultant integrations with respect to the number of molecules on substituting for dN_2 assume the form

$$\int \int \int Q V_{12}^{(n-5)/(n-1)} f_2(\xi_2, \eta_2, \zeta_2) \, d\xi_2 \, d\eta_2 \, d\zeta_2.$$

In general, a knowledge of f_2 under nonequilibrium conditions is necessary to perform the integrations. However, when $n = 5$, the relative velocity V_{12} disappears from the integral, and the final values can be written immediately as $\bar{Q} N_2$, where $\bar{Q}$ is the mean value of the quantity and N_2 is the total number of molecules of type 2. Physically, this simplification can be understood, as Boltzmann afterward pointed out, by noticing that the number of deflections through a given angle is a product of two factors, one of which (the cross section for scattering) decreases with V_{12} while the other (the number of collisions) increases with V_{12}. When $n = 5$, the two factors exactly balance. By introducing a new variable Φ such that $2 \cot^2 \Phi = b^4 \{V_{12}^2 M_1 M_2 / K(M_1 + M_2)\}^{4/n-1}$, Maxwell evaluated θ as $\pi/2 - \cos 2\Phi \, F_{\sin \Phi}$, where $F_{\sin \Phi}$ is the complete elliptic integral of the first kind. He tabulated values, drew a diagram of the scattering of molecules subject to inverse fifth-power repulsion from a point, and, most important, determined the numerical constants A_1 and A_2 corresponding to the integrals B_1 and B_2.

With this Maxwell was in a position to calculate physical properties of gases. Even with the simplification of an inverse fifth-power repulsion the mathematical task remained formidable, and an impressive feature of the paper is the notation he developed to keep track of different problems. It was, for example, necessary to calculate variations of quantities among a group of molecules arising from encounters with molecules of the same kind, molecules of a different kind, and the actions of external forces. Maxwell denoted the three processes by differential symbols $\delta_1, \delta_2, \delta_3$. Next he made a change of variables, using u, v, w to represent the mean translational velocity of the gas and ξ, η, ζ to represent deviations of individual molecules from the mean. He imagined a plane perpendicular to the x axis moving with a velocity whose x component is u', which in general is different from u. The expression for the quantity of Q that crosses the plane in unit time is

$$(u - u')\bar{Q}N + \overline{\xi Q}N, \tag{18}$$

where $\bar{Q} = \int Q \, dN$ is the average value of Q for all molecules of a given kind in the volume element.

With Q representing the mass M of a molecule, Eq. (18) gives the rate of transfer of matter across the plane; with Q representing momentum, Eq. (18) can be used to calculate systems of pressures in the gas; with Q, energy, it gives fluxes of heat.

Next Maxwell considered variations of Q within an element of volume of the gas through the actions of encounters or the external forces on molecules within the element or through the passage of molecules to or from the surrounding region. Denoting variations of the first kind by the symbol δ and variations of the second kind by ∂, Maxwell got from the theory a generalization of the standard equation of hydrodynamics:

$$\frac{\partial}{\delta t}\bar{Q}N + QN\left(\frac{du}{dx} + \frac{dv}{dy} + \frac{dw}{dz}\right) + \frac{d}{dx}(\overline{\xi Q}N) + \frac{d}{dy}(\overline{\eta Q}N) + \frac{d}{dz}(\overline{\zeta Q}N) = \frac{\delta}{\delta t}QN. \tag{19}$$

With Q as the mass M, Eq. (19) becomes the equation of continuity; that equation can then be substituted into the general version of Eq. (19), reducing it to a more useful form; with Q as the momentum, Eq. (19) becomes the equation of motion of the gas. From the equation of motion Maxwell derived Dalton's law of partial pressures, as he had in 1860, and formulas for the diffusion that could be compared with the experiments of Graham. With Q as the energy, Eq. (19) reduces to an equation that yields the law of equivalent volumes, formulas for specific heats, thermal effects of diffusion, and the coefficients of viscosity in simple and mixed gases. The viscosity equation for molecules subject to an inverse fifth-power repulsion, which replaces Eq. (9), is

$$\mu = \frac{1}{3}\frac{1}{kA_2}\frac{p}{\rho},\tag{20}$$

where $A_2 \cong 1.3682$, $k = \sqrt{K/2M^2}$, M is the mass of a molecule and K is the scaling constant for the intermolecular force. Because for an ideal gas $p/\rho \propto T$, μ is a linear function of temperature. Finally with Q a quantity of the third order in ξ, η, ζ, Eq. (20) gives effects of temperature gradients; from it Maxwell derived a formula relating the thermal conductivity of a gas to its viscosity, density, and specific heats ratio.

The calculations given by Maxwell in 1867 were the first mathematically exact investigations of transport properties in any gas model. Maxwell's viscosity equation, like that of Stokes, used a single coefficient to determine the dissipation of energy under both shear and compression. (For years, despite Stokes's own doubts, physicists applied the restricted equation to all fluids, until L. Tisza in 1942 pointed out that many discrepancies in the absorption of ultrasonic waves in gases and liquids disappear when the "Stokes condition" is abandoned.)[97] It is now known that the only fluids that obey the condition rigorously are systems of elastic spheres at low density. Thus, for monatomic gases at least, Maxwell's formulas are nearly correct. The grounds for his conclusion are therefore worth examining. At first sight he seems merely to slip in the Stokes condition by tacit assumption. Early in the paper he defines the mean pressure p at a point in the gas as $(1/3)(p_{xx} + p_{yy} + p_{zz})$, the average of the stresses along three orthogonal axes, and that definition automatically yields the condition. There is, however, an intriguing subsidiary argument given later on that can properly be used to justify the condition and that certainly strengthened Maxwell's conviction in its favor. This is the point already discussed: that the gas approximates at low pressures a quasisolid composed of stationary molecules governed by central forces. It is not hard to show that particles without internal degrees of freedom, such as those in a monatomic gas, indeed yield the same relation and satisfy Stokes's condition at higher pressures. Other gases depart from Stokes's condition through a variety of more complex molecular motions. Although Maxwell did not spell out all the implications, he left enough clues to indicate how a more complete analysis of the viscosity equation might proceed.

Again Maxwell's purpose in studying gases was not the theory of heat or that of gases but the theory of matter. Clearly his two theories of the transport properties of gases were based on completely different molecular models. Experimentalists, therefore, were not merely arguing over the relative merits of various experimental methods, data, and their analysis; they were probing for the first time the intimate details of molecular structure.

One of the more assiduous of these experimentalists was O. E. Meyer. In

1873, Meyer published some new results for a rather limited temperature range (0° to 21° C) that he said agreed with the formula

$$\mu = \mu_0(1 + \alpha t)^{3/4} = \mu_0(T/273)^{3/4}. \tag{21}$$

Despite his previous concession that Maxwell's experimental results had been superior to his own, Meyer now suggested that Maxwell's temperature dependence was incorrect because of a failure to control the temperature of the apparatus precisely enough.[98]

Maxwell's new viscosity theory was also disturbing on theoretical grounds. Having viewed Maxwell's first theory as being solidly based on the elastic-sphere model, Meyer could not easily follow Maxwell's switch to what he saw as a completely incompatible hypothesis, the model of the atom as a point center of force. Moreover, Meyer pointed out that the Joule-Thomson experiment indicated that the long-range forces between gas molecules must be attractive, not repulsive as Maxwell was now assuming. (Of course such an attraction was equally inconsistent with his philosophical commitment to elastic spheres, but Meyer did not seem to be able to resolve that problem.)

Although Meyer clearly would have preferred to retain the elastic-sphere model, he recognized that the $\sqrt{T}$ behavior predicted by this model did not agree with anyone's experimental data. He attributed this discrepancy to failure to take account of the internal motions of molecules. If the data is forced to fit the elastic-sphere formula, the diameter of the spheres must decrease with increasing temperature. This might seem surprising at first, Meyer said, because if the internal parts of the molecule acquire more energy, one would expect the size of the molecule to *increase* with temperature. But the contradiction can be resolved by noting that, at higher temperatures when two molecules collide, they can penetrate further into each other because there is more "open space" between their parts. Meyer was aware that Josef Stefan had already proposed a temperature-dependent diameter, but he did not accept Stefan's rationalization of this model in terms of ether envelopes.

Further experiments by Meyer, Puluj, von Obermayer, Warburg, and Barus gave temperature exponents in the range from 0.65 to 0.96 for various gases.[99] For a time there was a tendency for experimenters to fit their data with simple fractional exponents, presumably in the expectation that this would ultimately reveal some profound fact about the forces between gas molecules.

In his paper on rarefied gases (1879), Maxwell cited some of these experiments and stated that they "show that the viscosity of air varies according to a lower power of the absolute temperature than the first, probably the 0.77 power."[100] Maxwell ignored the already evident fact that this exponent would have significantly different values for some gases. Indeed he was not concerned about using a "realistic" model at this point; to complete his mathematical calculations he preferred to continue using his inverse fifth-power model, which implied a viscosity coefficient directly proportional to absolute temperature.

In 1900, Lord Rayleigh showed by dimensional analysis that if one assumes that the viscosity coefficient is independent of density, then there is a simple relation between the exponent n in the interatomic force law of the form $F \propto r^{-n}$ and the temperature dependence of the viscosity coefficient,

$$\mu \propto T^{(n+3)/(2n-2)}. \tag{22}$$

This includes as special cases Maxwell's two results for elastic spheres (considered as a limit, $n \to \infty$) and inverse fifth-power repulsion.[101] At that time no one had completed the calculations of the coefficient for any other values of n. This was first done in 1916 and 1917 by Chapman and Enskog, who generalized Rayleigh's formula.[102] On the other hand, even by 1900 it was already suspected that the formula $\mu \propto T^n$ might not be adequate even for a single gas for *any* value of n; a more complicated molecular force law might therefore be needed.[103]

Thermal conductivity is more difficult than viscosity to investigate, theoretically as well as experimentally, but Maxwell accomplished considerable progress when he showed that the two processes are closely related. His final result (1867) for the coefficient for thermal conductivity was

$$C = \frac{5}{3(\gamma - 1)} \frac{p_0}{\rho_0 - \theta_0} \frac{\mu}{s}, \tag{23}$$

where γ is the ratio of specific heats, μ is the viscosity, s is the specific gravity, and p_0, ρ_0, and θ_0 are the pressure, density, and temperature of a standard gas. Aside from the numerical factor 5/3, this formula is now considered correct for the special case of Maxwellian molecules and nearly correct for monatomic gases in general.

Maxwell's estimation of numerical values for particular gases was still rather casual (he was not aware of any experimental data and did not even cite the papers of Magnus). He estimated that "iron at 25° C. conducts heat 3525 times better than air at 16°.6 C.,"[104] and rather than give a direct comparison with Clausius's theoretical formula, he stated:

M. Clausius, from a different form of the theory, and from a different value of μ, found that lead should conduct heat 1400 times better than air. Now iron is twice as good a conductor of heat as lead, so that this estimate is not far different from that of M. Clausius in actual value.[104]

Since he assumed that the ideal gas law holds, the ratio $P_0/\rho_0\theta_0$ is a constant, and the thermal conductivity coefficient, like the viscosity coefficient, is independent of density but directly proportional to the temperature. On the other hand, Clausius's theory, like Maxwell's 1860 theory, was based on the elastic-sphere model and predicted a $\sqrt{T}$ law. So far there was no experimental data pertaining to this point. Maxwell noted at the end of this paper that oxygen, nitrogen, and carbonic acid should all have roughly the same conductivity as air; he did not mention the experiments indicating that hydrogen has a much greater conductivity, although this fact had been noted by Clausius in 1862 and was still the only confirmation of the theory available in 1867.

Responses to the implications of Maxwell's transport coefficient theory came mainly from experimentalists. Theoreticians were silent, except for Ludwig Boltzmann, who pursued kinetic theory for his own interests in thermodynamics.[105] Despite his different focus on the second law of thermodynamics and the microscopic structure of matter that would make that law sensible, in 1872 Boltzmann developed his own version of the kinetic theory based on an integrodifferential equation for the velocity distribution function.[106] His results for the transport coefficients of a gas of point particles with inverse fifth-power repulsive forces would have been identical with Maxwell's except that he discovered a mistake in Maxwell's calculation: the factor 5/3 in Eq. (23) should have been 5/2.

Boltzmann expressed the relation among heat conduction, viscosity, and specific heat in almost exactly the standard form that was adopted later:

$$k = \tfrac{5}{2}\mu c_v, \tag{24}$$

where k is the coefficient of thermal conductivity.

As Boltzmann remarked, "if one includes the intramolecular motion following Maxwell's method," i.e., by introducing the parameter β to represent the ratio of total molecular energy to translational energy, then the same formula should also apply to polyatomic gases. But Boltzmann does not accept this.

[T]his seems very arbitrary to me, and if one includes intramolecular motion in some other way, he can easily obtain significantly different values for the heat conduction constant. It appears that an exact calculation of this constant from the theory is impossible until we know more about the intramolecular motion.[107]

By this time Josef Stefan had published an experimental determination of the conductivity coefficients for several gases, so Boltzmann concluded that "our experimental knowledge is here much better than our theoretical knowledge." (That assessment suggests that Boltzmann was overly concerned with determining numerical values of coefficients and underrated the substantial advance in physical understanding achieved by Maxwell's theory.)

The first experimental study of heat conduction in gases designed to test the kinetic theory was published in 1871 by Friedrich Narr.[108] He cited Clausius's formula for heat conduction in the form

$$\frac{\text{velocity of cooling}}{\text{temperature of gradient}} = m \propto \frac{\gamma}{\sqrt{\sigma}}\varepsilon, \tag{25}$$

where γ is the ratio of specific heats, σ is the specific gravity, and ε is the mean free path. Assuming that ε is the same for all gases and taking $m = 100$ for air, Narr's computed values for m were close to those calculated from Clausius's formula for air, nitrogen, and carbon dioxide but almost twice that for hydrogen. He suggested that the discrepancy for hydrogen can be attributed to a longer mean free path because the hydrogen molecule is smaller than other molecules. Narr gave only relative values for the different gases, so his data cannot be directly related to the mean free paths found by other methods.

A more thorough investigation of heat conduction was undertaken by Josef Stefan. Stefan, one of Boltzmann's teachers in Vienna, had already published papers on various problems in gas theory and hydrodynamics, although he had not participated in the development of kinetic theory (aside from a brief note on the speed of sound in 1863). In 1872, he reviewed the various difficulties encountered by previous experimenters and claimed to have a more satisfactory method based on observing the cooling rates of thermometers immersed in gas. For the coefficient of thermal conductivity of air, he found a numerical value almost exactly the same as that deduced theoretically by Maxwell in 1867.

Few physical theories have produced such strikingly confirmed predictions, and one must indeed regard the dynamical theory of gases as one of the best founded physical theories.

Also another law, which is given by this theory, namely the independence of the conductivity from density, has been proved correct in a completely indisputable way by this experiment.

Similarly the relative values of the conductivities of different gases agree with the formulae of Maxwell.[109]

This idyllic state of perfect agreement between theory and experiment did not last long. By the time Stefan's paper appeared, Boltzmann had already discovered an arithmetic error in Maxwell's calculations, and when this error was corrected, the theoretical value was 50% larger.[110] Stefan suggested that if intramolecular motion does not contribute to heat conduction to the extent supposed in Maxwell's theory, the discrepancy might disappear.[111]

In 1875, Adolph Kundt and Emil Warburg published an important series of experiments on heat conduction, performed at low pressures in the belief that convection currents diminish, whereas if Maxwell's theory was valid, the conductivity maintains a constant value.[112] Their results indicated that the conductivity is indeed independent of pressure down to about 1 mm Hg for air and down to about 9 mm Hg for hydrogen. They also suggested that at lower pressures there is a temperature discontinuity at the interface between a hot solid and a cooler gas, analogous to the "slipping" (velocity discontinuity) that they found in their viscosity studies, but they were not able to establish this phenomenon experimentally.

Kundt and Warburg noted that, because air maintains its thermal conductivity down to such low pressures, Dulong and Petit's cooling law was invalid. The experimental proof of this law, by Dulong and Petit and subsequently by de la Provostaye and Desains, depended on the assumption that the heat loss from a thermometer into a "vacuum" depends entirely on radiation, aside from a small calculate correction. However, Dulong and Petit did not attain pressures lower than 0.5 or 1 mm Hg, and a recalculation (done by Kundt and Warburg) using contemporary data showed that the heat loss by conduction in their experiment would have been from 20% to 50% of that by radiation.[113]

The experimental difficulties facing all experimentalists trying to measure such small quantities as gaseous transport coefficients (so easily masked by other effects) and the real importance attached to these measurements because of their relation to molecular theories of matter can be seen in the work of A. Winkelmann. His first results on thermal conductivity roughly agreed with the theoretical formulas for elastic spheres and inverse fifth-power repulsive forces.[114] He then investigated the temperature dependence and found that the conductivity coefficient is directly proportional to the absolute temperature, as predicted by Maxwell for particles interacting with inverse fifth-power forces. However, Winkelmann argued that this theoretical model was untenable because, according to the results of the Joule-Thomson experiment, the long-range forces between molecules must be attractive not repulsive. He regarded his experimental data as a confirmation of Maxwell's temperature law but not of his atomic model. Winkelmann also noted that a modification of the Dulong-Petit cooling law was needed.[115]

In less than a year, however, Winkelmann retracted his result that thermal conductivity was proportional to the temperature.[116] He had not corrected for the decrease in the specific heat of mercury with temperature. With this correction the heat conduction coefficient varied as $T^{3/4}$, as did the viscosity.[117] His main concern was to establish the value of the ratio of the heat conduction coefficient to the product of the viscosity coefficient and specific heat at constant volume, $k/c_v\mu$. According to Maxwell's theory this ratio, later denoted by f and now usually called the Eucken factor, should be a constant characteristic of a given gas. Boltzman, however, argued that if only the translational motion of the molecules was transferred in collisions, then $f = 15(\gamma - 1)/4$, but if the total energy of the molecule was involved, $f = 2.5$ (Eq. (25)). Comparison with the experimental data of Stefan, Kundt and Warburg, and Winkelmann seemed to indicate that the actual conductivity lay somewhere in between these two values but was closer to the first one. Boltzmann proposed the semiempirical formula

$$f = \frac{3}{13}(2.5) + \frac{10}{13}\frac{15(\gamma - 1)}{4}$$

as the best fit to the data.[118]

The factor f was pursued further by Meyer. In his textbook he presented a new calculation of the heat conduction coefficient for a gas of elastic spheres for which he obtained $f = 1.537$. Later recalculations led to $f = 1.6027$.[119] Although Meyer's value of 1.6027 was in better agreement with some experimental data than the value of 2.5 that follows from Maxwell's theory, the subsequent work of Chapman and Enskog showed that it did not have as good a theoretical basis.[120]

For modern theorists, experimental results for monatomic gases are of much greater interest than those for polyatomic gases or mixtures, such as air. It had already been realized in 1876 that the values for polyatomic gases differed from theoretical predictions because of the transfer of internal molecular energy in collisions, even though the theory might be valid for point particles or elastic spheres.[121] The first known monatomic gas was mercury vapor. Schleiermacher measured its thermal conductivity by using Koch's earlier viscosity measurement and reported the value $f = 3.15$.[122] This result stood in the literature as an anomaly until 1959, when Zaitseva pointed out that if a more accuate value of the viscosity coefficient is used to reduce Schleiermacher's data, the value of f is 2.69.[123] In the meantime, Schwarze and Bannawitz had obtained values close to 2.500 for the monatomic rare gases argon, helium, and neon, and there was never much doubt about the validity of the theoretical value 2.5 for monatomic gases.[124] For polyatomic gases the situation was more complicated. Until fairly recently, kinetic theorists had not been able to improve on Eucken's 1913 semiempirical formula for f.[125] It must be admitted that the classical kinetic theory cannot give a completely satisfactory quantitative treatment of heat conduction without invoking some arbitrary postulate about the extent to which internal molecular energy is transferred in collisions. Aside from this defect, which affects only the numerical values of the coefficients, the kinetic theory of heat conduction has been reasonably successful in dealing with a property that has always presented rather severe experimental difficulties.[126]

Extensive work has been done on the temperature dependence of the coefficient of thermal conductivity. Originally it was hoped that experimental measurements could help to choose between the original elastic-sphere model (which predicts $k \propto \sqrt{T}$) and the inverse fifth-power force model (which predicts $k \propto T$). Despite a few early attempts to interpret the experimental data as confirming one or the other of these hypotheses, it soon became clear that for real gases the exponents of T are somewhere between 1/2 and 1, as in the case of viscosity, and the difficulty of getting accurate thermal conductivity data makes this approach less effective as a means of determining the intermolecular force law than the use of viscosity and equation of state data.[127]

Equally important for understanding the precise mechanical nature of the molecule was the third transport process, diffusion. In the early 1870s Maxwell conducted a detailed examination of experimental work and its impact on molecular models. His attention focused on the experiments of Josef Loschmidt, who had examined the temperature and pressure behavior of the diffusion coefficient D, predicted by Maxwell's theory to be

$$D = c\frac{\mu T}{p}, \tag{26}$$

where c is a constant that depends on molecular structure. For elastic spheres, $D \propto T^{3/2}$; for inverse fifth-power forces, $D \propto T^2$; and for both models $D \propto 1/\text{pressure}$.[128] Loschmidt, however, assumed that the kinetic theory predicted only $D \propto T^2$ without discriminating between molecular models.[129] Loschmidt was himself struck by the argeement between theoretical prediction and experimental results with respect to temperature and pressure behavior of the diffusion coefficient. Loschmidt suggested that for different gases the diffusion coefficients might vary with molecular mass according to the formula

$$D = D_0/\sqrt{m_1 m_2}, \tag{27}$$

but as Maxwell pointed out later,[130] this formula had no theoretical basis and did not even represent the experimental data well.

In 1871, Josef Stefan published an extensive discussion of the macroscopic theory of diffusion as related to kinetic theory,[131] and in another paper the following year he examined the problems of determining parameters for pairs of dissimilar molecules.[132] Maxwell's second theory (1867) led to an expression for the diffusion coefficient which could be put into the form

$$D = \frac{1}{A_{12}} \frac{p_0 p_0}{\rho_1 \rho_2} \frac{T^2}{T_0^2} \frac{1}{p}, \tag{28}$$

where p_0 is the normal (reference) pressure and ρ_1, ρ_2 are the densities of the two kinds of gas in the mixture.[133] Stefan used the notation A_{12} in place of Maxwell's A_1 to indicate that the constant must depend on the force between two different kinds of molecules. According to Maxwell's theory, this constant is independent of pressure and temperature, assuming an inverse fifth-power repulsion between the molecules; in this case D would be proportional to T^2/p, the law that Loschmidt had already confirmed by his experiments. If, Stefan pointed out, the inverse fifth-power model is adopted, there is no unambiguous way of deciding on a value of the force constant for dissimilar molecules, but if the elastic-sphere model is used, it can be assumed that the effective collision diameter is the arithmetic mean of the diameters of the two molecules:

$$d_{12} = (1/2)(d_1 + d_2).\tag{29}$$

From this Stefan deduced

$$A_{12} = \frac{4\pi d_{12}^2}{3}\frac{\sqrt{v_1^2 + v_2^2}}{m_1 + m_2}.\tag{30}$$

Because this formula depends on the molecular velocities, it appears that A_{12} is proportional to $\sqrt{T}$, and therefore $D \propto T^{3/2}$.[134]

Values of d could be estimated from the viscosity data of Maxwell and Meyer; Stefan found that when these values were substituted into Eq. (28) for D, using Eq. (30) for A_{12}, the results were roughly 5% to 10% greater than Loschmidt's experimental values; Stefan considered this to be satisfactory agreement.

Stefan also considered an alternative formula given by Clausius for the number of collisions between fast and slow molecules and derived an expression for A_{12} that gave equally good agreement with the experimental diffusion constants. He nevertheless rejected it because it was theoretically inferior. Stefan argued for the elastic-sphere model because it correlated the different properties of gases effectively, even though it could not include the internal structure and motions of polyatomic molecules. On the other hand, the $T^{3/2}$ law for the diffusion constant predicted by the elastic-sphere model appears to be incorrect. The difficulty disappears, however, if it is assumed that the molecular diameter depends on the velocities of the colliding molecules—a natural assumption if the molecules are really centers of repulsive force. With a velocity- or temperature-dependent molecular diameter, chosen so that the effective repulsive force varies inversely as the fifth power, the advantages of both of Maxwell's models can be combined (though probably at the cost of sacrificing Maxwell's standards of rigor).

Stefan, however, along with many of his colleagues in Germany, still accepted the atom as a solid nucleus surrounded by an atmosphere of ether (*Aetherhülle*). Since any other model, such as the elastic sphere or the point center of force, could be regarded only as an approximation to this picture, there could be no fundamental inconsistency in mixing the two models.

Maxwell's kinetic theories were dense with meanings for the mechanical molecular theory of matter of which Maxwell was much aware and to which he gave more and more attention after 1870. Despite his skepticism expressed in a letter to Stokes in October 1859 as he composed his first extended paper on kinetic theory,

I intend to arrange my propositions about the motions of elastic spheres in a manner independent of the speculations about gases, and I shall probably send them to the *Phil. Magazine,* which publishes a good deal about the dynamical theories of matter & heat,[135]

Maxwell was writing in the same era to William Thomson,

I shall be glad to know the max^m breadth of atoms. I suppose their length is not so easily limited, but if you could get a minimum breadth, you would go far to establish the existence of the atom.[136]

His more specific interest in quantitative estimates of molecular size appears to have been inspired by Loschmidt's work. Loschmidt, using the results of Clausius and Maxwell, was in fact the first scientist to derive a reasonable estimate of the size of a molecule from the kinetic theory of gases, and his experimental data on diffusion were used in turn by Maxwell for similar purposes. Maxwell's reasons for doing so were twofold. Consistent estimates of molecular diameter were added proof for his theory and, more importantly, confirmed his original goal for the theory itself, probing the structure of the molecule.[137]

Before the development of the kinetic theory, estimates of the diameter of molecules were not crucial to any physical or chemical theory, although one finds occasional order-of-magnitude guesses in the early literature. Probably the best founded of these was Thomas Young's estimate of the range of cohesive force involved in surface tension and capillarity, about a 250 millionth of an inch.[138]

In 1859 Clausius would only hazard a guess that the diameter of a molecule is much smaller than its mean free path, and even in 1860 Maxwell could only estimate the latter, but by 1870 four independent estimates of molecular diameter were available. The first, by Loschmidt in 1865, estimated molecular diameter explicitly from the condensation ratio between gases and liquids, assuming that the molecules in a liquid are nearly in contact, and then used Maxwell's value of l to estimate d. The result was a diameter of about 5 × 10^{-8} cm for a typical molecule.[139] Similar estimates were made independently by G. Johnstone Stoney, L. V. Lorenz, and William Thomson.[140]

Thomson's estimate was derived from several different phenomena, all of which could have been used to estimate molecular diameters *before* kinetic theory. Thomson was well aware that obtaining a quantitative value for the molecule's diameter brought the notion of the molecule out of the realm of metaphysics to which it had been sent by its vagueness and the unanswerable questions it presented to chemists. The atom had become "a real portion of matter occupying a finite space, and forming a not immeasurably small constituent of any palpable body."[141]

The determination of molecular size from the kinetic theory of gases permitted at the same time an estimate of the number of molecules in a given volume of gas. This led to what is now known as *Avogadro's number*, the number of molecules of a gas occupying 22.4 liters at standard pressure and temperature. Avogadro himself had no way of determining this number but simply proposed the hypothesis that the number is the same for all gases. *Loschmidt's number* is defined as the number of molecules in a cubic centimeter of gas under standard conditions (atmospheric pressure, $0°$ C) and was named for the person who first discovered how to make a reliable calculation of this number.[142]

Maxwell showed in his first kinetic theory paper that Avogadro's hypothesis follows from the equipartition theorem, and in fact this was one of the positive features of the theory which he pointed out to Stokes in his letter of 1859 (document III-1). He claimed that the kinetic theory thus gives an explanation of the "law of equivalent volumes" in gaseous reactions. Maxwell had the opportunity to make this point to British chemists at a meeting of the Chemical Society of London in June 1867, when he participated in the discussion of Sir Benjamin Brodie's proposal to replace the a*. ~nic theory by a symbolic calculus of chemical operations.[143] Maxwell noted that the chemical atomic theory was based on an assumption that could be deduced "from purely dynamical considerations" based on "the supposition advocated by Professor Clausius and others, that gases consist of molecules floating about in all directions, and producing pressure by their impact." That assumption (Maxwell called it a definition) was "that for every kind of substance the number of atoms, or molecules, in the gaseous state, occupying the space of a litre, at a temperature of 0 degrees, and of a pressure of 760 millimetres, must necessarily be the same."[144]

The fact that Avogadro's hypothesis was consistent with (and perhaps could be derived from) the kinetic theory was of some assistance to the chemists who were trying to establish this hypothesis. In the 1872 Faraday Lecture, Stanislao Cannizzaro told the Chemical Society of London that the kinetic theory based on Avogadro's hypothesis was the "corner-stone of the modern theory of molecules and atoms." He remarked that "the notion of a molecule, stripped of all the fantastic detail with which it has been encumbered, has acquired great credit amongst men devoted to mathematical and physical studies."[145] In support of this statement he mentioned Maxwell's "eloquent discourse" at the Liverpool meeting of the British Association and Tyndall's lecture on the "Scientific Use of the Imagination," delivered at the same meeting in 1870.[146]

Loschmidt pursued the study of diffusion to obtain a better estimate for molecular diameters. In 1873 Maxwell analyzed his results and carried the discussion a step further. He produced a revised theory of the elastic sphere gas, "using, however, the methods of my first paper, which are more difficult of application, and which led me into great confusion."[147] He derived a coefficient of diffusion somewhat similar to that given by Stefan:[148]

$$D = \frac{1}{2\sqrt{6\pi}} \sqrt{\frac{3p_0}{\rho_0}} \sqrt{\frac{1}{m_1} + \frac{1}{m_2}} \frac{1}{d_{12}^2}. \tag{31}$$

By applying Eq. (31) to the data, Maxwell obtained reasonably consistent values for d_{12} for different pairs of gases, and these values agreed with calculations based on viscosities. The case for regarding each molecule as an object of definite diameter was good. Numerical values ranged from 5.8 to 9.3 × 10^{-8} cm, about five times the modern estimates of molecular diameter. The discrepancy came from the assumption that molecules in liquids are in direct physical contact. The estimate of the number of molecules in a standard volume of gas was much nearer, being within 30% of the modern value.[149]

Maxwell did not make a systematic attempt to determine the strengths of intermolecular forces and their variation with distance, although several of his writings touched on this problem. His calculations in transport theory were based on models with purely repulsive forces ("elastic sphere" or inverse fifth power), but in his article on capillarity and elsewhere he recognized that several phenomena suggested the existence of attractive forces between molecules.[150] His discussion of van der Waals's theory of the continuity of liquids and gases also postulated that the intermolecular force changes from repulsion to attraction as the molecules move away from the region of direct contact.[151]

Of all the questions about molecules the most urgent concerned their internal structure. Maxwell was uneasy with the discrepancy between measured and calculated specific heats for diatomic and polyatomic gases. Uneasiness turned to alarm as Maxwell realized that the equipartition of energy included all degrees of freedom, just as clues to the rich variety of internal motions of molecules emerged from spectrum analysis.

From 1840 it had been suspected that the bright lines in spectra were related to the chemical composition of the substances from which they were emitted. The subject acquired fresh impetus in 1859 when Kirchhoff and Bunsen showed that the bright lines produced by known substances in the laboratory could be identified with the dark lines in the solar spectrum, the latter presumably indicating the absorption by those substances present in the sun's cooler outer atmosphere of radiation from the hotter interior.[152] The determination of the wavelengths at which an atom or molecule absorbs and emits radiation meant that the chemical composition of the sun and stars could be investigated. A further refinement of astronomical spectrum analysis, based on the Doppler shift, allowed the velocities of objects relative to the earth to be estimated.[153] In addition to the well-known cosmological consequences of this research, we note that it also provided definite confirmation of Maxwell's conclusion that Saturn's rings are composed of discrete particles.[154]

Since the characteristic spectrum of an atom or a molecule became one of its most important properties after 1859, it was obviously a matter of considerable interest to explain how this spectrum was related to its internal structure. During the 1860s a qualitative theory of spectra was worked out by R. B. Clifton, Johnstone Stoney, and Maxwell; they attributed the sharp lines to resonant vibrations of molecules excited by their mutual collisions.[155] The broad truth of the hypothesis seemed certain but it led, as Maxwell immediately saw, to two questions. First was the conclusion that an atom in Sirius and an atom in Arcturus have to be identical in all the details of their internal structure. There must be some universal dimensional constant determining vibration frequency: "each molecule ... throughout the universe bears impressed on it the stamp of metric system as distinctly as does the metre of the Archives of Paris, or the double royal cubit of the Temple of Karnac."[156] The royal cubit proved to be Planck's quantum of action discovered in 1900.

Maxwell assumed that the spectrum originated from the collisions of two molecules when the molecules would receive a jolt; yet the vibrations that produced the spectral lines obviously depended on the nature of the molecule, not on the collisions. The collision determined the amplitude of the vibrations but not their period.[157] Boltzman assumed that they originated at the moment of impact.[158] Whichever molecular model one chose, spectral data indicated that there were many more internal degrees of freedom to the molecule than those deduced from specific heats data. Initially each line of the spectrum was associated with a fundamental or overtone of a vibration of a particular molecule. Maxwell used the analogy of a church bell which when struck moves to produce many vibrations of different frequencies, all related and all depending on the particular bell.[159] Such a model of an elastic solid capable of many different elastic vibrations conflicted with specific heats data. The aural analogy remained popular until Arthur Schuster, using probability arguments, showed that it was untenable.[160] After Schuster's paper the search for a molecule whose fundamental vibration and harmonics could produce the observed spectral lines changed to an attempt to devise a mechanical model that could account for the spectral series of Balmer.[161]

One popular model in this new search was the vortex atom, and indeed this model is impressive when judged by the criteria of Victorian physicists for atomic explanations of phenomena.[162] It was commendably stable and had the characteristics of a "manufactured article," durable and interchangeable with others of the same kind.[163] Once set in motion, a vortex ring would retain its identity no matter how complex a system of interlocking rings it might become involved in. Unlike the center-of-force molecule, there were no arbitrary parameters to be fixed, and one did not have to postulate some artificial mechanism to keep several atomic rings together in a molecule as one did for the billiard-ball model.

Yet Maxwell saw a flaw in the vortex theory. The vortex was "only a mode of motion" and thus it was a mystery how it could have inertia, since "inertia is a property of matter not of modes of motion." Although vortex rings have definite momentum and energy, one must "show that bodies built up of vortex rings would have such momentum and energy as we know them to have [which] is, in the present state of the theory, a very difficult tast."[164]

The question of making the vortex model or any molecular model compatible with specific heats data was still open. But toward the end of the nineteenth century, with the acceptance of a purely electrical source for radiation and line spectra, this problem disappeared from kinetic theory. Oliver Lodge first suggested an electric source for radiation in the atom, and Stoney developed the first specifically electrical model for the origins of radiation.[165] By 1894 a qualitative theory offered some relief to the beleaguered defenders of kinetic theory. G. H. Bryan argued:

The electromagnetic theory of light relieves the kinetic theory from the burden which has been imposed on it by its opponents since (for example) if we regard the molecules of a gas as perfectly conducting spheres, spheroids or other bodies moving about in a dielectric "vacuum" ..., we shall be able to account for the spectra by means of electromagnetic oscillations determined by the surface harmonics of different orders without interfering with the assumptions required for explaining the specific heats of gases.[166]

Spectra and their production were purely electromagnetic problems that did not require a dynamical explanation. Specific heats data referred to only the dynamical, not the electrical, properties of molecules. Yet although the electromagnetic explanation removed the problem of spectra from the body of kinetic theory, one problem remained that was purely dynamical: explaining the experimental values for the specific heats ratio in terms of the equipartition theorem.

A molecular model, however, was absolutely necessary for the mechanical world view. A model of the molecule that exposed its inner structure was required to complete the new mechanical kinetic theory of matter. Thomson voiced his own dissatisfaction with the state of the theory of matter in 1871:

There can be no permanent satisfaction to the mind in explaining heat, light, elasticity, diffusion, electricity and magnetism in gases, liquids and solids, and describing precisely the relation of these different states of matter to one another by the statistics of great numbers of atoms, when the properties of the atom itself are simply assumed.[167]

The problem Thomson faced was to devise a mechanical molecular model that conformed to the demands of kinetic theory *and* to those of spectroscopy. These demands were totally incompatible as Thomson and his colleagues painfully learned. During the 1870s Maxwell rehearsed all the later difficulties that his contemporaries found when trying to meet these two sets of notions. In addition, Maxwell had philosophical criteria that any model had to fulfill. He quickly abandoned all the mechanical molecular models available to him, usually for a combination of physical and philosophical reasons. His caution in using any specific molecular model should be noted in any attempt to assess the importance of any one model for his physics. The trend of Maxwell's physics was away from any specific model no matter how useful it might have been in the generation of his early ideas in either gas or field theory.

Maxwell was never happy with any particular molecular model, although he pursued them all—billiard ball, center of force, and vortex. None of them could explain the range of known experimental facts that any molecular theory was expected to cover, even in the 1870s. One or another was useful in the explanation of a limited set of experimental facts, yet all had flaws of principle. If molecules were clusters of billiard balls or centers of force, one had to postulate a mechanism that kept the clusters together. Any model of a molecule that supposed the molecule to be elastic meant that it had many more internal degrees of freedom and was therefore not in keeping with indications from specific heats data. An elastic molecule could not explain the specific heats results, because (as Maxwell told the Chemical Society in 1875) such molecules would convert most of their energy of agitation into internal energy, and the specific heats ratio of a substance composed of them would be too great.[168]

Maxwell never liked any of the specific devices offered to resolve the specific heats problem, including Boltzmann and Bosanquet's "dumbbell" model proposed to account for the fact that some molecules seemed to have five mechanical degrees of freedom. Such a model still did not allow one to explain spectra unless it was also elastic, and such elastic molecules "were capable of internal vibrations and these of an infinite variety of types so that the body has an infinite number of degrees of freedom."[169] Boltzmann's alternative of postulating interactions with the ether did not solve the problem. Such models only added to the burden of the number of internal degrees of freedom that the molecule already possessed and thereby *decreased* a theoretical specific heats ratio, which was already too low.

The dilemma of explaining the specific heats data within the mechanical view of nature was not solved until the development of quantum theory. Although kinetic theory had led to a number of concrete results and demonstrated the real existence of molecules to the satisfaction of many nineteenth-century scientists, it could not lead to any consistent picture of the internal structure of that molecule, which the mechanical world view required for completeness. This was a puzzle for two generations of physicists after Maxwell's death. While some sensed the impossibility of solving it within the limitations of the mechanical world view, Maxwell's trial solutions were resurrected repeatedly during those thirty years.

Maxwell's correspondence does not add much to our knowledge of his views on atoms. There is a letter to Tait, dated November 13, 1867, which has a brief reference to the vortex theory:[170]

If you have any spare copies of your translation of Helmholtz on "Water Twists" I should be obliged to you if you could send me one.... I set the Helmholtz dogma to the Senate House in '66, and got it very nearly done by some men....

A letter to Pattison in 1868 contains the bald statement:[171]

I cannot admit any theory which considers matter as a system of points which are centers of force acting on other similar points, and admits nothing but these forces. For this does not account for the perseverance of matter in its state of motion and for the measure of matter.

Although Maxwell never used the vortex-atom in his work on kinetic theory, he did develop some of his early ideas about electromagnetic fields in terms of a model of "molecular vortices."[172] The source for this model is probably Rankine by way of William Thomson rather than Helmholtz;[173] Maxwell mentions the Helmholtz 1859 paper in a note but says he had already written the first part of his own paper before he saw Helmholtz's work.[174] Despite the adjective "molecular," Maxwell stated that "the size of the vortices is indeterminate, but is probably very small as compared with that of a complete molecule of ordinary matter."[175] Recalling this paper fifteen years later, Maxwell wrote to Tait (document II-24):

As for aether, when did I discuss the molecules thereof? Except perhaps in Phil. Mag. 1861–2 on molecular vortices, and these were *very* hypothetical....

Maxwell's intellectual interest spread well beyond the confines of science. His own interest and competency in philosophy is well known and he shared his contemporaries' sense of cultural life in which science was but one aspect. For Maxwell this cultural life included history and, in particular, the history of electrical ideas and kinetic theory.[176] Although Maxwell did not acknowledge the source of his first derivation of the distribution law, remarking merely that it was a law similar to the well-known law of errors, he did show considerable interest in the history of the physical ideas involved in the kinetic theory of gases. In his letter to Stokes of May 30, 1859, (document III-1) he refers to "Clausius' (or rather Herapath's) theory" perhaps having been led to Herapath's book *Mathematical Physics* (1848) by Joule. When a notice of a meeting of the British Association appeared in *The Athenaeum* announcing that Maxwell was to give a paper on the kinetic theory of gases, Herapath himself wrote a letter to *The Athenaeum* calling attention to his own previous research on the subject.[177] Probably Maxwell wrote directly to Herapath asking for further details of his work; in document II-3 we publish a letter from Herapath, probably to Maxwell (though there is no definite evidence of this), that appears to be an answer to such an inquiry.

In preparing his 1867 memoir on the dynamical theory of gases, Maxwell made a more extensive effort to review the previous history. He wrote to someone at Cambridge, probably H. A. J. Munro, a specialist in classical literature, to ask for confirmation of his interpretation of Lucretius (document II-4). In that letter we find a draft of a paragraph on Lucretius, which may be compared with the somewhat enlarged paragraph in the published version of the memoir.

The opinion that the observed properties of visible bodies apparently at rest are due to the action of invisible molecules in rapid motion is to be found in Lucretius. In the exposition which he gives of the theories of Democritus as modified by Epicurus, he describes the invisible atoms as all moving downwards with equal velocities, which, at quite uncertain times and places, suffer an imperceptible change, just enough to allow of occasional collisions taking place between the atoms. These atoms he supposes to set small bodies in motion by an action of which we may form some conception by looking at the motes in a sunbeam. The language of Lucretius must of course be interpreted according to the ideas of his age, but we need not wonder that it suggested to Le Sage the fundamental conception of his theory of gases, as well as his doctrine of ultramundane corpuscles.[178]

Maxwell gleaned further information about earlier kinetic theorists from a footnote that Clausius inserted in his paper on heat conduction in gases. P. Du Bois Reymond had discovered Bernoulli's kinetic theory of gases, published in Latin in his *Hydrodynamica* of 1738, and had published a German translation of the relevant portions. Clausius also mentions the works by Prevost and Lesage, *Deux Traités de Physique Mécanique* (1818), and he quoted Lesage's references to Lucretius, Gassendi, Boyle, Parent, Herman, and Daniel and Jean Bernoulli. Clausius concludes the footnote with the remark:

Amid the large number of authors who are now quoted with reference to this subject (a number which might perhaps be still increased—some of whom, however, I venture to think, although I have not read the passages referred to in the earlier ones, expressed themselves very likely somewhat vaguely), it would be difficult to indicate with any certainty the one to whom the first suggestion of this hypothesis is to be ascribed, and all we can do is to determine how much each one has contributed to develop the vague idea into an admissible physical theory.[179]

Maxwell seems to have followed up some of these references, so that he can make more definite judgements in his 1867 memoir:

Daniel Bernoulli ... distinctly explains the pressure of air by the impact of its particles on the sides of the vessel containing it.... Le Sage ... explains gravity by the impact of "ultramundane corpuscles" on bodies. These corpuscles also set in motion the particles of light and various ethereal media, which in their turn act on the molecules of gases and keep up their motions. His theory of impact is faulty, but his explanation of the expansive force of gases is essentially the same as in the dynamical theory as it now stands.... A more extensive application of the theory of moving molecules was made by Herapath. His theory of the collisions of perfectly hard bodies, such as he supposed the molecules to be, is faulty, inasmuch as it makes the result of impact depend on the absolute motion of the earth. [Cf. Herapath's reply to a similar objection about absolutely hard bodies (in document II-3); this is one reason for suspecting that the letter was written to Maxwell.] This author, however, has applied his theory to the numerical results of experiment in many cases, and his speculations are always ingenious, and often throw much real light on the questions treated. In particular, the theory of temperature and pressure in gases and the theory of diffusion are clearly pointed out.

Dr. Joule has also explained the pressure of gases by the impact of their molecules, and has calculated the velocity which they must have in order to produce the pressure observed in particular gases. [Maxwell does not realize that Herapath had previously computed molecular velocities.]

It is to Professor Clausius, of Zurich, that we owe the most complete dynamical theory of gases.... After reading his investigation of the distance described by each molecule between successive collisions, I published some propositions on the motions and collisions of perfectly elastic spheres, and deduced several properties of gases, especially the law of equivalent volumes, and the nature of gaseous friction. I also gave a theory of diffusion of gases, which I now know to be erroneous, and there were several errors in my theory of the conduction of heat in gases which M. Clausius has pointed out in an elaborate memoir on that subject.[180]

In 1871, Maxwell wrote out a summary of the history of the kinetic theory for William Thomson (document II-10).[181] This document was for Thomson's use when preparing his presidential address at the British Association meeting in Edinburgh. It gives a little more detail regarding the contributions of Lucretius, Herapath, and Clausius, and takes note of the recent estimates of atomic magnitudes by Loschmidt, Stoney, and Thomson. Maxwell suggests that the error that both Lesage and Herapath made in treating the impact of bodies may have been derived from Descartes' *Principia*. He also mentions unpublished work of his own on heat conduction and diffusion, done "in the old style" as a result of Clausius's critique of his 1859 paper (document III-12).

Our information about Maxwell's opinion of Boltzmann is curiously incomplete. The first published references are in 1873, in connection with the law of thermal equilibrium of gas subject to forces. In his address to the British Association, Maxwell mentions the theorem that asserts that in a mixture every molecule has the same average kinetic energy: "The proof of this dynamical theorem, in which I claim the priority, has recently been greatly developed and improved by Dr. Ludwig Boltzmann."[182] In the correspondence with Guthrie in *Nature* Maxwell makes a similar remark and continues, "I am greatly indebted to Boltzmann for the method used in the latter part of the sketch of the general investigation."[183] When Francis Guthrie raises further objections about the proof of the theorem, involving collisions of molecules in cycles, Maxwell dismisses the problem by saying, "For a far more elaborate theoretical treatment of the subject Prof. Guthrie is referred to the papers of Prof. Ludwig Boltzmann in the Vienna Transactions since 1868."[184]

We get some further insight into what Maxwell means when he tells someone to go read Boltzmann through a remark to Tait published by Knott (unfortunately the complete letter seems to have disappeared):

> By the study of Boltzmann I have been unable to understand him. He could not understand me on account of my shortness, and his length was and is an equal stumbling-block to me. Hence I am very much inclined to join the glorious company of supplanters and to put the whole business in about six lines.[185]

Maxwell refused to accept Boltzmann's hypothesis about the structure of polyatomic molecules (1876) and insisted that the specific heats problem could not satisfactorily be resolved in this way. Nevertheless he was impressed by Boltzmann's earlier work (1868) on thermal equilibrium, and when he returned to the equipartition theorem in 1879, he even referred to it as "Boltzmann's theorem."

Maxwell was still unaware of the work of Waterston, most of which was still in the Royal Society Archives, though a few brief accounts and references to it had been published. A letter from S. Tolver Preston to Maxwell in 1876 (document II-29) did bring Waterston's theory to Maxwell's attention in connection with the theory of sound. We know that Maxwell received Preston's letter and answered it because Preston refers to Maxwell's reply in a subsequently published paper. But the information came too late for inclusion in Maxwell's *Encyclopedia Britannica* article "Atom" (1875), and there is no reference to Waterston in any of Maxwell's publications or other writings. Had Maxwell followed up Preston's hint, he might have been able to rescue Waterston from obscurity before his death in 1883. As it turned out, Waterston did not receive recognition for his pioneering efforts in kinetic theory until 1892, when Lord Rayleigh read the same paper on sound that had caught Preston's eye and unearthed the 1845 manuscript from the Royal Society Archive for publication in the *Philosophical Transactions*.[186]

Notes

1. These connections have been emphasized recently by Eric Mendoza, "The Kinetic Theory of Matter, 1845–1855," *Ann. Sci.* 40 (1984), 184–220.

2. The engineering origin of thermodynamics is comprehensively described by D. S. Cardwell, *From Watt to Clausius: The Rise of Thermodynamics in the Early Industrial Age* (Ithaca: Cornell University Press, 1971). See also T. S. Kuhn, "Energy Conservation as an Example of Simultaneous Discovery," in *Critical Problems in the History of Science*, M. Clagett, ed. (Madison: University of Wisconsin Press, 1959), 321–356; Keith Hutchinson, "W. J. M. Rankine and the Rise of Thermodynamics," *British J. Hist. Sci.* 14 (1981), 1–26; Mendoza, op. cit. (note 1). The classic account of energy conservation as a "simultaneous discovery" is that of Kuhn, cited above; this has been criticized by Yehuda Elkana in his article "Helmholtz' 'Kraft': An Illustration of

Concepts in Flux," *Hist. Studies Phys. Sci.* 2 (1970), 263–298, and his book *The Discovery of the Conservation of Energy* (Cambridge, Mass.: Harvard University Press, 1974). Elkana gives greater emphasis to the role of Helmholtz and his changing use of the term *Kraft*. Elkana's interpretation has in turn been criticized by P. M. Heimann (now Harman), who also criticized some aspects of Kuhn's interpretation; see "Conversion of Forces and the Conservation of Energy," *Centaurus* 18 (1974), 147–161; "Helmholtz and Kant: The Metaphysical Foundations of *Über die Erhaltung der Kraft*," *Studies in Hist. Phil. Sci.* 5 (1974), 205–238. See also Clifford Truesdell, *The Tragicomical History of Thermodynamics, 1822–1854* (New York: Springer-Verlag, 1980). For the development of thermodynamics in Britain, Crosbie Smith sees mechanical philosophy and William Thomson as the key; see Crosbie Smith, "A New Chart for British Natural Philosophy: The Development of Energy Physics in the 19th Century," *Hist. Sci.* 16 (1978), 231–279, and "Mechanical Philosophy and the Emergence of Physics in Britain: 1800–1850," *Ann. Sci.* 33 (1976), 3–29. "William Thomson and the Creation of Thermodynamics, 1840–1855," *Arch. Hist. Exact Sci.* 16 (1977), 238–288, and, "Natural Philosophy and Thermodynamics: William Thomson and 'the Dynamical Theory of Heat,'" *British J. Hist. Sci.* 14 (1976), 293–319. The role of James Prescott Joule has been examined in Henry John Steffens, *James Prescott Joule and the Concept of Energy* (New York: Science History Publications, 1979); but see E. Mendoza and D. S. L. Cardwell, "On a Suggestion Concerning the Work of J. P. Joule," *British J. Hist. Sci.* 14 (1981), 177–180. Lest the German contributions be forgotten, see Herbert Breger, *Die Natur als arbeitende Maschine: der Physik 1840–1850* (New York: Campus Verlag, 1982).

3. Arnold Thackray, *John Dalton* (Cambridge, Mass.: Harvard University Press, 1972) and *Atoms and Powers* (Cambridge, Mass.: Harvard University Press, 1970) examines atomic theories in the eighteenth century. Stephen G. Brush, *Statistical Physics and the Atomic Theory of Matter* (Princeton: Princeton University Press, 1983) and *The Kind of Motion We Call Heat* (New York; North-Holland, 1976), 2 vols., treats aspects of atomic theory from Newton well into the twentieth century. Both contain bibliographies of recent specialist studies in the area.

4. See Robert Fox, *The Caloric Theory of Gases from Lavoisier to Regnault* (Oxford: Clarendon Press, 1971).

5. For a discussion of physics in the eighteenth century, see John Heilbron, *Elements of Early Modern Physics* (Berkeley: University of California Press, 1982).

6. The best known example of this is the caloric explanation of the velocity of sound. See T. S. Kuhn, "The Caloric Theory of Adiabatic Compression," *Isis* 49 (1958), 132–150. And, of course, there is Sadi Carnot's theory of the heat engine. See Robert Fox, *Sadi Carnot: Réflexions sur la puissance motrice du feu* (Paris, 1978), and Fox, op. cit. (note 4).

7. Problems with caloric theory have been examined in S. Lilley, "Attitudes to the Nature of Heat about the Beginning of the 19th Century," *Arch. Int. Hist. Sci.* 1 (1948), 630–639. Robert Kargon, "The Decline of the Caloric Theory of Heat, a Case Study," *Centaurus* 10 (1964), 35–39, and Fox, op. cit. (note 4). For a general discussion of theories of heat before 1840, see Brush, *The Kind of Motion We Call Heat*, vol. 1, 19–50.

8. See S. G. Brush, *The Kind of Motion We Call Heat*, vol. 2, 303–334, and earlier accounts of this theory by R. J. Morris, Jr., "Changing Concepts of Heat in the Early Nineteenth Century," *Proc. Oklahoma Acad. Sci.* 42 (1962), 195–199, and Charles Weiner, *Joseph Henry's Lectures on Natural Philosophy: Teaching and Research in Physics, 1832–1847*, Ph. D. Dissertation, Case Institute of Technology, 1965.

9. See E. S. Cornell, "Early Studies in Radiant Heat," *Ann. Sci.* 1 (1936), 217–225; A. Wolf, *A History of Science, Technology & Philosophy in the 18th Century* (London: George Allen & Unwin; New York: Macmillan, second edition, 1952), vol. 1, 206–212; E. S. Barr, "The Infrared Pioneers. I. Sir William Herschel," and "II. Macedonio Melloni," *Infrared Physics* 1 (1961), 1–4; 2 (1962), 67–73.

10. See S. G. Brush, "The Wave Theory of Heat: A Forgotten Stage in the Transition from the Caloric Theory to Thermodynamics," *British J. Hist. Sci.* 5 (1974), 145–167, and *On the Kind of Motion We Call Heat*, vol. 2, 303–334, esp. 325–329.

11. Mendoza, op. cit. (note 1), recently reemphasized the industrial and engineering concerns and concepts at the foundations of thermodynamics and kinetic theory. Also, on the gas laws and the laws of gaseous chemical reactions, see W. S. James, "The Discovery of the Gas Laws," *Science Progress* 23 (1929), 261–279; J. R. Partington, *An Advanced Treatise on Physical Chemistry*, vol. 1, *Fundamental Principles, The Properties of Gases* (London: Longmans, Green and Co., 1949); J. R. Partington, *A History of Chemistry*, vol. 4, (New York: St. Martin's Press, 1964); V. V. Raman, "Where Credit Is Due: The Gas Laws," *Physics Teacher* 77 (1973), 419–424; M. P. Crosland, "The Origins of Gay-Lussac's Law of Combining Volumes," *Ann. Sci.* 17 (1961 [pub. 1963]), 1–26. For a discussion of thermal conductivity and its relation to convection and radiation, see Brush, *Kind of Motion We Call Heat*, chap. 13; A. C. Burr, "Notes on the History of the Experimental Determination of the Thermal Conductivity of Gases," *Isis* 21 (1934), 169–186; and R. G. Olson, " Count Rumford, Sir John Leslie, and the Study of the Nature and Propagation of Heat at the Beginning of the Nineteenth Century," *Ann. Sci.* 26 (1970), 273–304. Graham's researches on diffusion and transpiration have been analyzed by E. A. Mason and Barbara Kronstadt, "Graham's Laws of Diffusion and Effusion," *J. Chem. Educ.* 44 (1967), 740–744; Mason, "Thomas Graham and the Kinetic Theory of Gases," *Phil. J.* 7 (1970), 99–115, and Mason and P. G. Wright, "Graham's Laws," *Contemporary Physics* 12 (1971). 179–186.

12. H. A. Lorentz, "On the Emission and Absorption by Metals of Rays of Light of Great Wavelengths," *Proc. Sec. Sci., K. Akad. Wet. Amsterdam* 5 (1903), 666–685; J. H. Jeans "The Mechanism of Radiation," *Phil. Mag.* [6] 2 (1901), 421–455; and J. D. van der Waals, Jr., "On the Relation between Radiation and Molecular Attraction," *Proc. Sec. Sci., K. Akad. Wet. Amsterdam* 3 (1901) 27–35.

13. Mendoza, op. cit. (note 1), 185.

14. D. Bernoulli, *Hydrodynamica* (Argentorati (Strassburg): Dulsecker, 1738), sectio decima; English translation of *Hydrodynamics* by Daniel Bernoulli and *Hydraulics* by Johann Bernoulli (New York: Dover Publications, 1968); an English translation of the section on kinetic theory is also in Brush, *Kinetic Theory* (New York: Pergamon, 1965), vol. 1, 57–65. For a discussion of the theories of Bernoulli and others in the eighteenth century, see C. Truesdell, "Rational Fluid Mechanics, 1687–1765," in L. Euler, *Opera Omnia* (Zürich: Orell Füssli, 1954), editor's introduction to vol. II.12; "Early Kinetic Theories of Gases," in *Essays in the History of Mechanics* (New York: Springer Verlag, 1968), 272–304; G. R. Talbot and A. J. Pacey, "Some Early Kinetic Theories of Gases: Herapath and His Predecessors," *British J. Hist. Sci.* 3 (1966), 113–149; A. J. Pacey and S. J. Fischer, "Daniel Bernoulli and the *vis viva* of Compressed Air," *British J. Hist. Sci.* 3 (1967), 388–392, and R. Hooykaas, "The First Kinetic Theory of Gases (1727)," *Arch. Int. Hist. Sci.* 27 (1948), 180–184.

15. J. Herapath, "On the Physical Properties of Gases," *Ann. Phil.* 8 (1816), 55–60; "A Mathematical Inquiry into the Causes, Laws and Principal Phenomena of Heat, Gases,

Gravitation, etc.," *Ann. Phil.* 1 (1821), 273–293, 340–351, 401–416, and several other papers reprinted in *Mathematical Physics* (London: Whittaker and Co., 1847), S. G. Brush, ed. (New York: Johnson Reprint Corp., 1970). For further discussion of Herapath's work, see S. G. Brush, "The Development of the Kinetic Theory of Gases. I. Herapath," *Ann. Sci.* 13 (1957), 188–198; "The Royal Society's First Rejection of the Kinetic Theory of Gases (1821), John Herapath versus Humphry Davy," *Notes Records R. Soc. London* 18 (1963), 161–180; G. R. Talbot and A. J. Pacey, op. cit. (note 14). Also see Brush, *The Kind of Motion We Call Heat*, vol. 1, 69–90, 107–133 for Herapath, 134–160, 335–339 for Waterston; J. J. Waterston, "On the Physics of Media that are Composed of Free and Perfectly Elastic Molecules in a State of Motion," *Phil. Trans.* 183A (1893), 5–79; an abstract was published in *Proc. R. Soc. London* 5 (1846), 605; "On a General Theory of Gases," *Rep. 21st Meeting BAAS* (1851), 6; "On the Theory of Sound," *Phil. Mag.* [4] 16 (1859), 481. These and other papers are reprinted in *The Collected Scientific Papers of John James Waterston*, J. S. Haldane, ed. (Edinburgh: Oliver and Boyd, 1928). That edition omits at least two other papers by Waterston: "On Dynamical Sequences in Kosmos," presented at the 23rd meeting of the BAAS in 1853, summary published in *Athenaeum* (1853) 1099–1100, and "Note of an Experiment on Voltaic Conduction," *Phil. Mag.* [4] 31 (1866), 83–84. On Waterston's life and work, see S. G. Brush, "The Development of the Kinetic Theory of Gases. II. Waterston," *Ann. Sci.* 13 (1957), 273–283; "John James Waterston and the Kinetic Theory of Gases," *Amer. Sci.* 49 (1961) 202–214; *Kinetic Theory*, vol. 1, 17–19.

16. L. Euler, "Tentamen explicationis phaenomenorum aeris," *Commentarii Academiae Scientiarum Imperialis Petropolitanae*, 2 (1727), 347. Molecules were vibrating or rotating about more or less fixed positions.

17. The story of the Royal Society's rejection of Herapath's theory and the loss of due historical consideration for Waterston's worth has been told elsewhere. See Brush, "The Royal Society's First Rejection of the Kinetic Theory of Gases, 1821," *Notes Records R. Soc. London* 18 (1963), 161, and *Kind of Motion We Call Heat*, vol. 1, 140–141.

18. J. P. Joule, "On the Mechanical Equivalent of Heat, and on the Constitution of Elastic Fluids," *Report 18th Meeting BAAS* (1848), pt. 2, 21 (abstr.); "Some Remarks on Heat, and the Constitution of Elastic Fluids," *Mem. Manchester Lit. Phil. Soc.* 9 (1851), 107–114 (read 1848), reprinted in *Phil. Mag.* 14 (1857), 211–216. On the relation between Herapath and Joule, see E. Mendoza, "The Surprising History of the Kinetic Theory of Gases," *Mem. Proc. Manchester Lit. and Phil. Soc.* 105 (1962–1963), no. 2, 1–13. Joule, Maxwell, and other nineteenth-century physicists generally used the word "velocity" rather than "speed" even in contexts in which no specific direction of motion was stated or implied. The distinction between "velocity" as a vector and "speed" as a scalar was adopted in the twentieth century, but even now is not universally followed since physicists still speak of the "velocity of light" as a scalar.

19. Clausius, "On the Mean Lengths of the Paths Described by the Separate Molecules of Gaseous Bodies," *Phil. Mag.* [4] 17 (1859), 81–91, translated from "Ueber die mittlere Länge der Wege, welche bei der Molecularbewegung gasförmigen Körper von den einzelnen Molecülen zurückgelegt werden, nebst einigen anderen Bemerkungen über die mechanische Wärmetheorie," *Ann. Phys.* [2] 105 (1858), 239–258. The English translation is reprinted in Brush, *Kinetic Theory*, vol. 1, 135–147.

On Clausius's contributions to kinetic theory, see E. W. Garber, "Clausius and Maxwell's Kinetic Theory of Gases," *Hist. Studies Phys. Sci.* 2 (1970), 299–319; E. E. Daub, "Waterston, Rankine, and Clausius on the Kinetic Theory of Gases," *Isis* 61

(1970), 105–106; Brush, *The Kind of Motion We Call Heat*, vol. 1, 168–182, 404–406; Ivo Schneider, "Rudolph Clausius' Beitrag zur Einführung wahrscheinlichkeits-theoretischer Methoden in die Physik der Gase nach 1856," *Arch. Hist. Exact Sci.* 14 (1975), 237–261; C. Truesdell, "Early Kinetic Theories of Gases," *Arch. Hist. Exact Sci.* 15 (1975), 1–66, esp. 20–30. Daub gives a concise sketch of Clausius's career in the *Dictionary of Scientific Biography* 3 (1971), 303–311.

20. Clausius, "Ueber die Art der Bewegung, welche wir Wärme nennen," *Ann. Phys.* [2] 100 (1857), 353–380. English translation in *Phil. Mag.* [4] 14 (1857), 108–127; reprinted in Brush, *Kinetic Theory*, vol. 1, 110–134.

21. A. K. Krönig, "Grundzuge einer Theorie der Gase," *Ann. Phys.* [2] 99 (1856), 315–322. See Grete Ronge, "Zur Geschichte der kinetische Wärmetheorie mit biographis-chen Notizen zu August Karl Krönig," *Gesnerus* 18 (1961), 45–70; E. E. Daub, "Waterston's influence on Krönig's Kinetic Theory of Gases," *Isis* 62 (1971), 512–515.

22. Such estimates of molecular velocity had already been made; see J. Herapath, "On the Physical Constitution of the World," *Railway Mag.* 1 (1836), 104–110 (written in 1832); *Mathematical Physics, or the Mathematical Principles of Natural Philosophy: with a Development of the Causes of Heat, Gaseous Elasticity, Gravitation, and other Great Phenomena of Nature* (London: Whittaker/Herapath's Railway Journal Office, 1847), reprinted with new introduction and bibliography of Herapath's publications by S. G. Brush (New York: Johnson Reprint Corp., 1972); J. P. Joule, "On the Mechanical Equivalent of Heat, and on the Constitution of Elastic Fluids," *Rep. 18th Meeting BAAS* (1848), 21–22; R. Clausius, op. cit. (note 20).

23. C. Truesdell, op. cit. (note 19), 19.

24 *Cambridge Dublin Math. J.* 8 (1853), 96, 270–272.

25. Clausius, op. cit. (note 20).

26. C. H. D. Buys-Ballot, "Ueber die Art von Bewegung, welche wir Wärme und Electricitat nennen," *Ann. Phys.* [2] 103 (1858), 240–259.

27. Clausius, op. cit. (note 19).

28. Maxwell, "Illustrations of the Dynamical Theory of Gases. I. On the Motions and Collisions of Perfectly Elastic Spheres," *Phil. Mag.* [4] 19 (1860), 19–32. "II. On the Process of Diffusion of Two or More Kinds of Moving Particles among one Another," *Phil. Mag.* [4] 20 (1860), 21–37. Reprinted in Maxwell's *Scientific Papers*, vol. 1, 377–409; reproduced in this book as document III-6. Maxwell read a paper on kinetic theory at the British Association meeting in 1859 (the same meeting that featured the historic confrontation between T. H. Huxley and Bishop Wilberforce on Darwin's theory of evolution) and continued his discussion in a paper on the transport coeffi-cients at the 1860 meeting. Neither of these reports is reprinted in the *Scientific Papers*; they are reproduced here as documents III-2 and III-8, respectively. For commentary see the publications of Brush, Garber, and Truesdell cited in notes 3 and 19.

   Maxwell sometimes adds the adjective "hard" or "rigid" to his "elastic spheres," presumably alluding to the ancient idea that atoms cannot be deformed. For a discussion of the apparent inconsistency involved in this description, see Wilson L. Scott, *The Conflict between Atomism and Conservation Theory 1644–1860* (New York: American Elsevier, 1970). The phrase "hard spheres" used in modern kinetic theory is equivalent to Maxwell's "elastic spheres."

29. Maxwell, "On the Dynamical Theory of Gases," *Phil. Trans. R. Soc. London* 157 (1867), 49–86; *Phil. Mag.* [4] 32 (1866), 390–393 (abstr.); 35 (1868), 129–145, 185–217;

Scientific Papers, vol. 2, 26–78 (esp. 43–45); document III-30. A detailed analysis and elaboration of the theory has been published by C. Truesdell and R. L. Muncaster, *Fundamentals of Maxwell's Kinetic Theory of a Simple Monatomic Gas Treated as a Branch of Rational Mechanics* (New York: Academic Press, 1980). On the origins of statistical mechanics, see Brush, *Kind of Motion We Call Heat*, 335–362, 419–420.

30. John Michell, "An Enquiry into the Probable Parallax and Magnitude of the Fixed Stars from the Quantity of Light which They Afford Us and the Particular Circumstances of Their Situation," *Phil. Trans. R. Soc. London* 57 (1767), 234–264; James Forbes, "On the Alleged Evidence for a Physical Connexion between Stars Forming Binary or Multiple Groups, Deduced from the Doctrine of Chances," *Phil. Mag.* 37 (1850), 401–427. For further details see Elizabeth Garber, "Aspects of the Introduction of Probability into Physics," *Centaurus* 17 (1972), 11–39; Barry Gower, "Astronomy and Probability: Forbes versus Michell on the Distribution of the Stars," *Ann. Sci.* 39 (1982), 145–160.

31. Pierre Simon Marquis de Laplace, *Essai Philosophique sur les Probabilités* (Paris: Courdier, 1814). George Boole, *An Investigation of the Laws of Thought on Which Are Founded the Mathematical Theories of Logic and Probabilities* (London: Macmillan, 1854).

32. John Frederick William Herschel, "Quetelet on Probabilities," *Edinburgh Rev.* 92 (1850), 1–57. Reprinted in *Essays from the "Edinburgh" and "Quarterly" Reviews* (London: Layton, 1857), 365–465.

33. Lewis Campbell and William Garnett, *The Life of James Clerk Maxwell* (London: Macmillan, 1882, second edition, 1884). First edition reprinted with a new preface by R. H. Kargon and appendix with letters (New York: Johnson Reprint Corp., 1969), 134–144, esp. 143.

34. Herschel, op. cit. (note 32) (*Essays*), 378.

35. Campbell and Garnett, op. cit. (note 33), letter to R. B. Litchfield, reprinted in Brush, Everitt, and Garber, *Maxwell on Saturn's Rings*, 64. The possible influence of Herschel's review was suggested by C. C. Gillispie, "Intellectual Factors in the Background of Analysis by Probabilities," in *Scientific Change*, A. C. Crombie, ed. (New York: Basic Books, 1963), 431–453. The significance of the letter to Campbell was pointed out by C. W. F. Everitt, as noted in Brush, *Kinetic Theory*, vol. 1, 30n. See also P. M. Heimann, "Maxwell and the Modes of Consistent Representation," *Arch. Hist. Exact Sci.* 6 (1970), 171–213; Heimann, "Molecular Forces, Statistical Representation and Maxwell's Demon," *Studies in Hist. Phil. Sci.* 1 (1970), 189–211.

36. See the comment in note 39 on Herschel's later qualification of this statement in his *Essays*.

37. Herschel, op. cit. (note 32) (*Edinburgh Rev.*), 19–20.

38. Garber, op. cit. (note 30), 21–23.

39. Herschel, op. cit. (note 32) (*Essays*), 400n. The italics are Herschel's. In a system of molecules with constant total energy, the fact that one molecule has an extremely high velocity *does* affect the probability that another molecule has a high velocity. This difficulty can be avoided either by taking the limit of an infinite number of molecules, in which case, as Boltzmann showed by an explicit calculation ("Studien über das Gleichgewicht der lebendigen Kraft zwischen bewegten materiellen Punkten," *Sitz. Math.-Naturwiss. Cl. Akad. Wiss., Wien*, Teil II, 58 (1868), 517–560), the Maxwell distribution is valid or by assuming that the system can exchange energy with its

surroundings in such a way that its *temperature* rather than its *energy* is fixed. (This would correspond to the canonical ensemble of Gibbs.)

40. Theodore M. Porter, "A Statistical Survey of Gases: Maxwell's Social Physics," *Hist. Studies Phys. Sci.* 12 (1981), 77–116. For an alternative view, see Ian Hacking, "Nineteenth Century Cracks in the Concept of Determinism," in *Probability and Conceptual Change in Scientific Thought*, M. Heidelberger and L. Kruger, eds. (Bielefeld: B. K. Verlag GmbH, 1982), 5–34.

41. Ibid., 79.

42. This issue is dealt with explicitly in Elizabeth Garber and Fred Weinstein, "History of Science as Social History," in *Studies in Psychoanalytic Sociology*, J. Rabow, G. Platt, and M. Goldmann, eds. (Melbourne, Florida: Robert Krieger, 1985).

43. Porter, op. cit. (note 40), 116.

44. Garber and Weinstein, op. cit. (note 42). For an explicit examination of subjectivity and the treatment of creativity by historians, see Fred Weinstein, "The Problem of Subjectivity in History," in *History and Psychology: Problems and Potential in Psychoanalytic Interpretation*, William McKinley Runyan, ed. (San Francisco: Guildford Press, 1985).

45. Clausius, "Ueber die Wärmeleitung gasförmiger Körper," *Ann. Phys.* [2] 115 (1862), 1–56 (English translation in *Phil. Mag.* [4] 23 (1862), 417–435, 512–524); "Ueber die Anwendung des Satzes von der Aequivalenz der Verwandlungen auf die inner Arbeit," *Ann. Phys.* [2] 116 (1862), 73–112 (English translation in *Phil. Mag.* [4] 24 (1862) 81–97, 201–213). A more detailed account of the history of the transport coefficients and Maxwell's contribution to their theoretical estimation and measurement is in Brush, *The Kind of Motion We Call Heat*, chaps. 5, 12, and 13.

46. G. G. Stokes, "On the Effect of the Internal Friction of Fluids on the Motion of Pendulums," *Trans. Cambridge Phil. Soc.* 9 (1850, pub. 1856), [8–106], reprinted in *Mathematical and Physical Papers by the Late Sir George Gabriel Stokes*, second edition, reprint (New York: Johnson Reprint Corp., 1966), vol. 3, 13.

47. S. G. Brush, C. W. F. Everitt, and E. W. Garber, *Maxwell on Saturn's Rings* (Cambridge, Mass.: MIT Press, 1983), 67, 138.

48. Stokes' reply to Maxwell's letter seems to be lost, but at the end of Part I of his 1860 paper Maxwell wrote, referring to his theoretical prediction that viscosity should be independent of pressure: "Such a consequence of a mathematical theory is very startling, and the only experiment I have met with on the subject does not seem to confirm it." He is probably referring to Stokes's account of Sabine's experiment; Stokes, op. cit. (note 46). On the subsequent clarification of this situation, see document III-10; O. E. Meyer, "Ueber die innere Reibung der Gase," *Ann. Phys.* [2] 127 (1865), 177–209, 401–420, 564–599; and material which Stokes added to his 1850 paper (1966 reprint, 13, 76–77, 137–141). In another paper ("Ueber die Reibung der Gase, Zweite Abhandlung. Ueber die Strömung der Gase durch Capillarröhren," *Ann. Phys.* [2] 127 (1866), 253–281, 353–382) Meyer argued that Graham's earlier experiments on gas diffusion could have been regarded as a confirmation of Maxwell's viscosity law. See also Mason, op. cit. (note 11).

49. Brush, *Kind of Motion We Call Heat*, chap. 13.

50. P. L. Dulong and A. T. Petit, "Recherches sur la mesure des temperatures, et sur les lois de la communication de la chaleur." *Ann. Chim. Phys.* 7 (1817), 113–154, 225–264, 337–367.

51. Maxwell, *Scientific Papers*, vol. 1, 405.

52. Brush, *Kind of Motion We Call Heat*, 515; J. Stefan, "Über die Beziehung zwischen der Wärmestrahlung und der Temperatur," *Sitz. Math.-Naturwiss. Cl. Akad. Wiss., Wien* 79 (1879), 391–428.

53. Maxwell, *Scientific Papers*, vol. 1, 403.

54. Ibid., 404–405.

55. Brush, *Kind of Motion We Call Heat*, 486.

56. Elizabeth Garber, "Clausius and Maxwell's Kinetic Theory of Gases," *Hist. Studies Phys. Sci.* 2 (1970), 299–319.

57. Thomas Graham, "A Short Account of Experimental Researches on the Diffusion of Gases through Each Other and Their Separation by Mechanical Means," *Quart. J. Sci.* 2 (1829), 74–83; "On the Law of the Diffusion of Gases," *Phil. Mag.* 2 (1833), 175–190, 269–276, 351–358; "On the Law of the Diffusion of Gases," *Trans. R. Soc. Edinburgh* 12 (1834), 222–258; "On the Motion of Gases," *Phil. Trans. R. Soc. London* 136 (1846), 573–632; 139 (1849), 349–392; these and other papers are reprinted in *Chemical and Physical Researches*, R. A. Smith, ed. (Edinburgh (privately printed), 1876).

58. Graham (1834), op. cit. (note 57), 224.

59. Graham (1849), op. cit. (note 57) cf. his earlier statement that his law of diffusion "is certainly not provided for in the corpuscular philosophy of the day," in "On the Law of Diffusion of Gases," *Phil. Mag.* 2 (1833), 357.

60. Graham (1849), op. cit. (note 57), 384–385.

61. Ibid., 385ff.

62. Herapath, op. cit. (note 15).

63. Thomas Graham, "On the Molecular Mobility of Gases," *Proc. R. Soc. London* 12 (1863), 611–623; *Phil. Trans. R. Soc. London* 153 (1863), 385–405.

64. Maxwell, *Scientific Papers*, vol. 1, 400.

65. Maxwell, "On Loschmidt's Experiments on Diffusion in Relation to the Kinetic Theory of Gases," *Nature* 8 (1873), 298–300; *Scientific Papers*, vol. 2, 345.

66. Maxwell, *Scientific Papers*, vol. 1, 409.

67. Brush, *Kind of Motion We Call Heat*, 139, 147.

68. Stanislao Cannizzaro, "Lettera al Prof. S. di Luca; sunto di un corso di Filosofia Chimica," *Il Nuovo Cimento* 7 (1858), 321–355. English translation, *Sketch of a Course of Chemical Philosophy*, Alembic Club Reprints no. 18, reissue edition (Edinburgh: Livingstone, 1961).

69. S. Cannizzaro, "Considerations on Some Points of the Theoretic Teaching of Chemistry," *J. Chem. Soc.* 10 (1872), 941–967.

70. Meyer, "Ueber die Reibung der Flüssigkeiten." *Ann. Phys.* [2] 113 (1861), 55–86, 193–238, 383–425.

71. Meyer (1865), op. cit. (note 48). The italics are Meyer's.

72. For a detailed discussion of Meyer's work see Brush, *The Kind of Motion We Call Heat*, chap. 12.

73. Meyer also published an elementary treatise on gases, *Die Kinetische Theorie des Gase* (Breslau: Maruschke & Berendt, 1877; second edition, 1899).

74. See his analysis of Graham's experiments of the flow of gases in tubes. Meyer (1866),

op. cit. (note 48). The experiments of Graham referred to are Graham (1846 and 1849), op. cit. (note 57).

75. J. Stefan, "Ueber die Bewegung fluessiger Koerper," *Sitz. Math.-Naturwiss. Cl. Akad. Wiss., Wien*, 46 (1862), 8–31, 495–520.

76. Maxwell, "On the Viscosity or Internal Friction of Air and Other Gases," *Phil. Trans. R. Soc. London* 156 (1866), 249–268; *Scientific Papers*, vol. 2, 1–25.

77. Shortly after the publication of Maxwell's experiments on viscosity, Meyer reconsidered his own work and launched a new series of experiments. Although his previous result, $\mu = 0.000275$ for the viscosity coefficient of air, was significantly higher than Maxwell's result ($\mu = 0.0001878(1 + \alpha t)$, where t = centigrade temperature), he had also deduced from Graham's observations a lower value, $\mu = 0.000178$. Since Maxwell's formula gives values within the limits of his own two results, "their correctness can scarcely be questioned." But he felt it necessary to repeat Maxwell's version of the observations with his own apparatus using round disks; there seemed to be some irregularities in the first set of results, so he made some minor changes in the setup. Meyer's new data, taken in 1868, yielded the following results for about $18°$ C: $\mu = 0.000197$ (from combining one series of observations) and $\mu = 0.000190$ (from a second series). Maxwell's formula would give 0.000200 for this temperature, which Meyer considered to be sufficiently good agreement. He also computed a revised table of transpiration coefficients and viscosities for other gases.

Meyer turned next to the problem of the temperature dependence of the viscosity coefficient. Maxwell's original mean free path theory based on the elastic-sphere model had predicted that the viscosity coefficient should be proportional to the square root of the absolute temperature, and Meyer had accepted this result, not necessarily as an accurate quantitative law but simply as a striking qualitative result contradicting commonsense knowledge of the properties of fluids. O. E. Meyer, "Pendelbeobachtungen," *Ann. Phys.* [2] 142 (1871), 481–524; "Ueber die pendelnde Bewegung einer Kugel unter dem Einflusse der inneren Reibung des umgebenden Mediums," *J. Reine Angew. Math.* 73 (1871), 14–26; "Ueber die innere Reibung der Gase. 3. Ueber Maxwell's Methode zur Bestimmung der Luftreibung," *Ann. Phys.* [2] 143 (1871), 14–26.

78. William Thomson's referee report to G. G. Stokes, April 11, 1866 (Royal Society Archives).

79. Clausius, op. cit. (note 45).

80. The Clausius formula for the mean free path was

$$l = \frac{1}{(4/3)\,d^2\,N}.$$

Maxwell's formula was the same except for a factor 2 in place of the (4/3). For the justification of Maxwell's factor, see Niven's note in Maxwell's *Scientific Papers*, vol. 1, 387.

81. Clausius, quoted from the English translation of his 1862 paper in *Phil. Mag.* (op. cit., note 45), 529.

82. Maxwell, op. cit. (note 28); *Scientific Papers*, vol. 1, 391.

83. For further details of Clausius's criticism and Maxwell's response, see Garber, op. cit. (note 56).

84. G. G. Stokes, "On the Theories of the Internal Friction of Fluids in Motion, and of the Equilibrium and Motion of Elastic Solids," *Trans. Cambridge Phil. Soc.* 8 (1849),

287–319, reprinted in *Mathematical and Physical Papers*, second edition (New York: Johnson Reprint Co., 1966), vol. 1.

85. Stokes, *Mathematical and Physical Papers*, vol. 1, 78.

86. S. D. Poisson, "Mémoire sur la propagation du mouvement dans les fluides élastiques," *Ann. Chim.* 22 (1823), 246–270, and "Mémoire sur l'équilibre et le mouvement des corps élastiques," *Mem. Acad. Sci. Paris* 8 (1829), 357–370, 623–627. For comment on this work, see Isaac Todhunter, *A History of the Theory of Elasticity and of the Strength of Materials from Galilei to Lord Kelvin*, (New York: Dover reprint of 1886–1893 edition, 1960), vol. 1, 220–224. See Clifford Truesdell, "Mechanical Foundations of Elasticity and Fluid Dynamics." *J. Rational Mech. Anal.* 2 (1953), 593–616. On experiments refuting Poisson's formula, see James F. Bell, "The Experimental Foundations of Solid Mechanics," *Handbuch der Physik*, S. Fluegge, ed. (New York: Springer-Verlag, 1973), vol. VI a/1, 1–813.

87. C. Truesdell, "The Mechanical Foundations of Elasticity and Fluid Dynamics," *J. Rational Mech. Anal.* 1 (1952), 125–300, esp. 228–231; "On the Viscosity of Fluids According to the Kinetic Theory," *Z. Phys.* 131 (1952), 273–289; S. G. Brush, "Theories of Liquid Viscosity" *Chem. Rev.* 62 (1962), 513–548, esp. 521–522.

88. See David B. Wilson, "George Gabriel Stokes on Stellar Aberration and the Luminiferous Ether," *British J. Hist. Sci.* 6 (1976), 57–72, for further discussion.

89. Forbes was Professor of Natural Philosophy at Edinburgh University and encouraged Maxwell's early scientific work and helped to persuade Maxwell's father to allow him to attend Cambridge. A list of Forbes's papers on glaciers is in the *Catalogue of Scientific Papers (1800–1863)*, compiled by the Royal Society of London (Metuchen, NJ: Scarecrow Reprint Corp., 1968), vol. 2, 660–662; see also J. S. Rowlinson, "The Theory of Glaciers," *Notes Records R. Soc. London* 26 (1971), 189–204.

90. Maxwell, "On the Equilibrium of Elastic Solids," *Trans. R. Soc. Edinburgh* 20 (1853), 87–120; reprinted in *Scientific Papers*, vol. 1, 31–75. David Brewster's paper is "On the Production of Regular Double Refraction in the Molecules of Bodies by Simple Pressure; with Observations on the Origin of the Double Refracting Structure," *Phil. Trans. R. Soc. London* (1830), 87–96, and *Edinburgh J. Sci.* 3 (1830), 328–337.

91. Maxwell, "On Faraday's Lines of Force," *Trans. Cambridge Phil. Soc.* 10 (1856), 27–83; reprinted in *Scientific Papers*, vol. 1, 155–229, esp. 160–175.

92. Maxwell, "On the Equilibrium of a Spherical Envelope," *Quart. J. Math.* 8 (1867), 325–333.

93. Maxwell, "The Dynamical Theory of the Electromagnetic Field," *Phil. Trans. R. Soc. London* 155 (1865), 459–512; reprinted in *Scientific Papers*, vol. 1, 526–597.

94. Maxwell, "On Double Refraction in a Viscous Fluid in Motion," *Proc. R. Soc. London* 22 (1873–1874), 46–47; reprinted in *Scientific papers*, vol. 2, 379.

95. Maxwell, "On the Dynamical Theory of Gases," *Phil. Trans. R. Soc. London* 157 (1867), 49–88; reprinted in *Scientific Papers*, vol. 2, 26–78, esp. 30.

96. Maxwell, "On Stresses in Rarefied Gases Arising from Inequalities of Temperature," *Phil. Trans. R. Soc. London* 170 (1880), 231–256. Also *Proc. R. Soc. London* 27 (1878), 304; reprinted in *Scientific Papers*, vol. 2, 681–712. Thomson's report, 15 June 1878, will be published in our next volume, containing Maxwell's papers on thermodynamics and statistical mechanics.

97. L. Tisza, "Supersonic Absorption and Stokes' Viscosity Relation," *Phys. Rev.* [3] 61 (1942), 531–536. For a modern discussion of Maxwell's theory, see C. Truesdell, "A

New Definition of a Fluid. II. The Maxwellian Fluid," *J. Math. Pures Appl.* 30 (1951), 111–158; E. Ikenberry and C. Truesdell, "On the Pressures and Flux of Energy in a Gas According to Maxwell's Kinetic Theory, I, II," *J. Rational Mech. Anal.* 5 (1956), 1–54, 55–128; Truesdell, "Une solution exacte des équations de Maxwell," *J. Math. Pures Appl.* 37 (1958), 119–133.

98. Meyer, "Ueber die innere Reibung der Gase. Vierte Abhandlung. Die Gultigkeit des Poiseuille'schen Gesetzes für die Transpiration der Gase," *Ann. Phys.* [2] 148 (1873), 1–144; "Ueber die innere Reibung der Gase. Funfte Abhandlung," *Ann. Phys.* [2] 148 (1873), 203–235; "Ueber die Bewegung einer Pendelkugel in der Luft," *J. Reine Angew. Math.* 75 (1873), 336–347; "Ueber die innere Reibung der Gase. Sechste Abhandlung. Ueber die Transpiration verschiedener Gase," *Ann. Phys.* [2] 148 (1873), 526–555.

99. Johann Puluj, "Ueber die Reibungsconstante der Luft als Function der Temperatur," *Sitz. Math.-Naturwiss. Cl. Akad. Wiss., Wien,* II, 69 (1874), 287–317; 70 (1875), 243–267; "Ueber die Abhaengigkeit der Reibung der Gase von der Temperatur," *Sitz. Math.-Naturwiss. Cl. Akad. Wiss., Wien,* II, 73 (1876), 589–628, and *Ann. Phys.* 1 (1876), 296–310. "Ueber die Abhängigkeit der Reibung der Gase von der Temperatur," *Sitz. Math.-Naturwiss. Cl. Akad. Wiss., Wien.* 73 (1876), 589–628; A. von Obermayer, "Ueber die Abhängigkeit der Reibungscoefficient der atmosphärischen Luft von der Temperatur," *Sitz. Math.-Naturwiss. Cl. Akad. Wiss., Wien.* 71 (1875), 281–308; "Ueber die Abhängigkeit der Coefficient der inneren Reibung der Gase von der Temperatur," *Sitz. Math.-Naturwiss. Cl. Akad. Wiss., Wien.* 73 (1876), 433–474. E. Warburg, "Ueber die Reibung und Wärme-leitung verdünnter Gase," *Berlin Akad. Monats.* (1875), 160–175; *Ann. Phys.* 155 (1875), 337–366, 525–550. Carl Barus, "Note on the Viscosity of Gases at High Temperature and on the Pyrometric Use of the Principle of Viscosity," *Amer. J. Sci.* 35 (1888), 407–410; "Maxwell's Theory of the Viscosity of Solids and Certain Features of Its Physical Verification," *Amer. J. Sci.* 36 (1888), 178–208.

100. Maxwell, op. cit. (note 96); *Scientific Papers,* vol. 2, 692.

101. Rayleigh, "On the Viscosity of Argon as Affected by Temperature," *Proc. R. Soc. London* 66 (1900), 68–74.

102. Brush, *Kinetic Theory,* op. cit. (note 14), vol. 3.

103. Meyer (1899), op. cit. (note 73), 218–222.

104. Maxwell, op. cit. (note 95); *Scientific Papers,* vol. 2, 77.

105. Most recent studies of Boltzmann's work concentrate on his contribution to clarifying the concept of entropy or the relation between his statistical mechanics and the later quantum physics. Brush has focused on Boltzmann's contributions to gas theory (*Kinetic Theory,* op. cit. (note 14), vol. 2, and *The Kind of Motion We Call Heat,* chaps. 6, 10, 12, 13) and has translated Boltzmann's treatise, *Lectures on Gas Theory* (Berkeley: University of California Press, 1964).

106. Boltzmann, "Weitere Studien ueber das Wärmegleichgewicht unter Gasmolekülen," *Sitz. Math.-Naturwiss. Cl. Akad. Wiss., Wien,* II, 66 (1872), 275–370. Translated in Brush, *Kinetic Theory,* op. cit. (note 14), vol. 2.

107. Brush, *Kinetic Theory,* op. cit. (note 14), vol. 2, 141.

108. Friedrich Narr, "Ueber die Erkaltung und Waermeleitung in Gasen," *Ann. Phys.* [2] 142 (1871), 123–158.

109. Stefan, "Untersuchungen ueber die Wärmeleitung in Gasen," *Sitz. Math.-Naturwiss. Cl. Akad. Wiss., Wien,* 65 (1872), 45–69. Abstract in *J. Chem. Soc.* 10 (1872), 591–592.

110. Stefan's thermal conductivity was 0.0000558 in cgs units for air.

111. Stefan, "Untersuchungen ueber Wärmeleitung in Gasen," *Sitz. Math.-Naturwiss. Cl. Akad. Wiss., Wien*, 72 (1876), 69–101. Abstract in *J. Chem. Soc.* 2 (1876), 37–38.

112. Kundt and Warburg, "Ueber Reibung und Wärmeleitung verdünnter Gase." *Ann. Phys.* [2] 155 (1875), 337–366, 525–550; 156 (1875), 177–211. English summary in *Phil. Mag.* [4] 50 (1875), 53–62.

113. Dulong and Petit, op. cit. (note 50).

114. Winkelmann, "Ueber die Wärmeleitung der Gase." *Ann. Phys.* [2] 153 (1875), 497–531.

115. Winkelmann, "Ueber die Waermeleitung der Gase, II," *Ann. Phys.* [2] 157 (1876), 497–555.

116. Winkelmann, "Ueber die Waermeleitung von Gasen und Daempfen und die Abhaengigkeit der specifischen Waerme derselben von der Temperatur," *Ann. Phys.* [2] 159 (1876), 177–197.

117. The decrease in specific heat of mercury with temperature was confirmed by many later studies; see J. R. Partington, *An Advanced Treatise on Physical Chemistry* (London: Longmans, Green and Co., 1951), vol. 2, 212. Thus a presupposition that had probably been used by many earlier experimenters on the basis of data suggesting an *increase* in the specific heat with temperature was demolished.

118. Boltzmann "Ueber die Natur der Gasmolecuele," *Sitz. Math.-Naturwiss. Cl. Akad. Wiss., Wien*, II, 74 (1877), 553–560.

119. Meyer (1877), op. cit. (note 73); the recalculation appears in the 1899 edition.

120. Brush, op. cit. (note 14).

121. See the discussion by Stefan, op. cit. (note 111).

122. Schleiermacher, "Ueber die Wärmeleitungsfähigkkeit des Quecksilberdampfes," *Ann. Phys.* [3] 36 (1889), 346–357.

123. L. S. Zaitseva, "An experimental investigation of thermal conductivity of monatomic gases in a wide range of temperatures," *Sov. Phys. Tech. Phys.* 4 (1959), 444–450.

124. See Brush, *Kind of Motion We Call Heat*, 503–507 for these references and further details of twentieth-century work.

125. A. Eucken, "Ueber das Waermeleitvermoegen, die spezifische Waerme und die innere Reibung der Gase," *Phys. Z.* 14 (1913), 324–332.

126. For a recent assessment of the theory, see J. A. Barker, M. V. Bobetic, and A. Pompe, "An Experimental Test of the Boltzmann Equation: Argon," *Molecular Physics* 20 (1971), 347–355. Values of thermal conductivity determined from viscosity data using the Chapman-Enskog theory are now considered more reliable than those found by direct experiments; see H. J. M. Hanley, R. D. McCarty, and H. Intemann, "The Viscosity and Thermal Conductivity of Dilute Gaseous Hydrogen from 15 to 5000 K," *J. Res. Nat. Bur. Stand.* 74A (1970), 331–353; and J. Kestin, S. T. Ro, and W. Wakeham, "An Extended Law of Corresponding States for the Equilibrium and Transport Properties of the Noble Gases," *Physica* 58 (1972), 165–221.

127. Brush, *Kinetic Theory*, vol. 3, 21–31.

128. Loschmidt, "Experimental-Untersuchungen ueber die Diffusion von Gasen ohne pörose Scheidewände," *Sitz. Math.-Naturwiss. Cl. Akad. Wiss., Wien* 61 (1870), 367–380; 62 (1870), 468–478.

129. In his 1867 paper Maxwell wrote that D "varies directly as the square of the

absolute temperature" without indicating that this law is restricted to the inverse fifth-power force model; not until 1873 did he state explicitly that $D \propto T^{3/2}$ for elastic spheres. The quotation is from Maxwell, op. cit. (note 95), reprinted in *Scientific Papers*, vol. 2, 60, and Maxwell, "On Loschmidt's Experiments on Diffusion in Relation to the Kinetic Theory of Gases," *Nature* 8 (1873), 298–300, reprinted in *Scientific Papers*, vol. 2, 343–350 (document III-42).

130. Maxwell, "On Loschmidt's Experiments," ibid., 344–345.

131. J. Stefan, "Ueber die Gleichgewicht und die Bewegung, in besondere die Diffusion von Gasmengen," *Sitz. Math.-Naturwiss. Cl. Akad. Wiss., Wien* 63 (1871), 63–124.

132. Stefan, "Ueber die dynamische Theorie der Diffusion der Gase," *Sitz. Math.-Naturwiss. Cl. Akad. Wiss., Wien* 65 (1872), 323–363.

133. Maxwell, op. cit. (note 95), *Scientific Papers*, vol. 2, 59.

134. This was the first time this result was stated explicitly in print. See note 129.

135. Maxwell to Stokes, October 8, 1859, in *Memoir and Scientific Correspondence of the late Sir George Gabriel Stokes*, Joseph Larmor, ed. (Cambridge: Cambridge University Press, 1909), vol. 2, 11–14. The letter is reproduced as document III-3.

136. Maxwell to William Thomson, December 17, 1861, in J. Larmor, "The origins of Clerk Maxwell's Electric Ideas, as Described in Familiar Letters to W. Thomson," *Proc. Cambridge Phil. Soc.* 32 (1936), 695–750, esp. 733.

137. For a more extensive discussion of Maxwell's work on molecular models, see Garber, "Molecular Science in Late-Nineteenth-Century Britain," *Hist. Studies in Phys. Sci.* 9 (1978), 265–297.

138. Young, in an article written in 1816, argued as follows:

"Now since there is reason to suppose the corpuscular force of a section of a square inch of water to be equivalent to the weight of a column about 750,000 feet high, at least if we allow the cohesion to be independent of the density, their magnitude will be expressed by 252.5 × 750,000 × 12 grains, which is to 9 as 252.5 × 1,000,000 to 1; consequently the extent of the cohesive force must be limited to about 250 millionth of an inch."— ("Cohesion," *Supplement to the Fourth, Fifth, and Sixth Editions of the Encyclopedia Britannica* (Edinburgh, 1824); reprinted in Young's *Miscellaneous Works* George Peacock, ed. (London; 1855), vol. 1, 461)

Young's method has been discussed by G. D. Scott and J. G. Macdonald, "Young's Estimate of the Size of Molecules," *Amer. J. Phys.* 33 (1965), 163–164; E. A. Mason, "Estimate of Molecular Sizes and Avogadro's Number from Surface Tension," *Amer. J. Phys.* 34 (1966), 1193; and A. French, "Earliest Estimates of Molecular Size," *Amer. J. Phys.* 35 (1967), 162–163.

139. Loschmidt, "Zur Grösse der Luftmolecule." *Sitz. Math.-Naturwiss. Cl. Akad. Wiss., Wien*, 52 (1865), 395–413.

140. G. J. Stoney, "The Internal Motions of Gases Compared with the Motions of Waves of Light," *Phil. Mag.* [4] 36 (1868), 132–141. L. V. Lorenz, "Zur Molecular-theorie und Elektricitätslehre," *Ann. Phys.* [2] 140 (1870), 644–647. William Thomson, "The Size of Atoms," *Nature*, 1 (1870), 551–553; "The Size of Molecules," *Nature* 2 (1870), 56–57. See notes i and j to document III-34.

141. Thomson, ibid., 553.

142. See Brush, *Kind of Motion We Call Heat*, 76 and references cited therein. For further discussion of Loschmidt's estimate, see R. M. Hawthorne, Jr., "Avogadro's Number: Early Values by Loschmidt and Others," *J. Chem. Educ.* 41 (1970), 751–755.

143. On Brodie's theory and the debate about chemical atomism in Britain at this time, see W. V. Farra, "Sir Benjamin Brodie and His Calculus of Chemical Operations," *Chymia* 9 (1964), 169–170; William H. Brock & D. M. Knight, "The Atomic Debates: Memorable and Interesting Evenings in the Life of the Chemical Society," *Isis* 56 (1965), 5–25; Brock, *The Atomic Debates: Brodie and the Rejection of the Atomic Theory* (New York: Humanities Press, 1967); Knight, *Atoms and Elements, a Study of Theories of Matter in England in the Nineteenth Century* (London: Hutchinson, 1967) and G. M. Fleck, "Atomism in Late Nineteenth Century Physical Chemistry," *J. Hist. Ideas* 24 (1963), 106–114.

144. Document II-5, reprinted with related papers in Knight, *Classical Scientific Papers: Chemistry* (New York: American Elsevier, 1968). Loschmidt's 1865 paper on molecular size apparently had not yet come to Maxwell's attention. His conflation of atoms and molecules here was repaired in later statements (see pp. 138, 140, 211).

145. Cannizzaro, op. cit. (note 68).

146. Cannizzaro, op. cit. (note 68). Maxwell, "Molecules," *Nature* 8 (1873), 437–441; reprinted in *Scientific Papers*, vol. 2, 361–377, and presented here in document II-16; and John Tyndall, *Fragments of Science* (New York: Appleton, 1897), sixth edition, vol. 2, 101–134.

147. Maxwell, op. cit. (note 129); reprinted in *Scientific Papers*, vol. 2, 345.

148. All the theoretical expressions for the diffusion coefficient derived up to this point made the coefficient independent of the composition of the mixture; in fact, the possibility that it might vary with composition does not seem to have been seriously considered. O. E. Meyer, op. cit. (note 73), later proposed a theory in which the diffusion coefficient did vary with composition, but extensive experiments in the early 1900s showed that while there is some variation, the Maxwell-Stefan theory is much more nearly correct. Further refinements in the kinetic theory by J. H. Jeans, D. Enskog, and S. Chapman confirmed this conclusion. See Jeans, *The Dynamical Theory of Gases*, fourth edition (Cambridge: Cambridge University Press, 1925), chap. 13; L. B. Loeb, *Kinetic Theory of Gases* (New York: McGraw-Hill, 1927), chap. 6, part 3; Sydney Chapman and T. G. Cowling, *The Mathematical Theory of Non-Uniform Gases* (Cambridge: Cambridge University Press, 1939).

149. Maxwell's final value for the mass of a hydrogen molecule is 4.6×10^{-24} g; for its diameter, 5.8×10^{-10} m; and for the number of molecules in a cubic centimeter of gas at $0°$ C and 760 mm pressure (now sometimes called "Loschmidt's number"), 19×10^{18}. Thomson reviewed and corrected some of Maxwell's estimates in his *Baltimore Lectures of Molecular Dynamics and the Wave Theory of Light* (London: Clay & Sons, 1904), Lecture 17. The modern values of these quantities are 3.34×10^{-24} g; approximately 2.6×10^{-10} m (not a very well-defined quantity according to modern physics), and 26.9×10^{18}.

150. Maxwell, "Capillarity," reprinted in *Scientific Papers*, vol. 2, 542. This article, first published in the *Encyclopedia Britannica*, is a substantial contribution to the subject that eliminates a number of errors from earlier theories of capillarity due to Laplace and Poisson.

151. Discussion of this will be presented in our next volume. Maxwell, "Van der Waals on the Continuity of the Gaseous and Liquid States," *Nature* 10 (1874), 477–480; reprinted in *Scientific Papers*, vol. 2, 407–415, esp. 412. On the history of interatomic forces see Brush, *Statistical Physics*, op. cit. (note 3), 204–217.

152. G. R. Kirchhoff, "Ueber die Fraunhofer'schen Linien," *Berlin Akad. Monats.*

(1859), 662–665; R. Bunsen and G. Kirchhoff, "Chemische Analyse durch Spectral-beobachten," *Ann. Phys.* [2] 110 (1860), 160–189. For additional references and discussion, see William McGucken, *Nineteenth-Century Spectroscopy: Development of the Understanding of Spectra 1802–1897* (Baltimore: Johns Hopkins Press, 1969), chap. 1.

153. J. C. Doppler, "Ueber das farbige Licht der Doppelsterne und einiger anderer Gesterne des Himmels," *Abh. Boehm. Ges. Wiss.* 2 (1842), 465–482. Joachim Thiele, "Zur Wirkungsgeschichte des Dopplerprinzips in neunzehnten Jahrhundert," *Ann. Sci.* 27 (1971), 393–407. On Friedrich von Hahn as a possible precursor of Doppler, see Peter Brosche, "Ein Vorlaeufer Christian Dopplers," *Phys. Blaetter* 32 (1977), 124–129. The acoustic Doppler shift was first demonstrated by C. H. D. Buys-Ballot, the same person whose criticism of kinetic theory led Clausius to introduce the mean free path concept.

154. *Maxwell on Saturn's Rings*, op. cit. (note 47). Donald E. Osterbrock, *James E. Keeler* (Cambridge: Cambridge University Press, 1984), 158–165.

155. R. B. Clifton, "An Attempt to Refer Some Phenomena Attending the Emission of Light to Mechanical Principles," *Proc. Manchester Lit. Phil. Soc.* 5 (1866), 24–28 (cited by McGucken, op. cit. (note 152), 41). G. J. Stoney, "The Internal Motions of Gases Compared with the Motions of Waves of Light," *Phil. Mag.* [4] 36 (1868), 132–141. Maxwell, *Theory of Heat* (London: Longmans, Green and Co., 1871 and later editions), chap. 22. The theories of Clifton, Stoney, Maxwell, and others are discussed by William McGucken, op. cit. (note 152).

156. Maxwell, op. cit. (note 146); reprinted in *Scientific Papers*, vol. 2, 376.

157. Maxwell, *Theory of Heat*, op. cit. (note 155), 306. This description remained the same throughout the editions that Maxwell personally saw through the press. A similar explanation is given in his *Britannica* article, "Atom," in *Scientific Papers*, vol. 2, 445–484, 463; presented here as document II-23.

158. Boltzmann, op. cit. (note 118).

159. Maxwell, "Atom," op. cit. (note 157), 471.

160. A Schuster, "On Harmonic Ratios in the Spectra of Gases," *Proc. R. Soc. London* 31 (1881), 337–347; McGucken, op. cit. (note 152), 122–125.

161. The details of the development of such models are in C. L. Maier, *The Role of Spectroscopy in the Acceptance of an Internally Structured Atom*, Ph.D. dissertation, University of Wisconsin, 1964 (New York: Arno Press, 1981); and McGucken, op. cit. (note 152).

162. Brush, *Kind of Motion We Call Heat*, 206–207, and references cited therein.

163. On the atom as a manufactured article, see Maxwell, op. cit. (note 146); reprinted in *Scientific Papers*, vol. 2, 376, 483, presented here in documents II-16 and II-23; J. Herschel, *A Preliminary Discourse on the Study of Natural Philosophy*, reprint of the 1830 edition (New York: Johnson Reprint Corp., 1966), 38; W. K. Clifford, *Lectures and Essays* (London: Macmillan, 1879), vol. 1, 222, 236.

164. Maxwell, "Atom," op. cit. (note 157); reprinted in *Scientific Papers*, vol. 2, 472.

165. For discussions of electrical models of the atom in spectroscopy, see McGucken, op. cit. (note 152), 184–198 and 209–213, and Maier, op. cit. (note 161), chap. 4. For Lodge's papers, see Oliver Lodge, "The Relation Between Electricity and Light," *Nature* 23 (1881), 302–304, "The Ether and Its Functions," *Nature* 27 (1882), 304–306, 328–330; and "The Modern Theory of Light," *Smithsonian Report* (1889), 441–448.

166. G. H. Bryan, "Report on the Present State of our Knowledge of Thermodynamics. Part II. The Laws of Distribution of Energy and their Limitation," *Rep. 64th Meeting BAAS* (1894), 64–106; "Prof. Boltzmann and the Kinetic Theory of Gases," *Nature* 51 (1894), 31.

167. William Thomson, "Presidential Address to the British Association Meeting at Edinburgh," *Rep. 41st Meeting BAAS* (1871), lxxxiv–cv.

168. See document II-24, especially pp. 230–232. Note that Maxwell rejected the assumption that a molecule of oxygen consists of two (point) atoms of oxygen (see pp. 140, 211, 232).

169. Maxwell, "The Kinetic Theory of Gases" (A review of Watson's *A Treatise on the Kinetic Theory of Gases*), *Nature* 18 (1877), 242–246, to be reprinted in our next volume.

170. C. G. Knott, *Life and Scientific Work of Peter Guthrie Tait* (Cambridge: Cambridge University Press, 1911), 106.

171. Maxwell to Mark Pattison, April 13, 1868; to be published in our next volume.

172. Maxwell, "On Physical Lines of Force," *Phil. Mag.* [4] 21 (1861), 161–175; 23 (1862), 281–291, 338–348; reprinted in *Scientific Papers*, vol. 1, 451–513.

173. C. W. F. Everitt, *James Clerk Maxwell: Physicist and Natural Philosopher* (New York: Scribner, 1975), 94–95.

174. Maxwell, op. cit. (note 172), 488.

175. Maxwell, op. cit. (note 172), 485.

176. Maxwell edited Henry Cavendish's electrical researches. He wrote an introductory essay and added extensive notes, *The Unpublished Electrical Writings of Hon. Henry Cavendish* (Cambridge: Cambridge University Press, 1879). For Maxwell's opinions on the history of science in general, see his 1856 Aberdeen lecture, published by R. V. Jones, "James Clerk Maxwell at Aberdeen 1856–1860," *Notes Records R. Soc. London* 23 (1973), 57–81. This work includes Maxwell's Inaugural Lecture, November 3, 1856. See also his 1860 lecture at Kings College, London, reprinted here in document II-2, and his inaugural lecture at Cambridge (1871), Maxwell, *Scientific Papers*, vol. 2, 241–255 (document II-11).

177. J. Herapath, "On the Dynamical Theory of Airs." *Athenaeum* (1860), 722; *Railway J.* 22 (1860), 552.

178. Maxwell, op. cit. (note 29).

179. Clausius, op. cit. (note 45).

180. Maxwell, op. cit. (note 29).

181. H. Bernstein, "J. Clerk Maxwell on the History of the Kinetic Theory of Gases, 1871," *Isis* 54 (1963), 206–216.

182. Maxwell, op. cit. (note 146); reprinted in *Scientific Papers*, vol. 2, 365–366.

183. Maxwell, "On the Equilibrium of Temperature of a Gaseous Column Subjected to Gravity," *Nature* 8 (1873), 527–528.

184. Maxwell, "On the Equilibrium of Temperature of a Gaseous Column," *Nature* 10 (1874), 123. This correspondence will be published in the next volume of this series.

185. Knott, op. cit. (note 170), 114.

186. See Strutt, *Life of John William Strutt, Third Baron Rayleigh* (London: Arnold, 1924). Reprinted with additions by J. N. Howard (Madison: University of Wisconsin

Press, 1968); Rayleigh, introduction to Waterston's paper as published in *Phil. Trans. R. Soc. London* 183A (1893), 5–79. In his introduction Rayleigh found it "singular that Waterston appears to have advanced no claim for subsequent publication, whether in the Transactions of the [Royal] Society, or through some other channel. At any time since 1860 reference would naturally have been made to Maxwell, and it cannot be doubted that he would have at once recommended that everything possible should be done to atone for the original failure of appreciation."

II   Documents on Atomic and Statistical Physics

1. "On the Properties of Matter"[a]

Lewis Campbell and William Garnett, *The Life of James Clerk Maxwell,* reprint of 1882 edition, (New York: Johnson Reprint Co., 1969), 109–111.

These properties are all relative to the three abstract entities connected with matter, namely, space, time, and force.

1. Since matter must be in some part of space, and in one part only at a time, it possesses the property of locality or position.

2. But matter has not only position but magnitude; this property is called extension.

3. And since it is not infinite it must have bounds, and therefore must possess figure.

These three properties belong both to matter and to imaginary geometrical figures, and may be called the geometrical properties of matter. The following properties do not necessarily belong to geometrical figures.

4. No part of space can contain at the same time more than one body, or no two bodies can coexist in the same space; this property is called impenetrability. It was thought by some that the converse of this was true, and that there was no part of space not filled with matter. If there be a vacuum, said they, that is empty space, it must be either a substance or an accident.

If a substance it must be created or uncreated.

If created it may be destroyed, while matter remains as it was, and thus length, breadth and thickness would be destroyed while the bodies remain at the same distance.

If uncreated, we are led into impiety.

If we say it is an accident, those who deny a vacuum challenge us to define it, and say that length, breadth and thickness belong exclusively to matter.

This is not true, for they belong also to geometric figures, which are forms of thought and not of matter; therefore the atomists maintain that empty space is an accident, and has not only a possible but a real existence, and that there is more empty space than full. This has been well stated by Lucretius.

5. Since there is a vacuum motion is possible; therefore we have a fifth property of matter called mobility.

And the impossibility of a body changing its state of motion or rest without some external force is called *inertia*.

Of forces acting between two particles of matter there are several kinds.

The first kind is independent of the quality of the particles and depends solely on their masses and their mutual distances. Of this kind is the attraction of gravitation and that repulsion which exists between the particles of matter which prevents any two from coming into contact.

The second kind depends on the quality of the particle; of this kind are the attractions of magnetism, electricity, and chemical affinity, which are all convertible into one another and affect all bodies.

The third kind acts between the particles of the same body, and tends to keep them at a certain distance from one another and in a certain configuration.

When this force is repulsive and inversely as the distance, the body is called gaseous.[b]

When it does not follow this law there are two cases.

There may be a force tending to preserve the figure of the body or not.

When this force vanishes the body is a liquid.

When it exists the body is a solid.

If it is small the body is soft; if great it is hard.

If it recovers its figure it is elastic; if not it is inelastic.

The forces in this third division depend almost entirely on heat.

The properties of bodies relative to heat and light are—
Transmission, Reflection, and Destruction,
and in the case of light these may be different for the three kinds of light, so that the properties of colour are—
Quality, Purity and Integrity; or
Hue, Tint, and Shade.

We come next to consider what properties of bodies may be perceived by the senses.

Now the only thing which can be directly perceived by the senses is Force, to which may be reduced heat, light, electricity, sound, and all the things which can be perceived by any sense.

In the sense of sight we perceive at the same time two spheres covered with different colours and shades. The pictures on these two spheres have a general resemblance, but are not exactly the same; and from a comparison of the two spheres we learn, by a kind of intuitive geometry, the position of external objects in three dimensions.

Thus, the object of the sense of sight is the impression made on the different parts of the retina by three kinds of light. By this sense we obtain the greater part of our practical knowledge of locality, extension, and figure as properties of bodies, and we actually perceive colour and angular dimension.

And if we take time into account (as we must always do, for no sense is instantaneous), we perceive relative angular motion.

By the sense of hearing we perceive the intensity, rapidity and quality of the vibrations of the surrounding medium.

By taste and smell we perceive the effects which liquids and aeriform bodies have on the nerves.

By touch we become acquainted with many conditions and qualities of bodies.

1. The actual dimensions of solid bodies in three dimensions, as compared with the dimensions of our own bodies.

2. The nature of the surface; its roughness or smoothness.

3. The state of the body with reference to heat.

To this is to be referred the sensation of wetness and dryness, on account of the close contact which fluids have with the skin.

By means of touch, combined with pressure and motion, we perceive—

1. Hardness and softness, comprehending elasticity, friability, tenacity, flexibility, rigidity, fluidity, etc.

2. Friction, vibration, weight, motion and the like.

The sensations of hunger and thirst, fatigue, and many others have no relations to the properties of bodies.

Lucretius on Empty Space

Nec tamen undique corporeâ stipata tementur
Omnia naturâ, namque est in rebus Inane.
Quod tibi cognôsse in multis erit utile rebus
Nec sinet errantem dubitare et quaere semper
De summa rerum; et nostris diffidere dictis.
Quapropter locus est intactus Inane Vacansque
Quod si non esset, nullâ ratione moveri
Res possent; namque officium quod corporis extat
Officere atque obstare, id in omni tempore adesset
Omnibus. Haud igitur quicquam procedere posset,
Principium quoniam cedendi nulla daret res,
At nunc per maria ac terras sublimaque caeli
Multa modis multis variâ ratione moveri
Cernimus ante oculus, quae, si non esset Inane,
Non tam sollicito motu privata carerent
Quam genita omninô nullâ ratione fuissent,
Undique materies quoniam stipata quiesset.[c]

a. Campbell and Garnett report that this essay by Maxwell was found among Sir William Hamilton's papers. They do not indicate whether it is an exercise based on notes taken in Hamilton's philosophy classes at the University of Edinburgh that Maxwell attended between 1847 and 1850 or notes for an original essay.

b. This refers to Newton's theory of gas pressure (*Principia*, Book 2, Section 5, Prop. XXIII), which was generally accepted until the early nineteenth century. Near the end of his life Maxwell made a more critical analysis of this theory; see documents II-31, 34, and 35.

c. Titus Lucretius Carus, *De Rerum Natura*, H. A. J. Munro, ed. (Cambridge, 1860), book 1, lines 329–345. Here is a standard translation by C. Bailey:

And yet all things are not held close pressed on every side by the nature of body; for there is void in things. To have learnt this will be of profit to you in dealing with many things; it will save you from wandering in doubt and always questioning about the sum of things, and distrusting my words. There is then a void, mere space, untouchable and empty. For if there were not, by no means could things move; for that which is the office of body, to offend and hinder, would at every moment beset all things; nothing, therefore, could advance, since nothing could make a start of yielding place. But as it is through seas and lands and the high tracts of heaven, we descry many things by many means moving in diverse ways before our eyes, which if there were not void would not so much be robbed and baulked of restless motion, but rather could in no way have been born at all, since matter would on every side be in closed-packed stillness. (*Lucretius on the Nature of Things* (Oxford: Clarendon Press, 1910), 38)

For more on Munro and Lucretius, see documents II-4 and 10. For the importance of Lucretius for Victorian culture, see Frank Turner, "Lucretius among the Victorians," *Vctorian Studies* 16 (1973), 329–348.

2. Lecture at Kings College London, 1860[a]

Cambridge University Library, Maxwell manuscripts, box 2.

Mr Principal and Gentlemen

The study of Natural Philosophy, when once entered on must preclude us from allowing our minds to dwell upon any ideas nobler than those of matter and motion. We must leave on one side all the questions which interest us as social and moral beings, & all the feelings which incline us to ⟨[?]⟩ take pleasure in what we see, without inquiring into what lies behind; and turn aside into a region where Force reigns supreme, and recognises matter as its only subject, and where the only theory of action is, that Might makes Right.

When we have commenced our course, it will be too late to speculate on the probable effect of such studies on the ultimate development of our minds, or even on the use of theoretical knowledge, as the prelude to a practical career. We must then give our whole minds to each subject as it comes before us, whether it be the contest of opposing forces, or the paths of moving bodies, and dismiss from our minds everything except what is involved in the problem we have to solve. If we require to make up our minds as to the relation of Natural Philosophy to other branches of knowledge, to our own education, or to the progress of mankind, we must do it today; for tomorrow our business will be not *about* Natural Philosophy, but will be Natural Philosophy itself. I shall now therefore endeavour to point out to you how some of these more general views of the subject appear to me, although I am sure, that after you have yourselves studied the subject, you will confess, that what you heard from me at first but faintly indicates what you learnt by experience at last.

Natural Philosophy is the name given in this country to a collection of sciences, consisting of two main groups. The first of these consists of Mechanics, and includes the general theory of motion and equilibrium, together with the application of the mechanical principles to the investigation of the phenomena of nature.

The second group of sciences is commonly called Physics, and includes at present the study of Light, Heat, Electricity and Magnetism; and, in general, of those phenomena which we have already referred to more general principles, but which we do not as yet contemplate as the result of known mechanical actions.

Natural Philosophy is bounded on the mechanical side by Mathematics, and on the Physical ⟨other⟩ side by Chemistry. Mechanics differs from Mathematics only by involving the ideas of matter, time, and force, in addition to those of Quantity and Space. The methods employed are the same as in mathematics, and the axioms, or laws of motion, upon which the science is founded are of the same kind as those of geometry.

Chemistry, the science which bounds us on the opposite or physical side, investigates those properties of matter by which one substance is distinguished from another, and it contemplates these substances as differing not merely in the degree in which they produce similar effects but in kind, in their very essence. In the physical sciences, we investigate general properties of matter, and refer them to causes which we ⟨express in⟩ conceive to operate on matter in general.

Thus the formation of water by the combination of oxygen and hydrogen is a chemical phenomenon, because it depends on the peculiar nature of those two gases, ⟨and⟩ on the proportions in which they are ⟨combined⟩ mixed, and on the facilities for combination. Chemistry professes to describe the properties of substances, to define the proportions in which they combine, and to state the conditions under which the union takes place, but it goes no further, it considers these as ultimate facts relating to each different substance.

It belongs to Physics to investigate the amount of heat produced by the union of the gases, & the effect of that heat in increasing the pressure, or expanding the combined gases, because the effects of heat, ⟨though different in degree⟩ on all substances, though different in degree, are the same in kind, and are subject to general laws, which apply to all substances.

The explosion may also produce the mechanical effect of bursting the vessel, blowing off the cover, or setting the air in vibration, and making a noise. These phenomena, considered as motions of various bodies, produced by forces of known amount, are to be investigated by mechanical methods.

I have taken three different classes of phenomena as illustrations of the subject matter of Chemistry, Physics and Mechanics. But the difference between these three sciences lies less in the subject studied than in the method of study. In Chemistry, we accept a number of elementary substances and study their properties. In Physics we accept certain great natural agencies, and study their effects on all kinds of matter. In Mechanics we accept nothing but matter and motion, and recognise no difference in matter except arrangement, and no energy in nature except motion.

If any one could explain, by means of known actions of heat, electricity or any other universal agent, the peculiar properties of oxygen and hydrogen, and the results of their combination, he would have offered a physical explanation of the nature of these substances, and he would have transferred certain phenomena from Chemistry into Physics.

Those who have attempted to conceive of the chemical elements as different arrangements of particles of one primitive kind of matter ⟨have⟩ would, if they had been successful, have reduced Chemistry to Mechanics.

No person, however, has hitherto been able to devise arrangements of particles by which chemical phenomena can be explained by the pure science of matter and motion, still less to prove that these arrangements will account for all the known properties of the substances. But though little has been accomplished in our attack on the chemical doctrine of elementary substances, something has been done towards the mechanical explanation of physical phenomena and though we cannot be said as yet to know scientifically the exact kind of motion to which such phenomena as heat and electricity are due, yet we have sufficient evidence to show that every labour we bestow in investigating such subjects by the aid of mechanical ideas will not be in vain.

Natural Philosophy, therefore, treats of those properties of matter which do not require us to conceive of different substances as essentially distinct—of the general properties of matter, as distinguished from the special properties of certain substances.

We receive from the Mathematicians the idea⟨s⟩ of Quantity and all the processes of pure Mathematics. We take possession of the field in which the mathematician is first trained in the exercise of his powers, the field of space with all the apparatus of geometry. We then introduce, in addition to the empty forms of geometry, our own idea of matter, and contemplate bodies placed in different parts of space. This arrangement we consider as subject to change, and thus we arrive at the notion of time and motion. In studying the causes of motion, we arrive at the idea of force, and its relation to the body moved. We then regard ⟨matter⟩ a body not merely as something occupying space and capable of motion, but as something requiring a definite amount of force to produce a definite motion, and capable of producing effects on other bodies depending on the amount of matter in the body moved. Thus we acquire the conception of Mass, or quantity of matter, as a measurable quantity, and also that of Energy, or the amount of work which a body is capable of doing, on account of its motion, or on account of any other state in which it is.

On these conceptions of matter, motion, force, and energy we found the mechanical sciences.

By considering the way in which a body must begin to move, we acquire distinct ideas of the nature of the forces which produce this motion; and we find that however numerous these forces may be, we can always ⟨reduce them⟩ find a very small number which would produce the same effect. Hence we arrive at the idea of equivalent systems of forces, and to the reduction of many forces to a smaller number. In such cases, the forces which we dismiss from our consideration are such as are so balanced as to produce no effect on the body. These balanced systems of forces are said to be in equilibrium, and the consideration of such systems forms the science of Statics.

In all cases in which the forces are not balanced, the science of Statics gives us the means of reducing them to the equivalent system of the form most convenient for our future operations, so that Statics is a necessary foundation for the ⟨whole of mechanics⟩ more general science of Dynamics.

We may also consider the possible motions of a body apart from the causes of that motion. The science of pure motion is called Cinematics or Kinetics, and forms the other foundation of Dynamics.

Dynamics considers the relation between force and motion. The forces are reduced by statics to their simplest form, and the possible motions are brought into a mathematical form by kinetics, and then these are brought into relation by Dynamics, which is the science of the motion of matter as produced by known forces.

Force is here considered as the cause of the motion of a body. But Force is always an action between *two* bodies, and is the result of some relation between them. The investigation of the particular relations between bodies which give rise to particular manifestations of force in nature forms a large portion of Experimental Physics, but there are certain general laws, regulating the amount of Energy arising from given conditions, and determining the total effect of the forces called into play, which are among the most important conclusions of physical science. The science founded on these laws is called Energetics. The application of these ⟨this science⟩ principles to natural phenomena is the special research which the present state of science points out as that from which ⟨most⟩ the greatest results are to be expected in the coming age. The work is only begun, and not till we have measured the energies of all known agents, can we hope to make any progress towards a mechanical explanation of their mode of action. Already in Astronomy & in the theory of Heat and Electricity have ⟨this⟩ the principles of Energetics led to new methods of research, to the discovery of unexpected relations between known properties of matter, and even to the ⟨discovery⟩ knowledge of new properties, which subsequent investigation has shown to exist.

The doctrine of the convertibility and equivalence of all forms of Energy may hereafter be made the basis of new inquiries, which, starting with the knowledge of the quantitative relations between mechanical energy and the other forms in which energy exists, such as heat, attraction &c. will proceed with the investigation of these special forms of energy, and discover, not only the *quantity*, but the *quality* of those forms of energy, and ascertain by what arrangements and motions of matter these different phenomena can be accounted for.

Statics, ⟨K⟩ Cinematics, Dynamics and Energetics may be regarded as the four great branches of abstract Mechanics. They may be applied to the equilibrium and motion of matter in various states of aggregation. The forms and dimensions of all known bodies are altered by the pressures which may be applied to them, and by the effects of Heat.

The theory of pressures, of changes of form, and of the relation between pressure and change of form constitutes an important branch of Mechanics and may be called the general theory of Elasticity. The theory of Pressure and of Elasticity, as applied to fluids, is much more simple than in the case of solids, and forms the sciences of Hydrostatics and Hydrodynamics.

The agency of Heat in producing similar changes is so intimately connected with that of pressure, that we are obliged to treat of the effects of heat and pressure at the same time. We are thus led to those practical applications of the theory of Energetics, which enable us to convert heat into mechanical energy, and form the basis of the theory of the Steam Engine.

The general theory of heat consists of four branches:

1. The laws of the production of heat by mechanical, chemical, or electric action, and the correlations of its transformation into other forms of energy.

2. The theory of the effects it produces on bodies by expanding them and changing their state.

3. The theory of the distribution of heat in bodies by conduction.

4. The theory of Radiant Heat.

The nature of radiant heat appears to differ in nothing from that of light. There is no doubt that the light which we see has all the properties of radiant heat, and that dark heat differs from light only in not being visible to our eyes. Radiant heat, then, being ⟨of⟩ the same ⟨nature⟩ thing as light, though perceived by a different sense, we prefer to use that organ which gives us most information, and call the radiation, *light*, and the *science*, Optics. From a few simple facts about reflexion and refraction, we are able to deduce a systematic science, which rivals the mechanical sciences in precision, and has this great advantage, as an educational science, that the elementary phenomena are easily observed. The science of Light, however, is one in which we have not only ⟨deduced⟩ explained phenomena by referring them to general laws, but in which those general laws have been explained by a mechanical theory.

To trace the steps by which the nature of the motion which we call Light has been ascertained, is one of the most instructive parts of Physical study, and is most likely to introduce the student into the right path for following out investigations in other parts of science.

Last of all we have the Electrical and Magnetic sciences, which treat of certain phenomena of attraction, heat, light and chemical action, depending on conditions of matter of which we have as yet only a partial and provisional knowledge. An immense mass of facts has been collected and these have been reduced to order, ⟨as the result⟩ and expressed as the results of a number of experimental laws, but the form under which these laws are ultimately to appear as deduced from central principles, is as yet uncertain. The present generation has no right to complain of the great discoveries already made, as if they left no room for further enterprize. They have only given Science a wider boundary and we have not only to reduce to order the regions already conquered, but to keep up constant operations on the frontier, on a continually increasing scale.

These are the main divisions of the science of matter and its forces. I must now speak of the method of study. We have seen that Physics differs from pure Mathematics in involving a greater number of ideas, while it agrees with mathematics in using these ideas as the foundation of systematic science. Our first duty must therefore be to require true ideas of the various kinds of quantity with which we have to deal. When we have done this, the application of these ideas to special cases will be comparatively easy. It is this which gives Physical science its peculiar value as a means of education. In all human knowledge, the acquisition of an idea brings with it, as a ⟨natural⟩ logical consequence, a certain system of truths dependent on it, but the mental process by which the idea is acquired is of a different order from that by which deductions are made from it. If a man understands what Force means, I have only to secure his attention, and I can prove to him as many propositions as I please, but if he has not the fundamental idea, no amount of demonstration will give it him. He must think for himself till he gets it.

Now in Natural Philosophy there are a great many different things which must be made our own, before we can have right notions upon what is to follow. And we have the great advantage over the student of many other sciences, that if we once go wrong, our errors become manifest as soon as we go a step forward, so that we have no fear of building complacently on a bad foundation for the whole will go to the ground as soon as we make the first practical application. We are therefore called upon, during our study of Natural Philosophy, to clear up our ideas of the fundamental truths upon which the science is built, and to test the success of this mental process by comparing the results with facts. ⟨It is this which makes our⟩

I shall not now enter upon the question whether the fundamental truths of Physics are to be regarded as mere facts discovered by experiment, or as necessary truths, which the mind must acknowledge as true as soon as its attention has been directed to them. Questions of this kind belong to Metaphysics. In this class we do not pretend to study Metaphysics in a formal and direct manner, but if by the careful study of the laws of nature and their dependence on each other we have been trained into watchfulness over the processes of our own minds, and clear habits of thought, we shall come all the better prepared for the study of higher problems, whether they are presented to us in a metaphysical shape, or as they occur sooner or later to every thinking being. If we have acquired, not the vague and popular notion that there are laws of Nature, but an acquaintance with some of those laws themselves, in their elementary form, and have been able to form some idea of their complication, when applied even to cases purposely simplified, we shall have learnt a lesson of caution, when we examine higher departments of nature, with the expectation of finding there laws of equal simplicity of expression, and equally agreeable to the present state of ⟨human [?]⟩ development of science.

Physical Science affords the exercise which has developed the powers of the greatest and most original thinkers. Bacon, though his supply of physical truth was scanty, had his mind fixed upon the discoveries of the future, and he draws both his wisdom and his eloquence from the contemplation of that new era of which he was himself the prophet. Descartes and Leibnitz need only to be named, to recall systems of metaphysics which were also systems of science. In fact the ideas discussed in metaphysics are so intimately connected with the foundations of Natural Philosophy, that we have only to read a few pages of a metaphysical work, if we wish to ascertain the precise limits of the authors knowledge of physical science.

In the course of your studies here, you will find abundant material for the most abstract speculation, but you must recollect that in physical speculation there must be nothing vague or indistinct. The truths with which we deal are far above the region of mist and storm which conceals them from the undisciplined mind, and yet they are solidly built on the very foundations of the world, and were established of old according to number and measure and weight. Nothing that we can say or think here can escape from the ordeal of the measuring rod and the balance. All quantities must be exact quantities, and all laws must be expressed with reference to exact quantities, so that we have a most effectual means of discovering error, and an absolute security against vagueness and ambiguity.

As we proceed in our course we shall see what part has been taken by experiment, demonstration, and hypothesis respectively in the advancement of science; and we shall have to distinguish between demonstrations founded on pure mathematical properties of space or motion, and those which start from a fact determined by experiment or from an hypothetical assumption. We shall also have to distinguish between experiments of illustration, which like the diagrams of Euclid, serve merely to direct the mind to the contemplation of the ⟨proper⟩ desired subject, and experiments of research, in which the thing sought is a quantity whose value could not be discovered without experiment. We shall also learn the value of hypotheses in the process of discovery, and by what method hypotheses may be made useful in relation to the present state of science, without forming an obstacle in the way of further discovery.

The student of physical science will find that the method or mode of procedure by which knowledge has been accumulated, and even the process by which he himself masters it from day to day, will furnish him with facts relating to the conditions of human knowledge, which he may take with him as guides to the study of other and more complicated subjects.

He will see as he advances that the laws of nature are not mere arbitrary and unconnected decisions of Supreme Power, but that they form essential parts of one universal system, in which infinite Power serves only to reveal unsearchable Wisdom and eternal Truth. When we examine the truths of science, and find that we can not only say "This *is* so" but "This *must be* so for otherwise it would not be consistent with the first principles of truth," or even when we can only say "This *ought to* be so, according to the analogy of nature," we should think what a great thing we are saying, when we pronounce sentence on the laws of creation, and say they are true, or right when judged by the principles of reason. Man has indeed a very partial knowledge of the simplest real thing. The nature of a drop of water has in it mysteries within mysteries utterly unknown to us at present, but what we do know, we know distinctly and scientifically, and we may expect that more will be understood in due time. Some facts we know in their first principles, others as experimentally true, we see more before us which we can only guess at as yet, but we are confident that there remains an inexhaustible inheritance of knowledge not revealed at once, lest we should become proud of our possession and despise patient inquiry, but so arranged, that as each new truth is unravelled, it becomes a clear, well established addition to science, quite free from the mystery which involves what lies beyond, and shows that every atom of creation is unfathomable in its perfection.

Objections have sometimes been raised to the study of physical science, on the ground of the supposed effect of exact science in making the mind unfitted to receive truths which it cannot fully comprehend. We shall find that it is the peculiar function of physical science to lead us, by the steps of rigid demonstration, to the confines of the incomprehensible, and to encourage us to apply our minds to that which we do not yet understand, since it is only to those who labour patiently and think steadily, that such mysteries are ever opened. The higher laws of nature are hid from us at present, but we, and those who come after us, will find in the search for them that which will prepare our minds for the next stage of human knowledge; and the discovery of each new law will not only open up new regions of science, but will alter mens [*sic*] expectation of what the laws of nature ought to be. New discoveries must be consistent with the old, for all *facts* are consistent with each other, but there is a vast mass of opinion about scientific facts, which I am certain is susceptible of modification, when the light of new truths is brought to bear upon it. We must beware of "anticipating nature" and reasoning from the supposed existence in ⟨an assumed⟩ particular form of laws of which we have no distinct idea, and assuming that the higher laws which we do not yet know are capable of being stated under the same forms as the lower ones which we do know.

When vague ideas are put into the form of physical arguments, we can expect nothing but vague conclusions, and great discredit to the mathematics or physics so desecrated.Vague ideas may possibly give picturesqueness to a declamation, but we must be very careful of them when they are disguised in the forms of exact science.

To avoid the vagueness ourselves, we must constantly make use of that method of expression which, by throwing away every idea but that of quantity, arrives at the utmost limit of distinctness. We cannot express physical facts except in a mathematical form. In this class, as I have said before, we have many *kinds* of quantities to deal with and therefore much of our time must be spent in becoming acquainted with these. This corresponds to that part of mathematics which is put in the form of definitions and axioms. But if geometry, or anything more than pure algebra, is mathematics, then Natural Philosophy is and ought to be mathematics, that is, the science ⟨in which⟩ of quantitative relations.

In this class, I hope you will all learn not only the mathematical accuracy of expression of which all physical facts are capable, but the mathematical necessity of their interdependence.

In this way we will carry with us, not merely results, or formulae applicable to cases that may possibly occur in our practice afterwards, but the principles upon which those formulae depend, and without which the formulae are mere mental rubbish, ⟨or s⟩

I know the tendency of the human mind to do anything rather than think. None of us expect to succeed without labour, and we all know that to learn any science requires mental labour, and I am sure we would all give a great deal of mental labour to get up our subjects. But mental labour is not thought, and those who have with great labour acquired the habit of application, often find it much easier to get up a formula than to master a principle. I shall endeavour to show you here, what you will find to be the case afterwards, that principles are fertile in results, but the mere results are barren, and that the man who has got up a formula is at the mercy of his memory, while the man who has thought out a principle may keep his mind clear of formulae, knowing that he could make any number of them when required.

I need hardly add, that though thought be a process from which the mind naturally recoils, yet that process once completed, the mind feels a power and an enjoyment which make it think little in future of the pains and throes which accompany the passage of the mind from one stage of development to another.

It is only by that mathematical training which enables us to see the consequences of the introduction of each new principle into science, that we can fully appreciate the value of these principles, and it is only by mathematical computation that the ultimate results can be compared with facts.

The intermediate portion of mathematical science, which consists of calculation and transformation of symbolical expressions, is most essential to physical science, but it is in reality *pure mathematics*. Everything connected with the original question may be dismissed from the mind during these operations, and the mathematician to whom they are referred may be ⟨ignorant⟩ doubtful whether his results are to be applied to solid geometry, to hydrostatics or to electricity. But as we are engaged in the study of Natural Philosophy we shall endeavour to put our calculations into such a form that every step may be capable of some physical interpretation, and thus we shall exercise powers far more useful than those of mere calculation—the application of principles, and the interpretation of results. In this class we profess to study Natural Philosophy as an applied science, and we hope in doing so to reap all the benefit to ourselves that can be drawn from the subject. If we studied first principles alone, we should become so familiar with the words and modes of expression of philosophy, that we should think ourselves masters of the subject, because we found no scruples in using language which ought to indicate something more than a talking acquaintance with high subjects.

If, dismissing all concern with principles, as too metaphysical for our taste, we were to accept them as mathematical data, embody them in formulae, and grind them together, till we get our x as one of the roots of an equation, we should have made ourselves a little more expert in algebra, but not otherwise wiser, especially if it should turn out that we could not determine whether x is a number of square feet or a velocity.

And if we had determined to be practical and instead of working out formulae, had taken the very case which we wished to solve, and had gone to a standard treatise, and copied out the formula which seemed most suitable, and then got out some result by substituting our numbers, we should *most probably*, by some little oversight get a perfectly useless result, and we should *certainly* leave our minds in a more confused state than when we began.

But if, while here, we can once acquire an intelligent conception of physical principles, and by first investigating the mathematical consequences of these in a few cases, and then applying the results to actual experiments or observations, convince ourselves that these principles are not mere abstractions made by philosophers, but the *key* by which we ourselves may interpret what we see every day, and the *charm* by which we may make the forces of nature do our bidding, then our minds will have received an extension and enlargement which will be *permanent*, for we have been forced to pass from philosophy to carpentry, and from the workshop to the *locus principiorum*, till we have learned by experience that the philosophical and the real is the same thing. If we can repeat the scientific expressions of physical facts in this class room, we shall have gained little, unless we are able to recognize those facts when we meet them out of doors, not dressed up for the lecture-table, but in that *natural retiring* form in which they escaped the notice of so many wise philosophers of old.

But if we can train our minds to see the physical significance of everything that happens, we shall be in the *first place* able to make use of our opportunities in the various professions to which we may be called, *secondly*, we shall never cease to seek, obtain, and enjoy additional knowledge of the world in which we are placed, and *thirdly*, all the skills and knowledge we lay up will round itself to a perfect whole of Wisdom, when all the elements of Science, from the matter which exhibits its modes of action, to the mind which perceives them, are felt to be mutually related parts of one great whole.

The whole course of history is full of examples showing how the neglect of scientific principles produces, in the first place, the certain failure of every enterprize; secondly, how the unscientific mind has been lead from one error to another up to the very pinnacle of absurdity, and thirdly how the want of observation, wrongheadedness, and superstition thus produced have generated systems of philosophy which, by beginning with the contradiction of physical facts, guarantee the thorough unreality of the whole superstructure. We have no time at present to study the history of physical science. We cannot understand the steps by which the human mind has ⟨passed⟩ advanced to its present state of knowledge, till we ourselves have some experience of what that state of knowledge is. When we have ⟨felt and⟩

encountered and overcome the resistance of our own minds to the acquisition of new ideas, we shall be the better able to appreciate the labours of those, who for the first time thought out those ideas, and transmitted them to us for a perpetual possession. While we feel the difficulties of acquiring knowledge which has been already discovered, and carefully put into the most convenient form for our acceptance, we must remember that the discoverers of that knowledge had to struggle against ⟨popular⟩ established notions, and the force of habits of thought to which they themselves had been trained, and to devote the energy of their lives to labours, the results of which *they* saw but dimly, and *others* not at all.

And now when discoveries have been made and results obtained, when public opinion has changed, and everything wearing the garb of science is respected, when, as in this institution, we are directed to study science with a view to success in life—shall we forget the men to whom all this is due? and ⟨appropriate⟩ accept the mere result of their labours, without entering into the spirit of their ideas? Shall our descendants be taught to repeat, like the Indian astronomers, statements of facts which had a meaning in former times, when men were found who could think? That condition I think need not be ours if we are careful to learn the higher lessons of science, while we study its facts. The conditions of knowledge are always the same, and we, while following ⟨other [?]⟩ out the discoveries of the leaders of science, must experience in some degree the same desire to know and the same joy in arriving at knowledge which encouraged and animated them. Do not check these feelings because you cannot expect mankind to sympathize with your triumph over some elementary proposition. None but yourselves can be partakers of the intellectual fruition which results from the comprehension of a principle, and you should beware how you ⟨slight⟩ throw away opportunities of a kind of enjoyment which neither university honours nor wordly reputation can ever afford. Do not be ashamed because the occasion seems small, but cherish any sensation of pleasure which you now feel in the opening up of the mind to the perception of truth. Nothing is easier than for the youngest of us, by repressing such natural feelings, to acquire a permanently contracted mind. The highest intellectual distinction at which man can aim, is to have preserved and nourished to maturity true liberality of thought, in a mind having all its actual knowledge in full and undoubted possession but always capable of advancing to higher and more comprehensive views of truth.

a. A slightly different transcription of this lecture has been published by C. Domb in the *American Journal of Physics*, 47 (1979), 928–933. An earlier version of Maxwell's views on some of these topics may be found in his 1856 Inaugural Lecture at Aberdeen, published by R. V. Jones in *Notes and Records of the Royal Society of London*, 28 (1973), 57–81.

3. Letter[a] from John Herapath to Maxwell, February 23, 1864

Catford Bridge,
Lewisham, Kent, S. E.
Feb. 23 1864

Dear Sir

Indifferent health my [?] many pressing demands on my time, will I hope plead for my delay in replying to your obliging note of the 8th inst.

I thank you most cordially for the trouble you have taken & the information you have collected respecting the molecular theory of gases, not a particle of which had I even heard of before I published my book, except a slight allusion very many years ago to Le Sage's ideas of gravitation,[b] & that was long after my theory of gases was perfected in 1814.

I think I published something about it in Dr Thompson's Annals of Philosophy in 1816, but if I did I have no account of it by me.[c] In 1821 however some papers of mine were published in the Annals on the collision of perfectly hard bodies & the theory of heat and gases.[d]

The theory of the collision of absolutely hard bodies rests on this principle, that each of two bodies feels equally the intensity of the collision on its [?] motion. If the bodies are moving in opposite directions with velocities a, b, the intensity of collision is the sum of their masses drawn into their respective velocities, that is $Aa + Bb$. Each therefore after collision retraces its path with the momentum the other had. If our A with the velocity a follows the other B, moving with the velocity b, then intensity of collision is $A(a - b)$, and the motion of B afterwards is $Aa - A(a - b) = Ab$, and that of B is $Bb + A(a - b)$.

This theory coincides with Wren & Huygens's mentioned in the Principia,[e] when & only when, the hard bodies are equal & moving with equal velocities in opposite directions.

With respect to the theory of heat in the Mathematical Physics,[f] the temperature is measured by the momentum of the particles of the body, and not, as I understand it in the old dynamical theory, simply by their velocity, which would do for measuring the different temperature of ⟨any⟩ the same body, but not to compare the temperature of ⟨this and⟩ different bodies.

As to the theory of gases, imagine an indefinite number of perfectly hard particles moving about & striking one another & the particles of the containing body in all directions, their motions being kept up & maintained by their motions, ⟨of⟩ when struck, of the particles of the containing material. The elasticity of the gas is measured by the number & intensity of the collisions on a given space in a unit of time. The intensity of the collisions on the sides of the vessel will depend on the momentum of the particles & the mean direction in which they strike the sides, which, as the particles move in all directions, will be as an average half a right angle.

I quite agree with you, that if there could be three or even one plane of hard solid matter put into a gas, it would in time destroy the motion of the particles of the gas. But is there any perfectly solid body, that is a body without particles? It is the [?] of the particles of the body on those of the gas which conserves their motions. A perfectly solid body could not in my opinion, except as a whole, have any motion, & therefore according to the dynamical theory of heat must be absolutely cold, which of itself would destroy the motion of the gaseous particles.

The theory of elastic particles is touched on in Scholium I pp. 245, 6 & 7 Vol. I M. P.,[g] but there are the other parts of my labours in relation to gases, and their chemical combination, &c, which I think, if I can ever find leisure to publish what ⟨by [?]⟩ I have had by me for many years, will change the ideas entertained respecting elastic particles in gases & some other points of Physics.

You will perceive that mine is strictly a mechanical system of Physics, based on the property of absolute hardness in the compound atoms or particles. The cause of magnetism I think I see, but Electricity, into which you have I perceive gone deeply, I could never satisfy myself about, though many years ago, when pursuing science, I tried very hard to do it.

Yours truly,
J. Herapath

a. For a copy of this letter we are indebted to its present owner, Professor C. Truesdell, Johns Hopkins University. The identification of the recipient as Maxwell is not certain, but is suggested by the reference to electricity in the last paragraph. See C. Truesdell, *Essays in the History of Mechanics* (New York: Springer-Verlag, 1968), 286n.

b. George-Louis Lesage's theory was that gravitation consisted in the bombardment bodies by streams of invisible particles whose net effect was the same as an attractive force between the bodies, as each shields the other from the stream. For an account of Lesage's theory see Samuel Aronson, "The Gravitational Theory of George-Louis Lesage," *Natural Philosopher*, 3 (1964), 51–74, and J. B. Gough, "Lesage, George-Louis," *Dictionary of Scientific Biography*, C. C. Gillispie, ed (New York: Scribners, 1973), vol. 8, 259–260. Additional references are given in S. G. Brush, *The Kind of Motion We Call Heat* (New York: North-Holland, 1976), 22, note 11.

c. John Herapath, "On the Physical Properties of Gases," *Ann. Phil.* 8 (1816), 56–60. This journal was edited by Thomas Thomson (without a *p*).

d. Herapath, "A Mathematical Enquiry into the Causes, Laws, and Principal Phenomena of Heat, Gases, Gravitation etc.," *Ann. Phil.* 1 (1821), 273–293, 340–351, 401–416; reprinted in Herapath, *Mathematical Physics* (London: Whittaker, 1847), 2 vols. In this paper is Herapath's deduction of the gas laws, theory of specific heats and latent heats, diffusion, velocity of sound, and the effect of the finite size of atoms.

For an account of his work, see Stephen G. Brush's introduction to the reprint (1972) of Herapath's papers (see note f below); Eric Mendoza, "A Critical Examination of Herapath's Dynamical Theory of Gases," *British J. Hist. Sci.* 8 (1975), 155–165, and "The Kinetic Theory of Matter," *Arch. Int. Hist. Sci.* 33 (1983), 186–220, esp. 187–190.

e. Isaac Newton, *Mathematical Principles of Natural Philosophy* (trans. Andrew Motte, 1729) (Berkeley: University of California Press, reprint, 1961), vol. 1, 25.

f. Herapath, *Mathematical Physics* (London Whittaker, 1847), 2 vols. *Mathematical Physics, Two Volumes in One, and Selected Papers by John Herapath*, Stephen G. Brush, ed. (New York: Johnson Reprint Corp., 1972).

g. Herapath is referrring to his *Mathematical Physics*, vol. I, 245–247 (op. cit., note f).

4. Letter from Maxwell to Unknown Correspondent,[a] February 7, 1866[b]

Trinity College, Cambridge, Add. Ms. c.111[10].

8 Palace Gardens Terrace
London W.
1866 Feb. 7

Dear Sir

I am writing about the Dynamical Theory of Gases and am making a short statement of those who have started or embraced similar theories before from Lucretius down to D. Bernouilli, Le Sage of Geneva under the name of Lucrèce Newtonien, and Clausius now professor at Zurich and myself.[c] The details of the mechanics are very different in these different writers on account of their different measure of acquaintance with the theory of collisions etc. With respect to those who flourished since the revival of science I can make out pretty well what they really meant but I am afraid to say anything of Lucretius because his words sometimes seem so appropriate that it is with great regret that one is compelled to cut off a great many marks from him for showing that he did not mean what he has already said so well.

Here is what I have written. Will you tell me if you think it unjust to Lucretius either in excess or defect.

"The notion of particles flying about in all directions, like the motes in a sunbeam, and causing by their impact the motion of larger bodies is to be found in the exposition of the theories of Democritus by Lucretius, but the nature if the impacts and the deviation produced in the paths of the particles are described in language which we must interpret according to the physical conceptions of the age of the author, that is we must get rid of every distinct physical idea which his words may have suggested to us."[d]

Is this saying it too severely about a clever and intelligent ancient?

In particular have the Lucretian atoms an original motion all the same and in the same (downward) direction and equally accelerated (lib. II 238, 239)[e] except insofar as they deviate, and so and only so come into collision (v 220 etc)[f] whereas (at v 90) spatium sine fine modoque est.

The words are such a good illustration of modern theory (v 11 lib II etc)[g] that it would be a pity if they meant something quite different.

The *great intervals* between the collisions in air are in fact about $\frac{1}{400\,000}$ of an inch but they are great compared to those in other media.

In my late capacity of Junior Moderator I have to obtain a poet for Tripos day 7th April. Finding no impulse towards poesy existing at Trinity Hall I asked Gray if the Trinity men were inclined that way. Can you tell me how the appointment is made, and whether Pollock or any other Trinity man would compose a hymn worthy of the day when the names of Colleges assume their Latin forms.

I remain Yours truly,

J. Clerk Maxwell

a. The unknown correspondent may well be Hugh Andrew Johnstone Munro (1819–1885), Fellow of Trinity College from 1843 until his death and a well-known classics scholar who published an important edition of Lucretius's poem *De Rerum Natura* (Cambridge, 1860), with the commentary and critical apparatus appearing in 1864. He was the British authority on Lucretius in this period. Also, he would be a likely candidate for advice on such customs associated with "Tripos day." From document II-10 it appears that Maxwell used the Munro edition in his account of the history of kinetic theory.

b. Courtesy of Thomas Simpson.

c. This history appears as document II-10 and has already been published. See Henry T, Bernstein, "J. Clerk Maxwell on the Kinetic Theory of Gases," *Isis* 54 (1963), 206–216.

d. This passage does not appear in Maxwell's final version of his history, see document II-10 and Bernstein, op. cit. (note c).

e. The lines to which Maxwell refers are in Munro's edition of *De Rerum Natura*, book 2, lines 238 and 239, "omnia quapropter debent per inane quietum aeque ponderibus non aequis concita fermi," translated as "wherefore all things must needs be borne on through the calm void moving at equal rate with unequal weights" (C. Bailey, *De Rerum Natura* (Oxford: Clarendon Press, 1947), vol. 1, 247).

f. [See] line 229, book 2 of the Munro edition. The lines 220–225 are:
tantium quod momen mutatem decen possis.
quod nisi declinare solerent, omnia deorsum.
imbris uti gattae, caderent per inane
profundum nec foret offersus natus nec plaga
creata principis.
The translation is "yet only just so much as you could call a change of trend. But if they were not used to swerve all things would fall downwards through the deep void

like drops of rain, nor could collision come to be nor a blow brought to pass for the first beginnings" (Bailey, op. cit. (note e), 243). See line 90. The line quoted is line 93 in the Munro edition, book 2: "since space is without bound or limit."

g. See line 11, book 2. This section of the poem does not refer to theories of matter but the life of contemplation and wisdom versus one of action and turmoil.
Lucretius's theory of beginnings and of matter starts book 2, line 80. Given the next lines in Maxwell's letter the most probable lines of the poem he had in mind were book 2, lines 105–107, Munro edition:
"cetera, quae porro magnum per inane vagortus,
paucula dissiliunt longe longeque recursaet
in magnis intervallis,"
translated as "Of the rest which wander through the great void a few leap apart and recoil afar with great spaces between" (Bailey, op. cit. (note e), 241).

5. [Discussion Remarks on Brodie's Theory at Chemical Society Meeting, 1867]

Chemical News 15 (1867), 303 (report on a meeting of the Chemical Society, London).

Professor CLERK MAXWELL said he confessed that when he came into the room his feelings received a wholesome shock from two of the statements in the diagrams—first, that space was a chemical substance, and second, that hydrogen and mercury were operations.[a] He now, however, understood what was meant. The present seemed to be an endeavour to cause the symbols of chemical substances to act in the formulae according to their own laws. The formulae at present used were made to express many valuable properties of chemical substances, just as a great many formulae were employed to represent the syllogism in logic, which required a logical mind to form them, to understand them, and to reason upon them. The only successful attempt to introduce a new system in the logical representation was that of Mr. Boole, who accomplished it by the metaphysical and mathematical conception that x^2 was equal to x. In Sir Benjamin Brodie's system α did not mean exactly "hydrogen," but "make hydrogen;" that is, take the cubic centimetre of space, and put hydrogen into it of the proper pressure and temperature. But if they were to compress into that space another volume of hydrogen, that would not be α^2, because it would increase the pressure to double what it was before. If it were possible to get α^2, they would require to combine two volumes together by a process unknown to chemists, keeping the pressure and temperature as before.[b] There was, in this respect, no doubt, an idea which differed from the mere collocation of symbols. The unit of ponderable matter described in the system was one which had been derived by chemists from chemical considerations alone. It had also been advocated by physicists from considerations derived from the theory of heat. In order to decide with certainty on the truth or falsehood of the atomic theory, it would be necessary to consider it from a dynamical point of view. He meant that kind of dynamics treated of in books on mechanics. It was worth while to direct the attention of chemists to the fact that a belief in atoms conducted necessarily to exactly the same definition as was given there—namely, that for every kind of substance the number of atoms, or molecules, in the gaseous state, occupying the space of a litre, at a temperature of 0 degrees, and of a pressure of 760 millimetres, must necessarily be the same. That was a consequence which could be deduced from purely dynamical considerations on the supposition advocated by Professor Clausius and others, that gases consist of molecules floating about in all directions, and producing pressure by their impact. That theory was now under probation among chemists, physicists, and others. The next step was one which might be far off—the finding of the number of these molecules. That number was a fixed one; and when it could be arrived at, we should have another unit of ponderable matter—that of a fixed molecule.[c]

a. This refers to Brodie's "operational" theory of chemistry; see references cited in chapter I, note 143.

b. According to D. M. Dallas, "Maxwell misunderstood the operational nature of the symbols, for he supposed that they symbolized physical operations, not chemical ones. Not all operations can be performed, only those which occur experimentally in reactions. The operation α^2 occurs in the following reaction:

$$CO + 2H_2 = CH_3OH,$$

$$\kappa\xi + 2\alpha = \kappa\alpha^2\xi,$$

but the operation α^2 is performed on carbon monoxide under certain conditions only, and only under these conditions will the use of α^2 be lawful. Maxwell may have been confusing this with 2α, which is two volumes of hydrogen at N.T.P." See "The Chemical Calculus of Sir Benjamin Brodie," in *The Atomic Debates*, W. H. Brock, ed. (Leicester, England: Leicester University Press, 1967), 31–90, esp. 50.

c. Maxwell also participated in a discussion of Brodie's theory at the British Association meeting at Dundee in September 1867. According to a letter from G. Griffiths (Assistant General Secretary of the B.A.) to Brodie, "Maxwell called on by the President, just made a few remarks, not however, of much importance." This quote appears in *The Atomic Debates*, op. cit. (note b), 132; note 112 to this quotation erroneously states that Maxwell's question to William Thomson, "Have you got anything about Sir B. Brodie...?" was before this meeting, but the contents of the letter shows that it could not have been written before 1871 (see document II-10).

6. Letter from Maxwell to Peter Guthrie Tait, November 13, 1867

Cambridge University Library, Maxwell manuscripts, Maxwell-Tait correspondence.

Glenlair,
Dalbeattie,
November 13, 1867

Dear Tait,

If you have any spare copies of your translation of Helmholtz on "Water Twists" I should be obliged to you if you could send me one.

I set the Helmholtz dogma to the Senate House in '66, and got it very nearly done by some men, completely as to the calculations nearly as to the interpretation. Thomson has set himself to spin the chains of destiny out of a fluid plenum as M. Scott set an eminent person to spin ropes from the sea sand, and I saw you had put your calculus in it too. May you both prosper and disentangle your formulae in proportion as you entangle your wurbles. But I fear that the simplest *indivisible* whorl is either two embracing whurles or a whorble embracing itself.

For a simple closed whorble[a] may be easily split and the parts separated [diagram 1] but two embracing wurbles preserve each others solidarity [diagram 2] thus though each may split into many every one of the one set must embrace every one of the other. So does a knotted one [diagram 3].

I send you one or two stereograms of Cyclides, ellipsoids and a parabolic hyperboloid. I have several more which I will send you when they are more perfect. I have got the wave surface and a magnified ⟨fue⟩ view of the point of adhesion also the surface of centres of an ellipsoid etc etc, but the engraver has to make some improvements on them.

Yours Truly,
J. Clerk Maxwell.

a. The term whorble probably refers to the German for a vortex, wirbel.

[Diagram 1]

[Diagram 2]

[Diagram 3]

7. Letter from Maxwell to Peter Guthrie Tait, 1867?[a]

C. G. Knott, *Life and Scientific Work of Peter Guthrie Tait* (Cambridge: Cambridge University Press, 1911), 195. Cambridge University Library, Maxwell manuscripts, Maxwell-Tait correspondence.

207.[b] Matter is *never* perceived by the senses. According to Torricelli, quoted by Berkeley 'Matter is nothing but an enchanted vase of Circe, fitted to receive Impulse and Energy, essences so subtle that nothing but the inmost nature of material bodies is able to contain them.' ...

208.[c] Newton's statement is meant to distinguish matter from space or volume, not to explain either matter or density.

Def. The mass of a body is that factor by which we must multiply the velocity to get the momentum, and by which we must multiply the half square of the velocity to get its energy.

Hence if we take the exchequer pound as unit of mass (which is made of platinum) and if we find a piece of copper such that when it and the exchequer pound move with equal velocity they have the same momentum (describe experiment) then the copper has a mass of one pound.

You may place the two masses in a common balance (which proves their *weights* equal), you may then cause the whole machine to move up or down. If the arm of the balance moves parallel to itself the *masses* must also be equal.

Some illustration of this sort (what you please) is good against heresy in the doctrine of the mass. Next show examples of things which are not matter, though they may be moved and acted on by forces, (1) The path of a body, (2) Its axis of rotation, (3) The form of a steady motion, (4) An undulation (sound or light), (5) Boscovich's centres of force. Next things which are matter such as the luminiferous aether, and if there be anything capable of momentum and kinetic energy.

a. Written soon after the publication of Thomson and Tait's *Treatise on Natural Philosophy*; Maxwell refers to paragraphs in this book.

b. This refers to Thomson and Tait's definition of matter to which Maxwell still took exception in his review of the *Treatise*. Cf. *Scientific Papers*, vol. 2, 776–785, esp. 779.

c. This refers to Thomson and Tait's definition of mass which they took from Newton, yet interpreted to mean density. Their definition contained no empirical demonstrations of the concept which Maxwell here suggests.

8. [Notes for British Association Address, 1870][a]

Cambridge University Library, Maxwell manuscripts, box 5.

In the case of the interdiffusion of two different gases, they can be separated again by chemical means but no natural process can be even thought of which will bring the individual molecules which are now in the upper part of the room into the upper part again after they have once been diffused among the lower particles.

This is a case of the irreversible diffusion of material bodies, but the conduction of heat is an example of the diffusion of energy, and it has been pointed out by Sir W. Thomson that this diffusion is not only irreversible, but that it is constantly diminishing that part of the stock of energy which exists in a form capable of being converted into mechanical work. This is Thomson's theory of the irreversible dissipation of energy, and is equivalent to Clausius' doctrine of the growth of what he calls Entropy.[b] The irreversible character of this process ⟨It⟩ is embodied in Fourier's theory of the conduction of heat, where the formulae themselves indicate a possible solution for all positive values of the time but ⟨become⟩ assume critical values when the time is made zero, and become absurd when the time is assumed to be negative.[c]

The idea which these researches impress on the mind when we follow the natural course of time, is that of an ultimate state of uniform diffusion of energy, which however is not actually reached in any finite time.

But if we reverse the process, and inquire into the former state of things by causing the symbol of time to diminish, we are led up to a state of things which cannot be conceived as the result of any previous state.

a. For the final published version of these paragraphs, see *Scientific Papers*, vol. 2, 226 (reprinted as document II-9).

b. William Thomson, "On a Universal Tendency in Nature to the Dissipation of Mechanical Energy," *Proc. R. Soc. Edinburgh* 3 (1851, pub. 1857), 139–142; *Phil, Mag*, [4] 4 (1852), 304–306. Rudolf Clausius, "Ueber verschiedene für die Anwendung bequeme Formen der Hauptgleichungen der mechanische Wärmetheorie," *Ann. Phys.* [2] 125 (1865), 353–400; English translation in R. Clausius, *The Mechanical Theory of Heat* (London: Macmillan, 1879).

c. Cf. Maxwell's letter to William Thomson, April 14, 1870, to be published in our next volume.

9. "Address to the Mathematical and Physical Sections of the British Association," September 15, 1870

Report of the 40th Meeting of the British Association for the Advancement of Science (1870), 1–9. Reprinted in *Scientific Papers*, vol. 2, 215–229.

[From the *British Association Report*, Vol. xl.]

XLI.   *Address to the Mathematical and Physical Sections of the British Association.*

[Liverpool, *September* 15, 1870.]

At several of the recent Meetings of the British Association the varied and important business of the Mathematical and Physical Section has been introduced by an Address, the subject of which has been left to the selection of the President for the time being. The perplexing duty of choosing a subject has not, however, fallen to me.

Professor Sylvester, the President of Section A at the Exeter Meeting, gave us a noble vindication of pure mathematics by laying bare, as it were, the very working of the mathematical mind, and setting before us, not the array of symbols and brackets which form the armoury of the mathematician, or the dry results which are only the monuments of his conquests, but the mathematician himself, with all his human faculties directed by his professional sagacity to the pursuit, apprehension, and exhibition of that ideal harmony which he feels to be the root of all knowledge, the fountain of all pleasure, and the condition of all action. The mathematician has, above all things, an eye for symmetry; and Professor Sylvester has not only recognized the symmetry formed by the combination of his own subject with those of the former Presidents, but has pointed out the duties of his successor in the following characteristic note:—

" Mr Spottiswoode favoured the Section, in his opening Address, with a combined history of the progress of Mathematics and Physics; Dr. Tyndall's address was virtually on the limits of Physical Philosophy; the one here in print," says Prof. Sylvester, " is an attempted faint adumbration of the nature of Mathematical Science in the abstract. What is wanting (like a fourth sphere resting on three others in contact) to build up the Ideal Pyramid is a discourse on the Relation of the two branches (Mathematics and Physics) to, their action and reaction upon, one another, a magnificent theme, with which it is to be hoped that some future President of Section A will crown the edifice and make the Tetralogy (symbolizable by $A + A'$, A, A', AA') complete."

The theme thus distinctly laid down for his successor by our late President is indeed a magnificent one, far too magnificent for any efforts of mine to realize.   I have endeavoured to follow Mr Spottiswoode, as with far-reaching vision he distinguishes the systems of science into which phenomena, our knowledge of which is still in the nebulous stage, are growing.   I have been carried by the penetrating insight and forcible expression of Dr Tyndall into that sanctuary of minuteness and of power where molecules obey the laws of their existence, clash together in fierce collision, or grapple in yet more fierce embrace, building up in secret the forms of visible things.   I have been guided by Prof. Sylvester towards those serene heights

> " Where never creeps a cloud, or moves a wind,
> Nor ever falls the least white star of snow,
> Nor ever lowest roll of thunder moans,
> Nor sound of human sorrow mounts to mar
> Their sacred everlasting calm."

But who will lead me into that still more hidden and dimmer region where Thought weds Fact, where the mental operation of the mathematician and the physical action of the molecules are seen in their true relation ? Does not the way to it pass through the very den of the metaphysician, strewed with the remains of former explorers, and abhorred by every man of science ? It would indeed be a foolhardy adventure for me to take up the valuable time of the Section by leading you into those speculations which require, as we know, thousands of years even to shape themselves intelligibly.

But we are met as cultivators of mathematics and physics. In our daily work we are led up to questions the same in kind with those of metaphysics ; and we approach them, not trusting to the native penetrating power of our own minds, but trained by a long-continued adjustment of our modes of thought to the facts of external nature.

As mathematicians, we perform certain mental operations on the symbols of number or of quantity, and, by proceeding step by step from more simple to more complex operations, we are enabled to express the same thing in many different forms. The equivalence of these different forms, though a necessary consequence of self-evident axioms, is not always, to our minds, self-evident ; but the mathematician, who by long practice has acquired a familiarity with many of these forms, and 'has become expert in the processes which lead from one to another, can often transform a perplexing expression into another which explains its meaning in more intelligible language.

As students of Physics we observe phenomena under varied circumstances, and endeavour to deduce the laws of their relations. Every natural phenomenon is, to our minds, the result of an infinitely complex system of conditions. What we set ourselves to do is to unravel these conditions, and by viewing the phenomenon in a way which is in itself partial and imperfect, to piece out its features one by one, beginning with that which strikes us first, and thus gradually learning how to look at the whole phenomenon so as to obtain a continually greater degree of clearness and distinctness. In this process, the feature which presents itself most forcibly to the untrained inquirer may not be that which is considered most fundamental by the experienced man of science ; for the success of any physical investigation depends on the judicious selection of what is to be observed as of primary importance, combined with a voluntary abstraction of the mind from those features which, however attractive they appear, we are not yet sufficiently advanced in science to investigate with profit.

Intellectual processes of this kind have been going on since the first formation of language, and are going on still. No doubt the feature which strikes us first and most forcibly in any phenomenon, is the pleasure or the pain which accompanies it, and the agreeable or disagreeable results which follow after it. A theory of nature from this point of view is embodied in many of our words and phrases, and is by no means extinct even in our deliberate opinions.

It was a great step in science when men became convinced that, in order to understand the nature of things, they must begin by asking, not whether a thing is good or bad, noxious or beneficial, but of what kind is it ? and how much is there of it ? Quality and Quantity were then first recognized as the primary features to be observed in scientific inquiry.

As science has been developed, the domain of quantity has everywhere encroached on that of quality, till the process of scientific inquiry seems to have become simply the measurement and registration of quantities, combined with a mathematical discussion of the numbers thus obtained. It is this scientific method of directing our attention to those features of phenomena which may be regarded as quantities which brings physical research under the influence of mathematical reasoning. In the work of the Section we shall have abundant examples of the successful application of this method to the most recent conquests of science; but I wish at present to direct your attention to some of the reciprocal effects of the progress of science on those elementary conceptions which are sometimes thought to be beyond the reach of change.

If the skill of the mathematician has enabled the experimentalist to see that the quantities which he has measured are connected by necessary relations, the discoveries of physics have revealed to the mathematician new forms of quantities which he could never have imagined for himself.

Of the methods by which the mathematician may make his labours most useful to the student of nature, that which I think is at present most important is the systematic classification of quantities.

The quantities which we study in mathematics and physics may be classified in two different ways.

The student who wishes to master any particular science must make himself familiar with the various kinds of quantities which belong to that science. When he understands all the relations between these quantities, he regards them as forming a connected system, and he classes the whole system of quantities together as belonging to that particular science. This classification is the most natural from a physical point of view, and it is generally the first in order of time.

But when the student has become acquainted with several different sciences, he finds that the mathematical processes and trains of reasoning in one science resemble those in another so much that his knowledge of the one science may be made a most useful help in the study of the other.

When he examines into the reason of this, he finds that in the two sciences he has been dealing with systems of quantities, in which the mathematical forms of the relations of the quantities are the same in both systems, though the physical nature of the quantities may be utterly different.

He is thus led to recognize a classification of quantities on a new principle, according to which the physical nature of the quantity is subordinated to its mathematical form. This is the point of view which is characteristic of the mathematician; but it stands second to the physical aspect in order of time, because the human mind, in order to conceive of different kinds of quantities, must have them presented to it by nature.

I do not here refer to the fact that all quantities, as such, are subject to the rules of arithmetic and algebra, and are therefore capable of being submitted to those dry calculations which represent, to so many minds, their only idea of mathematics.

The human mind is seldom satisfied, and is certainly never exercising its highest functions, when it is doing the work of a calculating machine. What the man of science, whether he is a mathematician or a physical inquirer, aims at is, to acquire and develope clear ideas of the things he deals with. For this purpose he is willing to enter on long calculations, and to be for a season a calculating machine, if he can only at last make his ideas clearer.

But if he finds that clear ideas are not to be obtained by means of processes the steps of which he is sure to forget before he has reached the conclusion, it is much better that he should turn to another method, and try to understand the subject by means of well-chosen illustrations derived from subjects with which he is more familiar.

We all know how much more popular the illustrative method of exposition is found, than that in which bare processes of reasoning and calculation form the principal subject of discourse.

Now a truly scientific illustration is a method to enable the mind to grasp some conception or law in one branch of science, by placing before it a conception or a law in a different branch of science, and directing the mind to lay hold of that mathematical form which is common to the corresponding ideas in the two sciences, leaving out of account for the present the difference between the physical nature of the real phenomena.

The correctness of such an illustration depends on whether the two systems of ideas which are compared together are really analogous in form, or whether, in other words, the corresponding physical quantities really belong to the same mathematical class. When this condition is fulfilled, the illustration is not only convenient for teaching science in a pleasant and easy manner, but the recognition of the formal analogy between the two systems of ideas leads to a knowledge of both, more profound than could be obtained by studying each system separately.

There are men who, when any relation or law, however complex, is put before them in a symbolical form, can grasp its full meaning as a relation among abstract quantities. Such men sometimes treat with indifference the further statement that quantities actually exist in nature which fulfil this relation. The mental image of the concrete reality seems rather to disturb than to assist their contemplations.

But the great majority of mankind are utterly unable, without long training, to retain in their minds the unembodied symbols of the pure mathematician, so that, if science is ever to become popular, and yet remain scientific, it must be by a profound study and a copious application of those principles of the mathematical classification of quantities which, as we have seen, lie at the root of every truly scientific illustration.

There are, as I have said, some minds which can go on contemplating with satisfaction pure quantities presented to the eye by symbols, and to the mind in a form which none but mathematicians can conceive.

There are others who feel more enjoyment in following geometrical forms, which they draw on paper, or build up in the empty space before them.

Others, again, are not content unless they can project their whole physical energies into the scene which they conjure up. They learn at what a rate the planets rush through space, and they experience a delightful feeling of exhilaration. They calculate the forces with which the heavenly bodies pull at one another, and they feel their own muscles straining with the effort.

To such men momentum, energy, mass are not mere abstract expressions of the results of scientific inquiry. They are words of power, which stir their souls like the memories of childhood.

For the sake of persons of these different types, scientific truth should be presented in different forms, and should be regarded as equally scientific, whether it appears in the robust form and the vivid colouring of a physical illustration, or in the tenuity and paleness of a symbolical expression.

Time would fail me if I were to attempt to illustrate by examples the scientific value of the classification of quantities. I shall only mention the name of that important class of magnitudes having direction in space which Hamilton has called vectors, and which form the subject-matter of the Calculus of Quaternions, a branch of mathematics which, when it shall have been thoroughly understood by men of the illustrative type, and clothed by them with physical imagery, will become, perhaps under some new name, a most powerful method of communicating truly scientific knowledge to persons apparently devoid of the calculating spirit.

The mutual action and reaction between the different departments of human thought is so interesting to the student of scientific progress, that, at the risk of still further encroaching on the valuable time of the Section, I shall say a few words on a branch of physics which not very long ago would have been considered rather a branch of metaphysics. I mean the atomic theory, or, as it is now called, the molecular theory of the constitution of bodies.

Not many years ago if we had been asked in what regions of physical science the advance of discovery was least apparent, we should have pointed to the hopelessly distant fixed stars on the one hand, and to the inscrutable delicacy of the texture of material bodies on the other.

Indeed, if we are to regard Comte as in any degree representing the scientific opinion of his time, the research into what takes place beyond our own solar system seemed then to be exceedingly unpromising, if not altogether illusory.

The opinion that the bodies which we see and handle, which we can set in motion or leave at rest, which we can break in pieces and destroy, are composed of smaller bodies which we cannot see or handle, which are always in motion, and which can neither be stopped nor broken in pieces, nor in any way destroyed or deprived of the least of their properties, was known by the name of the Atomic theory. It was associated with the names of Democritus, Epicurus, and Lucretius, and was commonly supposed to admit the existence only of atoms and void, to the exclusion of any other basis of things from the universe.

In many physical reasonings and mathematical calculations we are accustomed to argue as if such substances as air, water, or metal, which appear to our senses uniform and continuous, were strictly and mathematically uniform and continuous.

We know that we can divide a pint of water into many millions of portions, each of which is as fully endowed with all the properties of water as the whole pint was; and it seems only natural to conclude that we might go on subdividing the water for ever, just as we can never come to a limit in subdividing the space in which it is contained. We have heard how Faraday divided a grain of gold into an inconceivable number of separate particles, and we may see Dr Tyndall produce from a mere suspicion of nitrite of butyle an immense cloud, the minute visible portion of which is still cloud, and therefore must contain many molecules of nitrite of butyle.

But evidence from different and independent sources is now crowding in upon us which compels us to admit that if we could push the process of subdivision still further we should come to a limit, because each portion would then contain only one molecule, an individual body, one and indivisible, unalterable by any power in nature.

Even in our ordinary experiments on very finely divided matter we find that the substance is beginning to lose the properties which it exhibits when in a large mass, and that effects depending on the individual action of molecules are beginning to become prominent.

The study of these phenomena is at present the path which leads to the development of molecular science.

That superficial tension of liquids which is called capillary attraction is one of these phenomena. Another important class of phenomena are those which are due to that motion of agitation by which the molecules of a liquid or gas are continually working their way from one place to another, and continually changing their course, like people hustled in a crowd.

On this depends the rate of diffusion of gases and liquids through each other, to the study of which, as one of the keys of molecular science, that unwearied inquirer into nature's secrets, the late Prof. Graham, devoted such arduous labour.

The rate of electrolytic conduction is, according to Wiedemann's theory, influenced by the same cause ; and the conduction of heat in fluids depends probably on the same kind of action. In the case of gases, a molecular theory has been developed by Clausius and others, capable of mathematical treatment, and subjected to experimental investigation ; and by this theory nearly every known mechanical property of gases has been explained on dynamical principles ; so that the properties of individual gaseous molecules are in a fair way to become objects of scientific research.

Now Mr Stoney has pointed out* that the numerical results of experiments on gases render it probable that the mean distance of their particles at the ordinary temperature and pressure is a quantity of the same order of magnitude as a millionth of a millimetre, and Sir William Thomson has since† shewn, by several independent lines of argument, drawn from phenomena so different in themselves as the electrification of metals by contact, the tension of soap-bubbles, and the friction of air, that in ordinary solids and liquids the average distance between contiguous molecules is less than the hundred-millionth, and greater than the two-thousand-millionth of a centimetre.

* *Phil. Mag.* Aug. 1868.   † *Nature*, March 31, 1870.

These, of course, are exceedingly rough estimates, for they are derived from measurements some of which are still confessedly very rough; but if, at the present time, we can form even a rough plan for arriving at results of this kind, we may hope that, as our means of experimental inquiry become more accurate and more varied, our conception of a molecule will become more definite, so that we may be able at no distant period to estimate its weight with a greater degree of precision.

A theory, which Sir W. Thomson has founded on Helmholtz's splendid hydrodynamical theorems, seeks for the properties of molecules in the ring-vortices of a uniform, frictionless, incompressible fluid. Such whirling rings may be seen when an experienced smoker sends out a dexterous puff of smoke into the still air, but a more evanescent phenomenon it is difficult to conceive. This evanescence is owing to the viscosity of the air; but Helmholtz has shewn that in a perfect fluid such a whirling ring, if once generated, would go on whirling for ever, would always consist of the very same portion of the fluid which was first set whirling, and could never be cut in two by any natural cause. The generation of a ring-vortex is of course equally beyond the power of natural causes, but once generated, it has the properties of individuality, permanence in quantity, and indestructibility. It is also the recipient of impulse and of energy, which is all we can affirm of matter; and these ring-vortices are capable of such varied connexions and knotted self-involutions, that the properties of differently knotted vortices must be as different as those of different kinds of molecules can be.

If a theory of this kind should be found, after conquering the enormous mathematical difficulties of the subject, to represent in any degree the actual properties of molecules, it will stand in a very different scientific position from those theories of molecular action which are formed by investing the molecule with an arbitrary system of central forces invented expressly to account for the observed phenomena.

In the vortex theory we have nothing arbitrary, no central forces or occult properties of any other kind. We have nothing but matter and motion, and when the vortex is once started its properties are all determined from the original impetus, and no further assumptions are possible.

Even in the present undeveloped state of the theory, the contemplation of the individuality and indestructibility of a ring-vortex in a perfect fluid cannot fail to disturb the commonly received opinion that a molecule, in order to be permanent, must be a very hard body.

In fact one of the first conditions which a molecule must fulfil is, apparently, inconsistent with its being a single hard body. We know from those spectroscopic researches which have thrown so much light on different branches of science, that a molecule can be set into a state of internal vibration, in which it gives off to the surrounding medium light of definite refrangibility—light, that is, of definite wave-length and definite period of vibration. The fact that all the molecules (say, of hydrogen) which we can procure for our experiments, when agitated by heat or by the passage of an electric spark, vibrate precisely in the same periodic time, or, to speak more accurately, that their vibrations are composed of a system of simple vibrations having always the same periods, is a very remarkable fact.

I must leave it to others to describe the progress of that splendid series of spectroscopic discoveries by which the chemistry of the heavenly bodies has been brought within the range of human inquiry. I wish rather to direct your attention to the fact that, not only has every molecule of terrestrial hydrogen the same system of periods of free vibration, but that the spectroscopic examination of the light of the sun and stars shews that, in regions the distance of which we can only feebly imagine, there are molecules vibrating in as exact unison with the molecules of terrestrial hydrogen as two tuning-forks tuned to concert pitch, or two watches regulated to solar time.

Now this absolute equality in the magnitude of quantities, occurring in all parts of the universe, is worth our consideration.

The dimensions of individual natural bodies are either quite indeterminate, as in the case of planets, stones, trees, &c., or they vary within moderate limits, as in the case of seeds, eggs, &c.; but even in these cases small quantitative differences are met with which do not interfere with the essential properties of the body.

Even crystals, which are so definite in geometrical form, are variable with respect to their absolute dimensions.

Among the works of man we sometimes find a certain degree of uniformity.

There is a uniformity among the different bullets which are cast in the same mould, and the different copies of a book printed from the same type.

If we examine the coins, or the weights and measures, of a civilized country, we find a uniformity, which is produced by careful adjustment to standards made and provided by the state. The degree of uniformity of these national standards is a measure of that spirit of justice in the nation which has enacted laws to regulate them and appointed officers to test them.

This subject is one in which we, as a scientific body, take a warm interest; and you are all aware of the vast amount of scientific work which has been expended, and profitably expended, in providing weights and measures for commercial and scientific purposes.

The earth has been measured as a basis for a permanent standard of length, and every property of metals has been investigated to guard against any alteration of the material standards when made. To weigh or measure any thing with modern accuracy, requires a course of experiment and calculation in which almost every branch of physics and mathematics is brought into requisition.

Yet, after all, the dimensions of our earth and its time of rotation, though, relatively to our present means of comparison, very permanent, are not so by any physical necessity. The earth might contract by cooling, or it might be enlarged by a layer of meteorites falling on it, or its rate of revolution might slowly slacken, and yet it would continue to be as much a planet as before.

But a molecule, say of hydrogen, if either its mass or its time of vibration were to be altered in the least, would no longer be a molecule of hydrogen.

If, then, we wish to obtain standards of length, time, and mass which shall be absolutely permanent, we must seek them not in the dimensions, or the motion, or the mass of our planet, but in the wave-length, the period of vibration, and the absolute mass of these imperishable and unalterable and perfectly similar molecules.

When we find that here, and in the starry heavens, there are innumerable multitudes of little bodies of exactly the same mass, so many, and no more, to the grain, and vibrating in exactly the same time, so many times, and no more, in a second, and when we reflect that no power in nature can now alter in the least either the mass or the period of any one of them, we seem to have advanced along the path of natural knowledge to one of those points at which we must accept the guidance of that faith by which we understand that "that which is seen was not made of things which do appear."

One of the most remarkable results of the progress of molecular science is the light it has thrown on the nature of irreversible processes—processes, that is, which always tend towards and never away from a certain limiting state. Thus, if two gases be put into the same vessel, they become mixed, and the mixture tends continually to become more uniform. If two unequally heated portions of the same gas are put into the vessel, something of the kind takes place, and the whole tends to become of the same temperature. If two unequally heated solid bodies be placed in contact, a continual approximation of both to an intermediate temperature takes place.

In the case of the two gases, a separation may be effected by chemical means; but in the other two cases the former state of things cannot be restored by any natural process.

In the case of the conduction or diffusion of heat the process is not only irreversible, but it involves the irreversible diminution of that part of the whole stock of thermal energy which is capable of being converted into mechanical work.

This is Thomson's theory of the irreversible dissipation of energy, and it is equivalent to the doctrine of Clausius concerning the growth of what he calls Entropy.

The irreversible character of this process is strikingly embodied in Fourier's theory of the conduction of heat, where the formulæ themselves indicate, for all positive values of the time, a possible solution which continually tends to the form of a uniform diffusion of heat.

But if we attempt to ascend the stream of time by giving to its symbol continually diminishing values, we are led up to a state of things in which the formula has what is called a critical value; and if we inquire into the state of things the instant before, we find that the formula becomes absurd.

We thus arrive at the conception of a state of things which cannot be conceived as the physical result of a previous state of things, and we find that this critical condition actually existed at an epoch not in the utmost depths of a past eternity, but separated from the present time by a finite interval.

This idea of a beginning is one which the physical researches of recent times have brought home to us, more than any observer of the course of scientific thought in former times would have had reason to expect.

But the mind of man is not, like Fourier's heated body, continually settling down into an ultimate state of quiet uniformity, the character of which we can already predict; it is rather like a tree, shooting out branches which adapt themselves to the new aspects of the sky towards which they climb, and roots which contort themselves among the strange strata of the earth into which they delve. To us who breathe only the spirit of our own age, and know only the characteristics of contemporary thought, it is as impossible to predict the general tone of the science of the future as it is to anticipate the particular discoveries which it will make.

Physical research is continually revealing to us new features of natural processes, and we are thus compelled to search for new forms of thought appropriate to these features. Hence the importance of a careful study of those relations between Mathematics and Physics which determine the conditions under which the ideas derived from one department of physics may be safely used in forming ideas to be employed in a new department.

The figure of speech or of thought by which we transfer the language and ideas of a familiar science to one with which we are less acquainted may be called Scientific Metaphor.

Thus the words Velocity, Momentum, Force, &c. have acquired certain precise meanings in Elementary Dynamics. They are also employed in the Dynamics of a Connected System in a sense which, though perfectly analogous to the elementary sense, is wider and more general.

These generalized forms of elementary ideas may be called metaphorical terms in the sense in which every abstract term is metaphorical. The characteristic of a truly scientific system of metaphors is that each term in its metaphorical use retains all the formal relations to the other terms of the system which it had in its original use. The method is then truly scientific— that is, not only a legitimate product of science, but capable of generating science in its turn.

There are certain electrical phenomena, again, which are connected together by relations of the same form as those which connect dynamical phenomena. To apply to these the phrases of dynamics with proper distinctions and provisional reservations is an example of a metaphor of a bolder kind; but it is a legitimate metaphor if it conveys a true idea of the electrical relations to those who have been already trained in dynamics.

Suppose, then, that we have successfully introduced certain ideas belonging to an elementary science by applying them metaphorically to some new class of phenomena. It becomes an important philosophical question to determine in what degree the applicability of the old ideas to the new subject may be taken as evidence that the new phenomena are physically similar to the old.

The best instances for the determination of this question are those in which two different explanations have been given of the same thing.

The most celebrated case of this kind is that of the corpuscular and the undulatory theories of light. Up to a certain point the phenomena of light are equally well explained by both; beyond this point, one of them fails.

To understand the true relation of these theories in that part of the field where they seem equally applicable we must look at them in the light which Hamilton has thrown upon them by his discovery that to every brachistochrone problem there corresponds a problem of free motion, involving different velocities and times, but resulting in the same geometrical path. Professor Tait has written a very interesting paper on this subject.

According to a theory of electricity which is making great progress in Germany, two electrical particles act on one another directly at a distance, but with a force which, according to Weber, depends on their relative velocity, and according to a theory hinted at by Gauss, and developed by Riemann, Lorenz, and Neumann, acts not instantaneously, but after a time depending on the distance. The power with which this theory, in the hands of these eminent men, explains every kind of electrical phenomena must be studied in order to be appreciated.

Another theory of electricity, which I prefer, denies action at a distance and attributes electric action to tensions and pressures in an all-pervading medium, these stresses being the same in kind with those familiar to engineers, and the medium being identical with that in which light is supposed to be propagated.

Both these theories are found to explain not only the phenomena by the aid of which they were originally constructed, but other phenomena, which were not thought of or perhaps not known at the time; and both have independently arrived at the same numerical result, which gives the absolute velocity of light in terms of electrical quantities.

That theories apparently so fundamentally opposed should have so large a field of truth common to both is a fact the philosophical importance of which we cannot fully appreciate till we have reached a scientific altitude from which the true relation between hypotheses so different can be seen.

I shall only make one more remark on the relation between Mathematics and Physics. In themselves, one is an operation of the mind, the other is a dance of molecules. The molecules have laws of their own, some of which we select as most intelligible to us and most amenable to our calculation. We form a theory from these partial data, and we ascribe any deviation of the actual phenomena from this theory to disturbing causes. At the same time we confess that what we call disturbing causes are simply those parts of the true circumstances which we do not know or have neglected, and we endeavour in future to take account of them. We thus acknowledge that the so-called disturbance is a mere figment of the mind, not a fact of nature, and that in natural action there is no disturbance.

But this is not the only way in which the harmony of the material with the mental operation may be disturbed. The mind of the mathematician is subject to many disturbing causes, such as fatigue, loss of memory, and hasty conclusions; and it is found that, from these and other causes, mathematicians make mistakes.

I am not prepared to deny that, to some mind of a higher order than ours, each of these errors might be traced to the regular operation of the laws of actual thinking; in fact we ourselves often do detect, not only errors of calculation, but the causes of these errors. This, however, by no means alters our conviction that they are errors, and that one process of thought is right and another process wrong.

One of the most profound mathematicians and thinkers of our time, the late George Boole, when reflecting on the precise and almost mathematical character of the laws of right thinking as compared with the exceedingly perplexing though perhaps equally determinate laws of actual and fallible thinking, was led to another of those points of view from which Science seems to look out into a region beyond her own domain.

"We must admit," he says, "that there exist laws" (of thought) "which even the rigour of their mathematical forms does not preserve from violation. We must ascribe to them an authority, the essence of which does not consist in power, a supremacy which the analogy of the inviolable order of the natural world in no way assists us to comprehend."

10. Letter from Maxwell to William Thomson, 1871[a]

University of Glasgow, Kelvin Papers.

1 Democritus see Lucretius.

2 Lucretius α   His bodies are composed of a finite number of indivisible but invisible parts. β   These parts are in constant motion even when the motion of the body in mass is not perceived.

γ   The direction of this motion is *downward* and sensibly but not mathematically uniform. This is a strong point with Lucretius and the weak point of his theory.

δ   Irregularity of the deflections of the atoms introduced to account for free will etc. This is very important in T. L. Carus,[b]

Quare in seminibus quoque idem fateare necesse est
Esse aliam praeter plagas et pondera causum
motibus, unde haec est nobis innata potestas;
De nihilo quoniam fieri nil posse videmus,
Pondus enim prohibet, ne plagis omnia fiant,

Lib. II 284[c] Externa quasi vi: sed ne mens

ipsa necessum
Intestinum habeat cunctis in rebus agendis,
Et devicta quasi cogatur ferre patique:
Id facit exiguum clinamen principiorum
Nec regione loci certa, nec tempore certo.

3 Catena of upholders of intestine motion in hot bodies Bacon, Newton, Boyle, Cavendish, etc.

4 Dan. Bernoulli, not very definite but stated the theory of pressure produced by impact.[d]

5 Lesage of Geneva ⟨1787⟩ wrote an essay, Lucrèce Newtonien, deducing gravity from the impact of ultramundane corpuscles going *in all directions*, and maintaining that if Lucretius had possessed half the mathematical skill of his contemporary Euclid, of Alexandria he would have carried physical science far beyond the stage to which Newton advanced it. Lesage himself would have made a more important contribution to science, if, before calculating the results of the impact of his corpuscles, he had studied the few sentences in which Newton demonstrates the true laws of impact.[e]

6 Pierre Prèvost of Geneva (author of theory of Exchanges) published another treatise of Lesage and one of his own in which he ascribes the pressure of gases to the impact of their molecules against the sides of the vessel, but introduces the ultramundane corpuscles to maintain the motion of the gaseous molecules.[f]

7 Herapath in his Mathematical Physics, 1874 [i.e., 1847] gives still more extensive applications of the theory to gases flowing out of small holes, diffusing through each other etc. I think the notion of temperature being as the square of the velocity is his but he makes the "true temperature" the square root of what we call absolute temperature. This is a mere definition. He also gives $-480°$ F or $491°$ F as the "point of absolute cold. ["] It is remarkable that Lesage and Herapath should have independently fallen into similar errors about the impact of bodies, those errors being I believe known in the best [text?] books of their day. The only source from which these errors might have been derived is the Principia of Descartes.[g]

8 Joule in 1848 calculated with great exactness the velocity of the molecules of hydrogen and subjected the theory of the test of experiment.[h]

9 In 1856 Dr. Krönig directed attention to the kinetic theory of gases and showed how the gaseous laws may be deduced from the impact of perfectly elastic molecules. His conceptions of the arrangements and motions of the molecules are deficient in generality.[i]

10 The great development of the theory is due to Clausius.

α The arrangement of the molecules at any instant is perfectly general

β The impacts of the molecules against each other are taken fully into account

γ The relation between their diameter, the number in a given space and the mean free path is determined

δ Mathematical methods are introduced for dealing *statistically* with immense numbers of molecules by arranging them in groups according to their directions, velocities &c.

ε The slowness of diffusion is accounted for, and steps taken towards a complete theory

ζ Theory of evaporation and maximum density of vapours

η Theory of the change of partners among the molecules of compound bodies and the theory of electrolytic conduction under the smallest electromotive force &c. &c.

θ Internal energy of molecules[j]

11 Maxwell, 1860 α Clausius assumed the velocities of the molecules equal. (This is no essential part of his theory but may be regarded as a trial assumption) Maxwell showed that the velocities range through all values, being distributed according to the same law which prevails in the distribution of errors of observation and in general in all cases in which general uniformity exists in the mass amid apparent irregularity in individual cases. β When there are two or more kinds of molecules acting on one another by impact the average vis viva ["kinetic energy" is inserted

above "vis viva"] of a molecule is the same whatever its mass. Hence follows the dynamical interpretation of

1   Gay Lussac's law of equivalent volumes of gases
2   Dulong and Petit's law of specific heats of gases
I claim No 1 but am willing to distribute as regards No 2.

γ Theory of internal friction of gases and calculation of the mean length of path of the molecules from Stokes' theory of Baily's pendulum expts

δ Development of Clausius's theory of diffusion with errors and failures and a deduction of path from Graham's expts.

ε Theory of conduction of heat in gases obvious probably due to ever so many people, but comparison of conductivity of air and lead (erroneous) is my own.[k]

12 Clausius made objection No 1 to an integration founded on his theory of uniform velocity of molecules (This is the first commitment of Clausius to such a theory) As he was sure to be converted and I was lazy, I said 0. Objection No 2 &c to theory of diffusion and conduction were well founded, and in his paper on Conduction Clausius greatly advanced the methods of treatment and caused me to go through the subject still in the old style but improved (Not published)[l]

13 Oskar Emil Meyer made extensive experiments on internal friction and in 1865 made a more extensive theory of friction of gases, still on Maxwell's framework.[m]

14 Maxwell in 1865 made experiments on viscosity of gases proving that it is independent of pressure and proportional to absolute temperature, and that the ratios of the viscosity of air, carbonic acid and hydrogen agree with those given by Graham.[n] In 1866 he published a revised theory of gases in which the molecules are not regarded as hard elastic spheres but as acting on one another at various distances so as to produce an effect similar to that of a repulsive force varying inversely as the square of the distance. Mathematical methods altered and systematized.[o]

15 Prof. J. Loschmidt of Vienna 12 Oct 1865 communicated to the Imp. Acad. of Vienna a speculation on the size of the molecules of air deduced from Clausius's relation between the mean free path, the diameter and the number in unit of volume combined with an estimate of the volume occupied by the molecules themselves from a consideration of the volumes of various substances in the liquid state. Diameter of a molecule of air one millionth of a millimetre.[p]

16 Stoney in 1868 independently made an estimate of the same kind founded on the same data and leading to a similar result.[q]

17 W. Thomson's stereoscopic view of the same thing from several different directions.[r]

18 Loschmidt 1870 describes expts on diffusion of pairs of gases much more accurate than those of Graham.[s]

19 Gustav Hansemann of Eupen publishes 1871 "Die Atome und ihre Bewegungen, ein Versuch zur Verallgemeinerung der Krönig-Clausius'schen Theorie der Gase" pp. 191 and ranging from elastic spheres to the formation of the Tast-Geschmacks-und-Geruch-Organen and general theory of life, intellect and intellectual progress. But I only got this as a gift from the author a week ago and have not looked it in the mouth yet.[t]

I hope to see and hear you at Edinburgh. The printers are rather slow about Electricity and I have given up sending you proofs till you have served as an Ass.[u] Tait has been very useful about it. You should let the world know that the true source of mathematical methods applicable to physics is to be found in the Proceedings of the Edinburgh F. R. S. E.'s.[v] The volume- surface- and line-integrals of vectors and quaternions and their properties as in the course of being worked out by T′ is worth all that is going on in other seats of learning.[w]

Have you got anything about Sir B. Brodie or do you leave that to the Chemists? They have no right to it. Did you get my letters and proofs at the Athenaeum? I want to know if I may publish T's theorem as I have printed it.[x]

Yours

$$\frac{dp}{dt}$$

a. First published by H. T. Bernstein, "J. Clerk Maxwell on the Kinetic Theory of Gases," *Isis* 54 (1963), 206–216. Bernstein states that this letter was written by Maxwell to William Thomson in the first half of 1871.

b. Titus Lucretius Carus, *De Rerum Natura* (in six books). The edition Maxwell appears to use was that edited by Hugh A. J. Munro, (Cambridge: Cambridge University Press, 1860). Munro's translation, commentary, and method were published in 1864. He spent most of the rest of his life revising his text and translation, published in 1866 and in 1873. Munro collated all known copies of the text in Europe, but his text and translation were subject to much controversy during his life. The lines Maxwell quotes are lines 284–293, book 2, Munro edition, vol. 1. The translation is "Wherefore in the seeds too you must needs allow likewise that there is another cause of motion besides blows and weights, whence comes this power born in us, since we see that nothing comes from nothing. For weight prevents all things coming to pass by blows, as by some force without: But that the very mind feels not some necessity within in doing all things is not constrained like a conquered thing to bear and suffer, this is brought about by the tiny swerve of the first-beginnings in no determined direction of place and at no determined time" (Cyril Bailey, *De Rerum Natura* (Oxford: Clarendon Press, 1947) 251). Obviously neither Maxwell nor Thomson required any translation. The emphasis in the last two lines of the Latin is Maxwell's.

c. This is line 289 in the Munro edition.

d. On the theories of Daniel Bernoulli and other eighteenth-century scientists, see the works cited in chapter I, note 14.

e. George-Louis Lesage. The one major work that Lesage published in his life time was *Essai de Chimie Mécanique* (Rouen, 1758). Lesage's "Lucrèce Newtonien" appeared in Pierre Prevost, *Notice de la vie et des écrits de George-Louis Lesage de Génève* (Geneva, 1805), 189–502.

f. Prevost, *Deux Traités de physique mécanique* (Geneva: Paschoud, 1818).

g. René Descartes, *Principia Philosophiae* (Amsterdam, 1644), reprinted in his *Oeuvres* (Paris: Cerf, 1905), vol. 8; French translation (Paris, 1647), reprinted in *Oeuvres*, vol. 9. The first complete English translation, by V. R. Miller and R. P. Miller, is *Principles of Philosophy* (Boston: Reidel, 1984). The sections on collisions are translated by M. B. Hall, *Nature and Nature's Laws* (New York: Harper, 1970), 254–270. Descartes' theories of motion and collisions are discussed by Richard S. Westfall, *Force in Newton's Physics* (New York: American Elsevier, 1971), chap. 2; Richard J. Blackwell, "Descartes' Law of Motion," *Isis* 57 (1966), 220–234; Kurt Hübner, "Descartes' Rules of Impact and Their Criticism," in *Essays in Memory of Imre Lakatos*, R. S. Cohen *et al.*, eds. (Boston: Reidel, 1976), 299–310; Desmond N. Clarke, "The Impact Rules of Descartes' Physics," *Isis* 68 (1977), 55–66.

h. See chapter I, notes 18 and 22.

i. See chapter I, note 21.

j. See chapter I, notes 19 and 20.

k. See chapter I, note 28 and documents III-2, 6, and 8.

l. See chapter I, note 45; the unpublished account of heat conduction to which Maxwell refers is probably document III-12.

m. Meyer's papers are cited in chapter I, notes 48, 70 and 77. For a discussion of early research on gas transport properties, see J. R. Partington, *Advanced Treatise on Physical Chemistry* (London: Longmans, 1949), vol. 1.

n. Note 76, chapter I; documents II-17 and 18.

o. Note 29, chapter I; document III-30.

p. Notes 139 and 142, chapter I.

q. Note 140, chapter I.

r. William Thomson, "The Size of Atoms," *Nature* 1 (1870), 551–553. By stereoscopic view Maxwell is probably referring to the variety of data Thomson used to develop his various estimates, all of which were within reasonable range of one another.

s. Note 128, chapter I. Maxwell attached great importance to these experiments for estimates of molecular size; see documents III-34, 37, and 42.

t. Gustav Hansemann, *Die Atome und ihre Bewegungen: Ein Versuch zur Verallgemeinerung der Krönig-Clausius'schen Theorie der Gase* (Coln & Leipzig: Mayer, 1871).

u. Maxwell is probably referring to the page proofs of his *Treatise on Electricity and Magnetism* (Oxford, 1873). The slowness of the printers was exacerbated by the custom of Maxwell, Tait, and William Thomson to correct each other's page proofs. Thomson was always late returning such assignments. Thomson serving as an Ass probably refers to Thomson's being president of the British Association for the

Advancement of Science. His tenure began with a presidential lecture in 1871, reported in *Nature* 4 (1871), 262–270, and *Rep. 41st Meeting BAAS* (1871), lxxxiv–cv, for which Maxwell's help on the history of kinetic theory was asked.

v. Fellow of the Royal Society of Edinburgh in whose *Proceedings* both Thomson and Tait, especially the latter, published many short yet important papers on the mathematical methods of physics, usually as specific examples and solutions.

w. Peter Guthrie Tait, *Quaternions* (Oxford, 1875). In the correspondence among Maxwell, Tait, and Thomson, T' means Tait (see note a to document II-25).

x. In the correspondence T means Thomson; *dp/dt* means Maxwell, an allusion to the equation $dp/dt = JCM$. As explained by C. G. Knott, this equation is one form of the second law of thermodynamics, "as used by Thomson in his early papers and by Tait in his *Historic Sketch*, J being Joule's equivalent, C Carnot's function, and M the rate at which heat must be supplied per unit increase of volume, the temperature being constant." *Life and Scientific Work of Peter Guthrie Tait* (Cambridge: Cambridge University Press, 1911), 101.

Thomson's theorem is presented in Maxwell's *Treatise on Electricity and Magnetism* (New York: Dover, 1954), 138ff. This is a reprint of the third edition (1891), J. J. Thomson, ed.; this edition was adapted from the second, most of which was revised by Maxwell before his death in 1879. There are no indications of editorial changes in this section.

11. "Introductory Lecture on Experimental Physics," Cambridge, October 1871

Scientific Papers, vol. 2, 241–255.

XLIV. *Introductory Lecture on Experimental Physics.*

THE University of Cambridge, in accordance with that law of its evolution, by which, while maintaining the strictest continuity between the successive phases of its history, it adapts itself with more or less promptness to the requirements of the times, has lately instituted a course of Experimental Physics. This course of study, while it requires us to maintain in action all those powers of attention and analysis which have been so long cultivated in the University, calls on us to exercise our senses in observation, and our hands in manipulation. The familiar apparatus of pen, ink, and paper will no longer be sufficient for us, and we shall require more room than that afforded by a seat at a desk, and a wider area than that of the black board. We owe it to the munificence of our Chancellor, that, whatever be the character in other respects of the experiments which we hope hereafter to conduct, the material facilities for their full development will be upon a scale which has not hitherto been surpassed.

The main feature, therefore, of Experimental Physics at Cambridge is the Devonshire Physical Laboratory, and I think it desirable that on the present occasion, before we enter on the details of any special study, we should consider by what means we, the University of Cambridge, may, as a living body, appropriate and vitalise this new organ, the outward shell of which we expect soon to rise before us. The course of study at this University has always included Natural Philosophy, as well as Pure Mathematics. To diffuse a sound knowledge of Physics, and to imbue the minds of our students with correct dynamical principles, have been long regarded as among our highest functions, and very few of us can now place ourselves in the mental condition in which even such philosophers as the great Descartes were involved in the days before Newton had announced the true laws of the motion of bodies. Indeed the cultivation and diffusion of sound dynamical ideas has already effected a great change in the language and thoughts even of those who make no pretensions to science, and we are daily receiving fresh proofs that the popularisation of scientific doctrines is producing as great an alteration in the mental state of society as the material applications of science are effecting in its outward life. Such indeed is the respect paid to science, that the most absurd opinions may become current, provided they are expressed in language, the sound of which recals some well-known scientific phrase. If society is thus prepared to receive all kinds of scientific doctrines, it is our part to provide for the diffusion and cultivation, not only of true scientific principles, but of a spirit of sound criticism, founded on an examination of the evidences on which statements apparently scientific depend.

When we shall be able to employ in scientific education, not only the trained attention of the student, and his familiarity with symbols, but the keenness of his eye, the quickness of his ear, the delicacy of his touch, and the adroitness of his fingers, we shall not only extend our influence over a class of men who are not fond of cold abstractions, but, by opening at once all the gateways of knowledge, we shall ensure the association of the doctrines of science with those elementary sensations which form the obscure background of all our conscious thoughts, and which lend a vividness and relief to ideas, which, when presented as mere abstract terms, are apt to fade entirely from the memory.

In a course of Experimental Physics we may consider either the Physics or the Experiments as the leading feature. We may either employ the experiments to illustrate the phenomena of a particular branch of Physics, or we may make some physical research in order to exemplify a particular experimental method. In the order of time, we should begin, in the Lecture Room, with a course of lectures on some branch of Physics aided by experiments of illustration, and conclude, in the Laboratory, with a course of experiments of research.

Let me say a few words on these two classes of experiments,—Experiments of Illustration and Experiments of Research. The aim of an experiment of illustration is to throw light upon some scientific idea so that the student may be enabled to grasp it. The circumstances of the experiment are so arranged that the phenomenon which we wish to observe or to exhibit is brought into prominence, instead of being obscured and entangled among other phenomena, as it is when it occurs in the ordinary course of nature. To exhibit illustrative experiments, to encourage others to make them, and to cultivate in every way the ideas on which they throw light, forms an important part of our duty.

The simpler the materials of an illustrative experiment, and the more familiar they are to the student, the more thoroughly is he likely to acquire the idea which it is meant to illustrate. The educational value of such experiments is often inversely proportional to the complexity of the apparatus. The student who uses home-made apparatus, which is always going wrong, often learns more than one who has the use of carefully adjusted instruments, to which he is apt to trust, and which he dares not take to pieces.

It is very necessary that those who are trying to learn from books the facts of physical science should be enabled by the help of a few illustrative experiments to recognise these facts when they meet with them out of doors. Science appears to us with a very different aspect after we have found out that it is not in lecture rooms only, and by means of the electric light projected on a screen, that we may witness physical phenomena, but that we may find illustrations of the highest doctrines of science in games and gymnastics, in travelling by land and by water, in storms of the air and of the sea, and wherever there is matter in motion.

This habit of recognising principles amid the endless variety of their action can never degrade our sense of the sublimity of nature, or mar our enjoyment of its beauty. On the contrary, it tends to rescue our scientific ideas from that vague condition in which we too often leave them, buried among the other products of a lazy credulity, and to raise them into their proper position among the doctrines in which our faith is so assured, that we are ready at all times to act on them.

Experiments of illustration may be of very different kinds. Some may be adaptations of the commonest operations of ordinary life, others may be carefully arranged exhibitions of some phenomenon which occurs only under peculiar conditions. They all, however, agree in this, that their aim is to present some phenomenon to the senses of the student in such a way that he may associate with it the appropriate scientific idea. When he has grasped this idea, the experiment which illustrates it has served its purpose.

In an experiment of research, on the other hand, this is not the principal aim. It is true that an experiment, in which the principal aim is to see what happens under certain conditions, may be regarded as an experiment of research by those who are not yet familiar with the result, but in experimental researches, strictly so called, the ultimate object is to measure something which we have already seen—to obtain a numerical estimate of some magnitude.

Experiments of this class—those in which measurement of some kind is involved, are the proper work of a Physical Laboratory. In every experiment we have first to make our senses familiar with the phenomenon, but we must not stop here, we must find out which of its features are capable of measurement, and what measurements are required in order to make a complete specification of the phenomenon. We must then make these measurements, and deduce from them the result which we require to find.

This characteristic of modern experiments—that they consist principally of measurements,—is so prominent, that the opinion seems to have got abroad, that in a few years all the great physical constants will have been approximately estimated, and that the only occupation which will then be left to men of science will be to carry on these measurements to another place of decimals.

If this is really the state of things to which we are approaching, our Laboratory may perhaps become celebrated as a place of conscientious labour and consummate skill, but it will be out of place in the University, and ought rather to be classed with the other great workshops of our country, where equal ability is directed to more useful ends.

But we have no right to think thus of the unsearchable riches of creation, or of the untried fertility of those fresh minds into which these riches will continue to be poured. It may possibly be true that, in some of those fields of discovery which lie open to such rough observations as can be made without artificial methods, the great explorers of former times have appropriated most of what is valuable, and that the gleanings which remain are sought after, rather for their abstruseness, than for their intrinsic worth. But the history of science shews that even during that phase of her progress in which she devotes herself to improving the accuracy of the numerical measurement of quantities with which she has long been familiar, she is preparing the materials for the subjugation of new regions, which would have remained unknown if she had been contented with the rough methods of her early pioneers. I might bring forward instances gathered from every branch of science, shewing how the labour of careful measurement has been rewarded by the discovery of new fields of research, and by the development of new scientific ideas. But the history of the science of terrestrial magnetism affords us a sufficient example of what may be done by Experiments in Concert, such as we hope some day to perform in our Laboratory.

That celebrated traveller, Humboldt, was profoundly impressed with the scientific value of a combined effort to be made by the observers of all nations, to obtain accurate measurements of the magnetism of the earth; and we owe it mainly to his enthusiasm for science, his great reputation and his widespread influence, that not only private men of science, but the governments of most of the civilised nations, our own among the number, were induced to take part in the enterprise. But the actual working out of the scheme, and the arrangements by which the labours of the observers were so directed as to obtain the best results, we owe to the great mathematician Gauss, working along with Weber, the future founder of the science of electro-magnetic measurement, in the magnetic observatory of Göttingen, and aided by the skill of the instrument-maker Leyser. These men, however, did not work alone. Numbers of scientific men joined the Magnetic Union, learned the use of the new instruments and the new methods of reducing the observations; and in every city of Europe you might see them, at certain stated times, sitting, each in his cold wooden shed, with his eye fixed at the telescope, his ear attentive to the clock, and his pencil recording in his note-book the instantaneous position of the suspended magnet.

Bacon's conception of "Experiments in concert" was thus realised, the scattered forces of science were converted into a regular army, and emulation and jealousy became out of place, for the results obtained by any one observer were of no value till they were combined with those of the others.

The increase in the accuracy and completeness of magnetic observations which was obtained by the new method, opened up fields of research which were hardly suspected to exist by those whose observations of the magnetic needle had been conducted in a more primitive manner. We must reserve for its proper place in our course any detailed description of the disturbances to which the magnetism of our planet is found to be subject. Some of these disturbances are periodic, following the regular courses of the sun and moon. Others are sudden, and are called magnetic storms, but, like the storms of the atmosphere, they have their known seasons of frequency. The last and the most mysterious of these magnetic changes is that secular variation by which the whole character of the earth, as a great magnet, is being slowly modified, while the magnetic poles creep on, from century to century, along their winding track in the polar regions.

We have thus learned that the interior of the earth is subject to the influences of the heavenly bodies, but that besides this there is a constantly progressive change going on, the cause of which is entirely unknown. In each of the magnetic observatories throughout the world an arrangement is at work, by means of which a suspended magnet directs a ray of light on a prepared sheet of paper moved by clockwork. On that paper the never-resting heart of the earth is now tracing, in telegraphic symbols which will one day be interpreted, a record of its pulsations and its flutterings, as well as of that slow but mighty working which warns us that we must not suppose that the inner history of our planet is ended.

But this great experimental research on Terrestrial Magnetism produced lasting effects on the progress of science in general. I need only mention one or two instances. The new methods of measuring forces were successfully applied by Weber to the numerical determination of all the phenomena of electricity, and very soon afterwards the electric telegraph, by conferring a commercial value on exact numerical measurements, contributed largely to the advancement, as well as to the diffusion of scientific knowledge.

But it is not in these more modern branches of science alone that this influence is felt. It is to Gauss, to the Magnetic Union, and to magnetic observers in general, that we owe our deliverance from that absurd method of estimating forces by a variable standard which prevailed so long even among men of science. It was Gauss who first based the practical measurement of magnetic force (and therefore of every other force) on those long established principles, which, though they are embodied in every dynamical equation, have been so generally set aside, that these very equations, though correctly given in our Cambridge textbooks, are usually explained there by assuming, in addition to the variable standard of force, a variable, and therefore illegal, standard of mass.

Such, then, were some of the scientific results which followed in this case from bringing together mathematical power, experimental sagacity, and manipulative skill, to direct and assist the labours of a body of zealous observers. If therefore we desire, for our own advantage and for the honour of our University, that the Devonshire Laboratory should be successful, we must endeavour to maintain it in living union with the other organs and faculties of our learned body. We shall therefore first consider the relation in which we stand to those mathematical studies which have so long flourished among us, which deal with our own subjects, and which differ from our experimental studies only in the mode in which they are presented to the mind.

There is no more powerful method for introducing knowledge into the mind than that of presenting it in as many different ways as we can. When the ideas, after entering through different gateways, effect a junction in the citadel of the mind, the position they occupy becomes impregnable. Opticians tell us that the mental combination of the views of an object which we obtain from stations no further apart than our two eyes is sufficient to produce in our minds an impression of the solidity of the object seen; and we find that this impression is produced even when we are aware that we are really looking at two flat pictures placed in a stereoscope. It is therefore natural to expect that the knowledge of physical science obtained by the combined use of mathematical analysis and experimental research will be of a more solid, available, and enduring kind than that possessed by the mere mathematician or the mere experimenter.

But what will be the effect on the University, if men pursuing that course of reading which has produced so many distinguished Wranglers, turn aside to work experiments? Will not their attendance at the Laboratory count not merely as time withdrawn from their more legitimate studies, but as the introduction of a disturbing element, tainting their mathematical conceptions with material imagery, and sapping their faith in the formulæ of the textbooks? Besides this, we have already heard complaints of the undue extension of our studies, and of the strain put upon our questionists by the weight of learning which they try to carry with them into the Senate-House. If we now ask them to get up their subjects not only by books and writing, but at the same time by observation and manipulation, will they not break down altogether? The Physical Laboratory, we are told, may perhaps be useful to those who are going out in Natural Science, and who do not take in Mathematics, but to attempt to combine both kinds of study during the time of residence at the University is more than one mind can bear.

No doubt there is some reason for this feeling. Many of us have already overcome the initial difficulties of mathematical training. When we now go on with our study, we feel that it requires exertion and involves fatigue, but we are confident that if we only work hard our progress will be certain.

Some of us, on the other hand, may have had some experience of the routine of experimental work. As soon as we can read scales, observe times, focus telescopes, and so on, this kind of work ceases to require any great mental effort. We may perhaps tire our eyes and weary our backs, but we do not greatly fatigue our minds.

It is not till we attempt to bring the theoretical part of our training into contact with the practical that we begin to experience the full effect of what Faraday has called "mental inertia"—not only the difficulty of recognising, among the concrete objects before us, the abstract relation which we have learned from books, but the distracting pain of wrenching the mind away from the symbols to the objects, and from the objects back to the symbols. This however is the price we have to pay for new ideas.

But when we have overcome these difficulties, and successfully bridged over the gulph between the abstract and the concrete, it is not a mere piece of knowledge that we have obtained: we have acquired the rudiment of a permanent mental endowment. When, by a repetition of efforts of this kind, we have more fully developed the scientific faculty, the exercise of this faculty in detecting scientific principles in nature, and in directing practice by theory, is no longer irksome, but becomes an unfailing source of enjoyment, to which we return so often, that at last even our careless thoughts begin to run in a scientific channel.

I quite admit that our mental energy is limited in quantity, and I know that many zealous students try to do more than is good for them. But the question about the introduction of experimental study is not entirely one of quantity. It is to a great extent a question of distribution of energy. Some distributions of energy, we know, are more useful than others, because they are more available for those purposes which we desire to accomplish.

Now in the case of study, a great part of our fatigue often arises, not from those mental efforts by which we obtain the mastery of the subject, but from those which are spent in recalling our wandering thoughts; and these efforts of attention would be much less fatiguing if the disturbing force of mental distraction could be removed.

This is the reason why a man whose soul is in his work always makes more progress than one whose aim is something not immediately connected with his occupation. In the latter case the very motive of which he makes use to stimulate his flagging powers becomes the means of distracting his mind from the work before him.

There may be some mathematicians who pursue their studies entirely for their own sake. Most men, however, think that the chief use of mathematics is found in the interpretation of nature. Now a man who studies a piece of mathematics in order to understand some natural phenomenon which he has seen, or to calculate the best arrangement of some experiment which he means to make, is likely to meet with far less distraction of mind than if his sole aim had been to sharpen his mind for the successful practice of the Law, or to obtain a high place in the Mathematical Tripos.

I have known men, who when they were at school, never could see the good of mathematics, but who, when in after life they made this discovery, not only became eminent as scientific engineers, but made considerable progress in the study of abstract mathematics. If our experimental course should help any of you to see the good of mathematics, it will relieve us of much anxiety, for it will not only ensure the success of your future studies, but it will make it much less likely that they will prove injurious to your health.

But why should we labour to prove the advantage of practical science to the University? Let us rather speak of the help which the University may give to science, when men well trained in mathematics and enjoying the advantages of a well-appointed Laboratory, shall unite their efforts to carry out some experimental research which no solitary worker could attempt.

At first it is probable that our principal experimental work must be the illustration of particular branches of science, but as we go on we must add to this the study of scientific methods, the same method being sometimes illustrated by its application to researches belonging to different branches of science.

We might even imagine a course of experimental study the arrangement of which should be founded on a classification of methods, and not on that of the objects of investigation. A combination of the two plans seems to me better than either, and while we take every opportunity of studying methods, we shall take care not to dissociate the method from the scientific research to which it is applied, and to which it owes its value.

We shall therefore arrange our lectures according to the classification of the principal natural phenomena, such as heat, electricity, magnetism and so on.

In the laboratory, on the other hand, the place of the different instruments will be determined by a classification according to methods, such as weighing and measuring, observations of time, optical and electrical methods of observation, and so on.

The determination of the experiments to be performed at a particular time must often depend upon the means we have at command, and in the case of the more elaborate experiments, this may imply a long time of preparation, during which the instruments, the methods, and the observers themselves, are being gradually fitted for their work. When we have thus brought together the requisites, both material and intellectual, for a particular experiment, it may sometimes be desirable that before the instruments are dismounted and the observers dispersed, we should make some other experiment, requiring the same method, but dealing perhaps with an entirely different class of physical phenomena.

Our principal work, however, in the Laboratory must be to acquaint ourselves with all kinds of scientific methods, to compare them, and to estimate their value. It will, I think, be a result worthy of our University, and more likely to be accomplished here than in any private laboratory, if, by the free and full discussion of the relative value of different scientific procedures, we succeed in forming a school of scientific criticism, and in assisting the development of the doctrine of method.

But admitting that a practical acquaintance with the methods of Physical Science is an essential part of a mathematical and scientific education, we may be asked whether we are not attributing too much importance to science altogether as part of a liberal education.

Fortunately, there is no question here whether the University should continue to be a place of liberal education, or should devote itself to preparing young men for particular professions. Hence though some of us may, I hope, see reason to make the pursuit of science the main business of our lives, it must be one of our most constant aims to maintain a living connexion between our work and the other liberal studies of Cambridge, whether literary, philological, historical or philosophical.

There is a narrow professional spirit which may grow up among men of science, just as it does among men who practise any other special business. But surely a University is the very place where we should be able to overcome this tendency of men to become, as it were, granulated into small worlds, which are all the more worldly for their very smallness. We lose the advantage of having men of varied pursuits collected into one body, if we do not endeavour to imbibe some of the spirit even of those whose special branch of learning is different from our own.

It is not so long ago since any man who devoted himself to geometry, or to any science requiring continued application, was looked upon as necessarily a misanthrope, who must have abandoned all human interests, and betaken himself to abstractions so far removed from the world of life and action that he has become insensible alike to the attractions of pleasure and to the claims of duty.

In the present day, men of science are not looked upon with the same awe or with the same suspicion. They are supposed to be in league with the material spirit of the age, and to form a kind of advanced Radical party among men of learning.

We are not here to defend literary and historical studies. We admit that the proper study of mankind is man. But is the student of science to be withdrawn from the study of man, or cut off from every noble feeling, so long as he lives in intellectual fellowship with men who have devoted their lives to the discovery of truth, and the results of whose enquiries have impressed themselves on the ordinary speech and way of thinking of men who never heard their names? Or is the student of history and of man to omit from his consideration the history of the origin and diffusion of those ideas which have produced so great a difference between one age of the world and another?

It is true that the history of science is very different from the science of history. We are not studying or attempting to study the working of those blind forces which, we are told, are operating on crowds of obscure people, shaking principalities and powers, and compelling reasonable men to bring events to pass in an order laid down by philosophers.

The men whose names are found in the history of science are not mere hypothetical constituents of a crowd, to be reasoned upon only in masses. We recognise them as men like ourselves, and their actions and thoughts, being more free from the influence of passion, and recorded more accurately than those of other men, are all the better materials for the study of the calmer parts of human nature.

But the history of science is not restricted to the enumeration of successful investigations. It has to tell of unsuccessful inquiries, and to explain why some of the ablest men have failed to find the key of knowledge, and how the reputation of others has only given a firmer footing to the errors into which they fell.

The history of the development, whether normal or abnormal, of ideas is of all subjects that in which we, as thinking men, take the deepest interest. But when the action of the mind passes out of the intellectual stage, in which truth and error are the alternatives, into the more violently emotional states of anger and passion, malice and envy, fury and madness; the student of science, though he is obliged to recognise the powerful influence which these wild forces have exercised on mankind, is perhaps in some measure disqualified from pursuing the study of this part of human nature.

But then how few of us are capable of deriving profit from such studies. We cannot enter into full sympathy with these lower phases of our nature without losing some of that antipathy to them which is our surest safeguard against a reversion to a meaner type, and we gladly return to the company of those illustrious men who by aspiring to noble ends, whether intellectual or practical, have risen above the region of storms into a clearer atmosphere, where there is no misrepresentation of opinion, nor ambiguity of expression, but where one mind comes into closest contact with another at the point where both approach nearest to the truth.

I propose to lecture during this term on Heat, and, as our facilities for experimental work are not yet fully developed, I shall endeavour to place before you the relative position and scientific connexion of the different branches of the science, rather than to discuss the details of experimental methods.

We shall begin with Thermometry, or the registration of temperatures, and Calorimetry, or the measurement of quantities of heat. We shall then go on to Thermodynamics, which investigates the relations between the thermal properties of bodies and their other dynamical properties, in so far as these relations may be traced without any assumption as to the particular constitution of these bodies.

The principles of Thermodynamics throw great light on all the phenomena of nature, and it is probable that many valuable applications of these principles have yet to be made; but we shall have to point out the limits of this science, and to shew that many problems in nature, especially those in which the Dissipation of Energy comes into play, are not capable of solution by the principles of Thermodynamics alone, but that in order to understand them, we are obliged to form some more definite theory of the constitution of bodies.

Two theories of the constitution of bodies have struggled for victory with various fortunes since the earliest ages of speculation: one is the theory of a universal plenum, the other is that of atoms and void.

The theory of the plenum is associated with the doctrine of mathematical continuity, and its mathematical methods are those of the Differential

Calculus, which is the appropriate expression of the relations of continuous quantity.

The theory of atoms and void leads us to attach more importance to the doctrines of integral numbers and definite proportions; but, in applying dynamical principles to the motion of immense numbers of atoms, the limitation of our faculties forces us to abandon the attempt to express the exact history of each atom, and to be content with estimating the average condition of a group of atoms large enough to be visible. This method of dealing with groups of atoms, which I may call the statistical method, and which in the present state of our knowledge is the only available method of studying the properties of real bodies, involves an abandonment of strict dynamical principles, and an adoption of the mathematical methods belonging to the theory of probability. It is probable that important results will be obtained by the application of this method, which is as yet little known and is not familiar to our minds. If the actual history of Science had been different, and if the scientific doctrines most familiar to us had been those which must be expressed in this way, it is possible that we might have considered the existence of a certain kind of contingency a self-evident truth, and treated the doctrine of philosophical necessity as a mere sophism.

About the beginning of this century, the properties of bodies were investigated by several distinguished French mathematicians on the hypothesis that they are systems of molecules in equilibrium. The somewhat unsatisfactory nature of the results of these investigations produced, especially in this country, a reaction in favour of the opposite method of treating bodies as if they were, so far at least as our experiments are concerned, truly continuous. This method, in the hands of Green, Stokes, and others, has led to results, the value of which does not at all depend on what theory we adopt as to the ultimate constitution of bodies.

One very important result of the investigation of the properties of bodies on the hypothesis that they are truly continuous is that it furnishes us with a test by which we can ascertain, by experiments on a real body, to what degree of tenuity it must be reduced before it begins to give evidence that its properties are no longer the same as those of the body in mass. Investigations of this kind, combined with a study of various phenomena of diffusion and of dissipation of energy, have recently added greatly to the evidence in favour of the hypothesis that bodies are systems of molecules in motion.

254   INTRODUCTORY LECTURE ON EXPERIMENTAL PHYSICS.

I hope to be able to lay before you in the course of the term some of the evidence for the existence of molecules, considered as individual bodies having definite properties. The molecule, as it is presented to the scientific imagination, is a very different body from any of those with which experience has hitherto made us acquainted.

In the first place its mass, and the other constants which define its properties, are absolutely invariable; the individual molecule can neither grow nor decay, but remains unchanged amid all the changes of the bodies of which it may form a constituent.

In the second place it is not the only molecule of its kind, for there are innumerable other molecules, whose constants are not approximately, but absolutely identical with those of the first molecule, and this whether they are found on the earth, in the sun, or in the fixed stars.

By what process of evolution the philosophers of the future will attempt to account for this identity in the properties of such a multitude of bodies, each of them unchangeable in magnitude, and some of them separated from others by distances which Astronomy attempts in vain to measure, I cannot conjecture. My mind is limited in its power of speculation, and I am forced to believe that these molecules must have been made as they are from the beginning of their existence.

I also conclude that since none of the processes of nature, during their varied action on different individual molecules, have produced, in the course of ages, the slightest difference between the properties of one molecule and those of another, the history of whose combinations has been different, we cannot ascribe either their existence or the identity of their properties to the operation of any of those causes which we call natural.

Is it true then that our scientific speculations have really penetrated beneath the visible appearance of things, which seem to be subject to generation and corruption, and reached the entrance of that world of order and perfection, which continues this day as it was created, perfect in number and measure and weight?

We may be mistaken. No one has as yet seen or handled an individual molecule, and our molecular hypothesis may, in its turn, be supplanted by some new theory of the constitution of matter; but the idea of the existence of unnumbered individual things, all alike and all unchangeable, is one which cannot enter the human mind and remain without fruit.

INTRODUCTORY LECTURE ON EXPERIMENTAL PHYSICS.   255

But what if these molecules, indestructible as they are, turn out to be not substances themselves, but mere affections of some other substance?

According to Sir W. Thomson's theory of Vortex Atoms, the substance of which the molecule consists is a uniformly dense *plenum*, the properties of which are those of a perfect fluid, the molecule itself being nothing but a certain motion impressed on a portion of this fluid, and this motion is shewn, by a theorem due to Helmholtz, to be as indestructible as we believe a portion of matter to be.

If a theory of this kind is true, or even if it is conceivable, our idea of matter may have been introduced into our minds through our experience of those systems of vortices which we call bodies, but which are not substances, but motions of a substance; and yet the idea which we have thus acquired of matter, as a substance possessing inertia, may be truly applicable to that fluid of which the vortices are the motion, but of whose existence, apart from the vortical motion of some of its parts, our experience gives us no evidence whatever.

It has been asserted that metaphysical speculation is a thing of the past, and that physical science has extirpated it. The discussion of the categories of existence, however, does not appear to be in danger of coming to an end in our time, and the exercise of speculation continues as fascinating to every fresh mind as it was in the days of Thales.

12. Challis's "Mathematical Principles of Physics"

Review in *Nature* 8 (1873), 279–280; *Scientific Papers*, vol. 2, 338–342.

An Essay on the Mathematical Principles of Physics, &c. By the Rev. James Challis, M.A., F.R.S., F.R.A.S., Plumian Professor of Astronomy and Experimental Philosophy in the University of Cambridge, and Fellow of Trinity College. (Cambridge: Deighton, Bell, and Co., 1873.)

This essay is a sort of abstract or general account of the mathematical and physical researches on which the author has been so long engaged, portions of which have appeared from time to time in the *Philosophical Magazine*, and also in his larger work on the "Principles of Mathematics and Physics." [a] It is always desirable that mathematical results should be expressed in intelligible language, as well as in the symbolic form in which they were at first obtained, and we have to thank Professor Challis for this Essay, which though, or rather because, it hardly contains a single equation, sets forth his system more clearly than has been done in some of his previous mathematical papers.

The aim of this Essay, and of the author's long-continued labours, is to advance the theoretical study of Physics. He regards the material universe as "a vast and wonderful *mechanism*, of which not the least wonderful quality is, its being so constructed that we can understand it." The Book of Nature, in fact, contains elementary chapters, and, to those who know where to look for them, the mastery of one chapter is a preparation for the study of the next. The discovery of the calculation necessary to determine the acceleration of a particle whose position is given in terms of the time led to the Newtonian epoch of Natural Philosophy. The study from the cultivation of which our author looks for the "inauguration of a new scientific epoch," is that of the motion of fluids, commonly called Hydrodynamics. The scientific method which he recommends is that described by Newton as the "foundation of all philosophy," namely, that the properties which we attribute to the least parts of matter must be consistent with those of which experiments on sensible bodies have made us cognizant.

The world, according to Professor Challis, is made up of atoms and aether. The atoms are spheres, unalterable in magnitude, and endowed with inertia, but with no other property whatsoever. The aether is a perfect fluid endowed with inertia, and exerting a pressure proportional to its density. It is truly continuous (and therefore does not consist of atoms), and it fills up all the interstices of the atoms.

Here, then, we have set before us with perfect clearness the two constituents of the universe: the atoms, which we can picture in our minds as so many marbles; and the aether, which behaves exactly as air would do if Boyle's law were strictly accurate, if its temperature were invariable, if it were destitute of viscosity, and if gravity did not act on it.

We have no difficulty, therefore, in forming an adequate conception of the properties of the elements from which we have to construct a world. The hypothesis is at least an honest one. It attributes to the elements of things no properties except those which we can clearly define. It stands, therefore, on a different scientific level from those waxen hypotheses in which the atoms are endowed with a new system of attractive or repulsive forces whenever a new phenomena has to be explained.

But the task still before us is a herculean one. It is no less than to explain all actions between bodies or parts of bodies, whether in apparent contact or at stellar distances, by the motions of this all-embracing aether, and the pressure thence resulting.

One kind of motion of the aether is evidently a wave-motion, like that of sound-waves in air. How will such waves affect an atom? Will they propel it forward like the driftwood which is flung upon the shore, or will they draw it back like the shingle which is carried out by the returning wave? Or will they make it oscillate about a fixed position without any advance or recession on the whole?

We have no intention of going through the calculations necessary to solve this problem. They are not contained in this Essay, and Professor Challis admits that he has been unable to determine the absolute amount of the constant term which indicates the permanent effect of the waves on an atom. This is unfortunate, as it gives us no immediate prospect of making those numerical comparisons with observed facts which are necessary for the verification of the theory. Let us, however, suppose this purely mathematical difficulty surmounted, and let us admit with Professor Challis that if the wave-length of the undulations is very small compared with the diameter of the atom, the atom will be urged in the direction of wave-propagation, or in other words *repelled* from the origin of the waves. If on the other hand the wave-length is very great compared with the diameter of the atom, the atom will be urged in the direction opposite to that in which the waves travel, that is, it will be *attracted* towards the source of the waves.

The amount of this attraction or repulsion will depend on the mean of the square of the velocity of the periodic motion of the particles of the aether, and since the amplitude of a diverging wave is inversely as the distance from the centre of divergence, the force will be inversely as the square of this distance, according to Newton's law.

We must remember, however, that the problem is only imperfectly solved, as we do not know the absolute value of this force, and we have not yet arrived at an explanation of the fact that the attraction of gravitation is in exact proportion to the mass of the attracted body, whatever be its chemical nature. (See p. 36).

Admitting these results, and supposing the great ocean of aether to be traversed by waves, these waves impinge on the atoms, and are reflected in the form of diverging waves. These, in their turn, beat other atoms, and cause attraction or repulsion, according as their wave-length is great or small. Thus the waves of shortest period perform the office of repelling atom from atom, and rendering their collision for ever impossible. Other waves, somewhat longer, bind the atoms together in molecular groups. Others contribute to the elasticity of bodies of sensible size, while the long waves are the cause of universal gravitation, holding the planets in their courses, and preserving the most ancient heavens in all their freshness and strength. Then besides the waves of aether, our author contemplates its streams, spiral and otherwise, by which he accounts for electric, magnetic and galvanic phenomena.

Without pretending to have verified all or any of the calculations on which this theory is based, or to have compared the electric, magnetic, and galvanic phenomena, as described in the Essay, with those actually observed, we may venture to make a few remarks upon the theory of action at a distance here put forth.

The explanation of any action between distant bodies by means of a clearly conceivable process going on in the intervening medium is an achievement of the highest scientific value. Of all such actions, that of gravitation is the most universal and the most mysterious. Whatever theory of the constitution of bodies holds out a prospect of the ultimate explanation of the process by which gravitation is effected, men of science will be found ready to devote the whole remainder of their lives to the development of that theory.

The only theory hitherto put forth as a dynamical theory of gravitation is that of Lesage,[b] who adopts the Lucretian theory of atoms and void.

Gravitation on this theory is accounted for by the impact of atoms of incalculable minuteness, which are flying through the heavens with inconceivable velocity and in every possible direction. These "ultramundane corpuscles" falling on a solitary heavenly body would strike it on every side with equal impetus, and would have no effect upon it in the way of resultant force. If, however, another heavenly body were in existence, each would screen the other from a portion of the corpuscular bombardment, and the two bodies would be attracted to each other. The merits and defects of this theory have been recently pointed out by Sir W. Thomson.[c] If the corpuscles are perfectly elastic one body cannot protect the other from the storm, for it will reflect exactly as many corpuscles as it intercepts. If they are inelastic, as Lesage supposes, what becomes of them after collisions? Why are not bodies always growing by the perpetual accumulation of them? How do they get swept away? and what becomes of their energy? Why do they not volatilise the earth in a few minutes? I shall not enter on Sir W. Thomson's improvements of this theory, as it involves a different kind of hydro-dynamics from that cultivated in the Essay, but in whatever way we regard Lesage's theory, the cause of gravitation in the universe can be represented only as depending on an ever fresh supply of something *from without*.

Though Professor Challis has not, as far as we can see, stated in what manner his aethereal waves are originally produced, it would seem that on his theory also the primary waves, by whose action the waves diverging from the atoms are generated, must themselves be propagated from somewhere *outside* the world of stars.

On either theory, therefore, the universe is not even temporarily automatic, but must be fed from moment to moment by an agency external to itself.

If the corpuscles of the one theory, or the aethereal waves of the other, were from any cause to be supplied at a different rate, the value of every force in the universe would suffer change.

On both theories, too, the preservation of the universe is effected only by the unceasing expenditure of enormous quantities of work, so that the conservation of energy in physical operations, which has been the subject of so many measurements, and the study of which has led to so many discoveries, is apparent only, and is merely a kind of "moveable equilibrium" between supply and destruction.

It may seem a sort of anticlimax to descend from these highest heavens of invention down to the "equations of condition" of fluid motion. But it would not be right to pass by the fact that the fluids treated of in this Essay are not in all respects similar to those met with elsewhere. In all their motions they obey a law, which our author was the first to lay down, in addition— or perhaps in some cases in opposition—to those prescribed for them by Lagrange, Poisson, &c.

It is true that a perfect fluid, originally at rest, and afterwards acted on only by such forces as occur in nature, will freely obey this law, and that not only in the form laid down by Professor Challis, in which its rigour is partially relaxed by the introduction of an arbitrary factor, but in its original severe simplicity, as the condition of the existence of a velocity potential.[d]

But, on the one hand, problems in which the motion is assumed to violate this condition have been solved by Helmholtz[e] and Sir W. Thomson,[f] who tell us what the fluid will then do; and, on the other hand, Professor Challis's fluid is able, in virtue of the new equation, to transmit plane waves consisting of transverse displacements. As this is what takes place in the luminiferous aether, other physicists refuse to regard that aether as a fluid, because, according to their definition, the action between any contiguous portions of a fluid is entirely normal to the surface which separates them.

It is not necessary, however, for us to say any more on this subject, as the

Essay before us does not contain, in an explicit form, the equations referred to, but is devoted rather to the exposition of those wider theories of the constitution of matter and the phenomena of nature, some of which we have endeavoured to describe.[g]

a. James Challis (1803–1882) is best known in the history of science as the astronomer who failed to discover Neptune when John Couch Adams gave him the information needed to find it. Ironically, Challis later was responsible for deciding the subject for the Adams Prize and thus indirectly helped to launch Maxwell's career by stimulating his essay on Saturn's rings (see our previous volume, *Maxwell on Saturn's Rings*, 5–10).

In his brief article in the *Dictionary of Scientific Biography* ((1971), vol. 3, 186–187), Olin J. Eggen writes that "Challis was a spectacular failure as a scientist, and ironically, this failure has immortalized him." The previous "larger work" mentioned in this review was *Notes on the Principles of Pure and Applied Calculation; and Application of Mathematical Principles to Theories of the Physical Forces* (Cambridge: Deighton, Bell and Co., 1869); his numerous articles are listed in the Royal Society's *Catalogue of Scientific Papers*, vols. 1, 7, and 9. Challis attempted to reduce gravity, heat, and other phenomena to mechanical pressure through hydrodynamics; as Eggen remarks, "At a later time, or under less amiable circumstances, he would have been branded a charlatan." But his program, as Maxwell recognized, was in the grand tradition of Descartes, Lesage, Herapath, and the Helmholtz-Thomson vortex atom and could not be dismissed out of hand by a proponent of the kinetic theory of gases. Challis's hydrodynamic theory has not received adequate attention from modern historians of nineteenth-century physical science; the only serious treatment of which we are aware is the brief discussion by David B. Wilson in his paper, "George Gabriel Stokes on Stellar Aberration and the Luminiferous Ether," *British J. Hist. Sci.* 6 (1972), 57–72.

b. See note b to document II-3, and notes e and f to document II-10.

c. William Thomson, "On the Ultramundane Corpuscles of Le Sage," *Proc. R. Soc. Edinburgh* 7 (1872), 577–589; *Phil. Mag.* 45 (1873), 321–345. Thomson noted the relation of Le Sage's theory to the kinetic theory of gases.

d. The paper to which Maxwell refers is James Challis, "Discussion of a New Equation in Hydrodynamics," *Phil. Mag.* 20 (1842), 281–288.

e. Hermann von Helmholtz, "Ueber Integrale der hydrodynamischen Gleichungen welche den Wirbelbewegungen entsprechen," *J. Reine Angew. Math.* 55 (1858), 25–55.

f. William Thomson, "On Vortex Motion," *Trans. R. Soc. Edinburgh* 25 (1869), 217–260. R. H. Silliman, "William Thomson: Smoke Rings and Nineteenth-Century Atomism," *Isis* 54 (1963), 461–474.

g. The review was originally published anonymously, but its carefully worded yet devastating criticism of Challis led to a short exchange with Albert J. Mott (next two documents) in which "J. C. M." was revealed as the reviewer. Niven included the original review in his edition of Maxwell's *Scientific Papers* but omitted the response to Mott.

13. A. J. Mott, "Atoms and Ether"[a]

Nature 8 (1873), 322.

Attempts to dispense, in physics, with the ideas of direct attraction and repulsion, however interesting, lead generally to a *petitio principii*, and I fear Prof. Challis's view, to which attention is called in NATURE of August 7, cannot be received as an exception.

For an ether of which the density can be varied is a substance that can be compressed and expanded, and what idea is in our minds when we speak of compression and expansion in a really continuous substance? Continuity implies space, and space that is full. Can space be more than full? When we say that a fluid is compressible and elastic, do we mean anything else than that it is made of parts which can be pushed closer together, and which, being so pushed, will push each other back? But this is repulsion and action at a distance. We do not alter the fact by calling the substance ether, and relieving it from the influence of gravitation.

Is a continuous substance, which is capable of compression, conceivable? I think not; or if it is, the conception is at once more difficult and more opposed to sensible experience than that of attraction and repulsion.

The substance of a bar of iron is not continuous. If I draw one end of it towards me, why does the other end follow? What can be the relation between the movement of my end of the bar and the ethereal vibrations which must propel the other end and all intermediate parts in the same direction?

Albert J. Mott
Liverpool, Aug. 9

a. Albert J. Mott published a handful of papers on mechanics and gas theory and even one on the theory of evolution, but his career was based in Liverpool and his papers appear in the *Proceedings* of the Liverpool Literary and Philosophical Society. The Royal Society Catalogue does not list any obituary notices of him, and he is not mentioned in Pöggendorff.

14. "Atoms and Ether" [reply to Mott]

Nature 8 (1873), 361

I am not enough of a metaphysician to say whether a substance which can be compressed and expanded *necessarily* contains void spaces.

If so, the idea of air, furnished to a beginner by instruction in "Boyle's Law," is self-contradictory; and any molecular theory afterwards developed in order to account for "Boyle's Law," may claim not only ingenuity but necessity in order to abate a crying grievance to all right-minded persons.

I do not myself believe in Prof. Challis's aether, but at the same time I do not believe in the power of the human mind to pronounce that a continuous medium capable of being compressed is an impossibility.

But, on the other hand, I am sure that a medium consisting of molecules is essentially viscous; that is, any motions on a large scale which exist in it are always being converted into molecular agitation, otherwise called heat, so that every molecular medium is the seat of the dissipation of energy, and is getting hotter at the expense of the motions which it transmits. Hence no perfect fluid can be molecular. So far as I can see, Prof. Challis intends his aether to be a perfect fluid, and therefore continuous (see p. 16 of his Essay), though he does not himself pronounce upon its intimate constitution.[a]

Hansemann*[b] makes his aether molecular, and in fact a gas with the molecules immensely diminished in size.

With regard to Mr. Mott's iron bar, when he pulls one end he diminishes, in some unknown way, the pressure between the particles of the iron, and allows the pressure of the aether on the other end to produce its effect.

N.B. This is only the language of a theory, and that theory not mine; nevertheless, I think it is consistent with itself.

J. C. M.
Glenlair, Aug. 13

* Die Atome und ihre Bewegungen, von Gustav Hansemann, (E. H. Mayer: Coln, 1871.)

a. See the letter from Maxwell to Lewis Campbell, September 26, 1874: "In your letter you apply the word imponderable to a molecule. Don't do that again. It may also be worth knowing that the aether cannot be molecular. If it were, it would be a gas, and a pint of it would have the same properties as regards heat, etc., as a pint of air, except that it would not be so heavy." (Campbell and Garnett, *Life*, 390–391). D. J. Price has quoted a similar passage from an undated letter to Tait: "If aether is molecular be the molecules 1/1000 or 1/1000000 of those of hydrogen, the aether is a gas tending to equality of temperature with other bodies and having a capacity for heat not less than 3/5 that of H, O, N, &c. for equal volumes at the same temperature & pressure. Lesage's corpuscles also form a gas of great viscosity—for viscosity increases as the particles get smaller and therefore have longer free paths" ("The Cavendish Laboratory Archives," *Notes Records R. Soc. London* 10 (1953), 139–147).

b. Hansemann (1829–1902) retired as an industrialist in the Rhine to become a private scholar in Berlin in 1873. He wrote several papers on kinetic theory and applied the theory of evolution to social science (1901). See Maxwell's comments on this book in document II-10, item 19.

15. [Notes for "Molecules" Article][a]

Cambridge University Library, Maxwell papers.

[…] extreme slowness, whereas internal friction, or the lateral diffusion of momentum from one stratum to another, is rather favoured by the molecules being close together and the conduction of heat ⟨in⟩ which takes place in liquids not ⟨only⟩ so much by the transference of molecules ⟨but⟩ as by the communication of energy from one molecule to another ⟨on which it is act⟩ with which it is as it were in gear takes place ⟨most⟩ still more freely.

Another kind of diffusion which can be studied best in liquids is that which takes place under electric action.[b] Here is a solution of iodide of potassium through which an electric current passes. Iodine appears at one electrode and potassium at the other but ⟨the⟩ as potassium cannot exist in contact with water we have instead potass and hydrogen.

Clausius has thrown great light on this phenomenon which is called electrolysis by his theory that the molecules of iodide of potassium are always dancing about in the solution and that with such vigour that every now and then they bounce up against some other molecule with such force that the iodine and potassium part company and dance about through the crowd seeking partners among the other ⟨det⟩ dissociated molecules who ⟨may be also⟩ have suffered like things.[c] It is under these circumstances according to Clausius that the electromotive force ⟨exer⟩ produces the effect you have seen. As long as the molecules are in pairs, the electromotive force which pulls the two ⟨members of the pa⟩ molecules in opposite directions can do nothing, and by itself it is not sufficient to tear them asunder, but when by some violent shock the molecules are once parted the electromotive force exerts its guiding influence and ⟨leads⟩ bends the course of each of the unattached molecules ⟨also its prop⟩ each towards its proper electrode till the moment when meeting an unappropriated molecule of the opposite kind it enters into a new and closer alliance in which it is indifferent to mere electric suasion.

Another class of molecular phenomena from the study of which we may derive much light is the evaporation of liquids and the condensation of vapours. According to the theory of Clausius both these phenomena are always going on at the surface of every liquid.[d] Molecules are continually darting out of the liquid and other molecules from the gaseous mass above are darting into the liquid. When more molecules leave the liquid than enter it we call the ⟨pheno⟩ process evaporation, when more enter it than leave it we call it condensation.

The conditions under which evaporation and condensation take place have been long studied. Certain correct statements about them are to be found in every elementary book, so that we are in danger of thinking we know all that is to be known but we ⟨have⟩ need only to ⟨remember⟩ consider the real accession to our knowledge of the relations between gases and liquids which we owe to Dr Andrews^e to see that however well rounded our scientific doctrines may appear ⟨there they may be capable of a transformation into something higher⟩ their true interpretation may involve some ⟨look⟩ principle ⟨for more⟩ so profound that we are not even conscious that it yet remains to be discovered.

[End of this fragment. The next begins in mid-sentence with a brief passage that was moved to an earlier position in the published version (*Scientific Papers*, vol. 2, 369).]

[...] is very small. Roughly speaking it is about one tenth of a wave length of light.

From the mean path and the mean velocity it is easy to calculate the number of collisions which each molecule must undergo in a second in a gas of standard temperature and pressure. These numbers are also given in the tables. They are reckoned by thousands of millions in a second. The numbers of vibrations of light is only about a hundred thousand times greater.

Having advanced thus far in the study of molecules let us ⟨briefly for⟩ spend a moment taking account of the knowledge we have obtained.^f ⟨In the first place we have ascertained the velocity ⟨with which⟩ of the molecules ⟨must⟩ in the different gases.⟩

⟨We have also determined their relative masses.⟩

We have determined the relative masses of the molecules of different gases, their absolute velocities in metres per second and the amount of energy which takes the form of ⟨internal⟩ or rotary ⟨motion⟩ or vibratory motion of the parts of the molecule about its centre of gravity.

These ⟨molecular⟩ data of molecular science are obtained from those experiments on the pressure, density and specific heat of gases which have long been recognized as the regular business of a laboratory.

In the second rank we must place the determination of the relative size of the molecules of different gases; third the absolute values of their mean free paths and of the number of collisons a second. For these quantities we must have recourse to experiments of a more difficult order, on the diffusion of gases, their viscosity and their conductivity for heat. ⟨Great⟩ some progress has already been made in obtaining accuracy in researches of this kind, but we must remember that the numerical results of such experiments cannot as yet be regarded as so precise as those of the first rank.

But the results already obtained agree with each other sufficiently to show that the molecular theory of gases is worth the attention of scientific men.

There is another set of quantities which we must place in the third rank, as our knowledge of their numerical magnitude is neither precise, as in the first rank, nor approximate, as in the second; but is only as yet of the nature of a probable conjecture. These are the absolute mass and dimension of the molecules and the number of them in a cubic centimetre.

The most direct way of calculating these quantities is ⟨from⟩ by a comparison of the volume of the substance as a gas at standard temperature and pressure with its volume when the molecules are packed as closely together as possible. We cannot be certain that in any case the molecules are in actual contact. Indeed we are certain that in liquids they are not in contact, for the phenomena of diffusion show that they are always moving about even when the liquid is apparently at rest. But we have already shown that we can ascertain the relative dimensions of the molecules from experiments on diffusion ⟨and⟩ viscosity and conduction. ⟨Now⟩ We can also compare the volumes to which a cubic centimetre of different gases is reduced when in liquid form. In the case of those gases which have not been observed ⟨directly⟩ in the liquid form Kopp[g] has calculated their molecular volume from the volumes of their liquid compounds.

Now Loschmidt[h] and Lorenz Meyer[i] have shown that there is a remarkable though by no means perfect correspondence between the ⟨molecular volumes⟩ ratios of the volumes of the molecules deduced from experiments on viscosity and the molecular volumes of the substances in the liquid form. It would appear therefore that in most liquids the molecules are nearly in the same state of relative condensation so that the same proportional condensation would reduce them to the ideal state of maximum density in which no room at all is left between the molecules.

What the amount of this further condensation may be it is hard to say but we cannot suppose that any liquid is capable of any very great degree of condensation.

Now Loschmidt[j] has shown in a very simple way that if the mean path of a molecule were shortened in the same proportion as the volume of the gas would be diminished if it were reduced to its ideal condensation this shortened mean path would be about one eighth of the diameter of a molecule.

It was in this way that Loschmidt in 1866[k] first announced that the diameter of a molecule of either of the gases contained in air is about a millionth of a millimetre or ten tenth-metres. Independently of him and of each other Mr Stoney in 1868 and Sir W. Thomson in 1870[l] published results of a similar kind, those of Sir W. Thomson being deduced not only ⟨from⟩ by the ⟨method I have just⟩ consideration of the volume of a liquefied gas but from the phenomena of soap bubbles and the electric properties of metals.

According to an estimate which I have made on Loschmidt's plan the ⟨diameter⟩ size of the molecules of hydrogen is ⟨about⟩ such that about two million of them placed in a row would be about a millimetre long and a million

million million million of them would weigh about four grammes. In a cubic centimetre of gas at ⟨ordin⟩ standard pressure and temperature there are nineteen million million million molecules and if they were placed in regular cubical order the distance between consecutive molecules would be 37 tenth metres. We must remember that all these numbers which I have placed in the third rank are conjectural. ⟨The only⟩ In order to warrant us putting any confidence in numbers of this kind we should have to show that independent calculations founded on data ⟨given by mass⟩ obtained from experiments on many different gases, lead to consistent results.

But what we have already obtained is enough to show that we have some foundation for our conjectures about the weight and dimensions of molecules, and that our knowledge of the subject ⟨is in a very different⟩ has made much progress since the time when Graham[m] began to experiment on Diffusion and Transpiration. The adventurers who now undertake the quest of the ultimate atom ⟨have⟩ are at least aware of some of the phenomena which they have to study, and they may be sure that in following up the paths already known they will light on other hitherto unthought of
[space]

Some years ago the homeopathists in their endeavour to administer small enough doses of medicine prescribed them in the dilution (Here describe the process [space]

If the estimate we have formed of the size and weight of the molecules of simple substances be ⟨in any degree⟩ [a] just one, the chance of the patient receiving even one molecule of the drug would be but small.[n]

⟨Still more recently attempts have been made to express the results of the study of the hereditary transmission of features and peculiarities from one generation to another.⟩

There are other men who have made a study of the transmission of the characteristic features and peculiarities of individuals from one generation to another and of the remarkable manner in which these peculiarities sometimes reappear after having been apparently lost for one or more generations. Some of the results of these enquiries have been expressed in terms of the hypothesis that ⟨each⟩ a large number of
[End of fragment.]

a. These are fragments of notes for Maxwell's lecture on "Molecules," delivered at the British Association meeting in 1873. They are arranged in a sequence corresponding roughly to the final published version (document II-16).

b. Cf. the two paragraphs in the printed version which appear on pages 370–371 of *Scientific Papers*, vol. 2.

c. R. J. E. Clausius, "Ueber die Elektricitätsleitung in Elektrolyten," *Ann. Phys.* [2] 101 (1857), 338–360, translated as "On the Conduction of Electricity in Electrolytes," *Phil.*

Mag [4] 15 (1858), 94–109. This is sometimes called the "Williamson-Clausius hypothesis"; see J. R. Partington, *A History of Chemistry*, (New York: St. Martin's Press, 1972), vol. 4, 672. The detail Maxwell refers to is on p. 95 of Clausius's translated paper.

d. See the 1857 paper of Clausius cited in chapter I, note 20; a reprint appears in S. G. Brush, *Kinetic Theory* (New York: Pergamon Press, 1965), vol. 1, 118.

e. Thomas Andrews, "On the Continuity of the Gaseous and Liquid States of Matter," *Phil. Trans. R. Soc. London* 159 (1869), 575–590, is the main paper in which Andrews presents his experimental data.

f. This sentence resumes the material that follows the previous fragment in the published version; cf. *Scientific Papers*, vol. 2, 371 (document II-16).

g. Hermann Kopp, "Ueber die Volumänderung einiger Substanzen beim Erwärmen und Schmelzen," *Annalen der Chemie und Pharmacie* 93 (1855), 129–232; English summary, "On the Alteration of the Volume of Some Substances by Heating and Fusion," *Phil. Mag.* 9 (1855), 477–479; also "Atomwärme der Elemente und der starrer Korper," *Annalen der Chemie und Pharmacie* 3 (1864), 1–289.

h. See note 128, chapter I. Maxwell attached great importance to Loschmidt's experiments and from them derived his own estimates of molecular sizes. Calculations on these quantities using Loschmidt's data are scattered throughout his unpublished papers, and he also published his own assessment of the experiments in 1873 (document III-42).

i. Maxwell is probably referring to Julius Lothar Meyer (1830–1895), the brother of Oskar Emil Meyer (1834–1915) who made an important study of the viscosity of gases (see chapter I). Lothar Meyer published an investigation of the atomic weights and volumes of the elements, "Die Natur der chemische Elemente als Function ihrer Atomgewicht," *Annalen der Chemie* 7 (suppl. vol.) (1870), 354–364.

j. See note 139 to chapter I.

k. Maxwell is referring to the article cited in note i. Maxwell, following Stoney, used "tenth-metre" to mean 10^{-10} meter.

l. See note 140, chapter I.

m. Thomas Graham, *Elements of Chemistry* (London, 1846), collects the results of his previous experiments on the diffusion of gases.

n. See document III-44, note c, on homeopathic medicine.

16. "Molecules"

Nature, 8 (1873), 437–441; *Phil. Mag.* [4] 46 (1873), 453–469; Maxwell's *Scientific Papers*, vol. 2, 361–378.

[From *Nature*, Vol. VIII.]

LXII. *Molecules**.

AN atom is a body which cannot be cut in two. A molecule is the smallest possible portion of a particular substance. No one has ever seen or handled a single molecule. Molecular science, therefore, is one of those branches of study which deal with things invisible and imperceptible by our senses, and which cannot be subjected to direct experiment.

The mind of man has perplexed itself with many hard questions. Is space infinite, and if so in what sense ? Is the material world infinite in extent, and are all places within that extent equally full of matter ? Do atoms exist, or is matter infinitely divisible ?

The discussion of questions of this kind has been going on ever since men began to reason, and to each of us, as soon as we obtain the use of our faculties, the same old questions arise as fresh as ever. They form as essential a part of the science of the nineteenth century of our era, as of that of the fifth century before it.

We do not know much about the science organisation of Thrace twenty-two centuries ago, or of the machinery then employed for diffusing an interest in physical research. There were men, however, in those days, who devoted their lives to the pursuit of knowledge with an ardour worthy of the most distinguished members of the British Association ; and the lectures in which Democritus explained the atomic theory to his fellow-citizens of Abdera realised, not in golden opinions only, but in golden talents, a sum hardly equalled even in America.

To another very eminent philosopher, Anaxagoras, best known to the world as the teacher of Socrates, we are indebted for the most important service to

---

* A Lecture delivered before the British Association at Bradford.

the atomic theory, which, after its statement by Democritus, remained to be done.   Anaxagoras, in fact, stated a theory which so exactly contradicts the atomic theory of Democritus that the truth or falsehood of the one theory implies the falsehood or truth of the other.   The question of the existence or non-existence of atoms cannot be presented to us this evening with greater clearness than in the alternative theories of these two philosophers.

Take any portion of matter, say a drop of water, and observe its properties. Like every other portion of matter we have ever seen, it is divisible.   Divide it in two, each portion appears to retain all the properties of the original drop, and among others that of being divisible.   The parts are similar to the whole in every respect except in absolute size.

Now go on repeating the process of division till the separate portions of water are so small that we can no longer perceive or handle them.   Still we have no doubt that the sub-division might be carried further, if our senses were more acute and our instruments more delicate.   Thus far all are agreed, but now the question arises, Can this sub-division be repeated for ever ?

According to Democritus and the atomic school, we must answer in the negative.   After a certain number of sub-divisions, the drop would be divided into a number of parts each of which is incapable of further sub-division.   We should thus, in imagination, arrive at the atom, which, as its name literally signifies, cannot be cut in two.   This is the atomic doctrine of Democritus, Epicurus, and Lucretius, and, I may add, of your lecturer.

According to Anaxagoras, on the other hand, the parts into which the drop is divided are in all respects similar to the whole drop, the mere size of a body counting for nothing as regards the nature of its substance.   Hence if the whole drop is divisible, so are its parts down to the minutest sub-divisions, and that without end.

The essence of the doctrine of Anaxagoras is that parts of a body are in all respects similar to the whole.   It was, therefore, called the doctrine of Homoiomereia.   Anaxagoras did not of course assert this of the parts of organised bodies such as men and animals, but he maintained that those inorganic substances which appear to us homogeneous are really so, and that the universal experience of mankind testifies that every material body, without exception, is divisible.

The doctrine of atoms and that of homogeneity are thus in direct contradiction.

But we must now go on to molecules. Molecule is a modern word. It does not occur in *Johnson's Dictionary.* The ideas it embodies are those belonging to modern chemistry.

A drop of water, to return to our former example, may be divided into a certain number, and no more, of portions similar to each other. Each of these the modern chemist calls a molecule of water. But it is by no means an atom, for it contains two different substances, oxygen and hydrogen, and by a certain process the molecule may be actually divided into two parts, one consisting of oxygen and the other of hydrogen. According to the received doctrine, in each molecule of water there are two molecules of hydrogen and one of oxygen. Whether these are or are not ultimate atoms I shall not attempt to decide.

We now see what a molecule is, as distinguished from an atom.

A molecule of a substance is a small body such that if, on the one hand, a number of similar molecules were assembled together they would form a mass of that substance, while on the other hand, if any portion of this molecule were removed, it would no longer be able, along with an assemblage of other molecules similarly treated, to make up a mass of the original substance.

Every substance, simple or compound, has its own molecule. If this molecule be divided, its parts are molecules of a different substance or substances from that of which the whole is a molecule. An atom, if there is such a thing, must be a molecule of an elementary substance. Since, therefore, every molecule is not an atom, but every atom is a molecule, I shall use the word molecule as the more general term.

I have no intention of taking up your time by expounding the doctrines of modern chemistry with respect to the molecules of different substances. It is not the special but the universal interest of molecular science which encourages me to address you. It is not because we happen to be chemists or physicists or specialists of any kind that we are attracted towards this centre of all material existence, but because we all belong to a race endowed with faculties which urge us on to search deep and ever deeper into the nature of things.

We find that now, as in the days of the earliest physical speculations, all physical researches appear to converge towards the same point, and every inquirer, as he looks forward into the dim region towards which the path of discovery is leading him, sees, each according to his sight, the vision of the same Quest.

MOLECULES.

One may see the atom as a material point, invested and surrounded by potential forces. Another sees no garment of force, but only the bare and utter hardness of mere impenetrability.

But though many a speculator, as he has seen the vision recede before him into the innermost sanctuary of the inconceivably little, has had to confess that the quest was not for him, and though philosophers in every age have been exhorting each other to direct their minds to some more useful and attainable aim, each generation, from the earliest dawn of science to the present time, has contributed a due proportion of its ablest intellects to the quest of the ultimate atom.

Our business this evening is to describe some researches in molecular science, and in particular to place before you any definite information which has been obtained respecting the molecules themselves. The old atomic theory, as described by Lucretius and revived in modern times, asserts that the molecules of all bodies are in motion, even when the body itself appears to be at rest. These motions of molecules are in the case of solid bodies confined within so narrow a range that even with our best microscopes we cannot detect that they alter their places at all. In liquids and gases, however, the molecules are not confined within any definite limits, but work their way through the whole mass, even when that mass is not disturbed by any visible motion.

This process of diffusion, as it is called, which goes on in gases and liquids and even in some solids, can be subjected to experiment, and forms one of the most convincing proofs of the motion of molecules.

Now the recent progress of molecular science began with the study of the mechanical effect of the impact of these moving molecules when they strike against any solid body. Of course these flying molecules must beat against whatever is placed among them, and the constant succession of these strokes is, according to our theory, the sole cause of what is called the pressure of air and other gases.

This appears to have been first suspected by Daniel Bernoulli, but he had not the means which we now have of verifying the theory. The same theory was afterwards brought forward independently by Lesage, of Geneva, who, however, devoted most of his labour to the explanation of gravitation by the impact of atoms. Then Herapath, in his *Mathematical Physics*, published in 1847, made a much more extensive application of the theory to gases, and Dr Joule, whose absence from our meeting we must all regret, calculated the actual velocity of the molecules of hydrogen.

The further development of the theory is generally supposed to have begun with a paper by Krönig, which does not, however, so far as I can see, contain any improvement on what had gone before. It seems, however, to have drawn the attention of Professor Clausius to the subject, and to him we owe a very large part of what has been since accomplished.

We all know that air or any other gas placed in a vessel presses against the sides of the vessel, and against the surface of any body placed within it. On the kinetic theory this pressure is entirely due to the molecules striking against these surfaces, and thereby communicating to them a series of impulses which follow each other in such rapid succession that they produce an effect which cannot be distinguished from that of a continuous pressure.

If the velocity of the molecules is given, and the number varied, then since each molecule, on an average, strikes the sides of the vessel the same number of times, and with an impulse of the same magnitude, each will contribute an equal share to the whole pressure. The pressure in a vessel of given size is therefore proportional to the number of molecules in it, that is to the quantity of gas in it.

This is the complete dynamical explanation of the fact discovered by Robert Boyle, that the pressure of air is proportional to its density. It shews also that of different portions of gas forced into a vessel, each produces its own part of the pressure independently of the rest, and this whether these portions be of the same gas or not.

Let us next suppose that the velocity of the molecules is increased. Each molecule will now strike the sides of the vessel a greater number of times in a second, but, besides this, the impulse of each blow will be increased in the same proportion, so that the part of the pressure due to each molecule will vary as the *square* of the velocity. Now the increase of velocity corresponds, on our theory, to a rise of temperature, and in this way we can explain the effect of warming the gas, and also the law discovered by Charles that the proportional expansion of all gases between given temperatures is the same.

The dynamical theory also tells us what will happen if molecules of different masses are allowed to knock about together. The greater masses will go slower than the smaller ones, so that, on an average, every molecule, great or small, will have the same energy of motion.

The proof of this dynamical theorem, in which I claim the priority, has recently been greatly developed and improved by Dr Ludwig Boltzmann. The most important consequence which flows from it is that a cubic centimetre of every gas at standard temperature and pressure contains the same number of molecules. This is the dynamical explanation of Gay Lussac's law of the equivalent volumes of gases. But we must now descend to particulars, and calculate the actual velocity of a molecule of hydrogen.

A cubic centimetre of hydrogen, at the temperature of melting ice, and at a pressure of one atmosphere, weighs 0·00008954 grammes. We have to find at what rate this small mass must move (whether ·altogether or in separate molecules makes no difference) so as to produce the observed pressure on the sides of the cubic centimetre. This is the calculation which was first made by Dr Joule, and the result is 1,859 metres per second. This is what we are accustomed to call a great velocity. It is greater than any velocity obtained in artillery practice. The velocity of other gases is less, as you will see by the table, but in all cases it is very great as compared with that of bullets.

We have now to conceive the molecules of the air in this hall flying about in all directions, at a rate of about seventeen miles in a minute.

If all these molecules were flying in the same direction, they would constitute a wind blowing at the rate of seventeen miles a minute, and the only wind which approaches this velocity is that which proceeds from the mouth of a cannon. How, then, are you and I able to stand here? Only because the molecules happen to be flying in different directions, so that those which strike against our backs enable us to support the storm which is beating against our faces. Indeed, if this molecular bombardment were to cease, even for an instant, our veins would swell, our breath would leave us, and we should, literally, expire. But it is not only against us or against the walls of the hall that the molecules are striking. Consider the immense number of them, and the fact that they are flying in every possible direction, and you will see that they cannot avoid striking each other. Every time that two molecules come into collision, the paths of both are changed, and they go off in new directions. Thus each molecule is continually getting its course altered, so that in spite of its great velocity it may be a long time before it reaches any great distance from the point at which it set out.

I have here a bottle containing ammonia. Ammonia is a gas which you can recognise by its smell. Its molecules have a velocity of six hundred metres per second, so that if their course had not been interrupted by striking against the molecules of air in the hall, everyone in the most distant gallery would have smelt ammonia before I was able to pronounce the name of the gas. But instead of this, each molecule of ammonia is so jostled about by the molecules of air, that it is sometimes going one way and sometimes another, and like a hare which is always doubling, though it goes a great pace, it makes very little progress. Nevertheless, the smell of ammonia is now beginning to be perceptible at some distance from the bottle. The gas does diffuse itself through the air, though the process is a slow one, and if we could close up every opening of this hall so as to make it air-tight, and leave everything to itself for some weeks, the ammonia would become uniformly mixed through every part of the air in the hall.

This property of gases, that they diffuse through each other, was first remarked by Priestley. Dalton shewed that it takes place quite independently of any chemical action between the inter-diffusing gases. Graham, whose researches were especially directed towards those phenomena which seem to throw light on molecular motions, made a careful study of diffusion, and obtained the first results from which the rate of diffusion can be calculated.

Still more recently the rates of diffusion of gases into each other have been measured with great precision by Professor Loschmidt of Vienna.

He placed the two gases in two similar vertical tubes, the lighter gas being placed above the heavier, so as to avoid the formation of currents. He then opened a sliding valve, so as to make the two tubes into one, and after leaving the gases to themselves for an hour or so, he shut the valve, and determined how much of each gas had diffused into the other.

As most gases are invisible, I shall exhibit gaseous diffusion to you by means of two gases, ammonia and hydrochloric acid, which, when they meet, form a solid product. The ammonia, being the lighter gas, is placed above the hydrochloric acid, with a stratum of air between, but you will soon see that the gases can diffuse through this stratum of air, and produce a cloud of white smoke when they meet. During the whole of this process no currents or any other visible motion can be detected. Every part of the vessel appears as calm as a jar of undisturbed air.

But, according to our theory, the same kind of motion is going on in calm air as in the inter-diffusing gases, the only difference being that we can trace the molecules from one place to another more easily when they are of a different nature from those through which they are diffusing.

MOLECULES.

If we wish to form a mental representation of what is going on among the molecules in calm air, we cannot do better than observe a swarm of bees, when every individual bee is flying furiously, first in one direction and then in another, while the swarm, as a whole, either remains at rest, or sails slowly through the air.

In certain seasons, swarms of bees are apt to fly off to a great distance, and the owners, in order to identify their property when they find them on other people's ground, sometimes throw handfulls of flour at the swarm. Now let us suppose that the flour thrown at the flying swarm has whitened those bees only which happened to be in the lower half of the swarm, leaving those in the upper half free from flour.

If the bees still go on flying hither and thither in an irregular manner, the floury bees will be found in continually increasing proportions in the upper part of the swarm, till they have become equally diffused through every part of it. But the reason of this diffusion is not because the bees were marked with flour, but because they are flying about. The only effect of the marking is to enable us to identify certain bees.

We have no means of marking a select number of molecules of air, so as to trace them after they have become diffused among others, but we may communicate to them some property by which we may obtain evidence of their diffusion.

For instance, if a horizontal stratum of air is moving horizontally, molecules diffusing out of this stratum into those above and below will carry their horizontal motion with them, and so tend to communicate motion to the neighbouring strata, while molecules diffusing out of the neighbouring strata into the moving one will tend to bring it to rest. The action between the strata is somewhat like that of two rough surfaces, one of which slides over the other, rubbing on it. Friction is the name given to this action between solid bodies; in the case of fluids it is called internal friction, or viscosity.

It is, in fact, only another kind of diffusion—a lateral diffusion of momentum, and its amount can be calculated from data derived from observations of the first kind of diffusion, that of matter. The comparative values of the viscosity of different gases were determined by Graham in his researches on the transpiration of gases through long narrow tubes, and their absolute values have been deduced from experiments on the oscillation of discs by Oscar Meyer and myself.

Another way of tracing the diffusion of molecules through calm air is to heat the upper stratum of the air in a vessel, and to observe the rate at which this heat is communicated to the lower strata. This, in fact, is a third kind of diffusion—that of energy, and the rate at which it must take place was calculated from data derived from experiments on viscosity before any direct experiments on the conduction of heat had been made. Professor Stefan, of Vienna, has recently, by a very delicate method, succeeded in determining the conductivity of air, and he finds it, as he tells us, in striking agreement with the value predicted by the theory.

All these three kinds of diffusion—the diffusion of matter, of momentum, and of energy—are carried on by the motion of the molecules. The greater the velocity of the molecules and the further they travel before their paths are altered by collision with other molecules, the more rapid will be the diffusion. Now we know already the velocity of the molecules, and therefore, by experiments on diffusion, we can determine how far, on an average, a molecule travels without striking another. Professor Clausius, of Bonn, who first gave us precise ideas about the motion of agitation of molecules, calls this distance the mean path of a molecule. I have calculated, from Professor Loschmidt's diffusion experiments, the mean path of the molecules of four well-known gases. The average distance travelled by a molecule between one collision and another is given in the table. It is a very small distance, quite imperceptible to us even with our best microscopes. Roughly speaking, it is about the tenth part of the length of a wave of light, which you know is a very small quantity. Of course the time spent on so short a path by such swift molecules must be very small. I have calculated the number of collisions which each must undergo in a second. They are given in the table and are reckoned by thousands of millions. No wonder that the travelling power of the swiftest molecule is but small, when its course is completely changed thousands of millions of times in a second.

The three kinds of diffusion also take place in liquids, but the relation between the rates at which they take place is not so simple as in the case of gases. The dynamical theory of liquids is not so well understood as that of gases, but the principal difference between a gas and a liquid seems to be that in a gas each molecule spends the greater part of its time in describing its free path, and is for a very small portion of its time engaged in encounters with other molecules, whereas, in a liquid, the molecule has hardly any free path, and is always in a state of close encounter with other molecules.

Hence in a liquid the diffusion of motion from one molecule to another takes place much more rapidly than the diffusion of the molecules themselves, for the same reason that it is more expeditious in a dense crowd to pass on a letter from hand to hand than to give it to a special messenger to work his way through the crowd. I have here a jar, the lower part of which contains a solution of copper sulphate, while the upper part contains pure water. It has been standing here since Friday, and you see how little progress the blue liquid has made in diffusing itself through the water above. The rate of diffusion of a solution of sugar has been carefully observed by Voit. Comparing his results with those of Loschmidt on gases, we find that about as much diffusion takes place in a second in gases as requires a day in liquids.

The rate of diffusion of momentum is also slower in liquids than in gases, but by no means in the same proportion. The same amount of motion takes about ten times as long to subside in water as in air, as you will see by what takes place when I stir these two jars, one containing water and the other air. There is still less difference between the rates at which a rise of temperature is propagated through a liquid and through a gas.

In solids the molecules are still in motion, but their motions are confined within very narrow limits. Hence the diffusion of matter does not take place in solid bodies, though that of motion and heat takes place very freely. Nevertheless, certain liquids can diffuse through colloid solids, such as jelly and gum, and hydrogen can make its way through iron and palladium.

We have no time to do more than mention that most wonderful molecular motion which is called electrolysis. Here is an electric current passing through acidulated water, and causing oxygen to appear at one electrode and hydrogen at the other. In the space between, the water is perfectly calm; and yet two opposite currents of oxygen and of hydrogen must be passing through it. The physical theory of this process has been studied by Clausius, who has given reasons for asserting that in ordinary water the molecules are not only moving, but every now and then striking each other with such violence that the oxygen and hydrogen of the molecules part company, and dance about through the crowd, seeking partners which have become dissociated in the same way. In ordinary water these exchanges produce, on the whole, no observable effect; but no sooner does the electromotive force begin to act than it exerts its guiding influence on the unattached molecules, and bends the course of each toward its proper electrode, till the moment when, meeting with an unappropriated molecule of the opposite kind, it enters again into a more or less permanent union with it till it is again dissociated by another shock. Electrolysis, therefore, is a kind of diffusion assisted by electromotive force.

Another branch of molecular science is that which relates to the exchange of molecules between a liquid and a gas. It includes the theory of evaporation and condensation, in which the gas in question is the vapour of the liquid, and also the theory of the absorption of a gas by a liquid of a different substance. The researches of Dr Andrews on the relations between the liquid and the gaseous state have shewn us that though the statements in our elementary text-books may be so neatly expressed as to appear almost self-evident, their true interpretation may involve some principle so profound that, till the right man has laid hold of it, no one ever suspects that any thing is left to be discovered.

These, then, are some of the fields from which the data of molecular science are gathered. We may divide the ultimate results into three ranks, according to the completeness of our knowledge of them.

To the first rank belong the relative masses of the molecules of different gases, and their velocities in metres per second. These data are obtained from experiments on the pressure and density of gases, and are known to a high degree of precision.

In the second rank we must place the relative size of the molecules of different gases, the length of their mean paths, and the number of collisions in a second. These quantities are deduced from experiments on the three kinds of diffusion. Their received values must be regarded as rough approximations till the methods of experimenting are greatly improved.

There is another set of quantities which we must place in the third rank, because our knowledge of them is neither precise, as in the first rank, nor approximate, as in the second, but is only as yet of the nature of a probable conjecture. These are :—The absolute mass of a molecule, its absolute diameter, and the number of molecules in a cubic centimetre. We know the relative masses of different molecules with great accuracy, and we know their relative diameters approximately. From these we can deduce the relative densities of the molecules themselves. So far we are on firm ground.

The great resistance of liquids to compression makes it probable that their molecules must be at about the same distance from each other as that at which two molecules of the same substance in the gaseous form act on each other during an encounter. This conjecture has been put to the test by Lorenz Meyer, who has compared the densities of different liquids with the calculated relative densities of the molecules of their vapours, and has found a remarkable correspondence between them.

Now Loschmidt has deduced from the dynamical theory the following remarkable proportion :—As the volume of a gas is to the combined volume of all the molecules contained in it, so is the mean path of a molecule to one-eighth of the diameter of a molecule.

Assuming that the volume of the substance, when reduced to the liquid form, is not much greater than the combined volume of the molecules, we obtain from this proportion the diameter of a molecule. In this way Loschmidt, in 1865, made the first estimate of the diameter of a molecule. Independently of him and of each other, Mr Stoney in 1868, and Sir W. Thomson in 1870, published results of a similar kind, those of Thomson being deduced not only in this way, but from considerations derived from the thickness of soap-bubbles, and from the electric properties of metals.

According to the Table, which I have calculated from Loschmidt's data, the size of the molecules of hydrogen is such that about two millions of them in a row would occupy a millimetre, and a million million million million of them would weigh between four and five grammes.

In a cubic centimetre of any gas at standard pressure and temperature there are about nineteen million million million molecules. All these numbers of the third rank are, I need not tell you, to be regarded as at present conjectural. In order to warrant us in putting any confidence in numbers obtained in this way, we should have to compare together a greater number of independent data than we have as yet obtained, and to shew that they lead to consistent results.

Thus far we have been considering molecular science as an inquiry into natural phenomena. But though the professed aim of all scientific work is to unravel the secrets of nature, it has another effect, not less valuable, on the mind of the worker. It leaves him in possession of methods which nothing but scientific work could have led him to invent; and it places him in a position from which many regions of nature, besides that which he has been studying, appear under a new aspect.

The study of molecules has developed a method of its own, and it has also opened up new views of nature.

When Lucretius wishes us to form a mental representation of the motion of atoms, he tells us to look at a sunbeam shining through a darkened room (the same instrument of research by which Dr Tyndall makes visible to us the dust we breathe), and to observe the motes which chase each other, in all directions through it. This motion of the visible motes, he tells us, is but a result of the far more complicated motion of the invisible atoms which knock the motes about. In his dream of nature, as Tennyson tells us, he

> "Saw the flaring atom-streams
> And torrents of her myriad universe,
> Ruining along the illimitable inane,
> Fly on to clash together again, and make
> Another and another frame of things
> For ever."

And it is no wonder that he should have attempted to burst the bonds of Fate by making his atoms deviate from their courses at quite uncertain times and places, thus attributing to them a kind of irrational free will, which on his materialistic theory is the only explanation of that power of voluntary action of which we ourselves are conscious.

As long as we have to deal with only two molecules, and have all the data given us, we can calculate the result of their encounter; but when we have to deal with millions of molecules, each of which has millions of encounters in a second, the complexity of the problem seems to shut out all hope of a legitimate solution.

The modern atomists have therefore adopted a method which is, I believe, new in the department of mathematical physics, though it has long been in use in the section of Statistics. When the working members of Section F get hold of a report of the Census, or any other document containing the numerical data of Economic and Social Science, they begin by distributing the whole population into groups, according to age, income-tax, education, religious belief, or criminal convictions. The number of individuals is far too great to allow of their tracing the history of each separately, so that, in order to reduce their labour within human limits, they concentrate their attention on a small number of artificial groups. The varying number of individuals in each group, and not the varying state of each individual, is the primary datum from which they work.

This, of course, is not the only method of studying human nature. We may observe the conduct of individual men and compare it with that conduct which their previous character and their present circumstances, according to the best existing theory, would lead us to expect. Those who practise this method endeavour to improve their knowledge of the elements of human nature in much the same way as an astronomer corrects the elements of a planet by comparing its actual position with that deduced from the received elements. The study of human nature by parents and schoolmasters, by historians and statesmen, is therefore to be distinguished from that carried on by registrars and tabulators, and by those statesmen who put their faith in figures. The one may be called the historical, and the other the statistical method.

The equations of dynamics completely express the laws of the historical method as applied to matter, but the application of these equations implies a perfect knowledge of all the data. But the smallest portion of matter which we can subject to experiment consists of millions of molecules, not one of which ever becomes individually sensible to us. We cannot, therefore, ascertain the actual motion of any one of these molecules; so that we are obliged to abandon the strict historical method, and to adopt the statistical method of dealing with large groups of molecules.

The data of the statistical method as applied to molecular science are the sums of large numbers of molecular quantities. In studying the relations between quantities of this kind, we meet with a new kind of regularity, the regularity of averages, which we can depend upon quite sufficiently' for all practical purposes, but which can make no claim to that character of absolute precision which belongs to the laws of abstract dynamics.

Thus molecular science teaches us that our experiments can never give us anything more than statistical information, and that no law deduced from them can pretend to absolute precision. But when we pass from the contemplation of our experiments to that of the molecules themselves, we leave the world of chance and change, and enter a region where everything is certain and immutable.

The molecules are conformed to a constant type with a precision which is not to be found in the sensible properties of the bodies which they constitute. In the first place, the mass of each individual molecule, and all its other properties, are absolutely unalterable. In the second place, the properties of all molecules of the same kind are absolutely identical.

Let us consider the properties of two kinds of molecules, those of oxygen and those of hydrogen.

We can procure specimens of oxygen from very different sources—from the air, from water, from rocks of every geological epoch. The history of these specimens has been very different, and if, during thousands of years, difference of circumstances could produce difference of properties, these specimens of oxygen would shew it.

In like manner we may procure hydrogen from water, from coal, or, as Graham did, from meteoric iron. Take two litres of any specimen of hydrogen, it will combine with exactly one litre of any specimen of oxygen, and will form exactly two litres of the vapour of water.

Now if, during the whole previous history of either specimen, whether imprisoned in the rocks, flowing in the sea, or careering through unknown regions with the meteorites, any modification of the molecules had taken place, these relations would no longer be preserved.

But we have another and an entirely different method of comparing the properties of molecules. The molecule, though indestructible, is not a hard rigid body, but is capable of internal movements, and when these are excited, it emits rays, the wave-length of which is a measure of the time of vibration of the molecule.

By means of the spectroscope the wave-lengths of different kinds of light may be compared to within one ten-thousandth part. In this way it has been ascertained, not only that molecules taken from every specimen of hydrogen in our laboratories have the same set of periods of vibration, but that light, having the same set of periods of vibration, is emitted from the sun and from the fixed stars.

We are thus assured that molecules of the same nature as those of our hydrogen exist in those distant regions, or at least did exist when the light by which we see them was emitted.

From a comparison of the dimensions of the buildings of the Egyptians with those of the Greeks, it appears that they have a common measure. Hence, even if no ancient author had recorded the fact that the two nations employed the same cubit as a standard of length, we might prove it from the buildings themselves. We should also be justified in asserting that at some time or other a material standard of length must have been carried from one country to the other, or that both countries had obtained their standards from a common source.

But in the heavens we discover by their light, and by their light alone, stars so distant from each other that no material thing can ever have passed from one to another; and yet this light, which is to us the sole evidence of the existence of these distant worlds, tells us also that each of them is built up of molecules of the same kinds as those which we find on earth. A molecule of hydrogen, for example, whether in Sirius or in Arcturus, executes its vibrations in precisely the same time.

Each molecule, therefore, throughout the universe, bears impressed on it the stamp of a metric system as distinctly as does the metre of the Archives at Paris, or the double royal cubit of the Temple of Karnac.

No theory of evolution can be formed to account for the similarity of molecules, for evolution necessarily implies continuous change, and the molecule is incapable of growth or decay, of generation or destruction.

None of the processes of Nature, since the time when Nature began, have produced the slightest difference in the properties of any molecule. We are therefore unable to ascribe either the existence of the molecules or the identity of their properties to the operation of any of the causes which we call natural.

On the other hand, the exact equality of each molecule to all others of the same kind gives it, as Sir John Herschel has well said, the essential character of a manufactured article, and precludes the idea of its being eternal and self-existent.

Thus we have been led, along a strictly scientific path, very near to the point at which Science must stop. Not that Science is debarred from studying the internal mechanism of a molecule which she cannot take to pieces, any more than from investigating an organism which she cannot put together. But in tracing back the history of matter Science is arrested when she assures herself, on the one hand, that the molecule has been made, and on the other, that it has not been made by any of the processes we call natural.

Science is incompetent to reason upon the creation of matter itself out of nothing. We have reached the utmost limit of our thinking faculties when we have admitted that because matter cannot be eternal and self-existent it must have been created.

It is only when we contemplate, not matter in itself, but the form in which it actually exists, that our mind finds something on which it can lay hold.

That matter, as such, should have certain fundamental properties—that it should exist in space and be capable of motion, that its motion should be persistent, and so on, are truths which may, for anything we know, be of the kind which metaphysicians call necessary. We may use our knowledge of such truths for purposes of deduction, but we have no data for speculating as to their origin.

But that there should be exactly so much matter and no more in every molecule of hydrogen is a fact of a very different order. We have here a particular distribution of matter—a *collocation*—to use the expression of Dr Chalmers, of things which we have no difficulty in imagining to have been arranged otherwise.

The form and dimensions of the orbits of the planets, for instance, are not determined by any law of nature, but depend upon a particular collocation of matter. The same is the case with respect to the size of the earth, from which the standard of what is called the metrical system has been derived. But these astronomical and terrestrial magnitudes are far inferior in scientific importance to that most fundamental of all standards which forms the base of the molecular system. Natural causes, as we know, are at work, which tend to modify, if they do not at length destroy, all the arrangements and dimensions of the earth and the whole solar system. But though in the course of ages catastrophes have occurred and may yet occur in the heavens, though ancient systems may be dissolved and new systems evolved out of their ruins, the molecules out of which these systems are built—the foundation stones of the material universe—remain unbroken and unworn.

They continue this day as they were created—perfect in number and measure and weight, and from the ineffaceable characters impressed on them we may learn that those aspirations after accuracy in measurement, truth in statement, and justice in action, which we reckon among our noblest attributes as men, are ours because they are essential constituents of the image of Him who in the beginning created, not only the heaven and the earth, but the materials of which heaven and earth consist.

378   MOLECULES.

Table of Molecular Data.

		Hydrogen.	Oxygen.	Carbonic oxide.	Carbonic acid.
	Mass of molecule (hydrogen = 1)	1	16	14	22
Rank I.	Velocity (of mean square) metres per second at 0° C.	1859	465	497	396
	Mean path, tenth-metres	965	560	482	379
Rank II.	Collisions in a second (millions)	17750	7646	9489	9720
	Diameter, tenth-metres	5·8	7·6	8·3	9·3
Rank III.	Mass, twenty-fifth grammes	46	736	644	1012

Table of Diffusion. $\dfrac{(\text{Centimetre})^2}{\text{Second}}$ measure.

	Calculated.	Observed.	
H & O	0·7086	0·7214	
H & CO	0·6519	0·6422	
H & CO$_2$	0·5575	0·5558	Diffusion of matter observed by Loschmidt.
O & CO	0·1807	0·1802	
O & CO$_2$	0·1427	0·1409	
CO & CO$_2$	0·1386	0·1406	
H	1·2990	1·49	
O	0·1884	0·213	Diffusion of momentum (Graham and Meyer).
CO	0·1748	0·212	
CO$_2$	0·1087	0·117	
Air		0·256	
Copper		1·077	Diffusion of temperature observed by Stefan.
Iron		0·183	
Cane Sugar in water		0·00000365	Voit.
(Or in a day		0·3144)	
Salt in water		0·00000116	Fick.

17. Letter from Herbert Spencer[a] to Maxwell, December 4, 1873

Cambridge University Library, Maxwell Collection, Correspondence.

38 Queen's Gardens, 4th Dec. 1873

Dear Sir,

Sometime this year, Prof. Clifford[b] named to me a criticism you passed upon a certain hypothesis of mine respecting the process of nebular concentration, as tending to produce a hollow liquid spheroid during its closing stages. The moment he named to me this criticism, I saw that I had made a mistake. The process of condensation from the vaporous to the liquid form, I had at first considered in the case of Saturn's rings, where a precipitation into a denser form, recurring as the equatorial portion of the concentrating spheroid, might [?] produce a ring that would maintain its place; supposing the concentration to occur when the centripetal and centrifugal forces were balanced, I had inadvertently carried this conception to the cases of the other planets, where there could be no such balance of forces.

There has since occurred to me another hypothesis respecting the mode of condensation and the resulting structure. I have discussed this with my friends Tyndall[c] Hirst,[d] and Clifford; and while not committing themselves to it, they do not raise any objection against it. Prof. Clifford, however, expressing great faith in your intuitive insight into physical processes, recommended me to obtain, if possible, your opinion respecting the tenability of this hypothesis.[e]

I write to ask whether, after the close of the Cambridge term, you are likely to be in London; and whether, in that case, you could afford me half-an-hour's conversation; and further to ask whether, if you are not coming to London, you could grant me the same favour were I to come to Cambridge before the term ends.

I enclose the outline, in proof, of a speculation on another physical question, respecting which, also, I
[The rest of the letter is missing.]

a. Herbert Spencer (1820–1903) was already widely known in England and America for his writings on philosophy and evolution. His articles on physical astronomy also had some influence, in particular, "Recent Astronomy, and the Nebular Hypothesis," *Westminster Review* (July 1858); "Illogical Geology," *Universal Review* (July 1859); and "The Constitution of the Sun," *The Reader* (February 25, 1865); all these articles are reprinted in his *Essays: Scientific, Political, and Speculative* (London: Longman, Brown, Green, Longmans and Roberts, 1858–1874), 3 vols. Spencer supported Laplace's nebular hypothesis and attempted to construct various mechanisms to circumvent the difficulties faced by that hypothesis in accounting for details of planetary and satellite motions. He argued that as a hot gaseous body cools, forming a solid crust, it may never solidify completely but may retain a central core of highly compressed gas. The experiments of Thomas Andrews, showing that a gas could not be liquefied by pressure at temperatures above the critical point, seemed to support this argument, and in fact it was revived later in the nineteenth century on just this basis by A. Ritter, K. Zoeppritz, S. Arrhenius, and others.

Spencer incorporated the nebular hypothesis into his *First Principles* (London: Williams and Norgate, 1862) as a prime illustration of his general concept of "instability of the homogeneous"—the trend of evolution is not only from simplicity to complexity but from confusion to order; "not only a multiplication of unlike parts, but an increase in the distinctness with which these parts are marked off from one another" (fourth edition, §129).

b. William Kingdon Clifford (1845–1879), British mathematician, is now best known for his book, *The Common Sense of the Exact Sciences* (London: Kegan Paul, 1885). He was second wrangler at Cambridge and second Smith's prizeman in 1867. Clifford was appointed professor of Applied Mathematics at University College, London in 1868. His own scientific work was in geometry, in particular non-Euclidean geometry, and he was instrumental in introducing Riemann's ideas to British mathematicians. Together with Huxley and Tyndall, Clifford was an influential figure in the London scientific scene. There is no recent assessment of his mathematical work, but there is an appreciative biography by Leslie Stephen in the *Dictionary of National Biography*.

c. John Tyndall (1820–1893), by far the best known of this triumvirate, was equally involved in the X-Club and the Metaphysical Club and was best known as a politically powerful member of the scientific community in Britain in the late nineteenth century and as an evangelist for scientific naturalism. His public stance and the disputes he was involved in (with Forbes over glaciers and with Tait and Thomson over the history of thermodynamics) and his quarrels over government policies are much better known and have been more thoroughly investigated than even his committment to education in science or his extensive research interests. Tyndall was committed to an empirical and experimental physics in an era in which the mathematical development of theoretical ideas displaced this older tradition. Even so his scientific work in physics, as well as his impact on the teaching of science, deserve more careful assessment.

d. Thomas Arthur Hirst (1830–1892) was also a mathematician, appointed professor of Mathematical Physics and Mathematics at University College, London. His diaries constitute a candid record of the London scientific scene; see William H. Brock and Roy M. MacLeod, *Natural Knowledge in Social Context: The Journals of Thomas Arthur Hirst, FRS* (London: Mansell, 1980).

e. David Duncan, *Life and Letters of Herbert Spencer* (New York: Appleton and Co., 1908), vol. 2, 161ff, provides some further information on the impact of Maxwell's and Clifford's opinions on Spencer's work. Spencer wrote a postscript to his article "What is Electricity," *Reader*, 19 November, 1864. In a letter to Edward Livingstone Youmans (1821–1887), Spencer stated that he planned to send this postscript to "Tyndall and other authorities" for their comments. Duncan did not have Spencer's letter of December 4, 1873, which we publish here, and says only that "When sending the paper to Professor Clerk Maxwell reference seems to have been made to a remark made to Professor Kingdon Clifford regarding Spencer's views about nebular condensation."

In a later letter to Youmans, October 16, 1874, Spencer continues with the problem:

[Richard] Proctor in the last number of the *Cornhill*, has been drawing attention to the conclusions of your astronomer [Charles Augustus] Young that the sun is a hollow sphere.... His reasonings are in great measure the same as those set forth in my essay on the "Constitution of the Sun"—reasonings which I have been for the last year past intending to amend in respect of the particular process by which the precipitated matters form the molten shell. There are mechanical difficulties, named to Clifford by Clerk Maxwell, to the mode of formation as originally described. But, on pursuing the results of the process of precipitation into vapour and then into metallic rain, perpetually ascending and perpetually thickening as concentration goes on, I reached a conclusion respecting the formation of the shell, to which no objection has yet been made by the authorities with whom I have discussed it...." (p. 165).

Duncan continues, "He [Spencer] at once set about amending his reasonings in respect of the particular process by which the precipitated matters form the molten shell." A slip proof of the amended hypothesis was sent to Professor Clerk Maxwell, who, admitting that he did not "quite understand the principal features" of the hypothesis, adduced reasons to show that "a liquid shell supported by a nucleus of less density than itself, whether solid, liquid or gaseous, is essentially unstable. " On Professor Clerk Maxwell's letter (December 17, 1874) Spencer has pencilled, "This argument at first convinced me that my hypothesis was untenable. But subsequently the corollaries from [Thomas] Andrews's investigations concerning the critical point of gases, implying that a gas might become denser than a liquid and yet remain a gas, led me to readopt the hypothesis." Duncan here cites Spencer's *Essays*, vol. 1, 164. This was not the end of it, however, Duncan goes on, "This point with others is touched upon in correspondence with his French translator.

To. E. CAZELLES

12 May, 1875

I enclose impressions of some passages which will be substituted hereafter for certain parts of the essay on the 'Nebular Hypothesis.'"

In a note Duncan states that one of the alterations made was the "abandonment of an hypothesis which Professor Clerk Maxwell clearly proved to me as not tenable."

18. Letter from Maxwell to Herbert Spencer, December 5, 1873

David Duncan, *Life and Letters of Herbert Spencer* (New York: Appleton, 1908), vol. 2, 161–162.

I do not remember the particulars of what I said to Professor Clifford about nebular condensation. The occasion of it was I think a passage in an old edition of your *First Principles*,[a] and having since then made a little more acquaintance with your works, I regarded it merely as a temporary phase of the process of evolution which you have been carrying on within your own mind. Mathematicians by guiding their thoughts always along the same tracks, have converted the field of thought into a kind of railway system, and are apt to neglect cross-country speculations.

It is very seldom that any man who tries to form a system can prevent his system from forming round him before he is forty. Hence the wisdom of putting in some ingredients to check crystallization and keep the system in a colloidal condition. Candle-makers, I believe, use arsenic for this purpose.... But you seem to be able to retard the crystallization of parts of your system without stopping the process of evolution of the whole, and I therefore attach much more importance to the general scheme than to particular statements.[b]

a. Spencer's *First Principles* (London, 1862) was one of his developing attempts to apply the laws of biology and physics to analyze society. As their correspondence went on, Maxwell tried to give Spencer a physicist's interpretation of the second law of thermodynamics; this interpretation clearly contradicted Spencer's notion of the "instability of the homogeneous." In the latter case there is an innate tendency to change, and all change produces multiple effects so that there is ever-increasing heterogeneity.

Maxwell's real opinion of Spencer comes out in his correspondence with Tait. In a postcard to Tait, July 29th, 1876, Maxwell says, "Have you read Willard Gibbs on Equilibrium of Heterogeneous Substances. If not read him. Refreshing after H. Spencer on the Instability of the Homogeneous." (Letter in Cambridge University Library, Maxwell Collection, Maxwell–Tait Correspondence)

b. The rest of the letter is not available to us; Duncan indicates its contents as follows:

After describing several experiments, which he would not say were inconsistent with Spencer's theory, but which were very important and significant, Professor Clerk Maxwell continues "As I observe that you are always improving your phraseology I shall lay before you my notions on the nomenclature of molecular motions." One of the terms defined was "the motion of *agitation* of a molecule, namely 'that by which the actual velocity of an individual differs from the mean velocity of the group.'" (*Life and Letters*, 162)

19. Letter from Maxwell to Herbert Spencer, December 17, 1873

David Duncan, *Life and Letters of Herbert Spencer* (New York; Appleton, 1908), vol. 2, 162–163.

The reason for which I use the word "agitation" to distinguish the local motion of a molecule in relation to its neighbours is that I think with you that the word "agitation" conveys in a small degree, if at all, the notion of rhythm.

If motion is said to be rhythmic when the path is, on the whole, as much in one direction as in the opposite, then all motion is rhythmic when it is confined within a small region of space.

But if, as I understand the word rhythmic, it implies not only alteration, but regularity and periodicity, then the word "agitation" excludes the notion of rhythm, which was what I meant it to do.... A great scientific desideratum is a set of words of *little* meaning—words which mean no more than that a thing belongs to a very large class. Such words are much needed in the undulatory theory of light, in order to express fully what is proved by experiment, without conceding anything which is a mere hypothesis.[a]

a. Duncan shows Spencer's reaction to this letter in a letter, December 22, 1873 to John Tyndall in which Spencer comments:

I have had another letter from Clerk Maxwell, which considerably startled me by its views about molecular motion. I should like to talk to you about them. They seem to me to differ from those which I supposed you to hold, and which I supposed were held generally. (Duncan, *Spencer*, vol. 2, 161)

In the fourth edition of his *First Principles*, Spencer wrote, "After having for some years supposed myself alone in the belief that all motion is rhythmical, I discovered that my friend Professor Tyndall also held this doctrine." (p. 257)

20. Letter from Herbert Spencer to Maxwell, December 30, 1873

Cambridge University Library, Maxwell Collection, Correspondence.

38 Queen's Gardens, 30th Dec. 1873

Dear Sir,

I had no intention when I made my passing comment on the word *agitation* of drawing from you a second letter ⟨on the matter⟩. Nor do I now, in some remarks I propose to make, wish to do more than draw your attention to certain difficulties which occur to me, as presented by the hypothesis of molecular movements as you describe them.

I must confess that I was taken somewhat aback by the statement that you deliberately chose the word *agitation*, because it negatived the notion of rhythm. For I had hardly anticipated the tacit denial that the relative motions of molecules as wholes have rhythm. I feel fully the force of the reason for supposing that, when molecules are irregularly aggregated into a solid, the tensions due to their mutual actions will be so various as to produce great irregularity of motion; and I have, indeed, in the first part of the speculation concerning electricity, indicated this as a possible cause for the continuity of the spectrum of solids. But, admitting this, there seem to me two qualifying considerations. If, as shown in the lecture you were so kind to send me, molecules of different weights have different absolute velocities in the gaseous state; then, must it not happen that when such differently moving molecules are aggregated into solids, their *constitutional differences of mobility* will still show themselves? Such constitutional differences cannot well disappear without any result; and if they do not disappear, must there not result characteristic differences between their motions of agitation in the two solids they form—must not the two agitations differ in the *average periodicities* of the local motions constituting them? The second qualifying assumption which occurs to me is this. Though molecules, irregularly aggregated into a solid, may be expected to have motions more or less confused by the irregularities of the tensions; may we not say that, when they are *regularly* aggregated into a solid (as in a crystal), they will be subject to *regular* tensions, conducing to regular motions? Do not the formation and structure of a crystal imply that its units are all so homogeneously conditioned that they must have homogeneous motions?[a]

May I trespass on your patience by one further ⟨remark that⟩ comment respecting the constitution of gases—a comment which points in the same general direction? The conception of the molecules of a gas as having extremely rapid movements of translation, producing, by their multi-tudinous impacts, pressure upon the containing surfaces, becomes considerably modified, and approximated to the conception of rhythmical motion, when it is admitted that the individual molecules, instead of moving directly through space, make innumerable collisions with their neighbours in a second, and progress but slowly in any one direction. But it seems to me that the conception of something like a regular rhythm of the molecules as wholes, is implied by what takes place when one gas is diffused through another. For if it is true that after due lapse of time, this diffusion becomes complete through the whole chamber—if, in that case, the molecules of the diffused gas become at length homogeneously distributed with respect to the gas through which they are diffused—if, that is, they came finally to a state in which they are equidistantly dispersed throughout the whole mass, and if this is their state of equilibrium; then what must happen in such a state? They cannot pass out of such a state of equilibrium into a state ⟨not⟩ not of equilibrium: for this would imply change without a cause. They cannot change places with one another; since this would be a passing from a first state of equilibrium to a second state of equilibrium—a change without a cause or result. Hence such molecules must retain their places, and such motions as they have must be expended in performing regular oscillations. And if this is the condition of equilibrium of the molecules of the diffused gas, must it not also be that of the gas through which it is diffused?

I make these remarks simply with the view of showing the difficulties that occur to an outsider, and in the hope that outsiders may benefit by the further elucidations which these difficulties may suggest to you.

With many thanks

truly yours

Herbert Spencer

P. S.

Let me add another query which has suggested itself since writing the above. How is the motion of the molecules of a gas, supposing it a motion of translation, to be maintained? If the molecules are perpetually striking the sides of the containing vessel, and, in their movements through the occupied space, are perpetually coming in collision with one another, does not this imply a continual loss in the motion of translation? Part of the force of each impact with another molecule, must go to increase the vibrations within each; and similarly with the impacts
[The rest of the letter is missing.]

a. David Duncan, *Life and Letters of Herbert Spencer* (New York: Appleton, 1908), vol. 2, 163–164, quotes only this paragraph.

21. "Plateau on Soap-Bubbles"

Review in *Nature* 10 (1874), 119–121; *Scientific Papers*, vol. 2, 393–399.

[From *Nature*, Vol. x.]

LXVI. *Plateau on Soap-Bubbles*.

ON an Etruscan vase in the Louvre figures of children are seen blowing bubbles. Those children probably enjoyed their occupation just as modern children do. Our admiration of the beautiful and delicate forms, growing and developing themselves, the feeling that it is *our* breath which is turning dirty soap-suds into spheres of splendour, the fear lest by an irreverent touch we may cause the gorgeous vision to vanish with a sputter of soapy water in our eyes, our wistful gaze as we watch the perfected bubble when it sails away from the pipe's mouth to join, somewhere in the sky, all the other beautiful things that have vanished before it, assure us that, whatever our nominal age may be we are of the same family as those Etruscan children.

Here, for instance, we have a book, in two volumes, octavo, written by a distinguished man of science, and occupied for the most part with the theory and practice of bubble-blowing. Can the poetry of bubbles survive this? Will not the lovely visions which have floated before the eyes of untold generations collapse at the rude touch of Science, and "yield their place to cold material laws"? No, we need go no further than this book and its author to learn that the beauty and mystery of natural phenomena may make such an impression on a fresh and open mind that no physical obstacle can ever check the course of thought and study which it has once called forth.

M. Plateau in all his researches seems to have selected for his study those phenomena which exhibit some remarkable beauty of form or colour. In the zeal with which he devoted himself to the investigation of the laws of the subjective impressions of colour, he exposed his eyes to an excess of light, and

---

* *Statique expérimentale et théorique des Liquides soumis aux seules Forces moléculaires.* Par J. Plateau, Professeur à l'Université de Gand, &c. (Paris, Gauthier-Villars; London, Trübner & Co.; Gand et Leipzig, F. Clemm. 1873.)

has ever since been blind. But in spite of this great loss he has continued for many years to carry on experiments such as those described in this book, on the forms of liquid masses and films, which he himself can never either see or handle, but from which he gathers the materials of science as they are furnished to him by the hands, eyes, and minds of devoted friends.

So perfect has been the co-operation with which these experiments have been carried out, that there is hardly a single expression in the book to indicate that the measures which he took and the colours with which he was charmed were observed by him, not in the ordinary way, but through the mediation of other persons.

Which, now, is the more poetical idea—the Etruscan boy blowing bubbles for himself, or the blind man of science teaching his friends how to blow them, and making out by a tedious process of question and answer the conditions of the forms and tints which he can never see ?

But we must now attempt to follow our author as he passes from phenomena to ideas, from experiment to theory.

The surface which forms the boundary between a liquid and its vapour is the seat of phenomena on the careful study of which depends much of our future progress in the knowledge of the constitution of bodies. To take the simplest case, that of a liquid, say water, placed in a vessel which it does not fill, but which contains nothing else. The water lies at the bottom of the vessel, and the upper part, originally empty, becomes rapidly filled with the vapour of water. The temperature and the pressure—the quantities on which the thermal and statical relations of any body to external bodies depend—are the same for the water and its vapour, but the energy of a milligramme of the vapour greatly exceeds that of a milligramme of the water. Hence the energy of a milligramme of water-substance is much greater when it happens to be in the upper part of the vessel in the state of vapour, than when it happens to be in the lower part of the vessel in the state of water.

Now we find by experiment that there is no difference between the phenomena in one part of the liquid and those in another part except in a region close to the surface and not more than a thousandth or perhaps a millionth of a millimetre thick. In the vapour also, everything is the same, except perhaps in a very thin stratum close to the surface. The change in the value of the energy takes place in the very narrow region between water and vapour. Hence the energy of a milligramme of water is the same all through the mass of the water except in a thin stratum close to the surface, where it is some-what greater ; and the energy of a milligramme of vapour is the same all through the mass of vapour except close to the surface, where it is probably less.

The whole energy of the water is therefore, in the first place, that due to so many milligrammes of water; but besides this, since the water close to the surface has an excess of energy, a correction, depending on this excess, must be added. Thus we have, besides the energy of the water reckoned per milligramme, an additional energy to be reckoned per square millimetre of surface.

The energy of the vapour may be calculated in the same way at so much per milligramme, with a deduction of so much per square millimetre of surface. The quantity of vapour, however, which lies within the region in which the energy is beginning to change its value is so small that this deduction per square millimetre is always much smaller than the addition which has to be made on account of the liquid. Hence the whole energy of the system may be divided into three parts, one proportional to the mass of liquid, one to the mass of vapour, and the third proportional to the area of the surface which separates the liquid from the vapour.

If the system is displaced by an external agent in such a way that the area of the surface of the liquid is increased, the energy of the system is increased, and the only source of this increase of energy is the work done by the external agent. There is therefore a resistance to any motion which causes the extension of the surface of a liquid.

On the other hand, if the liquid moves in such a way that its surface diminishes, the energy of the system diminishes, and the diminution of energy appears in the form of work done on the external agent which allows the surface to diminish. Now a surface which tends to diminish in area, and which thus tends to draw together any solid framework which forms its boundary, is said to have surface-tension. Surface-tension is measured by the force acting on one millimetre of the boundary edge. In the case of water at 20° C., the tension is, according to M. Quincke, a force of 8·253 milligrammes weight per millimetre.

M. Plateau hardly enters into the theoretical deduction of the surface-tension from hypotheses respecting the constitution of bodies. We have there-fore thought it desirable to point out how the fact of surface-tension may be

50—2

deduced from the known fact that there is a difference in energy between a
liquid and its vapour, combined with the hypothesis, that as a milligramme of
the substance passes from the state of a liquid within the liquid mass, to that
of a vapour outside it, the change of its energy takes place, not instantaneously,
but in a continuous manner.

M. van der Waals, whose academic thesis, *Over de Continuiteit van den
Gas en Vloeistoftoestand* *, is a most valuable contribution to molecular physics,
has attempted to calculate approximately the thickness of the stratum within
which this continuous change of energy is accomplished, and finds it for water
about 0·0000003 millimetre.

Whatever we may think of these calculations, it is at least manifest that
the only path in which we may hope to arrive at a knowledge of the size
of the molecules of ordinary matter is to be traced among those phenomena
which come into prominence when the dimensions of bodies are greatly reduced,
as in the superficial layer of a liquid.

But it is in the experimental investigation of the effects of surface-tension
on the form of the surface of a liquid that the value of M. Plateau's book is to
be found. He uses two distinct methods. In the first he prepares a mixture
of alcohol and water which has the same density as olive oil, then introducing
some oil into the mixture and waiting till it has, by absorption of a small
portion of alcohol into itself, become accommodated to its position, he obtains a
mass of oil no longer under the action of gravity, but subject only to the
surface-tension of its boundary. Its form is therefore, when undisturbed,
spherical, but by means of rings, disks, &c., of iron, he draws out or com-
presses his mass of oil into a number of different figures, the equilibrium and
stability of which are here investigated, both experimentally and theoretically.

The other method is the old one of blowing soap-bubbles. M. Plateau,
however, has improved the art, first by finding out the best kind of soap and
the best proportion of water, and then by mixing his soapy water with
glycerine. Bubbles formed of this liquid will last for hours, and even days.

By forming a frame of iron wire and dipping it into this liquid he forms
a film, the figure of which is that of the surface of minimum area which has
the frame for its boundary. This is the case when the air is free on both
sides of the film. If, however, the portions of air on the two sides of the film

* Leiden : A. W. Sijthoff, 1873.

are not in continuous communication, the film is no longer the surface of absolute minimum area, but the surface which, with the given boundary, and inclosing a given volume, has a minimum area.

M. Plateau has gone at great length into the interesting but difficult question of the conditions of the persistence of liquid films. He shews that the surface of certain liquids has a species of viscosity distinct from the interior viscosity of the mass. This surface-viscosity is very remarkable in a solution of saponine. There can be no doubt that a property of this kind plays an important part in determining the persistence or collapse of liquid films. M. Plateau, however, considers that one of the agents of destruction is the surface-tension, and that the persistence mainly depends on the degree in which the surface-viscosity counteracts the surface-tension. It is plain, however, that it is rather the inequality of the surface-tension than the surface-tension itself which acts as a destroying force.

It has not yet been experimentally ascertained whether the tension varies according to the thickness of the film. The variation of tension is certainly insensible in those cases which have been observed.

If, as the theory seems to indicate, the tension diminishes when the thickness of the film diminishes, the film must be unstable, and its actual persistence would be unaccountable. On the other hand, the theory has not as yet been able to account for the tension increasing as the thickness diminishes.

One of the most remarkable phenomena of liquid films is undoubtedly the formation of the black spots, which were described in 1672 by Hooke, under the name of holes.

Fusinieri has given a very exact account of this phenomenon as he observed it in a vertical film protected from currents of air. As the film becomes thinner, owing to the gradual descent of the liquid of which it is formed, certain portions become thinner than the rest, and begin to shew the colours of thin plates. These little spots of colour immediately begin to ascend, dragging after them a sort of train like the tail of a tadpole. These tadpoles, as Fusinieri calls them, soon begin to accumulate near the top of the film, and to range themselves in horizontal bands according to their colours, those which have the colour corresponding to the smallest thickness ascending highest.

In this way the colours become arranged in horizontal bands in beautiful gradation, exhibiting all the colours of Newton's scale. When the frame of the film is made to oscillate, these bands oscillate like the strata formed by a series of liquids of different densities. This shews that the film is subject to dynamical conditions similar to those of such a liquid system. The liquid is subject to the condition that the volume of each portion of it is invariable, and the motion arises from the fact that by the descent of the denser portions (which is necessarily accompanied by the rise of the rarer portions) the gravitational potential energy of the system is diminished. In the case of the film, the condition which determines that the descent of the thicker portions shall entail the rise of the thinner portions must be that each portion of the film offers a special resistance to an increase or diminution of area. This resistance probably forms a large part of the superficial viscosity investigated by M. Plateau, which retards the motion of his magnetic needle, and evidently is far greater than the viscosity of figure, in virtue of which the film resists a shearing motion.

The coloured bands gradually descend from the top of the film, presenting at first a continuous gradation of colour, but soon a remarkable black, or nearly black, band begins to form at the top of the film, and gradually to extend itself downwards. The lower boundary of this black band is sharply defined. There is not a continuous gradation of colour according to the arrangement in Newton's table, but the black appears in immediate contact with the white or even the yellow of the first order, and M. Fusinieri has even observed it in contact with bands of the third order.

Nothing can shew more distinctly that there is some remarkable change in the physical properties of the film, when it is of a thickness somewhat greater than that of the black portion. And in fact the black part of the film is in many other respects different from the rest. It is easy, as Leidenfrost tells us, to pass a solid point through the thicker part of the film, and to withdraw it, without bursting the film, but if anything touches the black part, the film is shattered at once. The black portion does not appear to possess the mobility which is so apparent in the coloured parts. It behaves more like a brittle solid, such as a Prince Rupert's drop, than a fluid. Its edges are often very irregular, and when the curvature of the film is made to vary, the black portions sometimes seem to resist the change, so that their surface has no longer the same continuous curvature as the rest of the bubble. We have thus numerous indications of the great assistance which molecular science is likely to derive from the study of liquid films of extreme tenuity.

We have no time or inclination to discuss M. Plateau's work in a critical spirit. The directions for making the experiments are very precise, and if sometimes they appear tedious on account of repetitions, we must remember that it is by words, and words alone, that the author can learn the details of the experiment which he is performing by means of the hands of his friends, and that the repetition of phrases must in his case take the place of the ordinary routine of a careful experimenter. The description of the results of mathematical investigation, which is a most difficult but at the same time most useful species of literary composition, is a notable feature of this book, and could hardly be better done. The mathematical researches of Lindelöf, Lamarle, Scherk, Riemann, &c., on surfaces of minimum area, deserve to be known to others besides professed mathematicians, and M. Plateau deserves our thanks for giving us an intelligible account of them, and still more for shewing us how to make them visible with his improved soap-suds.

In the speculative part of the book, where the author treats of the causes of the phenomena, there is of course more room for improvement, as there always must be when a physicist is pushing his way into the unknown regions of molecular science. In such matters everything human, at least in our century, must be very imperfect, but for the same reason any real progress, however small, is of the greater value.

22. [Draft of "Atom" Article for the *Encyclopedia Britannica*]

Cambridge University Library, Maxwell collection.

We have mentioned the extreme forms of the theories of the constitution of bodies before describing the intermediate ones because it appears to us that most of these intermediate forms are mere hybrids arising from the union of inconsistent and incongruous ideas in the minds of men ⟨who could⟩ ⟨had imperfect⟩ ⟨of limited imaginat⟩ whose power of clearly imagining ⟨what they show⟩ their own fancies was defective.

In the first place come those theories which make the atom an extended body, having definite dimensions and figure. This is the nature of the atom as described by Lucretius. According to his description of the doctrine of Democritus the atoms are all in motion in a downward stream with an enormous ⟨but⟩ speed. ⟨There are⟩ It is the impact of these atoms on bodies which makes them heavy. At quite uncertain times and places these atoms ⟨chan⟩ are deflected from their vertical path. This causes them to jostle one another and by this fortuitous concourse they form visible bodies. The changes observed in these visible bodies are due to the motions of their invisible atoms and to the impact of other free atoms on them, and the uncertainty of the manner in which the atoms deviate from their rectilinear path introduces a corresponding uncertainty into everything that happens in the world.

⟨From⟩ This theory ⟨we may⟩ may be greatly improved by taking away one or two ⟨features⟩ of its [?] important elements ⟨introduced such as⟩ The downward direction of the flight of atoms ⟨which would not have been introduced if [?] after⟩ could only have had a meaning in those early times before it was known that the downward direction is different in different parts of the earth.

Lesage of Geneva replaced this rude conception by that of an infinity of streams of atoms flying through space and crossing each other in all directions. These he calls "ultramundane corpuscles" ⟨because he conceives them to come in all directions from regions far beyond that part of the visible system of the world which is known to us. A ⟨solitary⟩ body placed by itself in free space and exposed to the impacts of these corpuscles is bandied about by them in all directions but because, on the whole, it receives as many impacts on one side as on another it does not ⟨tend to change its⟩ acquire any perceptible velocity.⟩

But if there are two bodies in space, each of them will screen the other from a certain proportion of the corpuscles so that a smaller number of corpuscles will strike either body ⟨in the direction of the line from the other body to it⟩ on the side which is next the other body while the number of corpuscles which strike it in other directions remains the same. Each body will therefore be urged towards the other ⟨with⟩ by the effect of the excess of the impacts it receives on the side farthest from the other. The law of this action is remarkable. It is easy to show in the first place that ⟨the in this case⟩ the force urging *A* towards *B* is equal and opposite to that which urges *B* towards *A*. Hence we may assert that action and reaction are equal and opposite, even in the case in which the action and the reaction arise from the impacts of two different sets of corpuscles on two different bodies.

In the second place, if the ⟨diameter⟩ dimensions of the two bodies are small compared with the distance between them, the amount of the force will be inversely as the square of the distance. Here then is ⟨an opening⟩ a path, tending towards an explanation of the ⟨theory⟩ law of gravitation, which, if it only leads to ⟨consistent⟩ results in other respects consistent with facts, would be a royal road into the heart of science. But we must test it by means of a more complete knowledge of dynamical principles than was possessed either by Lucretius or Lesage. In the first place, are the impacts of the corpuscles on the body like those of perfectly elastic bodies or not. Is there or is there not a loss of kinetic energy on account of the impact. First let me suppose that there is no loss of energy, then on an average the velocity of a corpuscle will be the same after impact as it was before impact. But if this be the case the corpuscles rebounding from the body in any direction will be in number and velocity exactly equal to those which ⟨would⟩ are prevented from proceeding in the same direction by being deflected by the body, and it may be shown that this will be the case whatever be the shape of the body and however many other bodies be present in the field. Thus the rebounding molecules exactly make up for those which are deflected by the body, and ⟨the each of the⟩ there will be no excess of the impacts on any other body in one direction more than another. Thus the explanation of gravitation falls to the ground if the corpuscles are like perfectly elastic spheres and rebound with a velocity of separation equal to that of approach.

If on the other hand they rebound with a smaller velocity, the effect of attraction between the bodies will, no doubt, be produced and then we have to account for the loss of energy of motion. If we suppose it transformed into an irregular motion of agitation in the body or in the corpuscles there will be a constant generation of heat in all bodies, and when we come to calculate the amount of heat which would be thus generated in a second we find it so enormous that ⟨this hypothesis⟩ the whole world would soon be at a white heat.

It has been suggested by Sir W. Thomson that the corpuscle may before impact be moving without vibration or rotation but that after impact it may proceed with a smaller velocity of translation, the rest of the kinetic energy being in the form of rotation or if the corpuscule has parts it may be capable of vibration.

By the theory as thus modified we may preserve Lesage's explanation of gravitation consistently with the principle of the conservation of energy.

But several ⟨other⟩ points not yet mentioned were pointed out by Lesage himself as necessary to the theory.

In the first place we know that the intensity of gravitation depends ⟨not⟩ on the mass of a body and not on the extent of surface. A penny is equally heavy whether it be laid on the balance flat or on its edge. But if the penny were solid the number of corpuscules ⟨which⟩ moving downwards which would be caught by it would be greater when it is laid flat. Hence the penny, as regards the corpuscules is not a solid body. It must be regarded as made up of a system of molecules the inter ⟨stices⟩ spaces between which are so large compared with the size of the molecules ⟨when⟩ when the penny is placed edgewise to the stream of ⟨molecules⟩ corpuscules one molecule of the penny does not appreciably shelter another from the storm.

But the fact that the amount of gravitation does not depend on the configuration of a body is true of bodies as large as the earth and even the sun. We must therefore suppose that the earth is so porous, that its molecules are so wide apart compared with their actual dimensions that of the corpuscules which meet the earth in any direction only a very small proportion are stopped by striking its molecules, the far larger proportion of corpuscules passing right through the body of the earth without ever coming in contact with one of its molecules.

If it were not so then the sun and earth together would attract the moon less during an eclipse of the moon than at full moon when there is no eclipse; ⟨for during an eclipse a smaller number of corpuscules⟩ ⟨the sun an⟩ If a large proportion of ⟨molecules⟩ corpuscules falling on the sun were stopped before getting through his body then since during an eclipse of the moon the ⟨mole⟩ corpuscules which the earth screens from the moon are those which come to the earth from the sun, there will be a smaller number than usual to select from, and the effect of the earth on the moon will be less than if there had been no eclipse. Now no alteration of the moons orbit ⟨arises on account of⟩ is found to be produced during eclipses and this shows that the attraction of gravitation is not altered by ⟨the fact of two given attracting bodies the sun and the earth being in a straight lin⟩ one of the two attracting bodies being between the other and the moon. This leads us to ⟨form⟩ the contemplation [of] a ⟨very⟩ most formidable bombardment which is going on on all sides of us by the ultramundane corpuscules. It is but a very small fraction of them which are deflected by the interposition of the earth. Of these a fraction enormously smaller is intercepted by a small body, say a pound of matter at the earth's surface. And yet, by the abstraction of this fraction of a fraction of the bombardment, the body is impelled towards the earth with a force equal to the weight of a pound.

The velocity of the corpuscules must be enormously greater than that of any of the heavenly bodies, otherwise as may easily be shown, they would act as a resisting medium, opposing the motion of the ⟨heavenly bodies⟩ planets. Now the energy of a moving system is the half of its momentum multiplied by its velocity. Let us estimate the energy of that part of the stream of corpuscules which is bounded, say, by the circumference of a penny and which passes through this penny in one second. It is but a small fraction of the corpuscules which is stopped by the earth. ⟨As the diameter⟩ Find another fraction smaller than this in the ratio of the diameter of the earth to the thickness of a penny. This will represent the fraction stopped by the penny. The impulse of the molecules going downwards is greater than that of the molecules going upwards ⟨through⟩ and striking the penny in a second by the momentum which the penny falling freely would acquire in a second. To find the energy of these impacts we must multiply this by half the velocity of the corpuscules.

Hence to find the whole energy of the stream of ⟨molecules⟩ corpuscules flowing through the penny in a second we must multiply the energy of motion of the penny after falling for a second first by the ratio of the velocity of the corpuscules to that of the penny, next by the ratio of the diameter of the earth to the thickness of the penny and then by the square of the ratio of the number of molecules falling on the earth to the number stopped by it.

If these ratios could be estimated we should obtain an enormous number of horsepower as the rate of consumption of the energy of the universe ⟨for the purpose by which the weight of the penny is mai⟩ in order that the penny may be attracted towards the earth as we find it is.

It is needless to say that the mere magnitude of such a number ought not to astonish us. What is really astonishing is the theory which implies that forces of the magnitude [...] unless certain adjustments are always kept up so as to preserve the uniformity of this celestial bombardment, force of the magnitude of which we can form no adequate conception ⟨would⟩ and which even now are in existence though delicately balanced, would at once tear atom from atom and bring the universe to a white heat.

But we must now descend from the brilliant cosmical speculation of Lesage to the application of the atomic theory to the explanation of the properties of ⟨terrestr⟩ gross matter in the solid liquid or gaseous state. The properties of crystals—their mode of growth—their regular external figure—their cleavage planes—the ⟨action of⟩ existence of certain axes of direction within every portion of the crystal ⟨which determine its relations to the transmission of light⟩ ⟨mechanical stresses which act on it⟩ ⟨to the propag⟩ ⟨conduction of heat and electricity⟩ ⟨to magnetization to elasticity⟩ and ultimate[ly] which ⟨also⟩ determine its elasticity and ultimate strength in different directions and the mode in which it expands when heated and according to which the propagation of sound, the transmission of light, the conduction and heat and electricity and the process of magnetization take place differently in different directions, all these properties indicate that an apparently homogeneous and continuous solid may possess an internal structure. There are ⟨mathematical⟩ methods by which those properties of a substance which are different in different directions may be expressed mathematically so that the relations between these properties may be investigated analytically. Some of ⟨these⟩ ⟨in⟩ the investigation by which Poisson, Navier and the French mathematicians had obtained their mathematical results ⟨were obtained⟩ began with the assumption that the body is composed of atoms, whose configuration determines the properties of the body. But as the same results have been obtained by Stokes and others without any such hypothesis, this of itself is no evidence of atomic structure. The question of structure remains as before. ⟨Do the properties of crystals which we know to belong to the most minute parts of them which we can deal with arise from the configuration of certain ultimate ⟨small portions⟩ parts each of which is incapable of division or at least cannot be divided without losing the⟩ We know that the most minute portion of a crystal which we can deal with possesses the crystalline properties above mentioned. Now if the process of division were carried far enough would the ultimate portions still possess these properties (including that of cleavage into smaller portions) or would we arrive at portions not themselves possessing the crystalline properties but arranged in such a manner that the body which they constitute has these properties.

Cloth for example has certain properties related to direction. It can be stretched ⟨and torn⟩ more easily in one direction. It may be torn more easily in another. If we examine it we find it composed of threads. ⟨Each⟩ A single thread does not possess the properties which we observe in the cloth. The properties of the cloth arise from the arrangement or configuration of the threads. Should we or should we not in the process of dividing and subdividing a crystal arrive at a result analogous to that which we obtain when we have separated a piece of cloth into its component threads.

Again—when we cut ⟨cloth⟩ a very ⟨small piece⟩ narrow strip of cloth from the web we find that its properties when so cut differ from those it had when in the web. We have not reduced the strip to a single thread but we have brought these properties which depend on single threads into greater prominence by diminishing those which depend chiefly on the interfacing of innumerable threads. We might thus if our eyes were not able to distinguish the structure of the cloth ⟨were too minute to be seen⟩ obtain evidence that the structure of the cloth though too minute to be seen was not infinitely minute and we might even be able to make an estimate of the probable number of threads in the strip which we were examining.

23. "Atom"

Encyclopedia Britannica, ninth edition, vol. 3, 36–49 (1875); *Scientific Papers*, vol. 2, 445–484.

[From the *Encyclopœdia Britannica*.]

LXXIII. *Atom.*

Atom (ἄτομος) is a body which cannot be cut in two. The atomic theory is a theory of the constitution of bodies, which asserts that they are made up of atoms. The opposite theory is that of the homogeneity and continuity of bodies, and asserts, at least in the case of bodies having no apparent organisation, such, for instance, as water, that as we can divide a drop of water into two parts which are each of them drops of water, so we have reason to believe that these smaller drops can be divided again, and the theory goes on to assert that there is nothing in the nature of things to hinder this process of division from being repeated over and over again, times without end. This is the doctrine of the infinite divisibility of bodies, and it is in direct contradiction with the theory of atoms.

The atomists assert that after a certain number of such divisions the parts would be no longer divisible, because each of them would be an atom. The advocates of the continuity of matter assert that the smallest conceivable body has parts, and that whatever has parts may be divided.

In ancient times Democritus was the founder of the atomic theory, while Anaxagoras propounded that of continuity, under the name of the doctrine of homœomeria (Ὁμοιομέρια), or of the similarity of the parts of a body to the whole. The arguments of the atomists, and their replies to the objections of Anaxagoras, are to be found in Lucretius.

In modern times the study of nature has brought to light many properties of bodies which appear to depend on the magnitude and motions of their ultimate constituents, and the question of the existence of atoms has once more become conspicuous among scientific inquiries.

We shall begin by stating the opposing doctrines of atoms and of continuity before giving an outline of the state of molecular science as it now exists. In the earliest times the most ancient philosophers whose speculations

446   ATOM.

are known to us seem to have discussed the ideas of number and of continuous magnitude, of space and time, of matter and motion, with a native power of thought which has probably never been surpassed. Their actual knowledge, however, and their scientific experience were necessarily limited, because in their days the records of human thought were only beginning to accumulate. It is probable that the first exact notions of quantity were founded on the consideration of number. It is by the help of numbers that concrete quantities are practically measured and calculated. Now, number is discontinuous. We pass from one number to the next *per saltum*. The magnitudes, on the other hand, which we meet with in geometry, are essentially continuous. The attempt to apply numerical methods to the comparison of geometrical quantities led to the doctrine of incommensurables, and to that of the infinite divisibility of space. Meanwhile, the same considerations had not been applied to time, so that in the days of Zeno of Elea time was still regarded as made up of a finite number of "moments," while space was confessed to be divisible without limit. This was the state of opinion when the celebrated arguments against the possibility of motion, of which that of Achilles and the tortoise is a specimen, were propounded by Zeno, and such, apparently, continued to be the state of opinion till Aristotle pointed out that time is divisible without limit, in precisely the same sense that space is. And the slowness of the development of scientific ideas may be estimated from the fact that Bayle does not see any force in this statement of Aristotle, but continues to admire the paradox of Zeno. (Bayle's *Dictionary*, art. "Zeno".) Thus the direction of true scientific progress was for many ages towards the recognition of the infinite divisibility of space and time.

It was easy to attempt to apply similar arguments to matter. If matter is extended and fills space, the same mental operation by which we recognise the divisibility of space may be applied, in imagination at least, to the matter which occupies space. From this point of view the atomic doctrine might be regarded as a relic of the old numerical way of conceiving magnitude, and the opposite doctrine of the infinite divisibility of matter might appear for a time the most scientific. The atomists, on the other hand, asserted very strongly the distinction between matter and space. The atoms, they said, do not fill up the universe; there are void spaces between them. If it were not so, Lucretius tells us, there could be no motion, for the atom which gives way first must have some empty place to move into.

"Quapropter locus est intactus, inane, vacansque.
Quod si non esset, nulla ratione moveri
Res possent; namque, officium quod corporis exstat,
Officere atque obstare, id in omni tempore adesset
Omnibus: haud igitur quicquam procedere posset,
Principium quoniam cedendi nulla daret res." *De Rerum Natura,* I. 335.

The opposite school maintained then, as they have always done, that there is no vacuum—that every part of space is full of matter, that there is a universal plenum, and that all motion is like that of a fish in the water, which yields in front of the fish because the fish leaves room for it behind.

"Cedere squamigeris latices nitentibus aiunt
Et liquidas aperire vias, quia post loca pisces
Linquant, quo possint cedentes confluere undæ." I. 373.

In modern times Descartes held that, as it is of the essence of matter to be extended in length, breadth, and thickness, so it is of the essence of extension to be occupied by matter, for extension cannot be an extension of nothing.

"Ac proinde si quæratur quid fiet, si Deus auferat omne corpus quod in aliquo vase continetur, et nullum aliud in ablati locum venire permittat? respondendum est, vasis latera sibi invicem hoc ipso fore contigua. Cum enim inter duo corpora nihil interjacet, necesse est ut se mutuo tangant, ac manifeste repugnat ut distent, sive ut inter ipsa sit distantia, et tamen ut ista distantia sit nihil; quia omnis distantia est modus extensionis, et ideo sine substantia extensa esse non potest."
Principia, II. 18.

This identification of extension with substance runs through the whole of Descartes's works, and it forms one of the ultimate foundations of the system of Spinoza. Descartes, consistently with this doctrine, denied the existence of atoms as parts of matter, which by their own nature are indivisible. He seems to admit, however, that the Deity might make certain particles of matter indivisible in this sense, that no creature should be able to divide them. These particles, however, would be still divisible by their own nature, because the Deity cannot diminish his own power, and therefore must retain his power of dividing them. Leibnitz, on the other hand, regarded his monad as the ultimate element of everything.

There are thus two modes of thinking about the constitution of bodies, which have had their adherents both in ancient and in modern times. They correspond to the two methods of regarding quantity—the arithmetical and the geometrical. To the atomist the true method of estimating the quantity of matter in a body is to count the atoms in it. The void spaces between the atoms count for nothing. To those who identify matter with extension, the volume of space occupied by a body is the only measure of the quantity of matter in it.

Of the different forms of the atomic theory, that of Boscovich may be taken as an example of the purest monadism. According to Boscovich matter is made up of atoms. Each atom is an indivisible point, having position in space, capable of motion in a continuous path, and possessing a certain mass, whereby a certain amount of force is required to produce a given change of motion. Besides this the atom is endowed with potential force, that is to say, that any two atoms attract or repel each other with a force depending on their distance apart. The law of this force, for all distances greater than say the thousandth of an inch, is an attraction varying as the inverse square of the distance. For smaller distances the force is an attraction for one distance and a repulsion for another, according to some law not yet discovered. Boscovich himself, in order to obviate the possibility of two atoms ever being in the same place, asserts that the ultimate force is a repulsion which increases without limit as the distance diminishes without limit, so that two atoms can never coincide. But this seems an unwarrantable concession to the vulgar opinion that two bodies cannot co-exist in the same place. This opinion is deduced from our experience of the behaviour of bodies of sensible size, but we have no experimental evidence that two atoms may not sometimes coincide. For instance, if oxygen and hydrogen combine to form water, we have no experimental evidence that the molecule of oxygen is not in the very same place with the two molecules of hydrogen. Many persons cannot get rid of the opinion that all matter is extended in length, breadth, and depth. This is a prejudice of the same kind with the last, arising from our experience of bodies consisting of immense multitudes of atoms. The system of atoms, according to Boscovich, occupies a certain region of space in virtue of the forces acting between the component atoms of the system and any other atoms when brought near them. No other system of atoms can occupy the same region of space at the same time, because, before it could do so, the mutual action of the atoms would have caused a repulsion between the two systems insuperable by any force which we can command. Thus, a number of soldiers with firearms may occupy an extensive region to the exclusion of the enemy's armies, though the space filled by their bodies is but small. In this way Boscovich explained the apparent extension of bodies consisting of atoms, each of which is devoid of extension. According to Boscovich's theory, all action between bodies is action at a distance. There is no such thing in nature as actual contact between two bodies. When two bodies are said in ordinary language to be in contact, all that is meant is that they are so near together that the repulsion between the nearest pairs of atoms belonging to the two bodies is very great.

Thus, in Boscovich's theory, the atom has continuity of existence in time and space. At any instant of time it is at some point of space, and it is never in more than one place at a time. It passes from one place to another along a continuous path. It has a definite mass which cannot be increased or diminished. Atoms are endowed with the power of acting on one another by attraction or repulsion, the amount of the force depending on the distance between them. On the other hand, the atom itself has no parts or dimensions. In its geometrical aspect it is a mere geometrical point. It has no extension in space. It has not the so-called property of Impenetrability, for two atoms may exist in the same place. This we may regard as one extreme of the various opinions about the constitution of bodies.

The opposite extreme, that of Anaxagoras—the theory that bodies apparently homogeneous and continuous are so in reality—is, in its extreme form, a theory incapable of development. To explain the properties of any substance by this theory is impossible. We can only admit the observed properties of such substance as ultimate facts. There is a certain stage, however, of scientific progress in which a method corresponding to this theory is of service. In hydrostatics, for instance, we define a fluid by means of one of its known properties, and from this definition we make the system of deductions which constitutes the science of hydrostatics. In this way the science of hydrostatics may be built upon an experimental basis, without any consideration of the constitution of a fluid as to whether it is molecular or continuous. In like manner, after the French mathematicians had attempted, with more or less ingenuity, to construct a theory of elastic solids from the hypothesis that they consist of atoms in equilibrium under the action of their mutual forces, Stokes and others shewed that all the results of this hypothesis, so far at least as they agreed with facts, might be deduced from the postulate that elastic bodies exist, and from the hypothesis that the smallest portions into which we can divide them are sensibly homogeneous. In this way the principle of continuity, which is the basis of the method of Fluxions and the whole of modern mathematics, may

450   ATOM.

be applied to the analysis of problems connected with material bodies by assuming them, for the purpose of this analysis, to be homogeneous. All that is required to make the results applicable to the real case is that the smallest portions of the substance of which we take any notice shall be sensibly of the same kind. Thus, if a railway contractor has to make a tunnel through a hill of gravel, and if one cubic yard of the gravel is so like another cubic yard that for the purposes of the contract they may be taken as equivalent, then, in estimating the work required to remove the gravel from the tunnel, he may, without fear of error, make his calculations as if the gravel were a continuous substance. But if a worm has to make his way through the gravel, it makes the greatest possible difference to him whether he tries to push right against a piece of gravel, or directs his course through one of the intervals between the pieces ; to him, therefore, the gravel is by no means a homogeneous and continuous substance.

In the same way, a theory that some particular substance, say water, is homogeneous and continuous may be a good working theory up to a certain point, but may fail when we come to deal with quantities so minute or so attenuated that their heterogeneity of structure comes into prominence. Whether this heterogeneity of structure is or is not consistent with homogeneity and continuity of substance is another question.

The extreme form of the doctrine of continuity is that stated by Descartes, who maintains that the whole universe is equally full of matter, and that this matter is all of one kind, having no essential property besides that of extension. All the properties which we perceive in matter he reduces to its parts being movable among one another, and so capable of all the varieties which we can perceive to follow from the motion of its parts (*Principia*, II. 23). Descartes's own attempts to deduce the different qualities and actions of bodies in this way are not of much value. More than a century was required to invent methods of investigating the conditions of the motion of systems of bodies such as Descartes imagined. But the hydrodynamical discovery of Helmholtz that a vortex in a perfect liquid possesses certain permanent characteristics, has been applied by Sir W. Thomson to form a theory of vortex atoms in a homogeneous, incompressible, and frictionless liquid, to which we shall return at the proper time.

Outline of Modern Molecular Science, and in particular of the Molecular Theory of Gases.

We begin by assuming that bodies are made up of parts, each of which is capable of motion, and that these parts act on each other in a manner consistent with the principle of the conservation of energy. In making these assumptions, we are justified by the facts that bodies may be divided into smaller parts, and that all bodies with which we are acquainted are conservative systems, which would not be the case unless their parts were also conservative systems.

We may also assume that these small parts are in motion. This is the most general assumption we can make, for it includes, as a particular case, the theory that the small parts are at rest. The phenomena of the diffusion of gases and liquids through each other shew that there may be a motion of the small parts of a body which is not perceptible to us.

We make no assumption with respect to the nature of the small parts— whether they are all of one magnitude. We do not even assume them to have extension and figure. Each of them must be measured by its mass, and any two of them must, like visible bodies, have the power of acting on one another when they come near enough to do so. The properties of the body, or medium, are determined by the configuration and motion of its small parts.

The first step in the investigation is to determine the amount of motion which exists among the small parts, independent of the visible motion of the medium as a whole. For this purpose it is convenient to make use of a general theorem in dynamics due to Clausius.

When the motion of a material system is such that the time average of the quantity $\Sigma\,(mx^2)$ remains constant, the state of the system is said to be that of stationary motion. When the motion of a material system is such that the sum of the moments of inertia of the system, about three axes at right angles through its centre of mass, never varies by more than small quantities from a constant value, the system is said to be in a state of stationary motion.

The kinetic energy of a particle is half the product of its mass into the square of its velocity, and the kinetic energy of a system is the sum of the kinetic energy of all its parts.

452   ATOM.

When an attraction or repulsion exists between two points, half the product of this stress into the distance between the two points is called the *virial* of the stress, and is reckoned positive when the stress is an attraction, and negative when it is a repulsion. The virial of a system is the sum of the virials of the stresses which exist in it. If the system is subjected to the external stress of the pressure of the sides of a vessel in which it is contained, this stress will introduce an amount of virial $\frac{3}{2}pV$, where p is the pressure on unit of area and V is the volume of the vessel.

The theorem of Clausius may now be stated as follows :—In a material system in a state of stationary motion the time-average of the kinetic energy is equal to the time-average of the virial. In the case of a fluid enclosed in a vessel

$$\tfrac{1}{2}\Sigma\left(m\overline{v^2}\right) = \tfrac{3}{2}pV + \tfrac{1}{2}\Sigma\Sigma\left(Rr\right),$$

where the first term denotes the kinetic energy, and is half the sum of the product of each mass into the mean square of its velocity. In the second term, p is the pressure on unit of surface of the vessel, whose volume is V, and the third term expresses the virial due to the internal actions between the parts of the system. A double symbol of summation is used, because every pair of parts between which any action exists must be taken into account. We have next to shew that in gases the principal part of the pressure arises from the motion of the small parts of the medium, and not from a repulsion between them.

In the first place, if the pressure of a gas arises from the repulsion of its parts, the law of repulsion must be inversely as the distance. For, consider a cube filled with the gas at pressure p, and let the cube expand till each side is n times its former length. The pressure on unit of surface according to Boyle's law is now $\dfrac{p}{n^3}$, and since the area of a face of the cube is n^2 times what it was, the whole pressure on the face of the cube is $\dfrac{1}{n}$ of its original value. But since everything has been expanded symmetrically, the distance between corresponding parts of the air is now n times what it was, and the force is n times less than it was. Hence the force must vary inversely as the distance.

But Newton has shewn (*Principia*, Book I. Prop. 93) that this law is inadmissible, as it makes the effect of the distant parts of the medium on a particle greater than that of the neighbouring parts. Indeed, we should arrive at the conclusion that the pressure depends not only on the density of the air but on the form and dimensions of the vessel which contains it, which we know not to be the case.

If, on the other hand, we suppose the pressure to arise entirely from the motion of the molecules of the gas, the interpretation of Boyle's law becomes very simple. For, in this case

$$pV = \tfrac{1}{3}\Sigma\,(m\overline{v^2}).$$

The first term is the product of the pressure and the volume, which according to Boyle's law is constant for the same quantity of gas at the same temperature. The second term is two-thirds of the kinetic energy of the system, and we have every reason to believe that in gases when the temperature is constant the kinetic energy of unit of mass is also constant. If we admit that the kinetic energy of unit of mass is in a given gas proportional to the absolute temperature, this equation is the expression of the law of Charles as well as of that of Boyle, and may be written—

$$pV = R\theta,$$

where θ is the temperature reckoned from absolute zero, and R is a constant. The fact that this equation expresses with considerable accuracy the relation between the volume, pressure, and temperature of a gas when in an extremely rarified state, and that as the gas is more and more compressed the deviation from this equation becomes more apparent, shews that the pressure of a gas is due almost entirely to the motion of its molecules when the gas is rare, and that it is only when the density of the gas is considerably increased that the effect of direct action between the molecules becomes apparent.

The effect of the direct action of the molecules on each other depends on the number of pairs of molecules which at a given instant are near enough to act on one another. The number of such pairs is proportional to the square of the number of molecules in unit of volume, that is, to the square of the density of the gas. Hence, as long as the medium is so rare that the encounter between two molecules is not affected by the presence of others, the deviation from Boyle's law will be proportional to the square of the density. If the action between the molecules is on the whole repulsive, the pressure will be greater than that given by Boyle's law. If it is, on the whole, attractive, the pressure will be less than that given by Boyle's law. It appears, by the experiments of Regnault and others, that the pressure does deviate from Boyle's law when the density of the gas is increased. In the case of carbonic acid and other gases which are easily liquefied, this deviation is very great. In all cases, however, except that of hydrogen, the pressure is less than that given by Boyle's law, shewing that the virial is on the whole due to *attractive* forces between the molecules.

Another kind of evidence as to the nature of the action between the molecules is furnished by an experiment made by Dr Joule. Of two vessels, one was exhausted and the other filled with a gas at a pressure of 20 atmospheres; and both were placed side by side in a vessel of water, which was constantly stirred. The temperature of the whole was observed. Then a communication was opened between the vessels, the compressed gas expanded to twice its volume, and the work of expansion, which at first produced a strong current in the gas, was soon converted into heat by the internal friction of the gas. When all was again at rest, and the temperature uniform, the temperature was again observed. In Dr Joule's original experiments the observed temperature was the same as before. In a series of experiments, conducted by Dr Joule and Sir W. Thomson on a different plan, by which the thermal effect of free expansion can be more accurately measured, a slight cooling effect was observed in all the gases examined except hydrogen. Since the temperature depends on the velocity of agitation of the molecules, it appears that when a gas expands without doing external work the velocity of agitation is not much affected, but that in most cases it is slightly diminished. Now, if the molecules during their mutual separation act on each other, their velocity will increase or diminish according as the force is repulsive or attractive. It appears, therefore, from the experiments on the free expansion of gases, that the force between the molecules is small but, on the whole, attractive.

Having thus justified the hypothesis that a gas consists of molecules in motion, which act on each other only when they come very close together during an encounter, but which, during the intervals between their encounters which constitute the greater part of their existence, are describing free paths, and are not acted on by any molecular force, we proceed to investigate the motion of such a system.

The mathematical investigation of the properties of such a system of molecules in motion is the foundation of molecular science. Clausius was the first to express the relation between the density of the gas, the length of the free paths of its molecules, and the distance at which they encounter each other. He assumed, however, at least in his earlier investigations, that the velocities of all the molecules are equal. The mode in which the velocities are distributed was first investigated by the present writer, who shewed that in the moving system the velocities of the molecules range from zero to infinity, but that the number of molecules whose velocities lie within given limits can be expressed by a formula identical with that which expresses in the theory of errors the number of errors of observation lying within corresponding limits. The proof of this theorem has been carefully investigated by Boltzmann[1], who has strengthened it where it appeared weak, and to whom the method of taking into account the action of external forces is entirely due.

The mean kinetic energy of a molecule, however, has a definite value, which is easily expressed in terms of the quantities which enter into the expression for the distribution of velocities. The most important result of this investigation is that when several kinds of molecules are in motion and acting on one another, the mean kinetic energy of a molecule is the same whatever be its mass, the molecules of greater mass having smaller mean velocities. Now, when gases are mixed their temperatures become equal. Hence we conclude that the physical condition which determines that the temperature of two gases shall be the same is that the mean kinetic energies of agitation of the individual molecules of the two gases are equal. This result is of great importance in the theory of heat, though we are not yet able to establish any similar result for bodies in the liquid or solid state.

In the next place, we know that in the case in which the whole pressure of the medium is due to the motion of its molecules, the pressure on unit of area is numerically equal to two-thirds of the kinetic energy in unit of volume. Hence, if equal volumes of two gases are at equal pressures the kinetic energy is the same in each. If they are also at equal temperatures the mean kinetic energy of each molecule is the same in each. If, therefore, equal volumes of two gases are at equal temperatures and pressures, the number of molecules in each is the same, and therefore, the masses of the two kinds of molecules are in the same ratio as the densities of the gases to which they belong.

This statement has been believed by chemists since the time of Gay-Lussac, who first established that the weights of the chemical equivalents of

* *Sitzungsberichte der K. K. Akad.*, Wien, 8th Oct. 1868.

456   ATOM.

different substances are proportional to the densities of these substances when in the form of gas. The definition of the word molecule, however, as employed in the statement of Gay-Lussac's law is by no means identical with the definition of the same word as in the kinetic theory of gases. The chemists ascertain by experiment the ratios of the masses of the different substances in a compound. From these they deduce the chemical equivalents of the different substances, that of a particular substance, say hydrogen, being taken as unity. The only evidence made use of is that furnished by chemical combinations. It is also assumed, in order to account for the facts of combination, that the reason why substances combine in definite ratios is that the molecules of the substances are in the ratio of their chemical equivalents, and that what we call combination is an action which takes place by a union of a molecule of one substance to a molecule of the other.

This kind of reasoning, when presented in a proper form and sustained by proper evidence, has a high degree of cogency. But it is purely chemical reasoning; it is not dynamical reasoning. It is founded on chemical experience, not on the laws of motion.

Our definition of a molecule is purely dynamical. A molecule is that minute portion of a substance which moves about as a whole, so that its parts, if it has any, do not part company during the motion of agitation of the gas. The result of the kinetic theory, therefore, is to give us information about the relative masses of molecules considered as moving bodies. The consistency of this information with the deductions of chemists from the phenomena of combination, greatly strengthens the evidence in favour of the actual existence and motion of gaseous molecules.

Another confirmation of the theory of molecules is derived from the experiments of Dulong and Petit on the specific heat of gases, from which they deduced the law which bears their name, and which asserts that the specific heats of equal weights of gases are inversely as their combining weights, or, in other words, that the capacities for heat of the chemical equivalents of different gases are equal. We have seen that the temperature is determined by the kinetic energy of agitation of each molecule. The molecule has also a certain amount of energy of internal motion, whether of rotation or of vibration, but the hypothesis of Clausius, that the mean value of the internal energy always bears a proportion fixed for each gas to the energy of agitation, seems highly probable and consistent with experiment. The whole kinetic energy is therefore equal to the energy of agitation multiplied by a certain factor. Thus the energy communicated to a gas by heating it is divided in a certain proportion between the energy of agitation and that of the internal motion of each molecule. For a given rise of temperature the energy of agitation, say of a million molecules, is increased by the same amount whatever be the gas. The heat spent in raising the temperature is measured by the increase of the whole kinetic energy. The thermal capacities, therefore, of equal numbers of molecules of different gases are in the ratio of the factors by which the energy of agitation must be multiplied to obtain the whole energy. As this factor appears to be nearly the same for all gases of the same degree of atomicity, Dulong and Petit's law is true for such gases.

    Another result of this investigation is of considerable importance in relation to certain theories*, which assume the existence of æthers or rare media consisting of molecules very much smaller than those of ordinary gases. According to our result, such a medium would be neither more nor less than a gas. Supposing its molecules so small that they can penetrate between the molecules of solid substances such as glass, a so-called vacuum would be full of this rare gas at the observed temperature, and at the pressure, whatever it may be, of the ætherial medium in space. The specific heat, therefore, of the medium in the so-called vacuum will be equal to that of the same volume of any other gas at the same temperature and pressure. Now, the purpose for which this molecular æther is assumed in these theories is to act on bodies by its pressure, and for this purpose the pressure is generally assumed to be very great. Hence, according to these theories, we should find the specific heat of a so-called vacuum very considerable compared with that of a quantity of air filling the same space.

    We have now made a certain definite amount of progress towards a complete molecular theory of gases. We know the mean velocity of the molecules of each gas in metres per second, and we know the relative masses of the molecules of different gases. We also know that the molecules of one and the same gas are all equal in mass. For if they are not, the method of dialysis, as employed by Graham, would enable us to separate the molecules of smaller mass from those of greater, as they would stream through porous substances with greater velocity. We should thus be able to separate a gas, say hydrogen, into two portions, having different densities and other physical properties,

    * See Gustav Hansemann, *Die Atome und ihre Bewegungen.* 1871. (H. G. Mayer.)

458 ATOM.

different combining weights, and probably different chemical properties of other kinds. As no chemist has yet obtained specimens of hydrogen differing in this way from other specimens, we conclude that all the molecules of hydrogen are of sensibly the same mass, and not merely that their mean mass is a statistical constant of great stability.

But as yet we have not considered the phenomena which enable us to form an estimate of the actual mass and dimensions of a molecule. It is to Clausius that we owe the first definite conception of the free path of a molecule and of the mean distance travelled by a molecule between successive encounters. He shewed that the number of encounters of a molecule in a given time is proportional to the velocity, to the number of molecules in unit of volume, and to the square of the distance between the centres of two molecules when they act on one another so as to have an encounter. From this it appears that if we call this distance of the centres the diameter of a molecule, and the volume of a sphere having this diameter the volume of a molecule, and the sum of the volumes of all the molecules the molecular volume of the gas, then the diameter of a molecule is a certain multiple of the quantity obtained by diminishing the free path in the ratio of the molecular volume of the gas to the whole volume of the gas. The numerical value of this multiple differs slightly, according to the hypothesis we assume about the law of distribution of velocities. It also depends on the definition of an encounter. When the molecules are regarded as elastic spheres we know what is meant by an encounter, but if they act on each other at a distance by attractive or repulsive forces of finite magnitude, the distance of their centres varies during an encounter, and is not a definite quantity. Nevertheless, the above statement of Clausius enables us, if we know the length of the mean path and the molecular volume of gas, to form a tolerably near estimate of the diameter of the sphere of the intense action of a molecule, and thence of the number of molecules in unit of volume and the actual mass of each molecule. To complete the investigation we have, therefore, to determine the mean path and the molecular volume. The first numerical estimate of the mean path of a gaseous molecule was made by the present writer from data derived from the internal friction of air. There are three phenomena which depend on the length of the free path of the molecules of a gas. It is evident that the greater the free path the more rapidly will the molecules travel from one part of the medium to another, because their direction will not be so often altered by encounters with other molecules. If the molecules in different parts of the medium are of different kinds, their progress from one part of the medium to another can be easily traced by analysing portions of the medium taken from different places. The rate of diffusion thus found furnishes one method of estimating the length of the free path of a molecule. This kind of diffusion goes on not only between the molecules of different gases, but among the molecules of the same gas, only in the latter case the results of the diffusion cannot be traced by analysis. But the diffusing molecules carry with them in their free paths the momentum and the energy which they happen at a given instant to have. The diffusion of momentum tends to equalise the apparent motion of different parts of the medium, and constitutes the phenomenon called the internal friction or viscosity of gases. The diffusion of energy tends to equalise the temperature of different parts of the medium, and constitutes the phenomenon of the conduction of heat in gases.

These three phenomena—the diffusion of matter, of motion, and of heat in gases—have been experimentally investigated,—the diffusion of matter by Graham and Loschmidt, the diffusion of motion by Oscar Meyer and Clerk Maxwell, and that of heat by Stefan.

These three kinds of experiments give results which in the present imperfect state of the theory and the extreme difficulty of the experiments, especially those on the conduction of heat, may be regarded as tolerably consistent with each other. At the pressure of our atmosphere, and at the temperature of melting ice, the mean path of a molecule of hydrogen is about the 10,000th of a millimetre, or about the fifth part of a wave-length of green light. The mean path of the molecules of other gases is shorter than that of hydrogen.

The determination of the molecular volume of a gas is subject as yet to considerable uncertainty. The most obvious method is that of compressing the gas till it assumes the liquid form. It seems probable, from the great resistance of liquids to compression, that their molecules are about the same distance from each other as that at which two molecules of the same substance in the gaseous form act on each other during an encounter. If this is the case, the molecular volume of a gas is somewhat less than the volume of the liquid into which it would be condensed by pressure, or, in other words, the density of the molecules is somewhat greater than that of the liquid.

Now, we know the relative weights of different molecules with great

58—2

accuracy, and, from a knowledge of the mean path, we can calculate their relative diameters approximately. From these we can deduce the relative densities of different kinds of molecules. The relative densities so calculated have been compared by Lorenz Meyer with the observed densities of the liquids into which the gases may be condensed, and he finds a remarkable correspondence between them. There is considerable doubt, however, as to the relation between the molecules of a liquid and those of its vapour, so that till a larger number of comparisons have been made, we must not place too much reliance on the calculated densities of molecules. Another, and perhaps a more refined, method is that adopted by M. Van der Waals, who deduces the molecular volume from the deviations of the pressure from Boyle's law as the gas is compressed.

The first numerical estimate of the diameter of a molecule was that made by Loschmidt in 1865 from the mean path and the molecular volume. Independently of him and of each other, Mr Stoney, in 1868, and Sir W. Thomson, in 1870, published results of a similar kind—those of Thomson being deduced not only in this way, but from considerations derived from the thickness of soap bubbles, and from the electric action between zinc and copper.

The diameter and the mass of a molecule, as estimated by these methods, are, of course, very small, but by no means infinitely so. About two millions of molecules of hydrogen in a row would occupy a millimetre, and about two hundred million million million of them would weigh a milligramme. These numbers must be considered as exceedingly rough guesses; they must be corrected by more extensive and accurate experiments as science advances; but the main result, which appears to be well established, is that the determination of the mass of a molecule is a legitimate object of scientific research, and that this mass is by no means immeasurably small.

Loschmidt illustrates these molecular measurements by a comparison with the smallest magnitudes visible by means of a microscope. Nobert, he tells us, can draw 4000 lines in the breadth of a millimetre. The intervals between these lines can be observed with a good microscope. A cube, whose side is the 4000th of a millimetre, may be taken as the *minimum visibile* for observers of the present day. Such a cube would contain from 60 to 100 million molecules of oxygen or of nitrogen; but since the molecules of organised substances contain on an average about 50 of the more elementary atoms, we may assume that the smallest organised particle visible under the microscope contains about two million molecules of organic matter. At least half of every living organism consists of water, so that the smallest living being visible under the microscope does not contain more than about a million organic molecules. Some exceedingly simple organism may be supposed built up of not more than a million similar molecules. It is impossible, however, to conceive so small a number sufficient to form a being furnished with a whole system of specialised organs.

Thus molecular science sets us face to face with physiological theories. It forbids the physiologist from imagining that structural details of infinitely small dimensions can furnish an explanation of the infinite variety which exists in the properties and functions of the most minute organisms.

A microscopic germ is, we know, capable of development into a highly organised animal. Another germ, equally microscopic, becomes, when developed, an animal of a totally·different kind. Do all the differences, infinite in number, which distinguish the one animal from the other, arise each from some difference in the structure of the respective germs? Even if we admit this as possible, we shall be called upon by the advocates of Pangenesis to admit still greater marvels. For the microscopic germ, according to this theory, is no mere individual, but a representative body, containing members collected from every rank of the long-drawn ramification of the ancestral tree, the number of these members being amply sufficient not only to furnish the hereditary characteristics of every organ of the body and every habit of the animal from birth to death, but also to afford a stock of latent gemmules to be passed on in an inactive state from germ to germ, till at last the ancestral peculiarity which it represents is revived in some remote descendant.

Some of the exponents of this theory of heredity have attempted to elude the difficulty of placing a whole world of wonders within a body so small and so devoid of visible structure as a germ, by using the phrase structureless germs*. Now, one material system can differ from another only in the configuration and motion which it has at a given instant. To explain differences of function and development of a germ without assuming differences of structure is, therefore, to admit that the properties of a germ are not those of a purely material system.

The evidence as to the nature and motion of molecules, with which we have hitherto been occupied, has been derived from experiments upon gaseous

---

* See F. Galton, "On Blood Relationship," *Proc. Roy. Soc.*, June 13, 1872.

media, the smallest sensible portion of which contains millions of millions of molecules. The constancy and uniformity of the properties of the gaseous medium is the direct result of the inconceivable irregularity of the motion of agitation of its molecules. Any cause which could introduce regularity into the motion of agitation, and marshal the molecules into order and method in their evolutions, might check or even reverse that tendency to diffusion of matter, motion, and energy, which is one of the most invariable phenomena of nature, and to which Thomson has given the name of the dissipation of energy.

Thus, when a sound-wave is passing through a mass of air, this motion is of a certain definite type, and if left to itself the whole motion is passed on to other masses of air, and the sound-wave passes on, leaving the air behind it at rest. Heat, on the other hand, never passes out of a hot body except to enter a colder body, so that the energy of sound-waves, or any other form of energy which is propagated so as to pass wholly out of one portion of the medium and into another, cannot be called heat.

We have now to turn our attention to a class of molecular motions, which are as remarkable for their regularity as the motion of agitation is for its irregularity.

It has been found, by means of the spectroscope, that the light emitted by incandescent substances is different according to their state of condensation. When they are in an extremely rarefied condition the spectrum of their light consists of a set of sharply-defined bright lines. As the substance approaches a denser condition the spectrum tends to become continuous, either by the lines becoming broader and less defined, or by new lines and bands appearing between them, till the spectrum at length loses all its characteristics and becomes identical with that of solid bodies when raised to the same temperature.

Hence the vibrating systems, which are the source of the emitted light, must be vibrating in a different manner in these two cases. When the spectrum consists of a number of bright lines, the motion of the system must be compounded of a corresponding number of types of harmonic vibration.

In order that a bright line may be sharply defined, the vibratory motion which produces it must be kept up in a perfectly regular manner for some hundreds or thousands of vibrations. If the motion of each of the vibrating bodies is kept up only during a small number of vibrations, then, however regular may be the vibrations of each body while it lasts, the resultant disturbance of the luminiferous medium, when analysed by the prism, will be found to contain, besides the part due to the regular vibrations, other motions, depending on the starting and stopping of each particular vibrating body, which will become manifest as a diffused luminosity scattered over the whole length of the spectrum. A spectrum of bright lines, therefore, indicates that the vibrating bodies when set in motion are allowed to vibrate in accordance with the conditions of their internal structure for some time before they are again interfered with by external forces.

It appears, therefore, from spectroscopic evidence that each molecule of a rarefied gas is, during the greater part of its existence, at such a distance from all other molecules that it executes its vibrations in an undisturbed and regular manner. This is the same conclusion to which we were led by considerations of another kind at p. 452.

We may therefore regard the bright lines in the spectrum of a gas as the result of the vibrations executed by the molecules while describing their free paths. When two molecules separate from one another after an encounter, each of them is in a state of vibration, arising from the unequal action on different parts of the same molecule during the encounter. Hence, though the centre of mass of the molecule describing its free path moves with uniform velocity, the parts of the molecule have a vibratory motion with respect to the centre of mass of the whole molecule, and it is the disturbance of the luminiferous medium communicated to it by the vibrating molecules which constitutes the emitted light.

We may compare the vibrating molecule to a bell. When struck, the bell is set in motion. This motion is compounded of harmonic vibrations of many different periods, each of which acts on the air, producing notes of as many different pitches. As the bell communicates its motion to the air, these vibrations necessarily decay, some of them faster than others, so that the sound contains fewer and fewer notes, till at last it is reduced to the fundamental note of the bell*. If we suppose that there are a great many bells precisely similar to each other, and that they are struck, first one and then another, in a perfectly irregular manner, yet so that, on an average, as many bells are struck in one second of time as in another, and also in such a way

---

* Part of the energy of motion is, in the case of the bell, dissipated in the substance of the bell in virtue of the viscosity of the metal, and assumes the form of heat, but it is not necessary, for the purpose of illustration, to take this cause of the decay of vibrations into account.

that, on an average, any one bell is not again struck till it has ceased to vibrate, then the audible result will appear a continuous sound, composed of the sound emitted by bells in all states of vibration, from the clang of the actual stroke to the final hum of the dying fundamental tone.

But now let the number of bells be reduced while the same number of strokes are given in a second. Each bell will now be struck before it has ceased to vibrate, so that in the resulting sound there will be less of the fundamental tone and more of the original clang, till at last, when the peal is reduced to one bell, on which innumerable hammers are continually plying their strokes all out of time, the sound will become a mere noise, in which no musical note can be distinguished.

In the case of a gas we have an immense number of molecules, each of which is set in vibration when it encounters another molecule, and continues to vibrate as it describes its free path. The molecule is a material system, the parts of which are connected in some definite way, and from the fact that the bright lines of the emitted light have always the same wave-lengths, we learn that the vibrations corresponding to these lines are always executed in the same periodic time, and therefore the force tending to restore any part of the molecule to its position of equilibrium in the molecule must be proportional to its displacement relative to that position.

From the mathematical theory of the motion of such a system, it appears that the whole motion may be analysed into the following parts, which may be considered each independently of the others :—In the first place, the centre of mass of the system moves with uniform velocity in a straight line. This velocity may have any value. In the second place, there may be a motion of rotation, the angular momentum of the system about its centre of mass remaining during the free path constant in magnitude and direction. This angular momentum may have any value whatever, and its axis may have any direction. In the third place, the remainder of the motion is made up of a number of component motions, each of which is an harmonic vibration of a given type. In each type of vibration the periodic time of vibration is determined by the nature of the system, and is invariable for the same system. The relative amount of motion in different parts of the system is also determinate for each type, but the absolute amount of motion and the phase of the vibration of each type are determined by the particular circumstances of the last encounter, and may vary in any manner from one encounter to another.

The values of the periodic times of the different types of vibration are given by the roots of a certain equation, the form of which depends on the nature of the connections of the system. In certain exceptionally simple cases, as, for instance, in that of a uniform string stretched between two fixed points, the roots of the equation are connected by simple arithmetical relations, and if the internal structure of a molecule had an analogous kind of simplicity, we might expect to find in the spectrum of the molecule a series of bright lines, whose wave-lengths are in simple arithmetical ratios.

But if we suppose the molecule to be constituted according to some different type, as, for instance, if it is an elastic sphere, or if it consists of a finite number of atoms kept in their places by attractive and repulsive forces, the roots of the equation will not be connected with each other by any simple relations, but each may be made to vary independently of the others by a suitable change of the connections of the system. Hence, we have no right to expect any definite numerical relations among the wave-lengths of the bright lines of a gas.

The bright lines of the spectrum of an incandescent gas are therefore due to the harmonic vibrations of the molecules of the gas during their free paths. The only effect of the motion of the centre of mass of the molecule is to alter the time of vibration of the light as received by a stationary observer. When the molecule is coming towards the observer, each successive impulse will have a shorter distance to travel before it reaches his eye, and therefore the impulses will appear to succeed each other more rapidly than if the molecule were at rest, and the contrary will be the case if the molecule is receding from the observer. The bright line corresponding to the vibration will therefore be shifted in the spectrum towards the blue end when the molecule is approaching, and towards the red end when it is receding from the observer. By observations of the displacement of certain lines in the spectrum, Dr Huggins and others have measured the rate of approach or of recession of certain stars with respect to the earth, and Mr Lockyer has determined the rate of motion of tornadoes in the sun. But Lord Rayleigh has pointed out that according to the dynamical theory of gases the molecules are moving hither and thither with so great velocity that, however narrow and sharply-defined any bright line due to a single molecule may be, the displacement of the line towards the blue by the approaching molecules, and towards the red by the receding molecules, will produce a certain amount of widening and blurring of the line in the spectrum, so that there is a limit to the sharpness of definition of the lines of a gas. The widening of the lines due to this cause will be in proportion to the velocity of agitation of the molecules. It will be greatest for the molecules of smallest mass, as those of hydrogen, and it will increase with the temperature. Hence the measurement of the breadth of the hydrogen lines, such as C or F in the spectrum of the solar prominences, may furnish evidence that the temperature of the sun cannot exceed a certain value.

On the Theory of Vortex Atoms.

The equations which form the foundations of the mathematical theory of fluid motion were fully laid down by Lagrange and the great mathematicians of the end of last century, but the number of solutions of cases of fluid motion which had been actually worked out remained very small, and almost all of these belonged to a particular type of fluid motion, which has been since named the irrotational type. It had been shewn, indeed, by Lagrange, that a perfect fluid, if its motion is at any time irrotational, will continue in all time coming to move in an irrotational manner, so that, by assuming that the fluid was at one time at rest, the calculation of its subsequent motion may be very much simplified.

It was reserved for Helmholtz to point out the very remarkable properties of rotational motion in a homogeneous incompressible fluid devoid of all viscosity. We must first define the physical properties of such a fluid. In the first place, it is a material substance. Its motion is continuous in space and time, and if we follow any portion of it as it moves, the mass of that portion remains invariable. These properties it shares with all material substances. In the next place, it is incompressible. The form of a given portion of the fluid may change, but its volume remains invariable; in other words, the density of the fluid remains the same during its motion. Besides this, the fluid is homogeneous, or the density of all parts of the fluid is the same. It is also continuous, so that the mass of the fluid contained within any closed surface is always *exactly* proportional to the volume contained within that surface. This is equivalent to asserting that the fluid is not made up of molecules; for, if it were, the mass would vary in a discontinuous manner as the volume increases continuously, because first one and then another molecule would be included within the closed surface. Lastly, it is a perfect fluid, or, in other words, the stress between one portion and a contiguous portion is always normal to the surface which separates these portions, and this whether the fluid is at rest or in motion.

We have seen that in a molecular fluid the interdiffusion of the molecules causes an interdiffusion of motion of different parts of the fluid, so that the action between contiguous parts is no longer normal but in a direction tending to diminish their relative motion. Hence the perfect fluid cannot be molecular.

All that is necessary in order to form a correct mathematical theory of a material system is that its properties shall be clearly defined and shall be consistent with each other. This is essential; but whether a substance having such properties actually exists is a question which comes to be considered only when we propose to make some practical application of the results of the mathematical theory. The properties of our perfect liquid are clearly defined and consistent with each other, and from the mathematical theory we can deduce remarkable results, some of which may be illustrated in a rough way by means of fluids which are by no means perfect in the sense of not being viscous, such, for instance, as air and water.

The motion of a fluid is said to be irrotational when it is such that if a spherical portion of the fluid were suddenly solidified, the solid sphere so formed would not be rotating about any axis. When the motion of the fluid is rotational the axis and angular velocity of the rotation of any small part of the fluid are those of a *small* spherical portion suddenly solidified.

The mathematical expression of these definitions is as follows:—Let u, v, w be the components of the velocity of the fluid at the point (x, y, z), and let

$$a = \frac{dv}{dz} - \frac{dw}{dy}, \quad \beta = \frac{dw}{dx} - \frac{du}{dz}, \quad \gamma = \frac{du}{dy} - \frac{dv}{dx} \quad \text{...............} (1),$$

then a, β, γ are the components of the velocity of rotation of the fluid at the point (x, y, z). The axis of rotation is in the direction of the resultant of a, β, and γ, and the velocity of rotation, ω, is measured by this resultant.

A line drawn in the fluid, so that at every point of the line

$$\frac{1}{a}\frac{dx}{ds} = \frac{1}{\beta}\frac{dy}{ds} = \frac{1}{\gamma}\frac{dz}{ds} = \frac{1}{\omega} \quad \text{.........................} (2),$$

where s is the length of the line up to the point x, y, z, is called a vortex line. Its direction coincides at every point with that of the axis of rotation of the fluid.

We may now prove the theorem of Helmholtz, that the points of the fluid which at any instant lie in the same vortex line continue to lie in the same vortex line during the whole motion of the fluid.

The equations of motion of a fluid are of the form

$$\rho \,\frac{\delta u}{\delta t} + \frac{dp}{dx} + \rho \,\frac{dV}{dx} = 0 \quad\ldots\ldots\ldots\ldots\ldots\ldots\ldots (3),$$

when ρ is the density, which in the case of our homogeneous incompressible fluid we may assume to be unity, the operator $\frac{\delta}{\delta t}$ represents the rate of variation of the symbol to which it is prefixed at a point which is carried forward with the fluid, so that

$$\frac{\delta u}{\delta t} = \frac{du}{dt} + u\frac{du}{dx} + v\frac{du}{dy} + w\frac{du}{dz} \quad\ldots\ldots\ldots\ldots\ldots (4),$$

p is the pressure, and V is the potential of external forces. There are two other equations of similar form in y and z. Differentiating the equation in y with respect to z, and that in z with respect to y, and subtracting the second from the first, we find

$$\frac{d}{dz}\,\frac{\delta v}{\delta t} - \frac{d}{dy}\,\frac{\delta w}{\delta t} = 0 \quad\ldots\ldots\ldots\ldots\ldots\ldots\ldots (5).$$

Performing the differentiations and remembering equations (1) and also the condition of incompressibility,

$$\frac{du}{dx} + \frac{dv}{dy} + \frac{dw}{dz} = 0 \quad\ldots\ldots\ldots\ldots\ldots\ldots\ldots (6),$$

we find

$$\frac{\delta a}{\delta t} = a\frac{du}{dx} + \beta\frac{du}{dy} + \gamma\frac{du}{dz} \quad\ldots\ldots\ldots\ldots\ldots\ldots (7).$$

Now, let us suppose a vortex line drawn in the fluid so as always to begin at the same particle of the fluid. The components of the velocity of this point are u, v, w. Let us find those of a point on the moving vortex line at a distance ds from this point where

$$ds = \omega d\sigma \quad\ldots\ldots\ldots\ldots\ldots\ldots\ldots\ldots\ldots (8).$$

The co-ordinates of this point are

$$x + a\, d\sigma, \quad y + \beta\, d\sigma, \quad z + \gamma\, d\sigma \dotfill (9),$$

and the components of its velocity are

$$u + \frac{\delta a}{\delta t}\, d\sigma, \quad v + \frac{\delta \beta}{\delta t}\, d\sigma, \quad w + \frac{\delta \gamma}{\delta t}\, d\sigma \dotfill (10).$$

Consider the first of these components. In virtue of equation (7) we may write it

$$u + \frac{du}{dx}\, a\, d\sigma + \frac{du}{dy}\, \beta\, d\sigma + \frac{du}{dz}\, \gamma\, d\sigma \dotfill (11),$$

or

$$u + \frac{du}{dx}\frac{dx}{d\sigma}\, d\sigma + \frac{du}{dy}\frac{dy}{d\sigma}\, d\sigma + \frac{du}{dz}\frac{dz}{d\sigma}\, d\sigma \dotfill (12),$$

or

$$u + \frac{du}{d\sigma}\, d\sigma \dotfill (13).$$

But this represents the value of the component u of the velocity of the fluid itself at the same point, and the same thing may be proved of the other components.

Hence the velocity of the second point on the vortex line is identical with that of the fluid at that point. In other words, the vortex line swims along with the fluid, and is always formed of the same row of fluid particles. The vortex line is therefore no mere mathematical symbol, but has a physical existence continuous in time and space.

By differentiating equations (1) with respect to x, y, and z respectively, and adding the results, we obtain the equation—

$$\frac{da}{dx} + \frac{d\beta}{dy} + \frac{d\gamma}{dz} = 0 \dotfill (14).$$

This is an equation of the same form with (6), which expresses the condition of flow of a fluid of invariable density. Hence, if we imagine a fluid, quite independent of the original fluid, whose components of velocity are a, β, γ, this imaginary fluid will flow without altering its density.

Now, consider a closed curve in space, and let vortex lines be drawn in both directions from every point of this curve. These vortex lines will form a tubular surface, which is called a vortex tube or a vortex filament. Since the imaginary fluid flows along the vortex lines without change of density, the quantity which in unit of time flows through any section of the same vortex tube must be the same. Hence, at any section of a vortex tube the product of the area of the section into the mean velocity of rotation is the same. This quantity is called the *strength* of the vortex tube.

A vortex tube cannot begin or end within the fluid; for, if it did, the imaginary fluid, whose velocity components are a, β, γ, would be generated from nothing at the beginning of the tube, and reduced to nothing at the end of it. Hence, if the tube has a beginning and an end, they must lie on the surface of the fluid mass. If the fluid is infinite the vortex tube must be infinite, or else it must return into itself.

We have thus arrived at the following remarkable theorems relating to a finite vortex tube in an infinite fluid :—(1) It returns into itself, forming a closed ring. We may therefore describe it as a vortex *ring*. (2) It always consists of the same portion of the fluid. Hence its volume is invariable. (3) Its strength remains always the same. Hence the velocity of rotation at any section varies inversely as the area of that section, and that of any segment varies directly as the length of that segment. (4) No part of the fluid which is not originally in a state of rotational motion can ever enter into that state, and no part of the fluid whose motion is rotational can ever cease to move rotationally. (5) No vortex tube can ever pass through any other vortex tube, or through any of its own convolutions. Hence, if two vortex tubes are linked together, they can never be separated, and if a single vortex tube is knotted on itself, it can never become untied. (6) The motion at any instant of every part of the fluid, including the vortex rings themselves, may be accurately represented by conceiving an electric current to occupy the place of each vortex ring, the strength of the current being proportional to that of the ring. The magnetic force at any point of space will then represent in direction and magnitude the velocity of the fluid at the corresponding point of the fluid.

These properties of vortex rings suggested to Sir William Thomson* the possibility of founding on them a new form of the atomic theory. The conditions which must be satisfied by an atom are—permanence in magnitude, capability of internal motion or vibration, and a sufficient amount of possible characteristics to account for the difference between atoms of different kinds.

The small hard body imagined by Lucretius, and adopted by Newton, was invented for the express purpose of accounting for the permanence of the pro-

---

* "On Vortex Atoms," *Proc. Roy. Soc. Edin.*, 18th February, 1867.

perties of bodies. But it fails to account for the vibrations of a molecule as revealed by the spectroscope. We may indeed suppose the atom elastic, but this is to endow it with the very property for the explanation of which, as exhibited in aggregate bodies, the atomic constitution was originally assumed. The massive centres of force imagined by Boscovich may have more to recommend them to the mathematician, who has no scruple in supposing them to be invested with the power of attracting and repelling according to any law of the distance which it may please him to assign. Such centres of force are no doubt in their own nature indivisible, but then they are also, singly, incapable of vibration. To obtain vibrations we must imagine molecules consisting of many such centres, but, in so doing, the possibility of these centres being separated altogether is again introduced. Besides, it is in questionable scientific taste, after using atoms so freely to get rid of forces acting at sensible distances, to make the whole function of the atoms an action at insensible distances.

On the other hand, the vortex ring of Helmholtz, imagined as the true form of the atom by Thomson, satisfies more of the conditions than any atom hitherto imagined. In the first place, it is quantitatively permanent, as regards its volume and its strength,—two independent quantities. It is also qualitatively permanent as regards its degree of implication, whether "knottedness" on itself or "linkedness" with other vortex rings. At the same time, it is capable of infinite changes of form, and may execute vibrations of different periods, as we know that molecules do. And the number of essentially different implications of vortex rings may be very great without supposing the degree of implication of any of them very high.

But the greatest recommendation of this theory, from a philosophical point of view, is that its success in explaining phenomena does not depend on the ingenuity with which its contrivers "save appearances," by introducing first one hypothetical force and then another. When the vortex atom is once set in motion, all its properties are absolutely fixed and determined by the laws of motion of the primitive fluid, which are fully expressed in the fundamental equations. The disciple of Lucretius may cut and carve his solid atoms in the hope of getting them to combine into worlds; the follower of Boscovich may imagine new laws of force to meet the requirements of each new phenomenon; but he who dares to plant his feet in the path opened up by Helmholtz and Thomson has no such resources. His primitive fluid has no other properties than inertia, invariable density, and perfect mobility, and the method by which the

472   ATOM.

motion of this fluid is to be traced is pure mathematical analysis. The difficulties of this method are enormous, but the glory of surmounting them would be unique.

There seems to be little doubt that an encounter between two vortex atoms would be in its general character similar to those which we have already described. Indeed, the encounter between two smoke rings in air gives a very lively illustration of the elasticity of vortex rings.

But one of the first, if not the very first desideratum in a complete theory of matter is to explain—first, mass, and second, gravitation. To explain mass may seem an absurd achievement. We generally suppose that it is of the essence of matter to be the receptacle of momentum and energy, and even Thomson, in his definition of his primitive fluid, attributes to it the possession of mass. But according to Thomson, though the primitive fluid is the only true matter, yet that which we call matter is not the primitive fluid itself, but a mode of motion of that primitive fluid. It is the mode of motion which constitutes the vortex rings, and which furnishes us with examples of that permanence and continuity of existence which we are accustomed to attribute to matter itself. The primitive fluid, the only true matter, entirely eludes our perceptions when it is not endued with the mode of motion which converts certain portions of it into vortex rings, and thus renders it molecular.

In Thomson's theory, therefore, the mass of bodies requires explanation. We have to explain the inertia of what is only a mode of motion, and inertia is a property of matter, not of modes of motion. It is true that a vortex ring at any given instant has a definite momentum and a definite energy, but to shew that bodies built up of vortex rings would have such momentum and energy as we know them to have is, in the present state of the theory, a very difficult task.

It may seem hard to say of an infant theory that it is bound to explain gravitation. Since the time of Newton, the doctrine of gravitation has been admitted and expounded, till it has gradually acquired the character rather of an ultimate fact than of a fact to be explained.

It seems doubtful whether Lucretius considers gravitation to be an essential property of matter, as he seems to assert in the very remarkable lines—

> "Nam si tantundem est in lanæ glomere, quantum
> Corporis in plumbo est, tantundem pendere par est :
> Corporis officium est quoniam premere omnia deorsum."—*De Rerum Natura*, I. 361.

If this is the true opinion of Lucretius, and if the downward flight of the atoms arises, in his view, from their own gravity, it seems very doubtful whether he attributed the weight of sensible bodies to the impact of the atoms. The latter opinion is that of Le Sage, of Geneva, propounded in his *Lucrèce New-tonien*, and in his *Traité de Physique Mécanique*, published, along with a second treatise of his own, by Pierre Prevost, of Geneva, in 1818*. The theory of Le Sage is that the gravitation of bodies towards each other is caused by the impact of streams of atoms flying in all directions through space. These atoms he calls ultramundane corpuscules, because he conceives them to come in all directions from regions far beyond that part of the system of the world which is in any way known to us. He supposes each of them to be so small that a collision with another ultramundane corpuscule is an event of very rare occurrence. It is by striking against the molecules of gross matter that they discharge their function of drawing bodies towards each other. A body placed by itself in free space and exposed to the impacts of these corpuscules would be bandied about by them in all directions, but because, on the whole, it receives as many blows on one side as on another, it cannot thereby acquire any sensible velocity. But if there are two bodies in space, each of them will screen the other from a certain proportion of the corpuscular bombardment, so that a smaller number of corpuscules will strike either body on that side which is next the other body, while the number of corpuscules which strike it in other directions remains the same.

Each body will therefore be urged towards the other by the effect of the excess of the impacts it receives on the side furthest from the other. If we take account of the impacts of those corpuscules only which come directly from infinite space, and leave out of consideration those which have already struck mundane bodies, it is easy to calculate the result on the two bodies, supposing their dimensions small compared with the distance between them.

The force of attraction would vary directly as the product of the areas of the sections of the bodies taken normal to the distance and inversely as the square of the distance between them.

Now, the attraction of gravitation varies as the product of the *masses* of the bodies between which it acts, and inversely as the square of the distance between them. If, then, we can imagine a constitution of bodies such that

* See also *Constitution de la Matière*, &c., par le P. Leray, Paris, 1869.

the effective areas of the bodies are proportional to their masses, we shall make the two laws coincide. Here, then, seems to be a path leading towards an explanation of the law of gravitation, which, if it can be shewn to be in other respects consistent with facts, may turn out to be a royal road into the very arcana of science.

Le Sage himself shews that, in order to make the effective area of a body, in virtue of which it acts as a screen to the streams of ultramundane corpuscules, proportional to the mass of the body, whether the body be large or small, we must admit that the size of the solid atoms of the body is exceedingly small compared with the distances between them, so that a very small proportion of the corpuscules are stopped even by the densest and largest bodies. We may picture to ourselves the streams of corpuscules coming in every direction, like light from a uniformly illuminated sky. We may imagine a material body to consist of a congeries of atoms at considerable distances from each other, and we may represent this by a swarm of insects flying in the air. To an observer at a distance this swarm will be visible as a slight darkening of the sky in a certain quarter. This darkening will represent the action of the material body in stopping the flight of the corpuscules. Now, if the proportion of light stopped by the swarm is very small, two such swarms will stop nearly the same amount of light, whether they are in a line with the eye or not, but if one of them stops an appreciable proportion of light, there will not be so much left to be stopped by the other, and the effect of two swarms in a line with the eye will be less than the sum of the two effects separately.

Now, we know that the effect of the attraction of the sun and earth on the moon is not appreciably different when the moon is eclipsed than on other occasions when full moon occurs without an eclipse. This shews that the number of the corpuscules which are stopped by bodies of the size and mass of the earth, and even the sun, is very small compared with the number which pass straight through the earth or the sun without striking a single molecule. To the streams of corpuscules the earth and the sun are mere systems of atoms scattered in space, which present far more openings than obstacles to their rectilinear flight.

Such is the ingenious doctrine of Le Sage, by which he endeavours to explain universal gravitation. Let us try to form some estimate of this continual bombardment of ultramundane corpuscules which is being kept up on all sides of us.

We have seen that the sun stops but a very small fraction of the corpuscules which enter it. The earth, being a smaller body, stops a still smaller proportion of them. The proportion of those which are stopped by a small body, say a 1 lb. shot, must be smaller still in an enormous degree, because its thickness is exceedingly small compared with that of the earth.

Now, the weight of the ball, or its tendency towards the earth, is produced, according to this theory, by the excess of the impacts of the corpuscules which come from above over the impacts of those which come from below, and have passed through the earth. Either of these quantities is an exceedingly small fraction of the momentum of the whole number of corpuscules which pass through the ball in a second, and their difference is a small fraction of either, and yet it is equivalent to the weight of a pound. The velocity of the corpuscules must be enormously greater than that of any of the heavenly bodies, otherwise, as may easily be shewn, they would act as a resisting medium opposing the motion of the planets. Now, the energy of a moving system is half the product of its momentum into its velocity. Hence the energy of the corpuscules, which by their impacts on the ball during one second urge it towards the earth, must be a number of foot-pounds equal to the number of feet over which a corpuscule travels in a second, that is to say, not less than thousands of millions. But this is only a small fraction of the energy of all the impacts which the atoms of the ball receive from the innumerable streams of corpuscules which fall upon it in all directions.

Hence the rate at which the energy of the corpuscules is spent in order to maintain the gravitating property of a single pound, is at least millions of millions of foot-pounds per second.

What becomes of this enormous quantity of energy? If the corpuscules, after striking the atoms, fly off with a velocity equal to that which they had before, they will carry their energy away with them into the ultramundane regions. But if this be the case, then the corpuscules rebounding from the body in any given direction will be both in number and in velocity exactly equivalent to those which are prevented from proceeding in that direction by being deflected by the body, and it may be shewn that this will be the case whatever be the shape of the body, and however many bodies may be present in the field. Thus, the rebounding corpuscules exactly make up for those which are deflected by the body, and there will be no excess of the impacts on any other body in one direction or another.

60—2

The explanation of gravitation, therefore, falls to the ground if the corpuscules are like perfectly elastic spheres, and rebound with a velocity of separation equal to that of approach. If, on the other hand, they rebound with a smaller velocity, the effect of attraction between the bodies will no doubt be produced, but then we have to find what becomes of the energy which the molecules have brought with them but have not carried away.

If any appreciable fraction of this energy is communicated to the body in the form of heat, the amount of heat so generated would in a few seconds raise it, and in like manner the whole material universe, to a white heat.

It has been suggested by Sir W. Thomson that the corpuscules may be so constructed as to carry off their energy with them, provided that part of their kinetic energy is transformed, during impact, from energy of translation to energy of rotation or vibration. For this purpose the corpuscules must be material systems, not mere points. Thomson suggests that they are vortex atoms, which are set into a state of vibration at impact, and go off with a smaller velocity of translation, but in a state of violent vibration. He has also suggested the possibility of the vortex corpuscule regaining its swiftness and losing part of its vibratory agitation by communion with its kindred corpuscules in infinite space.

We have devoted more space to this theory than it seems to deserve, because it is ingenious, and because it is the only theory of the cause of gravitation which has been so far developed as to be capable of being attacked and defended. It does not appear to us that it can account for the temperature of bodies remaining moderate while their atoms are exposed to the bombardment. The temperature of bodies must tend to approach that at which the average kinetic energy of a molecule of the body would be equal to the average kinetic energy of an ultramundane corpuscule.

Now, suppose a plane surface to exist which stops *all* the corpuscules. The pressure on this plane will be $p = NMu^2$ where M is the mass of a corpuscule, N the number in unit of volume, and u its velocity normal to the plane. Now, we know that the very greatest pressure existing in the universe must be much less than the pressure p, which would be exerted against a body which stops all the corpuscules. We are also tolerably certain that N, the number of corpuscules which are at any one time within unit of volume, is small compared with the value of N for the molecules of ordinary bodies. Hence, Mu^2 must be enormous compared with the corresponding quantity for ordinary bodies, and it follows that the impact of the corpuscules would raise all bodies to an enormous temperature. We may also observe that according to this theory the habitable universe, which we are accustomed to regard as the scene of a magnificent illustration of the conservation of energy as the fundamental principle of all nature, is in reality maintained in working order only by an enormous expenditure of external power, which would be nothing less than ruinous if the supply were drawn from anywhere else than from the infinitude of space, and which, if the contrivances of the most eminent mathematicians should be found in any respect defective, might at any moment tear the whole universe atom from atom.

We must now leave these speculations about the nature of molecules and the cause of gravitation, and contemplate the material universe as made up of molecules. Every molecule, so far as we know, belongs to one of a definite number of species. The list of chemical elements may be taken as representing the known species which have been examined in the laboratories. Several of these have been discovered by means of the spectroscope, and more may yet remain to be discovered in the same way. The spectroscope has also been applied to analyse the light of the sun, the brighter stars, and some of the nebulæ and comets, and has shewn that the character of the light emitted by these bodies is similar in some cases to that emitted by terrestrial molecules, and in others to light from which the molecules have absorbed certain rays. In this way a considerable number of coincidences have been traced between the systems of lines belonging to particular terrestrial substances and corresponding lines in the spectra of the heavenly bodies.

The value of the evidence furnished by such coincidences may be estimated by considering the degree of accuracy with which one such coincidence may be observed. The interval between the two lines which form Fraunhofer's line D is about the five hundredth part of the interval between B and G on Kirchhoff's scale. A discordance between the positions of two lines amounting to the tenth part of this interval, that is to say, the five thousandth part of the length of the bright part of the spectrum, would be very perceptible in a spectroscope of moderate power. We may define the power of the spectroscope to be the number of times which the smallest measurable interval is contained in the length of the visible spectrum. Let us denote this by p. In the case we have supposed p will be about 5000.

If the spectrum of the sun contains n lines of a certain degree of inten-

sity, the probability that any one line of the spectrum of a gas will coincide
with one of these n lines is

$$1 - \left(1 - \frac{1}{p}\right)^n = \frac{n}{p}\left(1 - \frac{n-1}{2}\frac{1}{p} + \&c.\right),$$

and when p is large compared with n, this becomes nearly $\frac{n}{p}$. If there are
r lines in the spectrum of the gas, the probability that each and every one
shall coincide with a line in the solar spectrum is approximately $\frac{n^r}{p^r}$. Hence,
in the case of a gas whose spectrum contains several lines, we have to compare the results of two hypotheses. If a large amount of the gas exists in
the sun, we have the strongest reason for expecting to find all the r lines in
the solar spectrum. If it does not exist, the probability that r lines out of
the n observed lines shall coincide with the lines of the gas is exceedingly
small. If, then, we find all the r lines in their proper places in the solar
spectrum, we have very strong grounds for believing that the gas exists in the
sun. The probability that the gas exists in the sun is greatly strengthened
if the character of the lines as to relative intensity and breadth is found to
correspond in the two spectra.

The absence of one or more lines of the gas in the solar spectrum tends
of course to weaken the probability, but the amount to be deducted from the
probability must depend on what we know of the variation in the relative
intensity of the lines when the temperature and the pressure of the gas are
made to vary.

Coincidences observed, in the case of several terrestrial substances, with
several systems of lines in the spectra of the heavenly bodies, tend to increase
the evidence for the doctrine that terrestrial substances exist in the heavenly
bodies, while the discovery of particular lines in a celestial spectrum which do
not coincide with any line in a terrestrial spectrum does not much weaken
the general argument, but rather indicates either that a substance exists in the
heavenly body not yet detected by chemists on earth, or that the temperature
of the heavenly body is such that some substance, undecomposable by our
methods, is there split up into components unknown to us in their separate
state.

We are thus led to believe that in widely-separated parts of the visible
universe molecules exist of various kinds, the molecules of each kind having

their various periods of vibration either identical, or so nearly identical that our spectroscopes cannot distinguish them. We might argue from this that these molecules are alike in all other respects, as, for instance, in mass. But it is sufficient for our present purpose to observe that the same kind of molecule, say that of hydrogen, has the same set of periods of vibration, whether we procure the hydrogen from water, from coal, or from meteoric iron, and that light, having the same set of periods of vibration, comes to us from the sun, from Sirius, and from Arcturus.

The same kind of reasoning which led us to believe that hydrogen exists in the sun and stars, also leads us to believe that the molecules of hydrogen in all these bodies had a common origin. For a material system capable of vibration may have for its periods of vibration any set of values whatever. The probability, therefore, that two material systems, quite independent of each other, shall have, to the degree of accuracy of modern spectroscopic measurements, the same set of periods of vibration, is so very small that we are forced to believe that the two systems are not independent of each other. When, instead of two such systems, we have innumerable multitudes all having the same set of periods, the argument is immensely strengthened.

Admitting, then, that there is a real relation between any two molecules of hydrogen, let us consider what this relation may be.

We may conceive of a mutual action between one body and another tending to assimilate them. Two clocks, for instance, will keep time with each other if connected by a wooden rod, though they have different rates if they were disconnected. But even if the properties of a molecule were as capable of modification as those of a clock, there is no physical connection of a sufficient kind between Sirius and Arcturus.

There are also methods by which a large number of bodies differing from each other may be sorted into sets, so that those in each set more or less resemble each other. In the manufacture of small shot this is done by making the shot roll down an inclined plane. The largest specimens acquire the greatest velocities, and are projected farther than the smaller ones. In this way the various pellets, which differ both in size and in roundness, are sorted into different kinds, those belonging to each kind being nearly of the same size, and those which are not tolerably spherical being rejected altogether.

If the molecules were originally as various as these leaden pellets, and were afterwards sorted into kinds, we should have to account for the dis-

ATOM.

appearance of all the molecules which did not fall under one of the very limited number of kinds known to us; and to get rid of a number of indestructible bodies, exceeding by far the number of the molecules of all the recognised kinds, would be one of the severest labours ever proposed to a cosmogonist.

It is well known that living beings may be grouped into a certain number of species, defined with more or less precision, and that it is difficult or impossible to find a series of individuals forming the links of a continuous chain between one species and another. In the case of living beings, however, the generation of individuals is always going on, each individual differing more or less from its parents. Each individual during its whole life is undergoing modification, and it either survives and propagates its species, or dies early, accordingly as it is more or less adapted to the circumstances of its environment. Hence, it has been found possible to frame a theory of the distribution of organisms into species by means of generation, variation, and discriminative destruction. But a theory of evolution of this kind cannot be applied to the case of molecules, for the individual molecules neither are born nor die, they have neither parents nor offspring, and so far from being modified by their environment, we find that two molecules of the same kind, say of hydrogen, have the same properties, though one has been compounded with carbon and buried in the earth as coal for untold ages, while the other has been "occluded" in the iron of a meteorite, and after unknown wanderings in the heavens has at last fallen into the hands of some terrestrial chemist.

The process by which the molecules become distributed into distinct species is not one of which we know any instances going on at present, or of which we have as yet been able to form any mental representation. If we suppose that the molecules known to us are built up each of some moderate number of atoms, these atoms being all of them exactly alike, then we may attribute the limited number of molecular species to the limited number of ways in which the primitive atoms may be combined so as to form a permanent system.

But though this hypothesis gets rid of the difficulty of accounting for the independent origin of different species of molecules, it merely transfers the difficulty from the known molecules to the primitive atoms. How did the atoms come to be all alike in those properties which are in themselves capable of assuming any value?

If we adopt the theory of Boscovich, and assert that the primitive atom is a mere centre of force, having a certain definite mass, we may get over the difficulty about the equality of the mass of all atoms by laying it down as a doctrine which cannot be disproved by experiment, that mass is not a quantity capable of continuous increase or diminution, but that it is in its own nature discontinuous, like number, the atom being the unit, and all masses being multiples of that unit. We have no evidence that it is possible for the ratio of two masses to be an incommensurable quantity, for the incommensurable quantities in geometry are supposed to be traced out in a continuous medium. If matter is atomic, and therefore discontinuous, it is unfitted for the construction of perfect geometrical models, but in other respects it may fulfil its functions.

But even if we adopt a theory which makes the equality of the mass of different atoms a result depending on the nature of mass rather than on any quantitative adjustment, the correspondence of the periods of vibration of actual molecules is a fact of a different order.

We know that radiations exist having periods of vibration of every value between those corresponding to the limits of the visible spectrum, and probably far beyond these limits on both sides. The most powerful spectroscope can detect no gap or discontinuity in the spectrum of the light emitted by incandescent lime.

The period of vibration of a luminous particle is therefore a quantity which in itself is capable of assuming any one of a series of values, which, if not mathematically continuous, is such that consecutive observed values differ from each other by less than the ten thousandth part of either. There is, therefore, nothing in the nature of time itself to prevent the period of vibration of a molecule from assuming any one of many thousand different observable values. That which determines the period of any particular kind of vibration is the relation which subsists between the corresponding type of displacement and the force of restitution thereby called into play, a relation involving constants of space and time as well as of mass.

It is the equality of these space- and time-constants for all molecules of the same kind which we have next to consider. We have seen that the very different circumstances in which different molecules of the same kind have been placed have not, even in the course of many ages, produced any appreciable difference in the values of these constants. If, then, the various processes of nature to which these molecules have been subjected since the world began have not been able in all that time to produce any appreciable difference between the constants of one molecule and those of another, we are forced to conclude that it is not to the operation of any of these processes that the uniformity of the constants is due.

The formation of the molecule is therefore an event not belonging to that order of nature under which we live. It is an operation of a kind which is not, so far as we are aware, going on on earth or in the sun or the stars, either now or since these bodies began to be formed. It must be referred to the epoch, not of the formation of the earth or of the solar system, but of the establishment of the existing order of nature, and till not only these worlds and systems, but the very order of nature itself is dissolved, we have no reason to expect the occurrence of any operation of a similar kind.

In the present state of science, therefore, we have strong reasons for believing that in a molecule, or if not in a molecule, in one of its component atoms, we have something which has existed either from eternity or at least from times anterior to the existing order of nature. But besides this atom, there are immense numbers of other atoms of the same kind, and the constants of each of these atoms are incapable of adjustment by any process now in action. Each is physically independent of all the others.

Whether or not the conception of a multitude of beings existing from all eternity is in itself self-contradictory, the conception becomes palpably absurd when we attribute a relation of quantitative equality to all these beings. We are then forced to look beyond them to some common cause or common origin to explain why this singular relation of equality exists, rather than any one of the infinite number of possible relations of inequality.

Science is incompetent to reason upon the creation of matter itself out of nothing. We have reached the utmost limit of our thinking faculties when we have admitted that, because matter cannot be eternal and self-existent, it must have been created. It is only when we contemplate not matter in itself, but the form in which it actually exists, that our mind finds something on which it can lay hold.

That matter, as such, should have certain fundamental properties, that it should have a continuous existence in space and time, that all action should be between two portions of matter, and so on, are truths which may, for aught we know, be of the kind which metaphysicians call necessary. We may use our knowledge of such truths for purposes of deduction, but we have no data for speculating on their origin.

But the equality of the constants of the molecules is a fact of a very different order. It arises from a particular distribution of matter, a *collocation*, to use the expression of Dr Chalmers, of things which we have no difficulty in imagining to have been arranged otherwise. But many of the ordinary instances of collocation are adjustments of constants, which are not only arbitrary in their own nature, but in which variations actually occur; and when it is pointed out that these adjustments are beneficial to living beings, and are therefore instances of benevolent design, it is replied that those variations which are not conducive to the growth and multiplication of living beings tend to their destruction, and to the removal thereby of the evidence of any adjustment not beneficial.

The constitution of an atom, however, is such as to render it, so far as we can judge, independent of all the dangers arising from the struggle for existence. Plausible reasons may, no doubt, be assigned for believing that if the constants had varied from atom to atom through any sensible range, the bodies formed by aggregates of such atoms would not have been so well fitted for the construction of the world as the bodies which actually exist. But as we have no experience of bodies formed of such variable atoms this must remain a bare conjecture.

Atoms have been compared by Sir J. Herschel to manufactured articles, on account of their uniformity. The uniformity of manufactured articles may be traced to very different motives on the part of the manufacturer. In certain cases it is found to be less expensive as regards trouble, as well as cost, to make a great many objects exactly alike than to adapt each to its special requirements. Thus, shoes for soldiers are made in large numbers without any designed adaptation to the feet of particular men. In another class of cases the uniformity is intentional, and is designed to make the manufactured article more valuable. Thus, Whitworth's bolts are made in a certain number of sizes, so that if one bolt is lost, another may be got at once, and accurately fitted to its place. The identity of the arrangement of the words in the different copies of a document or book is a matter of great practical importance, and it is more perfectly secured by the process of printing than by that of manuscript copying.

In a third class not a part only but the whole of the value of the object arises from its exact conformity to a given standard. Weights and measures belong to this class, and the existence of many well-adjusted material standards

61—2

of weight and measure in any country furnishes evidence of the existence of a system of law regulating the transactions of the inhabitants, and enjoining in all professed measures a conformity to the national standard.

There are thus three kinds of usefulness in manufactured articles—cheapness, serviceableness, and quantitative accuracy. Which of these was present to the mind of Sir J. Herschel we cannot now positively affirm, but it was at least as likely to have been the last as the first, though it seems more probable that he meant to assert that a number of exactly similar things cannot be each of them eternal and self-existent, and must therefore have been made, and that he used the phrase "manufactured article" to suggest the idea of their being made in great numbers.

24. "On the Dynamical Evidence of the Molecular Constitution of Bodies"

Nature 11 (1875), 357–359, 374–377; *Journal of the Chemical Society* 13 (1875), 493–508; *Scientific Papers*, vol. 2, 418–438.

[From *Nature*, Vol. XI.]

LXXI.   *On the Dynamical Evidence of the Molecular Constitution of Bodies*.*

WHEN any phenomenon can be described as an example of some general principle which is applicable to other phenomena, that phenomenon is said to be explained. Explanations, however, are of very various orders, according to the degree of generality of the principle which is made use of. Thus the person who first observed the effect of throwing water into a fire would feel a certain amount of mental satisfaction when he found that the results were always similar, and that they did not depend on any temporary and capricious antipathy between the water and the fire. This is an explanation of the lowest order, in which the class to which the phenomenon is referred consists of other phenomena which can only be distinguished from it by the place and time of their occurrence, and the principle involved is the very general one that place and time are not among the conditions which determine natural processes. On the other hand, when a physical phenomenon can be completely described as a change in the configuration and motion of a material system, the dynamical explanation of that phenomenon is said to be complete. We cannot conceive any further explanation to be either necessary, desirable, or possible, for as soon as we know what is meant by the words configuration, motion, mass, and force, we see that the ideas which they represent are so elementary that they cannot be explained by means of anything else.

The phenomena studied by chemists are, for the most part, such as have not received a complete dynamical explanation.

Many diagrams and models of compound molecules have been constructed. These are the records of the efforts of chemists to imagine configurations of material systems by the geometrical relations of which chemical phenomena may

---

* A lecture delivered at the Chemical Society, Feb. 18, by Prof. Clerk-Maxwell, F.R.S.

be illustrated or explained. No chemist, however, professes to see in these diagrams anything more than symbolic representations of the various degrees of closeness with which the different components of the molecule are bound together.

In astronomy, on the other hand, the configurations and motions of the heavenly bodies are on such a scale that we can ascertain them by direct observation. Newton proved that the observed motions indicate a continual tendency of all bodies to approach each other, and the doctrine of universal gravitation which he established not only explains the observed motions of our system, but enables us to calculate the motions of a system in which the astronomical elements may have any values whatever.

When we pass from astronomical to electrical science, we can still observe the configuration and motion of electrified bodies, and thence, following the strict Newtonian path, deduce the forces with which they act on each other; but these forces are found to depend on the distribution of what we call electricity. To form what Gauss called a "construirbar Vorstellung" of the invisible process of electric action is the great desideratum in this part of science.

In attempting the extension of dynamical methods to the explanation of chemical phenomena, we have to form an idea of the configuration and motion of a number of material systems, each of which is so small that it cannot be directly observed. We have, in fact, to determine, from the observed external actions of an unseen piece of machinery, its internal construction.

The method which has been for the most part employed in conducting such inquiries is that of forming an hypothesis, and calculating what would happen if the hypothesis were true. If these results agree with the actual phenomena, the hypothesis is said to be verified, so long, at least, as some one else does not invent another hypothesis which agrees still better with the phenomena.

The reason why so many of our physical theories have been built up by the method of hypothesis is that the speculators have not been provided with methods and terms sufficiently general to express the results of their induction in its early stages. They were thus compelled either to leave their ideas vague and therefore useless, or to present them in a form the details of which could be supplied only by the illegitimate use of the imagination.

In the meantime the mathematicians, guided by that instinct which teaches them to store up for others the irrepressible secretions of their own minds, had developed with the utmost generality the dynamical theory of a material system.

Of all hypotheses as to the constitution of bodies, that is surely the most warrantable which assumes no more than that they are material systems, and proposes to deduce from the observed phenomena just as much information about the conditions and connections of the material system as these phenomena can legitimately furnish.

When examples of this method of physical speculation have been properly set forth and explained, we shall hear fewer complaints of the looseness of the reasoning of men of science, and the method of inductive philosophy will no longer be derided as mere guess-work.

It is only a small part of the theory of the constitution of bodies which has as yet been reduced to the form of accurate deductions from known facts. To conduct the operations of science in a perfectly legitimate manner, by means of methodised experiment and strict demonstration, requires a strategic skill which we must not look for, even among those to whom science is most indebted for original observations and fertile suggestions. It does not detract from the merit of the pioneers of science that their advances, being made on unknown ground, are often cut off, for a time, from that system of communications with an established base of operations, which is the only security for any permanent extension of science.

In studying the constitution of bodies we are forced from the very beginning to deal with particles which we cannot observe. For whatever may be our ultimate conclusions as to molecules and atoms, we have experimental proof that bodies may be divided into parts so small that we cannot perceive them.

Hence, if we are careful to remember that the word particle means a small part of a body, and that it does not involve any hypothesis as to the ultimate divisibility of matter, we may consider a body as made up of particles, and we may also assert that in bodies or parts of bodies of measurable dimensions, the number of particles is very great indeed.

The next thing required is a dynamical method of studying a material system consisting of an immense number of particles, by forming an idea of their configuration and motion, and of the forces acting on the particles, and deducing from the dynamical theory those phenomena which, though depending on the configuration and motion of the invisible particles, are capable of being observed in visible portions of the system.

The dynamical principles necessary for this study were developed by the fathers of dynamics, from Galileo and Newton to Lagrange and Laplace; but the special adaptation of these principles to molecular studies has been to a great extent the work of Prof. Clausius of Bonn, who has recently laid us under still deeper obligations by giving us, in addition to the results of his elaborate calculations, a new dynamical idea, by the aid of which I hope we shall be able to establish several important conclusions without much symbolical calculation.

The equation of Clausius, to which I must now call your attention, is of the following form:

$$pV = \tfrac{2}{3}T - \tfrac{2}{3}\Sigma\Sigma(\tfrac{1}{2}Rr).$$

Here p denotes the pressure of a fluid, and V the volume of the vessel which contains it. The product pV, in the case of gases at constant temperature, remains, as Boyle's Law tells us, nearly constant for different volumes and pressures. This member of the equation, therefore, is the product of two quantities, each of which can be directly measured.

The other member of the equation consists of two terms, the first depending on the motion of the particles, and the second on the forces with which they act on each other.

The quantity T is the kinetic energy of the system, or, in other words, that part of the energy which is due to the motion of the parts of the system.

The kinetic energy of a particle is half the product of its mass into the square of its velocity, and the kinetic energy of the system is the sum of the kinetic energy of its parts.

In the second term, r is the distance between any two particles, and R is the attraction between them. (If the force is a repulsion or a pressure, R is to be reckoned negative.)

The quantity $\tfrac{1}{2}Rr$, or half the product of the attraction into the distance across which the attraction is exerted, is defined by Clausius as the virial of the attraction. (In the case of pressure or repulsion, the virial is negative.)

The importance of this quantity was first pointed out by Clausius, who, by giving it a name, has greatly facilitated the application of his method to physical exposition.

The virial of the system is the sum of the virials belonging to every pair of particles which exist in the system. This is expressed by the double sum

$\Sigma\Sigma\left(\frac{1}{2}Rr\right)$, which indicates that the value of $\frac{1}{2}Rr$ is to be found for every pair of particles, and the results added together.

Clausius has established this equation by a very simple mathematical process, with which I need not trouble you, as we are not studying mathematics to-night. We may see, however, that it indicates two causes which may affect the pressure of the fluid on the vessel which contains it: the motion of its particles, which tends to increase the pressure, and the attraction of its particles, which tends to increase the pressure.

We may therefore attribute the pressure of a fluid either to the motion of its particles or to a repulsion between them.

Let us test by means of this result of Clausius the theory that the pressure of a gas arises entirely from the repulsion which one particle exerts on another, these particles, in the case of gas in a fixed vessel, being really at rest.

In this case the virial must be negative, and since by Boyle's Law the product of pressure and volume is constant, the virial also must be constant, whatever the volume, in the same quantity of gas at constant temperature. It follows from this that Rr, the product of the repulsion of two particles into the distance between them, must be constant, or in other words that the repulsion must be inversely as the distance, a law which Newton has shewn to be inadmissible in the case of molecular forces, as it would make the action of the distant parts of bodies greater than that of contiguous parts. In fact, we have only to observe that if Rr is constant, the virial of every pair of particles must be the same, so that the virial of the system must be proportional to the number of pairs of particles in the system—that is, to the square of the number of particles, or in other words to the square of the quantity of gas in the vessel. The pressure, according to this law, would not be the same in different vessels of gas at the same density, but would be greater in a large vessel than in a small one, and greater in the open air than in any ordinary vessel.

The pressure of a gas cannot therefore be explained by assuming repulsive forces between the particles.

It must therefore depend, in whole or in part, on the motion of the particles.

If we suppose the particles not to act on each other at all, there will be no virial, and the equation will be reduced to the form

$$Vp = \tfrac{2}{3}T.$$

If M is the mass of the whole quantity of gas, and c is the mean square of the velocity of a particle, we may write the equation—

$$Vp = \tfrac{1}{3}Mc^2$$

or in words, the product of the volume and the pressure is one-third of the mass multiplied by the mean square of the velocity. If we now assume, what we shall afterwards prove by an independent process, that the mean square of the velocity depends only on the temperature, this equation exactly represents Boyle's Law.

But we know that most ordinary gases deviate from Boyle's Law, especially at low temperatures and great densities. Let us see whether the hypothesis of forces between the particles, which we rejected when brought forward as the sole cause of gaseous pressure, may not be consistent with experiment when considered as the cause of this deviation from Boyle's Law.

When a gas is in an extremely rarefied condition, the number of particles within a given distance of any one particle will be proportional to the density of the gas. Hence the virial arising from the action of one particle on the rest will vary as the density, and the whole virial in unit of volume will vary as the square of the density.

Calling the density ρ, and dividing the equation by V, we get—

$$p = \tfrac{1}{3}\rho c^2 - \tfrac{2}{3}A\rho^2$$

where A is a quantity which is nearly constant for small densities.

Now, the experiments of Regnault shew that in most gases, as the density increases the pressure falls below the value calculated by Boyle's Law. Hence the virial must be positive; that is to say, the mutual action of the particles must be in the main attractive, and the effect of this action in diminishing the pressure must be at first very nearly as the square of the density.

On the other hand, when the pressure is made still greater the substance at length reaches a state in which an enormous increase of pressure produces but a very small increase of density. This indicates that the virial is now negative, or, in other words, the action between the particles is now, in the main, repulsive. We may therefore conclude that the action between two particles at any sensible distance is quite insensible. As the particles approach each other the action first shews itself as an attraction, which reaches a maximum, then diminishes, and at length becomes a repulsion so great that no attainable force can reduce the distance of the particles to zero.

The relation between pressure and density arising from such an action between the particles is of this kind.

As the density increases from zero, the pressure at first depends almost entirely on the motion of the particles, and therefore varies almost exactly as the pressure, according to Boyle's Law. As the density continues to increase, the effect of the mutual attraction of the particles becomes sensible, and this causes the rise of pressure to be less than that given by Boyle's Law. If the temperature is low, the effect of attraction may become so large in proportion to the effect of motion that the pressure, instead of always rising as the density increases, may reach a maximum, and then begin to diminish.

At length, however, as the average distance of the particles is still further diminished, the effect of repulsion will prevail over that of attraction, and the pressure will increase so as not only to be greater than that given by Boyle's Law, but so that an exceedingly small increase of density will produce an enormous increase of pressure.

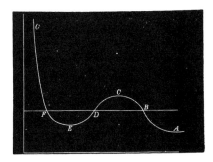

Hence the relation between pressure and volume may be represented by the curve *ABCDEFG*, where the horizontal ordinate represents the volume, and the vertical ordinate represents the pressure.

As the volume diminishes, the pressure increases up to the point *C*, then diminishes to the point *E*, and finally increases without limit as the volume diminishes.

We have hitherto supposed the experiment to be conducted in such a way that the density is the same in every part of the medium. This, however, is impossible in practice, as the only condition we can impose on the medium from without is that the whole of the medium shall be contained within a certain vessel. Hence, if it is possible for the medium to arrange itself so that part has one density and part another, we cannot prevent it from doing so.

Now the points B and F represent two states of the medium in which the pressure is the same but the density very different. The whole of the medium may pass from the state B to the state F, not through the intermediate states CDE, but by small successive portions passing directly from the state B to the state F. In this way the successive states of the medium as a whole will be represented by points on the straight line BF, the point B representing it when entirely in the rarefied state, and F representing it when entirely condensed. This is what takes place when a gas or vapour is liquefied.

Under ordinary circumstances, therefore, the relation between pressure and volume at constant temperature is represented by the broken line $ABFG$. If, however, the medium when liquefied is carefully kept from contact with vapour, it may be preserved in the liquid condition and brought into states represented by the portion of the curve between F and E. It is also possible that methods may be devised whereby the vapour may be prevented from condensing, and brought into states represented by points in BC.

The portion of the hypothetical curve from C to E represents states which are essentially unstable, and which cannot therefore be realised.

Now let us suppose the medium to pass from B to F along the hypothetical curve $BCDEF$ in a state always homogeneous, and to return along the straight line FB in the form of a mixture of liquid and vapour. Since the temperature has been constant throughout, no heat can have been transformed into work. Now the heat transformed into work is represented by the excess of the area FDE over BCD. Hence the condition which determines the maximum pressure of the vapour at given temperature is that the line BF cuts off equal areas from the curve above and below.

The higher the temperature, the greater the part of the pressure which depends on motion, as compared with that which depends on forces between the particles. Hence, as the temperature rises, the dip in the curve becomes less marked, and at a certain temperature the curve, instead of dipping, merely becomes horizontal at a certain point, and then slopes upward as before. This point is called the critical point. It has been determined for carbonic acid by the masterly researches of Andrews. It corresponds to a definite temperature, pressure and density.

At higher temperatures the curve slopes upwards throughout, and there is nothing corresponding to liquefaction in passing from the rarest to the densest state.

The molecular theory of the continuity of the liquid and gaseous states forms the subject of an exceedingly ingenious thesis by Mr Johannes Diderick van der Waals*, a graduate of Leyden. There are certain points in which I think he has fallen into mathematical errors, and his final result is certainly not a complete expression for the interaction of real molecules, but his attack on this difficult question is so able and so brave, that it cannot fail to give a notable impulse to molecular science. It has certainly directed the attention of more than one inquirer to the study of the Low-Dutch language in which it is written.

The purely thermodynamical relations of the different states of matter do not belong to our subject, as they are independent of particular theories about molecules. I must not, however, omit to mention a most important American contribution to this part of thermodynamics by Prof. Willard Gibbs†, of Yale College, U.S., who has given us a remarkably simple and thoroughly satisfactory method of representing the relations of the different states of matter by means of a model. By means of this model, problems which had long resisted the efforts of myself and others may be solved at once.

Let us now return to the case of a highly rarefied gas in which the pressure is due entirely to the motion of its particles. It is easy to calculate the mean square of the velocity of the particles from the equation of Clausius, since the volume, the pressure, and the mass are all measurable quantities. Supposing the velocity of every particle the same, the velocity of a molecule of oxygen would be 461 metres per second, of nitrogen 492, and of hydrogen 1844, at the temperature of 0° C.

The explanation of the pressure of a gas on the vessel which contains it by the impact of its particles on the surface of the vessel has been suggested at various times by various writers. The fact, however, that gases are not observed to disseminate themselves through the atmosphere with velocities at all approaching those just mentioned, remained unexplained, till Clausius, by a

---

* *Over de continuiteit van den gas en vloeistoftoestand.* (Leiden : A. W. Sijthoff, 1873.)

† "A method of geometrical representation of the thermodynamic properties of substances by means of surfaces." *Transactions of the Connecticut Academy of Arts and Sciences*, Vol. II. Part 2.

thorough study of the motions of an immense number of particles, developed the methods and ideas of modern molecular science.

To him we are indebted for the conception of the mean length of the path of a molecule of a gas between its successive encounters with other molecules. As soon as it was seen how each molecule, after describing an exceedingly short path, encounters another, and then describes a new path in a quite different direction, it became evident that the rate of diffusion of gases depends not merely on the velocity of the molecules, but on the distance they travel between each encounter.

I shall have more to say about the special contributions of Clausius to molecular science. The main fact, however, is, that he opened up a new field of mathematical physics by shewing how to deal mathematically with moving systems of innumerable molecules.

Clausius, in his earlier investigations at least, did not attempt to determine whether the velocities of all the molecules of the same gas are equal, or whether, if unequal, there is any law according to which they are distributed. He therefore, as a first hypothesis, seems to have assumed that the velocities are equal. But it is easy to see that if encounters take place among a great number of molecules, their velocities, even if originally equal, will become unequal, for, except under conditions which can be only rarely satisfied, two molecules having equal velocities before their encounter will acquire unequal velocities after the encounter. By distributing the molecules into groups according to their velocities, we may substitute for the impossible task of following every individual molecule through all its encounters, that of registering the increase or decrease of the number of molecules in the different groups.

By following this method, which is the only one available either experimentally or mathematically, we pass from the methods of strict dynamics to those of statistics and probability.

When an encounter takes place between two molecules, they are transferred from one pair of groups to another, but by the time that a great many encounters have taken place, the number which enter each group is, on an average, neither more nor less than the number which leave it during the same time. When the system has reached this state, the numbers in each group must be distributed according to some definite law.

As soon as I became acquainted with the investigations of Clausius, I endeavoured to ascertain this law.

The result which I published in 1860 has since been subjected to a more strict investigation by Dr Ludwig Boltzmann, who has also applied his method to the study of the motion of compound molecules. The mathematical investigation, though, like all parts of the science of probabilities and statistics, it is somewhat difficult, does not appear faulty. On the physical side, however, it leads to consequences, some of which, being manifestly true, seem to indicate that the hypotheses are well chosen, while others seem to be so irreconcilable with known experimental results, that we are compelled to admit that something essential to the complete statement of the physical theory of molecular encounters must have hitherto escaped us.

I must now attempt to give you some account of the present state of these investigations, without, however, entering into their mathematical demonstration.

I must begin by stating the general law of the distribution of velocity among molecules of the same kind.

If we take a fixed point in this diagram and draw from this point a line representing in direction and magnitude the velocity of a molecule, and make a dot at the end of the line, the position of the dot will indicate the state of motion of the molecule.

If we do the same for all the other molecules, the diagram will be dotted all over, the dots being more numerous in certain places than in others.

The law of distribution of the dots may be shewn to be the same as that which prevails among errors of observation or of adjustment.

The dots in the diagram before you may be taken to represent the velocities of molecules, the different observations of the position of the same star, or the bullet-holes round the bull's-eye of a target, all of which are distributed in the same manner.

The velocities of the molecules have values ranging from zero to infinity, so that in speaking of the average velocity of the molecules we must define what we mean.

The most useful quantity for purposes of comparison and calculation is called the "velocity of mean square." It is that velocity whose square is the average of the squares of the velocities of all the molecules.

This is the velocity given above as calculated from the properties of different gases. A molecule moving with the velocity of mean square has a kinetic energy equal to the average kinetic energy of all the molecules in the medium, and if a single mass equal to that of the whole quantity of gas were moving

with this velocity, it would have the same kinetic energy as the gas actually
has, only it would be in a visible form and directly available for doing work.

If in the same vessel there are different kinds of molecules, some of greater
mass than others, it appears from this investigation that their velocities will be
so distributed that the average kinetic energy of a molecule will be the same,
whether its mass be great or small.

Diagram of Velocities.

Here we have perhaps the most important application which has yet been
made of dynamical methods to chemical science.   For, suppose that we have
two gases in the same vessel.   The ultimate distribution of agitation among the
molecules is such that the average kinetic energy of an individual molecule is
the same in either gas.   This ultimate state is also, as we know, a state of
equal temperature.   Hence the condition that two gases shall have the same
temperature is that the average kinetic energy of a single molecule shall be the
same in the two gases.

Now, we have already shewn that the pressure of a gas is two-thirds of the kinetic energy in unit of volume. Hence, if the pressure as well as the temperature be the same in the two gases, the kinetic energy per unit of volume is the same, as well as the kinetic energy per molecule. There must, therefore, be the same number of molecules in unit of volume in the two gases.

This result coincides with the law of equivalent volumes established by Gay Lussac. This law, however, has hitherto rested on purely chemical evidence, the relative masses of the molecules of different substances having been deduced from the proportions in which the substances enter into chemical combination. It is now demonstrated on dynamical principles. The molecule is defined as that small portion of the substance which moves as one lump during the motion of agitation. This is a purely dynamical definition, independent of any experiments on combination.

The density of a gaseous medium, at standard temperature and pressure, is proportional to the mass of one of its molecules ,as thus defined.

We have thus a safe method of estimating the relative masses of molecules of different substances when in the gaseous state. This method is more to be depended on than those founded on electrolysis or on specific heat, because our knowledge of the conditions of the motion of agitation is more complete than our knowledge of electrolysis, or of the internal motions of the constituents of a molecule.

I must now say something about these internal motions, because the greatest difficulty which the kinetic theory of gases has yet encountered belongs to this part of the subject.

We have hitherto considered only the motion of the centre of mass of the molecule. We have now to consider the motion of the constituents of the molecule relative to the centre of mass.

If we suppose that the constituents of a molecule are atoms, and that each atom is what is called a material point, then each atom may move in three different and independent ways, corresponding to the three dimensions of space, so that the number of variables required to determine the position and configuration of all the atoms of the molecule is three times the number of atoms.

It is not essential, however, to the mathematical investigation to assume that the molecule is made up of atoms. All that is assumed is that the position and configuration of the molecule can be completely expressed by a certain number of variables.

Let us call this number n.

Of these variables, three are required to determine the position of the centre of mass of the molecule, and the remaining $n-3$ to determine its configuration relative to its centre of mass.

To each of the n variables corresponds a different kind of motion.

The motion of translation of the centre of mass has three components.

The motions of the parts relative to the centre of mass have $n-3$ components.

The kinetic energy of the molecule may be regarded as made up of two parts—that of the mass of the molecule supposed to be concentrated at its centre of mass, and that of the motions of the parts relative to the centre of mass. The first part is called the energy of translation, the second that of rotation and vibration. The sum of these is the whole energy of motion of the molecule.

The pressure of the gas depends, as we have seen, on the energy of translation alone. The specific heat depends on the rate at which the whole energy, kinetic and potential, increases as the temperature rises.

Clausius had long ago pointed out that the ratio of the increment of the whole energy to that of the energy of translation may be determined if we know by experiment the ratio of the specific heat at constant pressure to that at constant volume.

He did not however, attempt to determine *à priori* the ratio of the two parts of the energy, though he suggested, as an extremely probable hypothesis, that the average values of the two parts of the energy in a given substance always adjust themselves to the same ratio. He left the numerical value of this ratio to be determined by experiment.

In 1860 I investigated the ratio of the two parts of the energy on the hypothesis that the molecules are elastic bodies of invariable form. I found, to my great surprise, that whatever be the shape of the molecules, provided they are not perfectly smooth and spherical, the ratio of the two parts of the energy must be always the same, the two parts being in fact equal.

This result is confirmed by the researches of Boltzmann, who has worked out the general case of a molecule having n variables.

He finds that while the average energy of translation is the same for molecules of all kinds at the same temperature, the whole energy of motion is to the energy of translation as n to 3.

For a rigid body $n = 6$, which makes the whole energy of motion twice the energy of translation.

But if the molecule is capable of changing its form under the action of impressed forces, it must be capable of storing up potential energy, and if the forces are such as to ensure the stability of the molecule, the average potential energy will increase when the average energy of internal motion increases.

Hence, as the temperature rises, the increments of the energy of translation, the energy of internal motion, and the potential energy are as 3, $(n-3)$, and e respectively, where e is a positive quantity of unknown value depending on the law of the force which binds together the constituents of the molecule.

When the volume of the substance is maintained constant, the effect of the application of heat is to increase the whole energy. We thus find for the specific heat of a gas at constant volume—

$$\frac{1}{2J} \frac{p_0 V_0}{273^\circ} (n + e).$$

where p_0 and V_0 are the pressure and volume of unit of mass at zero centigrade, or 273° absolute temperature, and J is the dynamic equivalent of heat. The specific heat at constant pressure is

$$\frac{1}{2J} \frac{p_0 V_0}{273^\circ} (n + 2 + e).$$

In gases whose molecules have the same degree of complexity the value of n is the same, and that of e *may* be the same.

If this is the case, the specific heat is inversely as the specific gravity, according to the law of Dulong and Petit, which is, to a certain degree of approximation, verified by experiment.

But if we take the actual values of the specific heat as found by Regnault and compare them with this formula, we find that $n + e$ for air and several other gases cannot be more than 4·9. For carbonic acid and steam it is greater. We obtain the same result if we compare the ratio of the calculated specific heats

$$\frac{2 + n + e}{n + e}$$

with the ratio as determined by experiment for various gases, namely, 1·408.

And here we are brought face to face with the greatest difficulty which the molecular theory has yet encountered, namely, the interpretation of the equation $n + e = 4\cdot9$.

If we suppose that the molecules are atoms—mere material points, incapable of rotatory energy or internal motion—then n is 3 and e is zero, and the ratio of the specific heats is $1\cdot66$, which is too great for any real gas.

But we learn from the spectroscope that a molecule can execute vibrations of constant period. It cannot therefore be a mere material point, but a system capable of changing its form. Such a system cannot have less than six variables. This would make the greatest value of the ratio of the specific heats $1\cdot33$, which is too small for hydrogen, oxygen, nitrogen, carbonic oxide, nitrous oxide, and hydrochloric acid.

But the spectroscope tells us that some molecules can execute a great many different kinds of vibrations. They must therefore be systems of a very considerable degree of complexity, having far more than six variables. Now, every additional variable introduces an additional amount of capacity for internal motion without affecting the external pressure. Every additional variable, therefore, increases the specific heat, whether reckoned at constant pressure or at constant volume.

So does any capacity which the molecule may have for storing up energy in the potential form. But the calculated specific heat is already too great when we suppose the molecule to consist of two atoms only. Hence every additional degree of complexity which we attribute to the molecule can only increase the difficulty of reconciling the observed with the calculated value of the specific heat.

I have now put before you what I consider to be the greatest difficulty yet encountered by the molecular theory. Boltzmann has suggested that we are to look for the explanation in the mutual action between the molecules and the ætherial medium which surrounds them. I am afraid, however, that if we call in the help of this medium, we shall only increase the calculated specific heat, which is already too great.

The theorem of Boltzmann may be applied not only to determine the distribution of velocity among the molecules, but to determine the distribution of the molecules themselves in a region in which they are acted on by external forces. It tells us that the density of distribution of the molecules at a point where the potential energy of a molecule is ψ, is proportional to $e^{-\frac{\psi}{\kappa\theta}}$ where θ is the absolute temperature, and κ is a constant for all gases. It follows from this, that if several gases in the same vessel are subject to an external force like that of gravity, the distribution of each gas is the same as if no other gas were present. This result agrees with the law assumed by Dalton, according to which the atmosphere may be regarded as consisting of two independent atmospheres, one of oxygen, and the other of nitrogen ; the density of the oxygen diminishing faster than that of the nitrogen, as we ascend.

This would be the case if the atmosphere were never disturbed, but the effect of winds is to mix up the atmosphere and to render its composition more uniform than it would be if left at rest.

Another consequence of Boltzmann's theorem is, that the temperature tends to become equal throughout a vertical column of gas at rest.

In the case of the atmosphere, the effect of wind is to cause the temperature to vary as that of a mass of air would do if it were carried vertically upwards, expanding and cooling as it ascends.

But besides these results, which I had already obtained by a less elegant method and published in 1866, Boltzmann's theorem seems to open up a path into a region more purely chemical. For if the gas consists of a number of similar systems, each of which may assume different states having different amounts of energy, the theorem tells us that the number in each state is proportional to $e^{-\frac{\psi}{\kappa\theta}}$ where ψ is the energy, θ the absolute temperature, and κ a constant.

It is easy to see that this result ought to be applied to the theory of the states of combination which occur in a mixture of different substances. But as it is only during the present week that I have made any attempt to do so, I shall not trouble you with my crude calculations.

I have confined my remarks to a very small part of the field of molecular investigation. I have said nothing about the molecular theory of the diffusion of matter, motion, and energy, for though the results, especially in the diffusion of matter and the transpiration of fluids are of great interest to many chemists, and though from them we deduce important molecular data, they belong to a part of our study the data of which, depending on the conditions of the encounter of two molecules, are necessarily very hypothetical. I have thought it better to exhibit the evidence that the parts of fluids are in motion, and to describe the manner in which that motion is distributed among molecules of different masses.

To shew that all the molecules of the same substance are equal in mass, we may refer to the methods of dialysis introduced by Graham, by which two gases of different densities may be separated by percolation through a porous plug.

If in a single gas there were molecules of different masses, the same process of dialysis, repeated a sufficient number of times, would furnish us with two portions of the gas, in one of which the average mass of the molecules would be greater than in the other. The density and the combining weight of these two portions would be different. Now, it may be said that no one has carried out this experiment in a sufficiently elaborate manner for every chemical substance. But the processes of nature are continually carrying out experiments of the same kind; and if there were molecules of the same substance nearly alike, but differing slightly in mass, the greater molecules would be selected in preference to form one compound, and the smaller to form another. But hydrogen is of the same density, whether we obtain it from water or from a hydrocarbon, so that neither oxygen nor carbon can find in hydrogen molecules greater or smaller than the average.

The estimates which have been made of the actual size of molecules are founded on a comparison of the volumes of bodies in the liquid or solid state, with their volumes in the gaseous state. In the study of molecular volumes we meet with many difficulties, but at the same time there are a sufficient number of consistent results to make the study a hopeful one.

The theory of the possible vibrations of a molecule has not yet been studied as it ought, with the help of a continual comparison between the dynamical theory and the evidence of the spectroscope. An intelligent student, armed with the calculus and the spectroscope, can hardly fail to discover some important fact about the internal constitution of a molecule.

The observed transparency of gases may seem hardly consistent with the results of molecular investigations.

A model of the molecules of a gas consisting of marbles scattered at distances bearing the proper proportion to their diameters, would allow very little light to penetrate through a hundred feet.

But if we remember the small size of the molecules compared with the length of a wave of light, we may apply certain theoretical investigations of Lord Rayleigh's about the mutual action between waves and small spheres, which shew that the transparency of the atmosphere, if affected only by the presence of molecules, would be far greater than we have any reason to believe it to be.

A much more difficult investigation, which has hardly yet been attempted, relates to the electric properties of gases. No one has yet explained why dense gases are such good insulators, and why, when rarefied or heated, they permit the discharge of electricity, whereas a perfect vacuum is the best of all insulators.

It is true that the diffusion of molecules goes on faster in a rarefied gas, because the mean path of a molecule is inversely as the density. But the electrical difference between dense and rare gas appears to be too great to be accounted for in this way.

But while I think it right to point out the hitherto unconquered difficulties of this molecular theory, I must not forget to remind you of the numerous facts which it satisfactorily explains. We have already mentioned the gaseous laws, as they are called, which express the relations between volume, pressure, and temperature, and Gay Lussac's very important law of equivalent volumes. The explanation of these may be regarded as complete. The law of molecular specific heats is less accurately verified by experiment, and its full explanation depends on a more perfect knowledge of the internal structure of a molecule than we as yet possess.

But the most important result of these inquiries is a more distinct conception of thermal phenomena. In the first place, the temperature of the medium is measured by the average kinetic energy of translation of a single molecule of the medium. In two media placed in thermal communication, the temperature as thus measured tends to become equal.

In the next place, we learn how to distinguish that kind of motion which we call heat from other kinds of motion. The peculiarity of the motion called heat is that it is perfectly irregular; that is to say, that the direction and magnitude of the velocity of a molecule at a given time cannot be expressed as depending on the present position of the molecule and the time.

In the visible motion of a body, on the other hand, the velocity of the centre of mass of all the molecules in any visible portion of the body is the observed velocity of that portion, though the molecules may have also an irregular depending agitation on account of the body being hot.

In the transmission of sound, too, the different portions of the body have a motion which is generally too minute and too rapidly alternating to be directly observed. But in the motion which constitutes the physical phenomenon of sound, the velocity of each portion of the medium at any time can be expressed as depending on the position and the time elapsed ; so that the motion of a medium during the passage of a sound-wave is regular, and must be distinguished from that which we call heat.

If, however, the sound-wave, instead of travelling onwards in an orderly manner and leaving the medium behind it at rest, meets with resistances which fritter away its motion into irregular agitations, this irregular molecular motion becomes no longer capable of being propagated swiftly in one direction as sound, but lingers in the medium in the form of heat till it is communicated to colder parts of the medium by the slow process of conduction.

The motion which we call light, though still more minute and rapidly alternating than that of sound, is, like that of sound, perfectly regular, and therefore is not heat. What was formerly called Radiant Heat is a phenomenon physically identical with light.

When the radiation arrives at a certain portion of the medium, it enters it and passes through it, emerging at the other side. As long as the medium is engaged in transmitting the radiation it is in a certain state of motion, but as soon as the radiation has passed through it, the medium returns to its former state, the motion being entirely transferred to a new portion of the medium.

Now, the motion which we call heat can never of itself pass from one body to another unless the first body is, during the whole process, hotter than the second. The motion of radiation, therefore, which passes entirely out of one portion of the medium and enters another, cannot be properly called heat.

We may apply the molecular theory of gases to test those hypotheses about the luminiferous æther which assume it to consist of atoms or molecules.

Those who have ventured to describe the constitution of the luminiferous æther have sometimes assumed it to consist of atoms or molecules.

The application of the molecular theory to such hypotheses leads to rather startling results.

In the first place, a molecular æther would be neither more nor less than a gas. We may, if we please, assume that its molecules are each of them equal to the thousandth or the millionth part of a molecule of hydrogen, and that they can traverse freely the interspaces of all ordinary molecules. But, as we have seen, an equilibrium will establish itself between the agitation of the ordinary molecules and those of the æther. In other words, the æther and the bodies in it will tend to equality of temperature, and the æther will be subject to the ordinary gaseous laws as to pressure and temperature.

Among other properties of a gas, it will have that established by Dulong and Petit, so that the capacity for heat of unit of volume of the æther must be equal to that of unit of volume of any ordinary gas at the same pressure. Its presence, therefore, could not fail to be detected in our experiments on specific heat, and we may therefore assert that the constitution of the æther is not molecular.

25. Postcard from Maxwell to Peter Guthrie Tait, March 19, 1875

Tait Letters at Cambridge University.

Glenlair
Dalbeattie
19 March 1875
OT'[a]

The sooner A. C. B.[b] d____ds my notions the better as they require
d____d'ing. Of course it is something more than about the comparative
claims of Gay Lussac, Avogadro, &c. Also—distinguish between the mire of
Macademic molecules and the concrete or dust of atoms.[c] As for the aether,
when did I discuss the molecules thereof? Except perhaps in Phil. Mag.
1861–2 on molecular vortices,[d] and these were *very* hypothetical. Of course
Thomson's Ur-fluid is not molecular for molecules have to be created out of
it each by its own peculiar spin.[e]

If this be creation we can form a tolerably distinct idea of the process.
That is we can conceive a thing at rest and then in motion and the setting in
motion may occupy any required time. So, speaking kinematically we know
the whole process.

That the process, as thus conceived, cannot be done in accordance with
certain prevalent laws of force may be unfavourable to our expectations of
its occurrence within the present year but it by no means interferes with the
distinctness of our conception of what happens.

I have looked up the titles of the officers of the R. S. E.[f] and find that the
personage referred to is the Keeper. A list of defaulters is kept I believe and
also a list of persons who have received their Transactions and Proceedings.
To this M[r] Cockburn may refer, and if he will send me the publications due
to me and inform me if my name is written in the list and is not blackened
with ink, he will greatly oblige me.

Address of books
Glenlair
Care of Corsock Postman
Dalbeattie

Yours

$$\frac{dp}{dt}$$

Here is a transformation of rectangular coords, which I invented yesterday
in the railway. No doubt it is old and it is very Cartesian.

$$x'(1 + \alpha^2 + \beta^2 + \gamma^2) = x(1 + \alpha^2 - \beta^2 - \gamma^2) + 2y(\alpha\beta - \gamma) + 2z(\alpha\gamma + \beta)$$

$$y'(1 + \alpha^2 + \beta^2 + \gamma^2) = 2x(\beta\alpha + \gamma) + y(1 + \beta^2 - \gamma^2 - \alpha^2) + 2z(\beta\gamma - \alpha)$$

$$z'(1 + \alpha^2 + \beta^2 + \gamma^2) = 2x(\gamma\alpha - \beta) + 2y(\gamma\beta + \alpha) + z(1 + \gamma^2 - \alpha^2 - \beta^2)$$

Here $\alpha\beta\gamma$ are any numerical quantities. The direction of the vector whose components are α β γ is that of the axis of rotation and the length of the vector is $\tan\frac{1}{2}\varphi$, φ being the angle of rotation.

I have not had time to compare this with your quaternion method of expressing the angular position of a body.

a. T′ is Tait. This derives from Tait's association with William Thomson in writing *A Treatise on Natural Philosophy*. The book became known as "Thomson and Tait," then T and T′ or T & T′. Thomson became T and Tait T′ in Maxwell's and their later correspondence.

b. Possibly Alexander Crum Brown, Tait's brother-in-law.

c. "Macadam" roads were introduced around 1820 by John McAdam (1756–1836).

d. Maxwell, "On Physical Lines of Force," *Phil. Mag.* [4] 21 (1861), 161–175, 281–291; 23 (1862), 12–24; *Scientific Papers*, vol. 1, 451–513 (esp. 489–513). On the question of whether the ether is molecular, see document II-14 and note b to it.

e. William Thomson, "On Vortex Atoms," *Phil. Mag.* [4] 34 (1867), 15–24, and later papers on this subject.

f. Royal Society of Edinburgh.

26. Letter from Maxwell to George Gabriel Stokes, September 25, 1875

Memoir and Scientific Correspondence of the Late Sir George Gabriel Stokes, selected and arranged by Joseph Larmor (Cambridge: Cambridge University Press, 1907; reprint, New York: Johnson Reprint Corp., 1971), vol. 2, 36–37.

Glenlair, Dalbeattie
25 Sept. 1875

My dear Stokes,

I quite concur with you that Mr Gore's paper should now be printed *in extenso* in the proceedings.

Mrs Maxwell is keeping pretty well and has been able to get out on her pony.

Would you agree with the following statements about elasticity and viscosity, as related to a molecular theory?

When after being strained the groups of molecules in a body tend to return to the same stable configuration as when unstrained, the body is elastic. If at corresponding stages of the straining and restitution the stresses are the same, the body is perfectly elastic. If the stress during the restitution is less than that during deformation the elasticity is imperfect.

If, when the strain exceeds a certain value, complete restitution does not occur, this value is called the limiting strain and the stress the limiting stress of elasticity.

If, when the stress is removed, the body does not completely return to its original form the body is said to be plastic or viscous. A viscous body, if kept strained long enough, loses all tendency to change its form. A plastic body does not permanently change its form unless the stress exceeds a certain value, and if kept strained it never loses all stress.

THEORY

In an elastic solid the thermal agitation of the molecules does not carry them beyond the limits of oscillating about stable configurations.

But as the thermal agitation increases so many molecules per second are thrown out and oscillate about a new configuration, the nature of which is determined by the present form of the body and not by its unstrained form.[a]

The greater the strain it is probable that more molecules will be so thrown out; but the number is not proportional to the strain, but varies very little for small strains, and suddenly rises enormously for breaking strains.

If $1/l$ of molecules are thrown out in unit of time from a state of strain represented by e, f, g, a, b, c, see Thomson and Tait, and if these molecules enter into the state $(1/3)\theta, (1/3)\theta, (1/3)\theta, 0, 0, 0$ where $\theta = e + f + g$, then[b]

$$\frac{de}{dt} = -\frac{1}{l}(e - (1/3)\theta) + \frac{du}{dx}$$

$$\frac{du}{dt} = -\frac{1}{l}a + \frac{dv}{dz} + \frac{dw}{dy}$$

and if k and n are the coefficients of elasticity so that,

$$P = \left(k - \frac{2n}{3}\right)\theta + 2ne \quad \text{and} \quad S = na$$

the equations of motion will be of the form

$$\frac{1}{\frac{1}{l} + \frac{d}{dt}}\left\{\left[\frac{k}{l} + \left(k + \frac{n}{3}\right)\frac{d}{dt}\right]\frac{d\theta}{dx} - n\nabla^2 u\right\} + X - \rho\frac{Du}{dt} = 0$$

with 2 others and

$$\frac{d\theta}{dt} = \frac{du}{dx} + \frac{dv}{dy} + \frac{dw}{dz}$$

If l is small we may put

$$p = C - k\theta \quad \text{and} \quad nl = \mu$$

and the equation becomes

$$-\frac{dp}{dx} - \mu\nabla^2 u + X - \rho\frac{Du}{dt} = 0$$

Yours very truly,
J. Clerk Maxwell.

a. Note by J. Larmor: "Cf. Maxwell's article "Constitution of Bodies," *Enc. Brit.* 9th ed., *Collected* [i.e., *Scientific*] *Papers*, vol. ii, p.624 (Document II-30], referred to by Rayleigh in connexion with Ewing's model of magnetic hysteresis; the next page gives a development of his ideas." "The next page" means the paragraph in Maxwell's letter starting "If $1/l$ of the molecules..." which begins page 37 in the Stokes *Memoir*.

b. Note by J. Larmor: "Here (u, v, w) is velocity. The signs of the terms involving $\nabla^2 u$ below should be changed unless $\nabla^2 u$ itself has the quaternionic sign."

27. Letter from the Right Rev. C. J. Ellicott, D. D., Lord Bishop of Gloucester and Bristol, to Maxwell, November 21, 1876

Lewis Campbell and William Garnett, *The Life of James Clerk Maxwell* (London: Macmillan and Co., 1882), 392.

Palace, Gloucester, 21st Nov. 1876.

MY DEAR SIR—Will you kindly pardon a great liberty? I have quoted in a forthcoming charge a remarkable expression of yours that atoms are "manufactured articles." Could you in your kindness give me the proper title and reference to the paper and the page? I am now, alas, far from libraries, and have, in matters scientific especially, to ask the aid of others. Will you excuse me asking this further question?

Are you, as a scientific man, able to accept the statement that is often made on the theological side, viz. that the creation of the sun posterior to light involves no serious difficulty,—the creation of light being the establishment of the primal vibrations, generally, the creation of the sun, the primal formation of an origin, whence vibrations would be propagated earthward?

My own mind,—far from a scientific one,—is not clear on this point. I surmise, then, that the scientific mind might not only not be clear as to the explanation, but equitably bound to say that it was no explanation at all. Excuse the trouble I am giving you, for the truth's sake, and believe me, very faithfully yours,

C. J. GLOUCESTER AND BRISTOL.

28. Letter from Maxwell to the Right Rev. C. J. Ellicott, D. D., Lord Bishop of Gloucester and Bristol, November 22, 1876

Lewis Campbell and William Garnett, *The Life of James Clerk Maxwell* (London: Macmillan and Co., 1882), 393–395.

---

11 Scroope Terrace, Cambridge, 22d Nov. 1876.

MY LORD BISHOP—The comparison of atoms or of molecules to "manu-factured articles," was originally made by Sir J. F. D. Herschel in his "Preliminary Discourse on the Study of Natural Philosophy," Art. 28, p. 38 (ed. 1851, Longmans).

I send you by book post several papers in which I have directed attention to certain kinds of equality among all molecules of the same substance, and to the bearing of this fact on speculations as to their origin.

The comparison to "manufactured articles" was criticised (I think in a letter to *Nature*) by Mr. C. J. Monro [*Nature*, x. 481, 15th October 1874], and the latter part of the *Encyc. Brit.*, Article "Atom," is intended to meet this criticism, which points out that in some cases the uniformity among manufactured articles is evidence of want of power in the manufacturer to adapt each article to its special use.

What I thought of was not so much that uniformity of result which is due to uniformity in the process of formation, as a uniformity intended and accomplished by the same wisdom and power of which uniformity, accuracy, symmetry, consistency, and continuity of plan are as important attributes as the contrivance of the special utility of each individual thing.

With respect to your second question, there is a statement printed in most commentaries that the fact of light being created before the sun is in striking agreement with the last results of science (I quote from memory).

I have often wished to ascertain the date of the original appearance of this statement, as this would be the only way of finding what "last result of science" it referred to. It is certainly older than the time when any notions of the undulatory theory became prevalent among men of science or commentators.

If it were necessary to provide an interpretation of the text in accordance with the science of 1876 (which may not agree with that of 1896), it would be very tempting to say that the light of the first day means the all-embracing aether, the vehicle of radiation, and not actual light, whether from the sun or from any other source. But I cannot suppose that this was the very idea meant to be conveyed by the original author of the book to those for whom he was writing. He tells us of a previous darkness. Both light and darkness imply a being who can see if there is light, but not if it is dark, and the words are always understood so. That light and darkness are terms relative to the creature only is recognised in Ps. cxxxix. 12.

As a mere matter of conjectural cosmogony, however, we naturally suppose those things most primeval which we find least subject to change.

Now the aether or material substance which fills all the interspace between world and world, without a gap or flaw of $\frac{1}{100\,000}$ inch anywhere, and which probably penetrates through all grosser matters, is the largest, most uniform and apparently most permanent object we know, and we are therefore inclined to suppose that it existed before the formation of the systems of gross matter which now exist within it, just as we suppose the sea older than the individual fishes in it.

But I should be very sorry if an interpretation founded on a most conjectural scientific hypothesis were to get fastened to the text in Genesis, even if by so doing it got rid of the old statement of the commentators which has long ceased to be intelligible. The rate of change of scientific hypothesis is naturally much more rapid than that of Biblical interpretations, so that if an interpretation is founded on such an hypothesis, it may help to keep the hypothesis above ground long after it ought to be buried and forgotten.

At the same time I think that each individual man should do all he can to impress his own mind with the extent, the order, and the unity of the universe, and should carry these ideas with him as he reads such passages as the 1st Chap. of the Ep. to Colossians (see Lightfoot on Colossians, p. 182), just as enlarged conceptions of the extent and unity of the world of life may be of service to us in reading Psalm viii.; Heb. ii. 6, etc. Believe me, yours faithfully,

J. Clerk Maxwell.

29. Letter from Samuel Tolver Preston[a] to Maxwell, December 5, 1876

Cambridge University Library, Maxwell Collection, Miscellaneous correspondence.

Dear Sir

Knowing the great interest you take in questions of Molecular Physics, I venture to send you the enclosed paper which is the result of much attention and study. It appeared to me an interesting question as to how the propagation of Sound (its velocity etc.) could be explained by means of the Kinetic Theory of Gases. The perusal of a paper by Mr. J. J. Waterston in the Philosophical Mag. for 1859 "On the Theory of Sound" formed the starting point of the enclosed paper.[b]

I do not wish to claim any undue originality & therefore enclose a copy of Mr. Waterston's paper taken from the Phil. Mag. On page 489 Mr. Waterston gives a method of illustrating the propagation of pulses or waves in a gas (constituted according to the kinetic theory), by means of a system of balls or spheres moving in a special manner. Mr. Waterston you will observe gives no special explanation as to how the ⟨special⟩ mode of motion assigned by him to spheres can properly represent the character of the motion of the molecules of a gas in its normal state, according to the kinetic theory. This therefore is the point I set myself to investigate, and the result arrived at is contained in the first part of my paper to which I would direct your kind attention & I should particularly value your opinion. It is possible that Mr. Waterston may have undertaken some such investigation but it does not appear in his paper. I am not a mathematician myself & if I should happen to arrive at any original idea, it is probably due rather to the time I give to the consideration of these subjects than to anything else. Of course the enclosed paper is at your disposal to make what use you please of. It appears to me a somewhat remarkable fact that Mr. Waterston should already in 1845 (as appears from his paper) have investigated to some extent mathematically the kinetic theory of gases.[c]

Yours very truly

S. Tolver Preston

P. S. You will excuse any apparent superfluity of explanation in the enclosed paper, as I intended it might also serve for a general reader. I am of course quite open to the contingency that you (without my knowledge) may already have turned your attention to this special subject, & thus find nothing new in my paper.

a. Samuel Tolver Preston (1844–?) was a telegraph engineer who retired to Germany and received a Ph.D. at Munich in 1894. He continued to work in physics as a private scholar. He disappears from the records after 1904. Among his interests were theories of the ether, gases, light, and gravitation.

b. John James Waterston, "On the Theory of Sound," *Phil. Mag.* [4] 16 (1859), 481–495, reprinted in *The Collected Scientific Papers of John James Waterston*, J. S. Haldane, ed. (Edinburgh: Oliver and Boyd, 1928), 345–362.

c. Maxwell's reply is missing but Preston alludes to it in his published paper, "Mode of the Propagation of Sound, and the Physical Condition Determining Its Velocity on the Basis of the Kinetic Theory of Gases," *Phil. Mag.* [5] 3 (1877), 441–453.

Preston says that Maxwell, to whom the paper was sent, showed that the ratio of the speed of sound to the molecular velocity for a gas of spherical atoms is $\sqrt{5/3} =$ 0.745. See also further papers on this subject by Preston, "On the Equilibrium of Pressure in a Gas," *Phil. Mag.* [5] 4 (1877), 77, and "On the View of the Propagation of Sound Demanded by the Acceptance of the Kinetic Theory of Gases," *Nature* 18 (1878), 253–255. In the latter paper Preston says Maxwell's result has to be corrected if the molecules have rotational motion but does not say precisely how; he does not seem to be aware of the connection with the ratio of specific heats. Preston also says that he discussed sound propagation in his book *Physics of the Ether* (London: Spon, 1875).

Waterston's work remained generally unknown until after his death; see note 186, chapter I, and Brush, *Kind of Motion We Call Heat*, 134–159.

30. "Constitution of Bodies"

Encyclopedia Britannica, ninth edition, vol. 6, 310–313; *Scientific Papers*, vol. 2, 616–624.

[From the *Encyclopædia Britannica.*]

LXXXVIII.   *Constitution of Bodies.*

THE question whether the smallest parts of which bodies are composed are finite in number, or whether, on the other hand, bodies are infinitely divisible, relates to the *ultimate* constitution of bodies, and is treated of in the article ATOM.

The mode in which elementary substances combine to form compound substances is called the *chemical* constitution of bodies, and is treated of in CHEMISTRY.

The mode in which sensible quantities of matter, whether elementary or compound, are aggregated together so as to form a mass having certain observed properties, is called the *physical* constitution of bodies.

Bodies may be classed in relation to their physical constitution by considering the effects of internal stress in changing their dimensions. When a body can exist in equilibrium under the action of a stress which is not uniform in all directions it is said to be solid.

When a body is such that it cannot be in equilibrium unless the stress at every point is uniform in all directions, it is said to be fluid.

There are certain fluids, any portion of which, however small, is capable of expanding indefinitely, so as to fill any vessel, however large. These are called gases. There are other fluids, a small portion of which, when placed in a large vessel, does not at once expand so as to fill the vessel uniformly, but remains in a collected mass at the bottom, even when the pressure is removed. These fluids are called liquids.

When a liquid is placed in a vessel so large that it only occupies a part of it, part of the liquid begins to evaporate, or in other words it passes into the state of a gas, and this process goes on either till the whole of the liquid is evaporated, or till the density of the gaseous part of the substance has reached a certain limit. The liquid and the gaseous portions of the substance are then in equilibrium. If the volume of the vessel be now made smaller, part of the gas will be condensed as a liquid, and if it be made larger, part of the liquid will be evaporated as a gas.

The processes of evaporation and condensation, by which the substance passes from the liquid to the gaseous, and from the gaseous to the liquid state, are discontinuous processes, that is to say, the properties of the substance are very different just before and just after the change has been effected. But this difference is less in all respects the higher the temperature at which the change takes place, and Cagniard de la Tour in 1822* first shewed that several substances, such as ether, alcohol, bisulphide of carbon, and water, when heated to a temperature sufficiently high, pass into a state which differs from the ordinary gaseous state as much as from the liquid state. Dr Andrews has since† made a complete investigation of the properties of carbonic acid both below and above the temperature at which the phenomena of condensation and evaporation cease to take place, and has thus explored as well as established the continuity of the liquid and gaseous states of matter.

For carbonic acid at a temperature, say of 0° C., and at the ordinary pressure of the atmosphere, is a gas. If the gas be compressed till the pressure rises to about 40 atmospheres, condensation takes place, that is to say, the substance passes in successive portions from the gaseous to the liquid condition.

If we examine the substance when part of it is condensed, we find that the liquid carbonic acid at the bottom of the vessel has all the properties of a liquid, and is separated by a distinct surface from the gaseous carbonic acid which occupies the upper part of the vessel.

But we may transform gaseous carbonic acid at 0° C. into liquid carbonic acid at 0° C. without any abrupt change, by first raising the temperature of the gas above 30°.92 C. which is the *critical* temperature, then raising the pressure to about 80 atmospheres, and then cooling the substance, still at high pressure, to zero.

During the whole of this process the substance remains perfectly homogeneous. There is no surface of separation between two forms of the substance, nor can any sudden change be observed like that which takes place when the gas is condensed into a liquid at low temperatures; but at the end of the

* *Annales de Chimie,* 2ᵐᵉ série, XXI. et XXII.
† *Phil. Trans.* 1869, p. 575.

process the substance is undoubtedly in the liquid state, for if we now diminish the pressure to somewhat less than 40 atmospheres the substance will exhibit the ordinary distinction between the liquid and the gaseous state, that is to say, part of it will evaporate, leaving the rest at the bottom of the vessel, with a distinct surface of separation between the gaseous and the liquid parts.

The passage of a substance between the liquid and the solid state takes place with various degrees of abruptness. Some substances, such as some of the more crystalline metals, seem to pass from a completely fluid to a completely solid state very suddenly. In some cases the melted matter appears to become thicker before it solidifies, but this may arise from a multitude of solid crystals being formed in the still liquid mass, so that the consistency of the mass becomes like that of a mixture of sand and water, till the melted matter in which the crystals are swimming becomes all solid.

There are other substances, most of them colloidal, such that when the melted substance cools it becomes more and more viscous, passing into the solid state with hardly any discontinuity. This is the case with pitch.

The theory of the consistency of solid bodies will be discussed in the article ELASTICITY, but the manner in which a solid behaves when acted on by stress furnishes us with a system of names of different degrees and kinds of solidity.

A fluid, as we have seen, can support a stress only when it is uniform in all directions, that is to say, when it is of the nature of a hydrostatic pressure.

There are a great many substances which so far correspond to this definition of a fluid that they cannot remain in permanent equilibrium if the stress within them is not uniform in all directions.

In all existing fluids, however, when their motion is such that the shape of any small portion is continually changing, the internal stress is not uniform in all directions, but is of such a kind as to tend to check the relative motion of the parts of the fluid.

This capacity of having inequality of stress called into play by inequality of motion is called viscosity. All real fluids are viscous, from treacle and tar to water and ether and air and hydrogen.

When the viscosity is very small the fluid is said to be mobile, like water and ether.

When the viscosity is so great that a considerable inequality of stress, though it produces a continuously increasing displacement, produces it so slowly that we can hardly see it, we are often inclined to call the substance a solid, and even a hard solid. Thus the viscosity of cold pitch or of asphalt is so great that the substance will break rather than yield to any sudden blow, and yet if it is left for a sufficient time it will be found unable to remain in equilibrium under the slight inequality of stress produced by its own weight, but will flow like a fluid till its surface becomes level.

If, therefore, we define a fluid as a substance which cannot remain *in permanent equilibrium* under a stress not equal in all directions, we must call these substances fluids, though they are so viscous that we can walk on them without leaving any footprints.

If a body, after having its form altered by the application of stress, tends to recover its original form when the stress is removed, the body is said to be elastic.

The ratio of the numerical value of the stress to the numerical value of the strain produced by it is called the *coefficient of elasticity*, and the ratio of the strain to the stress is called the *coefficient of pliability*.

There are as many kinds of these coefficients as there are kinds of stress and of strains or components of strains produced by them.

If, then, the values of the coefficients of elasticity were to increase without limit, the body would approximate to the condition of a rigid body.

We may form an elastic body of great pliability by dissolving gelatine or isinglass in hot water and allowing the solution to cool into a jelly. By diminishing the proportion of gelatine the coefficient of elasticity of the jelly may be diminished, so that a very small force is required to produce a large change of form in the substance.

When the deformation of an elastic body is pushed beyond certain limits depending on the nature of the substance, it is found that when the stress is removed it does not return exactly to its original shape, but remains permanently deformed. These limits of the different kinds of strain are called the limits of perfect elasticity.

There are other limits which may be called the limits of cohesion or of tenacity, such that when the deformation of the body reaches these limits the body breaks, tears asunder, or otherwise gives way, and the continuity of its substance is destroyed.

A body which can have its form permanently changed without any flaw or break taking place is called *mild*. When the force required is small the

78—2

body is said to be *soft*; when it is great the body is said to be *tough*. A body which becomes flawed or broken before it can be permanently deformed is called *brittle*. When the force required is great the body is said to be *hard*.

The stiffness of a body is measured by the force required to produce a given amount of deformation.

Its strength is measured by the force required to break or crush it.

We may conceive a solid body to approximate to the condition of a fluid in several different ways.

If we knead fine clay with water, the more water we add the softer does the mixture become till at last we have water with particles of clay slowly subsiding through it. This is an instance of a mechanical mixture the constituents of which separate of themselves. But if we mix bees-wax with oil, or rosin with turpentine, we may form permanent mixtures of all degrees of softness, and so pass from the solid to the fluid state through all degrees of viscosity.

We may also begin with an elastic and somewhat brittle substance like gelatine, and add more and more water till we form a very weak jelly which opposes a very feeble resistance to the motion of a solid body, such as a spoon, through it. But even such a weak jelly may not be a true fluid, for it may be able to withstand a very small force, such as the weight of a small mote. If a small mote or seed is enclosed in the jelly, and if its specific gravity is different from that of the jelly, it will tend to rise to the top or sink to the bottom. If it does not do so we conclude that the jelly is not a fluid but a solid body, very weak, indeed, but able to sustain the force with which the mote tends to move.

It appears, therefore, that the passage from the solid to the fluid state may be conceived to take place by the diminution without limit either of the coefficient of rigidity, or of the ultimate strength against rupture, as well as by the diminution of the viscosity. But whereas the body is not a true fluid till the ultimate strength, or the coefficient of rigidity, is reduced to zero, it is not a true solid as long as the viscosity is not infinite.

Solids, however, which are not viscous in the sense of being capable of an unlimited amount of change of form, are yet subject to alterations depending on the time during which stress has acted on them. In other words, the stress at any given instant depends, not only on the strain at that instant,

CONSTITUTION OF BODIES.   621

but on the previous history of the body. Thus the stress is somewhat greater when the strain is increasing than when it is diminishing, and if the strain is continued for a long time, the body, when left to itself, does not at once return to its original shape, but appears to have taken a set, which, however, is not a permanent set, for the body slowly creeps back towards its original shape with a motion which may be observed to go on for hours and even weeks after the body is left to itself.

Phenomena of this kind were pointed out by Weber and Kohlrausch (*Pogg. Ann.* Bd. 54, 119 and 128), and have been described by O. E. Meyer (*Pogg. Ann.* Bd. 131, 108), and by Maxwell (*Phil. Trans.* 1866, p. 249), and a theory of the phenomena has been proposed by Dr L. Boltzmann (*Wiener Sitzungsberichte*, 8th October 1874).

The German writers refer to the phenomena by the name of "elastische Nachwirkung," which might be translated "elastic reaction" if the word reaction were not already used in a different sense. Sir W. Thomson speaks of the viscosity of elastic bodies.

The phenomena are most easily observed by twisting a fine wire suspended from a fixed support, and having a small mirror suspended from the lower end, the position of which can be observed in the usual way by means of a telescope and scale. If the lower end of the wire is turned round through an angle not too great, and then left to itself, the mirror makes oscillations, the extent of which may be read off on the scale. These oscillations decay much more rapidly than if the only retarding force were the resistance of the air, shewing that the force of torsion in the wire must be greater when the twist is increasing than when it is diminishing. This is the phenomenon described by Sir W. Thomson under the name of the viscosity of elastic solids. But we may also ascertain the middle point of these oscillations, or the point of temporary equilibrium when the oscillations have subsided, and trace the variations of its position.

If we begin by keeping the wire twisted, say for a minute or an hour, and then leave it to itself, we find that the point of temporary equilibrium is displaced in the direction of twisting, and that this displacement is greater the longer the wire has been kept twisted. But this displacement of the point of equilibrium is not of the nature of a permanent set, for the wire, if left to itself, creeps back towards its original position, but always slower and slower. This slow motion has been observed by the writer going on for more than a week, and he also found that if the wire was set in vibration the motion of the point of equilibrium was more rapid than when the wire was not in vibration.

We may produce a very complicated series of motions of the lower end of the wire by previously subjecting the wire to a series of twists. For instance, we may first twist it in the positive direction, and keep it twisted for a day, then in the negative direction for an hour, and then in the positive direction for a minute. When the wire is left to itself the displacement, at first positive, becomes negative in a few seconds, and this negative displacement increases for some time. It then diminishes, and the displacement becomes positive, and lasts a longer time, till it too finally dies away.

The phenomena are in some respects analogous to the variations of the surface temperature of a very large ball of iron which has been heated in a furnace for a day, then placed in melting ice for an hour, then in boiling water for a minute, and then exposed to the air; but a still more perfect analogy may be found in the variations of potential of a Leyden jar which has been charged positively for a day, negatively for an hour, and positively again for a minute*.

The effects of successive magnetization on iron and steel are also in many respects analogous to those of strain and electrification†.

The method proposed by Boltzmann for representing such phenomena mathematically is to express the actual stress, $L_{(t)}$, in terms not only of the actual strain, $\theta_{(t)}$, but of the strains to which the body has been subjected during all previous time.

His equation is of the form

$$L_t = K\theta_t - \int_0^\infty \psi(\omega)\,\theta_{t-\omega}\,d\omega,$$

where ω is the interval of time reckoned backwards from the actual time t to the time $t - \omega$, when the strain $\theta_{t-\omega}$ existed, and $\psi(\omega)$ is some function of that interval.

We may describe this method of deducing the actual state from the previous states as the historical method, because it involves a knowledge of the previous history of the body. But this method may be transformed into another,

* See Dr Hopkinson, "On the Residual Charge of the Leyden Jar," *Proc. R. S.* xxiv. 408, March 30, 1876.

† See Wiedemann's *Galvanismus*, vol. ii. p. 567.

in which the present state is not regarded as influenced by any state which has ceased to exist. For if we expand $\theta_{t-\omega}$ by Taylor's theorem,

$$\theta_{t-\omega} = \theta_t - \omega \frac{d\theta}{dt} + \frac{\omega^2}{1 \cdot 2} \frac{d^2\theta}{dt^2} - \&\mathrm{c}.$$

and if we also write

$$A = \int_0^\infty \psi(\omega)\, d\omega, \quad B = \int_0^\infty \omega\psi(\omega)\, d\omega, \quad C = \int_0^\infty \frac{\omega^2}{1 \cdot 2} \psi(\omega)\, d\omega, \ \&\mathrm{c}.$$

then equation (1) becomes

$$L = (K - A)\, \theta + B \frac{d\theta}{dt} - C \frac{d^2\theta}{dt^2} + \&\mathrm{c}.$$

where no symbols of time are subscribed, because all the quantities refer to the present time.

This expression of Boltzmann's, however, is not in any sense a physical theory of the phenomena; it is merely a mathematical formula which, though it represents some of the observed phenomena, fails to express the phenomenon of permanent deformation. Now we know that several substances, such as gutta-percha, India-rubber, &c., may be permanently stretched when cold, and yet when afterwards heated to a certain temperature they recover their original form. Gelatine also may be dried when in a state of strain, and may recover its form by absorbing water.

We know that the molecules of all bodies are in motion. In gases and liquids the motion is such that there is nothing to prevent any molecule from passing from any part of the mass to any other part; but in solids we must suppose that some, at least, of the molecules merely oscillate about a certain mean position, so that, if we consider a certain group of molecules, its configuration is never very different from a certain stable configuration, about which it oscillates.

This will be the case even when the solid is in a state of strain, provided the amplitude of the oscillations does not exceed a certain limit, but if it exceeds this limit the group does not tend to return to its former configuration, but begins to oscillate about a new configuration of stability, the strain in which is either zero, or at least less than in the original configuration.

The condition of this breaking up of a configuration must depend partly on the amplitude of the oscillations, and partly on the amount of strain in the original configuration; and we may suppose that different groups of molecules, even in a homogeneous solid, are not in similar circumstances in this respect.

Thus we may suppose that in a certain number of groups the ordinary agitation of the molecules is liable to accumulate so much that every now and then the configuration of one of the groups breaks up, and this whether it is in a state of strain or not. We may in this case assume that in every second a certain proportion of these groups break up, and assume configurations corresponding to a strain uniform in all directions.

If all the groups were of this kind, the medium would be a viscous fluid.

But we may suppose that there are other groups, the configuration of which is so stable that they will not break up under the ordinary agitation of the molecules unless the average strain exceeds a certain limit, and this limit may be different for different systems of these groups.

Now if such groups of greater stability are disseminated through the substance in such abundance as to build up a solid framework, the substance will be a solid, which will not be permanently deformed except by a stress greater than a certain given stress.

But if the solid also contains groups of smaller stability and also groups of the first kind which break up of themselves, then when a strain is applied the resistance to it will gradually diminish as the groups of the first kind break up, and this will go on till the stress is reduced to that due to the more permanent groups. If the body is now left to itself, it will not at once return to its original form, but will only do so when the groups of the first kind have broken up so often as to get back to their original state of strain.

This view of the constitution of a solid, as consisting of groups of molecules some of which are in different circumstances from others, also helps to explain the state of the solid after a permanent deformation has been given to it. In this case some of the less stable groups have broken up and assumed new configurations, but it is quite possible that others, more stable, may still retain their original configurations, so that the form of the body is determined by the equilibrium between these two sets of groups; but if, on account of rise of temperature, increase of moisture, violent vibration, or any other cause, the breaking up of the less stable groups is facilitated, the more stable groups may again assert their sway, and tend to restore the body to the shape it had before its deformation.

31. Letter from Maxwell to Peter Guthrie Tait, August 7, 1878

Tait Letters, Cambridge University.

Glenlair
18 78.7.8

O. T'   Much obliged for Macfarlane and his pictures, also for θ H [thermoelectric] diagrams.

What of those demons Nickel & Cobalt? are they electrolytically deposited?

I do not see Carbon.

I have no objection to lynx prooves[a] if they are not 4[nions] [quaternions]

In respect of Macfarlane—was it you or some German who taught him to make empirical equations?

The equation between V and p for constant length of spark shows no symptom (within the range of the expt[s]) of cocking its tail up at zero as we know it does. But perhaps it was difficult to exhaust so large a vessel with so large a mouth and such wide lips.

Can you explain the following?

When the law of electric force is supposed to be r^{-n}, and if n is not less than 2, then

$$\rho = C\left(1 - \left(\frac{x^2}{a^2} + \frac{y^2}{b^2} + \frac{z^2}{c^2}\right)\right)^{(4-n)/2} \tag{1}$$

gives a distribution in the ellipsoid (abc) which is in equilibrium.

The whole quantity in the ellipsoid is

$$E = 2\pi Cabc\frac{\Gamma\left(\dfrac{n-2}{2}\right)\Gamma\left(\dfrac{3}{2}\right)}{\Gamma\left(\dfrac{n+1}{2}\right)} \tag{2}$$

which is finite when n is greater than 2. When $n = 2$ and E is given, $C = 0$, or $\rho = 0$ within the ellipsoid and the whole charge is on the surface as it ought to be.

But when n is less than 2

$$\iiint \rho \, dx \, dy \, dz$$

is infinite, more so, indeed than when $n = 2$. Hence we might in this case also expect the whole charge to be on the surface and $\rho = 0$ within. But if so there would be a finite force within the ellipsoid *from the centre* as

Cavendish shows,[b] and this would draw fluid out of it till it was so much undercharged in every part as just to balance the effect of the surface-charge.

If you now shut your eyes and open Green [=ᵘᵐ of Fluids, Ferrers p.   ][c] you will find that he gets a result of exactly this kind by the simple process of determining $\Gamma \dfrac{n-2}{2}$ by the equation

$$\Gamma \frac{n-2}{2} = \frac{2}{n-2} \Gamma \frac{n+1}{2} \tag{3}$$

which, when $n < 2$, is of course negative. He thus very neatly gets a definite distribution of negative electricity throughout the interior even as Cavendish prophesied, though the exact law was hid from him. The volume-integral of this, however, is not E but $-\infty$, so in order to put it right, he spreads a charge $E + \infty$ over the surface, and then he insinuates that all is well. But in the first place (1) is not a solution when $n < 2$ because it gives an infinite ⟨negati⟩ skin which produces an infinite force at any point inside except the centre which cannot be balanced by a finite distribution inside. In the second place the $\Gamma \dfrac{n-2}{2}$ which occurs in (2) is not the finite negative quantity determined by (3) but the $\int_0^\infty x^{(n-4)/2} e^{-x} dx$ which is $+\infty$.

If these two mistakes really balance one another and lead to a true result, it is one of the most remarkable instances of design in the whole of mathematics.

Meanwhile I take some consolation in reflecting that no one has hitherto been able to produce a fluid the particles of which repel one another with a force inversely as a power lower than the square. Sir Isaac Newton indeed showed that the particles of air could not repel each other according to any other law than r^{-1} but at the same time he assures us that it is not so, for in that case a large quantity of air would ⟨have⟩ exert more pressure than a small one of the same density, so it follows that if the air is composed of particles repelling one another the repulsion is confined to the nearest particles. But he adds that whether the air consists of particles repelling each other is a Physical Question, with which remark he closes the discussion.[d]

Now if particles wish to repel each other with a force inversely as the distance and to confine this repulsion to their next neighbour, let them refuse to act except at close quarters, and then let them act with ferocity. The result will be a general bombardment in which the time-average of the repulsion is the mean impulse of a collision divided by the time between the collisions. It is therefore inversely as the distance between those particles which collide, and though these are not exactly next neighbours, they are neighbours in the antievangelical sense of falling foul of each other.

Princip. II. Prop. 23.

$$\frac{dp}{dt}$$

I have got Tait and Dewar on Vacua in Nature.[e]

There are some beautiful misprints.

a. Lynxing prooves (i.e., proofs). The common phrase in the correspondence of Maxwell, Tait, and William Thomson for checking the proofs of each other's books or articles, perhaps in allusion to the Italian Accademia dei Lincei, the earliest of the European scientific societies, whose founders chose the title Lincei (Lynxes) to signify the sharpsightedness of science from the proverbially keen sight of the lynx.

b. Maxwell was at this time editing the electrical papers of Henry Cavendish (1731–1810). For further discussion of the topic of this letter, see document II-34.

c. George Green, "Mathematical Investigations Concerning the Laws of the Equilibrium of Fluids Analogous to the Electric Fluid, with Other Similar Researches," *Trans. Cambridge Phil. Soc.* 5 (1835), 1–64; reprinted in *Mathematical Papers of George Green*, N. M. Ferrers, ed. (London, 1871). Maxwell is citing the Ferrers edition.

d. Newton, *Philosophiae Naturalis Principia Mathematica* (London, 1687; third edition, 1726), Book 2, Section 5; English translation of the proposition is reprinted in Brush, *Kinetic Theory* (1965), vol. 1, 54–56.

e. J. Dewar and P. G. Tait, "Charcoal Vacua," *Nature* 12 (1875), 217–218. This paper will be discussed in connection with the theory of the radiometer effect in our next volume.

32. [Notes for a Lecture on Kinetic Theory]

Cambridge University Library, Maxwell Collection.

⟨Thus far we have been following ⟨what⟩ the progress of science along a particular track, that is to say. the⟩

Thus far we have been describing scientific work from the point of view of the worker himself. The aim we have kept in view is the study of the constitution of bodies and we have estimated the value of experimental methods and theories according to the help they have afforded us in this study.

But the effect of scientific work is twofold. Its proposed object is to unravel ⟨reveals⟩ the secrets of nature ⟨on the one hand and⟩ but it also reacts on the mind of the worker. ⟨The man⟩ He becomes
[space]

But though the professed aim of scientific work is to unravel the secrets of nature, it has another ⟨effect⟩ result hardly less important in its reaction on the mind of the worker. It leaves him in possession of methods which he could never have developed ⟨if the⟩ without the incitement of the pursuit of natural knowledge and it places him in a position from which ⟨his a great part of his previous⟩ much of ⟨his⟩ his ⟨takes a view of nature⟩ many regions of nature besides that which he has been studying appear under a different aspect. The study of molecules has developed a method of its own and it has also opened up new views of nature.

To describe mathematical methods is at all times ⟨difficult⟩ tedious and I have no intenton of making use of a black board. But I think that the method which we are forced to adopt in the study of molecular phenomena affords one of the ⟨most str⟩ best illustrations of the nature and limitations of a large part of human knowledge.

The molecular theory of gases resolves itself into that of the motions and encounters of the molecules. As long as we have to deal only with one pair of molecules and to determine from their known motion before the encounter what will be their motion after the encounter we require no methods but those of elementary dynamics.

When air is compressed the sides of the vessel are moving to meet the molecules ⟨and the molecules⟩ like a cricket bat swung forward to meet the ball, and the molecules like the cricket ball rebound with a greater velocity than before. When air is allowed to expand the sides of the vessel are retreating from the molecules like the cricketer's hands when he is stopping the ball and the molecules rebound with diminished velocity ⟨as the ball does when it is.⟩ Hence air becomes warmer when compressed and cooler when allowed to expand. The observed amount of this heating and cooling however is less than it would be if the only motion of the molecules were that of their centres of inertia. Hence as Clausius has shown the molecules must have other motions such as rotation and vibration as well as motion of simple translation, and a definite portion of the energy of the gas arises from these rotatory or vibratory motions of the molecule, which have no effect on the pressure of the gas.[a]

Thus far any person acquainted with the most elementary dynamics may pursue the kinetic theory of gases. It is easy to understand the effects of the collisons of the ⟨partic⟩ molecules against the sides of the vessel. But when we attempt to trace the motion of each molecule among innumerable others and to determine the effects of all the collisons which must occur in the flying crowd we find that special methods are requisite for the treatment of so intricate a problem.

Lucretius has given ⟨a striking illustration⟩ us a hint how to form a mental representation of the dance of atoms by looking at the motes which we see chasing each other through a sunbeam in a darkened room. Their number their minuteness and the variety and perpetual alteration of the motion of each atom might well confuse the

[There are one or two pages missing here.]

data of Economic or Social Science ⟨the results they obtain are⟩ they obtain their results by distributing the whole population of a country into groups according to age, income tax, reading and writing, number of convictions and so forth. They do so because the number of individuals is far too great to allow of their tracing the history of each separately, so that in order to reduce their labour within human compass they concentrate their attention on artificial groups ⟨and⟩ so that the varying numbers in each group and not the ⟨names⟩ varying state of each individual is the primary datum from which they work.

⟨This may not be the way in which Shakespeare ⟨and⟩ obtained his knowledge of human nature.⟩

⟨This method of studying human nature, the principles of which have been⟩

This is of course not the only method of studying human nature. ⟨We may compare the actual conduct of individuals with the conduct which we should have expected from them, taking into account all our previous knowledge of them and all our⟩

We may observe the conduct of individual men and compare it with that conduct which their previous character, as judged according to ⟨those maxims and generalisations which we⟩ the best attainable rules and maxims would lead us to expect. In this way we may correct and improve these rules and maxims just as an astronomer corrects the elements of a planet by comparing its actual position with that deduced from received elements.

This method which we may call the astronomical or dynamical method is evidently of a higher order than the statistical method, but we can employ it only in the case of the limited number of individuals of whose history we have sufficient knowledge.

Now in our ⟨experiments⟩ physical researches every portion of matter which we can subject to experiment consists of millions and millions of molecules not one of which ever becomes individually sensible to us. Hence the statistical method is the only available one.

What is the nature of our data. We can take a drop of water and weigh it. We thus obtain a measure of the sum of the masses of all the molecules it contains, though we may be ignorant of the mass of any one of them. Again by multiplying this mass by the apparent velocity of the drop we obtain its momentum and this is the resultant of ⟨all⟩ the momenta of all the individual molecules though in point of fact these molecules may be moving in every direction within the drop.

a. There is a single stroke down the page indicating that all of this paragraph and the first two sentences of the next are to be omitted.

33. [Fragment on Statistical Regularity]

Cambridge University Library, Maxwell Collection.

[...] and the complete solution of the problem would enable us to determine at any given instant the position and motion of any given molecule from a knowledge of the positions and motions of all the molecules in their initial state.[a]

According to the statistical method ⟨we abandon the attempt to follow every molecule. We⟩ the state of the system at any instant is ascertained by distributing the molecules into groups, the definition of each group being founded on some variable property of the molecules. Each individual molecule is sometimes in one of these groups and sometimes in another but we make no attempt to follow it. We simply take account of the number of molecules which at a given instant belong to each group.

Thus we may consider as a group those molecules which at a given instant lie within a given region in space. Molecules may pass into or out of this region, but we confine our attention to the increase or diminution of the number of molecules within it. In the same way the population of a water-place, considered as a mere number, varies in the same way whether its visitors ⟨frequent it⟩ return to it season after season or whether ⟨they consist of persons who each new [?] consists of⟩ the annual [?] consists each year of ⟨new different⟩ fresh individuals. We may also form our groups out of those molecules which at a given instant have velocities lying within given limits. When a molecule has an encounter and changes its velocity it passes out of one of these groups and enters another, but as other molecules are also changing their velocities the number of molecules in each group ⟨is⟩ varies very little from a certain average value.

We thus meet with a new kind of regularity, the regularity of averages—a regularity which when we are dealing with millions of millions of individuals is so unvarying that we are almost in danger of confounding it with absolute uniformity.

Laplace in his theory of Probability[b] has given many examples of this kind of statistical regularity and has shown how this regularity is consistent with the utmost irregularity among the individual instances which are enumerated in making up the results.[c] ⟨The very same facts are held by Mr Buckle and⟩ In the hands of Mr Buckle ⟨these same⟩ facts ⟨become evidence⟩ of the same kind were brought forward as instances of the unalterable character of natural law.[d]

But the stability of the average of large numbers of ⟨independent⟩ variable events must be carefully distinguished from the absolute uniformity of sequence according to which we suppose that every individual ⟨thing has its motion determined by⟩ event is determined by its antecedents. For instance if a quantity of air is enclosed in a vessel and left to itself we may be ⟨perfectly⟩ morally certain that whenever we choose to examine it we shall find the pressure uniform in horizontal strata and greater below than above, that the temperature will be uniform throughout, and that there will be no sensible currents of air in the vessel.

But there is nothing inconsistent with the laws of motion in supposing that in a particular case a very different event might occur. For instance if at a given instant ⟨the result⟩ a certain number ⟨one half⟩ of the molecules should each of them encounter ⟨a molecule⟩ one of the remaining molecules and if in each case one of the molecules after the encounter should be moving vertically upwards and if in addition the molecules above them happened to be ⟨so [?] out of the way⟩ not to get into the way of these upward moving molecules,—the result would be a sort of explosion by which a mass of air would be projected upwards with the velocity of a cannon ball while a larger mass would be blown downwards with an equivalent momentum.

We are morally certain that such an event will not take place within the air of the vessel however long we leave it. What are the grounds of this certainty.

The explosion will certainly happen if certain conditions are satisfied. Each of this conditions by itself is not only possible but is in the common course of events as often satisfied as not. But as the number of conditions which must be satisfied at once is to be counted by millions of millions the improbability of the concurrence of all these conditions amounts to what we are unable to distinguish from an impossibility. ⟨Nonetheless the explosion which we have described is no more improbable in itself than say any other event ⟨in w⟩ say that which actually occurs provided that both events are described with complete accuracy defining the⟩

Nevertheless it is no more improbable that at a given instant the molecules should be arranged in one definite manner than in any other definite manner. We are as certain that the exact arrangement which the molecules have at ⟨this⟩ the present instant will never again be repeated as that the arrangement which would bring about the explosion will never occur.

a. This and the following four paragraphs are nearly identical with the middle section of the introduction of H. W. Watson, *A Treatise on the Kinetic Theory of Gases* (Oxford: Clarendon Press, 1876), vi–vii.

b. Pierre Simon, Marquis de Laplace published a series of memoirs on probability from 1773 onwards. Maxwell is probably referring to *Théorie analytique des probabilités* (Paris, 1820); this is vol. 7 of *Oeuvres complètes de Laplace* (Paris, 1886). Laplace's *Essai philosophique sur les probabilités* (Paris, 1814) was also widely known, not only as a popular exposition of the statistical approach but also for its statement of

"Laplacean Determinism." For additional references, see Brush, *Kind of Motion We Call Heat*, 583–595.

c. End of section used in Watson's preface; there was a footnote here, "MS. notes by Professor Clerk Maxwell." At the end of his preface (p. iv), Watson says, "To Professor Clerk Maxwell I am indebted for much kind assistance, and especially for access to some of his manuscript notes on this subject, from which I have taken many valuable suggestions." Our next volume will include Watson's letter to Maxwell (October 29, 1875), in which Watson asks Maxwell's advice on some technical points in kinetic theory, and Maxwell's review of Watson's book (published in *Nature*, 1877).

d. Henry Thomas Buckle (1821–1862), *History of Civilisation in England* (London, 1851–1861), 4 vols. The author of this popular history, known for its "scientific" approach, was brought up in a conservative and evangelical home but rebelled to become a radical and freethinker. See Campbell and Garnett, *Life*, 295, for Maxwell's first impression.

34. "Gases, Kinetic Theory of[a] [Draft of Editorial Note for Papers of Henry Cavendish]

Cambridge University Library, Maxwell Collection.

Newton, in the preface to the Principia, pointed out the true path of physical investigation in the following words.

Omnis enim philosophiae difficultas in eo versari videtur, ut a phaenomenis motuum investigemus vires naturae, deinde ab his viribus demonstremus phaemonena reliqua.

⟨He himself did⟩ To do this for the motions of the heavenly bodies is the aim of the Principia, but he proceeds—

Utinam caetera naturae phaenomena ex principiis mechanicis eodem argumentandi genere derivare liceret. Nam multa me movent, ut nonnihil suspicer ea omnia ex viribus quibusdam pendere posse, quibus corporum particulae per causas nondum cognitas vel in se mutuo impelluntur et secundum figuras regulares cohaerent, vel ab invicem fugantur et recedunt: quibus viribus ignotis, philosophi hactenus naturam frustra tentarunt. Spero autem quod vel huic philosophandi modo, vel veriori alicui, principia hic posita lucem aliquam praebebunt.[b]

The application of this method of investigation is comparatively easy when we are able to observe directly the motions of the bodies between which the forces to be discovered are acting, as in the case of the heavenly bodies as treated by Newton himself or in that of electrified bodies as treated by Cavendish and Coulomb. It becomes more difficult when the bodies themselves are so small, so numerous, and so variable in their motions that we are precluded from all direct observation as in the phaenomena of material bodies composed of exceedingly small particles. Newton however has given an example of the treatment of a question of this kind in the Principia II, 23 where he shows that if an elastic fluid is composed of ⟨particles at rest⟩ mutually repelling particles at rest, and if the pressure is proportional to the density, then the repulsion between two particles must be inversely as the distance between them.[c] He supposes a constant quantity of air enclosed in a cubical vessel ⟨the dimensions of⟩ which ⟨are⟩ is made to vary ⟨while it⟩ so as to become a cube of greater or smaller dimensions. Then since by Boyles law the product of the pressure of the air on unit of surface into the volume of the cube is constant, and since the volume of the cube is the product of the area of a face into the edge perpendicular to it it follows that the product of the total pressure on a face of the cube into the edge of the cube is constant.

Now if an imaginary plane be drawn through the cube parallel to one of its faces the mutual pressure between the portions of air on opposite sides of this plane is equal to the pressure on a face of the cube.

Consider two states of the cubical vessel. The number of molecules is the same in both states and their configurations are geometrically similar. Hence the distance between any two given molecules must vary as the edge of the cube and the ⟨force⟩ repulsion between these two molecules must vary as the ⟨total⟩ resultant force between the two sets of molecules separated by the imaginary plane. Hence the product of the repulsion between two molecules into the distance between them must be constant, or the repulsion must vary inversely as the distance.

This demonstration applies only to the case in which a definite number of molecules are made to occupy a larger or smaller vessel and are not acted on by other molecules. Indeed Newton points out in his Scholium that ⟨it would⟩ if the same law of repulsion hold for every pair of molecules it would require a greater pressure to produce the same density in a larger mass of air which is contrary to experience.

We must therefore suppose that the repulsion exists not between every pair of molecules but only between each molecule and a certain definite number of other molecules, which we may supposed to be defined as the *n* molecules nearest to the given molecule.

Newton gives as an example of such a kind of action the attraction of a magnet, the field of which is contracted when a plate of iron is interposed, as if the attractive power was bounded by the nearest body attracted.

Henry Cavendish, in his first paper on electricity[d] ⟨in the Phil⟩ points out ⟨that⟩ if the repulsion were limited to those pairs of molecules which are at a certain *distance* from each other, the pressure arising from such a repulsion would vary nearly as the square of the density provided a large number of molecules are always within the sphere of repulsion of any one molecule. Hence this hypothesis will not account for the pressure varying as the density.

On the other hand, if the repulsion were limited to particular pairs of molecules, then since the molecules are free to move about among each other, their particular pairs would soon get out of each others reach till only those particles were near each other between which the repulsive force is supposed not to exist, and the fluid ⟨not⟩ cease to be elastic ⟨at all⟩. It would appear, therefore, that the hypothesis stated by Newton and adopted by Cavendish— that the repulsive force is inversely as the distance but exists only between each molecule and a certain number of those nearest to it—is the only statical hypothesis which is admissable.

There is one obvious objection, however, to the definition of the *n* nearest molecules, for it is quite possible that *A* may be one of the *n* molecules nearest to *B* while *B* is not one of the *n* molecules nearest to *A* and this would make it impossible to decide whether *A* and *B* are to repel each other or not.

Newton's own conclusion with respect to this statical molecular hypothesis is as follows

An vero fluida elastica ex particulis se mutuo fugentibus constent, quaestio physica est. Nos proprietatem fluidorum ex ejusmodi particulis constantium mathematica demonstravimus, ut philosophis ausam praebeamus quaestionem illum tractandi.[e]

The conditions of the question, as thus pointed out by Newton, are satisfied by the kinetic theory in a much more natural manner than by any modification of the statical theory.

On the kinetic theory, as will be presently shown, the pressure of a gas may still be said in a certain sense to arise from the repulsive force between its molecules, only instead of this repulsive force being in constant action, it is called into play only during the encounters between two molecules.

⟨Though the intensity of the impulse is not the same for all encounters, yet as it does not depend on the interval between successive encounters, we may consider its mean value as constant.⟩

The average value of the repulsion between two molecules which encounter each other may be estimated by dividing the impulse of the encounter by half the interval of time between the encounter ⟨immediately⟩ preceding and the encounter following the given encounter.

Neither the velocity of the molecules nor the impulse of their encounters are constant, but as they do not depend on the distance between the molecules, we may when considering the variation of that distance take the mean values of the velocity and of the impulse.

We thus find that the repulsion exists only between those pairs of molecules which encounter each other, and that its average value varies inversely as the distance travelled by the molecule between two ⟨successive encounters⟩ consecutive encounters.

This is as close an approach to the conditions indicated by Newton as we could expect without introducing highly strained hypotheses.

a. This is a draft of "On the Molecular Constitution of Air" (next document).

b. Isaac Newton, *Philosophiae Naturalis Principia Mathematica* (London, 1687; third edition, 1726). The standard Motte-Cajori translation of this passage is somewhat misleading, according to Prof. C. Truesdell (letter to S. G. Brush, January 30, 1970). The following translation is based on Prof. Truesdell's suggestions (except for the last sentence):

All the difficulty of Philosophy seems to lie in this, that from the phenomena of motions we should investigate the forces of nature, and thereupon from these forces we should demonstrate the rest of the phenomena. [We omit here three sentences dealing with the explanation of planetary motions by means of gravitational forces] Would it were possible to derive the rest of the phenomena of nature from mechanical principles by the same kind of reasoning! For many things move me to suspect somewhat that they all may depend upon certain forces, by which the little parts of bodies through causes not yet known are either driven against one another and crowd together according to regular figures, or are driven away from each other and drawn apart: these forces being unknown, philosophers have hitherto attempted the search of nature in vain. But I hope the principles here laid down will afford some light either to this or some truer method of philosophy.

On the problems of translating this passage, see C. Truesdell, *Essays in the History of Mechanics* (New York: Springer-Verlag, 1968), 178–179.

c. See the English translation of this passage, reprinted in Brush, *Kinetic Theory*, vol. 1, 54–56.

d. (Maxwell's footnote) *Phil. Trans.* 1771 or *Cavendishes Electrical Researches,* art 97. (*The Electrical Researches of the Honourable Henry Cavendish,* J. Clerk Maxwell, ed. (Cambridge: At the University Press, 1879), 43.)

e. "But whether elastic fluids do really consist of particles so repelling each other, is a physical question. We have here demonstrated mathematically the property of fluids consisting of particles of this kind, that hence philosophers may take occasion to discuss that question" (*Mathematical Principles of Natural Philosophy,* trans. A. Motte (1729) (Berkeley: University of California Press, 1962), vol. 1, 302]

35. "On the Molecular Constitution of Air"

The Electrical Researches of the Honourable Henry Cavendish, J. Clerk Maxwell, ed. (Cambridge: At the University Press, 1879; reprinted, London: Frank Cass and Co., Ltd., 1967), 380–382.

Note 6, Art. 97, p. 43.

On the Molecular Constitution of Air.

The theory of Sir Isaac Newton here referred to is given in the *Principia*, Lib. II., Prop. XXIII.

Newton supposes a constant quantity of air enclosed in a cubical vessel which is made to vary so as to become a cube of greater or smaller dimensions. Then since by Boyle's law the product of the pressure of the air on unit of surface into the volume of the cube is constant; and since the volume of the cube is the product of the area of a face into the edge perpendicular to it, it follows that the product of the total pressure on a face of the cube into the edge of the cube is constant, or the total pressure on a face is inversely as the edge of the cube.

Now if an imaginary plane be drawn through the cube parallel to one of its faces, the mutual pressure between the portions of air on opposite sides of this plane is equal to the pressure on a face of the cube. But the number of particles is the same, and their configuration is geometrically similar whether the cube is large or small. Hence the distance between any two given molecules must vary as the edge of the cube, and the force between the two molecules must vary as the total force between the sets of molecules separated by the imaginary plane, and therefore the product of the repulsion between two given molecules into the distance between them must be constant, in other words the repulsion varies inversely as the distance.

In this demonstration the repulsion considered is that between two *given* molecules, and it is shown that this must vary inversely as the distance between them in order to account for Boyle's law of the elasticity of air.

If, however, we suppose the same law of repulsion to hold for every pair of molecules, Newton shows in his Scholium that it would require a greater pressure to poduce the same density in a larger mass of air.

We must therefore suppose that the repulsion exists, not between every pair of molecules, but only between each molecule and a certain definite number of other molecules, which we may suppose to be defined as those nearest to the given molecules. Newton gives as an example of such a kind of action the attraction of a magnet, the field of which is contracted when a plate of iron is interposed, so that the attractive power appears to be bounded by the nearest body attracted.

If the repulsion were confined to those molecules which are within a certain *distance* of each other, then, as Cavendish points out, the pressure arising from this repulsion would vary nearly as the square of the density, provided a large number of molecules are within this distance. Hence the hypothesis will not explain the fact that the pressure varies as the density.

On the other hand, if the repulsion were limited to particular pairs of particles, then since the particles are free to move, these pairs of particles would move away from each other till only those particles were near each other between which the repulsive force is supposed not to exist.

It would appear therefore that the hypothesis stated by Newton and adopted by Cavendish is the only admissible one, namely, that the repulsive force is inversely as the distance, but is exerted only between the nearest molecules.

Newton's own conclusion to his investigation of the properties of air on the statical molecular hypothesis is as follows:—"An vero Fluida Elastica ex particulis se mutuo fugantibus constent, Quaestio Physica est. Nos proprietatem Fluidorum ex ejusmodi particulis constantium mathematice demonstravimus, ut Philosophis ansam praebeamus Quaestionem illam tractandi."

The theory that the molecules of elastic fluids are in motion satisfies the conditions of the question as pointed out by Newton in a much more natural manner than any modification of the statical hypothesis.

According to the kinetic theory of gases, each molecule is in motion, and this motion is during the greater part of its course undisturbed by the action of other molecules, and is therefore uniform and in a straight line. When however it comes very near another molecule, the two molecules act on each other for a very short time, the courses of both are changed and they go on in the new courses till they encounter other molecules.

It would appear from the observed properties of gases that the mutual action between two molecules is insensible at all sensible distances. As the molecules approach, the action is at first attractive, but soon changes to a repulsive force of far greater magnitude, so that the general character of the encounter depends mainly on the repulsive force.

On this theory, the elasticity of the gas may still be said in a certain sense to arise from the repulsive force between its molecules, only instead of this repulsive force being in constant action, it is called into play only during the encounters between two molecules. The intensity of the impulse is not the same for all encounters, but as it does not depend on the interval between the encounters, we may consider its mean value as constant. The average value of the force between two molecules is in this case the value of the impulse divided by the time between two encounters. Hence we may say that the force is inversely as the distance between the molecules, and that it acts between those molecules only which encounter each other.

For an earlier investigation by Cavendish of the properties of an elastic fluid, see Note 18.

36. "Physical Sciences"[a]

Encyclopedia Britannica, ninth edition, vol. 19, 1–3.

According to the original meaning of the word, physical science would be that knowledge which is conversant with the order of nature—that is, with the regular succession of events whether mechanical or vital—in so far as it has been reduced to a scientific form. The Greek word "physical" would thus be the exact equivalent of the Latin word "natural." In the actual development, however, of modern science and its terminology these two words have come to be restricted each to one of the two great branches into which the knowledge of nature is divided according to its subject-matter. Natural science is now understood to refer to the study of organized bodies and their development, while physical science investigates those phenomena primarily which are observed in things without life, though it does not give up its claim to pursue this investigation when the same phenomena take place in the body of a living being. In forming a classification of sciences the aim must be to determine the best arrangement of them in the state in which they now exist. We therefore make no attempt to map out a scheme for the science of future ages. We can no more lay down beforehand the plan according to which science will be developed by our successors than we can anticipate the particular discoveries which they will make. Still less can we found our classification on the order in time according to which different sciences have been developed. This would be no more scientific than the classification of the properties of matter according to the senses by which we have become acquainted with their existence.

It is manifest that there are some sciences, of which we may take arithmetic as the type, in which the subject-matter is abstract, capable of exact definition, and incapable of any variation arising from causes unknown to us which would in the slightest degree alter its properties. Thus in arithmetic the properties of numbers depend entirely on the definitions of these numbers, and these definitions may be perfectly understood by any person who will attend to them. The same is true of theoretical geometry, though, as this science is associated in our minds with practical geometry, it is difficult to avoid thinking of the probability of error arising from unknown causes affecting the actual measurement of the quantities. There are other sciences, again, of which we may take biology as the type, in which the subject-matter is concrete, not capable of exact definition, and subject to the influence of many causes quite unknown to us. Thus in biology many abstract words such as "species," "generation," &c., may be employed, but the only thing which we can define is the concrete individual, and the ideas which the most accomplished biologist attaches to such words as "species" or "generation" have a very different degree of exactness from those which mathematicians associate, say, with the class or order of a surface, or with the unbilical generation of conicoids. Sciences of this kind are rich in facts, and will be well occupied for ages to come in the co-ordination of these facts, though their cultivators may be cheered in the meantime by the hope of the discovery of laws like those of the more abstract sciences, and may indulge their fancy in the contemplation of a state of scientific knowledge when maxims cast in the same mould as those which apply to our present ideas of dead matter will regulate all our thoughts about living things.

What is commonly called "physical science" occupies a position intermediate between the abstract sciences of arithmetic, algebra, and geometry and the morphological and biological sciences. The principal physical sciences are as follows.

A. *The Fundamental Science of Dynamics, or the doctrine of the motion of bodies as affected by force.*—The divisions of dynamics are the following. (1) Kinematics, or the investigation of the kinds of motion of which a body or system of bodies is capable, without reference to the cause of these motions. This science differs from ordinary geometry only in introducing the idea of motion,—that is, change of position going on continuously in space and time. Kinematics includes, of course, geometry, but in every existing system of geometry the idea of motion is freely introduced to explain the tracing of lines, the sweeping out of surfaces, and the generation of solids. (2) Statics, or the investigation of the equilibrium of forces,—that is to say, the conditions under which a system of forces may exist without producing motion of the body to which they are applied. Statics includes the discussion of systems of forces which are equivalent to each other. (3) Kinetics, or the relations between the motions of material bodies and the forces which act on them. Here the idea of matter as something capable of being set in motion by force, and requiring a certain force to generate a given motion, is first introduced into physical science. (4) Energetics, or the investigation of the force which acts between two bodies or parts of a body, as dependent on the conditions under which action takes place between one body or part of a body and another so as to transfer energy from one to the other.

The science of dynamics may be divided in a different manner with respect to the nature of the body whose motion is studied. This forms a cross division. (1) Dynamics of a particle; including its kinematics or the theory of the tracing of curves, its statics or the doctrine of forces acting at a point, its kinetics or the elementary equations of motion of a particle, and its energetics, including, as examples, the theory of collision and that of central forces. (2) Dynamics of a connected system, including the same subdivisions. This is the most important section in the whole of physical science, as every dynamical theory of natural phenomena must be founded on it. The subdivisions of this, again, are—*a*. dynamics of a rigid system, or a body of invariable form; *b*. dynamics of a fluid, including the discussion (α) of its possible motion, (β) of the conditions of its equilibrium (hydrostatics), (γ) of the action of force in producing motion (hydrodynamics, not so unsatisfactory since Helmholtz, Stokes, and Thomson's investigations), and (δ) of the forces called into play by change of volume; *c*. dynamics of an elastic body; *d*. dynamics of a viscous body.

B. The Secondary Physical Sciences—Each of these sciences consists of two divisions or stages. In the elementary state it is occupied in deducing from the observed phenomena certain general laws, and then employing these laws in the calculation of all varieties of the phenomena. In the dynamical stage the general laws already discovered are analysed and shown to be equivalent to certain forms of the dynamical relations of a connected system (A, 2), and the attempt is made to discover the nature of the dynamical system of which the observed phenomena are the motions. This dynamical stage includes, of course, several other stages rising one above the other; for we may successfully account for a certain phenomenon, say the turning of a weathercock towards the direction of the wind, by assuming the existence of a force having a particular direction and tending to turn the tail of the cock in that direction. In this way we may account not only for the setting of the weathercock but for its oscillations about its final position. This, therefore, is entitled to rank as a dynamical theory. But we may go on and discover a new fact, that the air exerts a pressure and that there is a greater pressure on that side of the cock on which the wind blows. This is a further development of the theory, as it tends to account for the force already discovered. We may go on and explain the dynamical connexion between this inequality of pressure and the motion of the air regarded as a fluid. Finally, we may explain the pressure of the air on the hypothesis that the air consists of molecules in motion, which strike against each other and against the surface of any body exposed to the air.

The dynamical theories of the different physical sciences are in very different stages of development, and in almost all of them a sound knowledge of the subject is best acquired by adopting, at least at first, the method which we have called " elementary,"—that is to say, the study of the connexion of the phenomena peculiar to the science without reference to any dynamical explanations or hypotheses. Thus we have—

(1) Theory of gravitation, with discussion of the weight and motion of bodies near the earth, of the whole of physical astronomy, and of the figure of the earth. There is a great deal of dynamics here, but we can hardly say that there is even a beginning of a dynamical theory of the method by which bodies gravitate towards each other.

(2) Theory of the action of pressure and heat in changing in the dimensions and state of bodies. This is a very large subject and might be divided into two parts, one treating of the action of pressure and the other of heat. But it is much more instructive to study the action of both causes together, because they produce effects of the same kind, and therefore mutually influence each other. Hence the term "thermodynamics" might be extended to the whole subject were it not that it is already restricted to a very important department relating to the transformation of energy from the thermal to the mechanical form and the reverse. The divisions of the subject are seven. (*a*) Physical states of a substance,—gaseous, liquid, and solid; elasticity of volume in all three states; elasticity of figure in the solid state; viscosity in all three states; plasticity in the solid state; surface-tension, or capillarity; tenacity of solids; cohesion of liquids; adhesion of gases to liquids and solids. (*b*) Effects of heat in raising temperature, altering size and form, changing physical state. (*c*) Thermometry. (*d*) Calorimetry. (*e*) Thermodynamics, or the mutual convertibility of heat and work. (*f*) Dissipation of energy by diffusion of matter by mixture, diffusion

of motion by internal friction of fluids, diffusion of heat by conduction. (*g*) Theory of propagation of sound, vibrations of strings, rods, and other bodies.

(3) Theory of radiance. (*a*) Geometrical optics; theory of conjugate foci and of instruments. (*b*) Velocity of light in different media. (*c*) Prismatic analysis of light,—spectroscopy, radiant heat, visible radiance, ultra-violet rays, calorescence, &c., fluorescence, &c. (*d*) Colours of thin plates, diffraction, &c. (*d'*) Proof of the existence of wave-lengths and wave-periods (preparation for dynamical theory). (*e*) Polarized light, radiant heat, &c. (*e'*) The disturbance is transverse to the ray. (*f*) Quantity of energy in the total radiation from a hot body; Prévost's theory of exchanges, &c. (*g*) Theory of three primary colours.

(4) Electricity and magnetism. (*a*) Electrostatics, or distribution and effects of electricity in equilibrium. (*b*) Electrokinematics, or distribution of currents in conductors. (*c*) Magnetism and magnetic induction (diamagnetism, &c.). (*d*) Electromagnetism, or the effects of an electric current at a distance. Under (*b*) we may discuss electrochemistry, or the theory of electrolysis; under (*c*) terrestrial magnetism and ship's magnetism; and after (*d*) comes electrokinetics, or electromagnetic phenomena considered with reference to the fundamental science of dynamics. There is also Faraday's discovery of the effect of magnetism on light and the electromagnetic theory of light.

Chemistry is not included in this list, because, though dynamical science is continually reclaiming large tracts of good ground from the one side of chemistry, chemistry is extending with still greater rapidity on the other side into regions where the dynamics of the present day must put her hand upon her mouth. Chemistry, however, is a physical science, and a physical science which occupies a very high rank.   (J. C. M.)

a. Published with the following footnote by the editor of the *Britannica*: "The paper of the late Professor J. Clerk Maxwell which is presented to the reader under this head was prepared at the time when the ninth edition of the *Encyclopedia Britannica* was being planned, and bore in his MS, the title 'Remarks on the Classification of the Physical Sciences.'"

37. Letter from Maxwell to Simon Newcomb, May 31 and June 2, 1879

Library of Congress, Manuscript Division, Newcomb Papers.

Cavendish Laboratory
Cambridge
England

31 May 1879

Professor S. Newcomb[a]
Washington

Dear Sir

I have received your very interesting letter on the conditions which determine the mean temperature of the earth. You have laid on me the somewhat difficult task of judging in what degree you share in that ignorance of which we all are partakers. It seems to me that if your letter shows any, it is of an historical rather than a physical kind, and is therefore all the more easily removed, and consequently i[t]s removal is all the more called for in the event of publication.

In short it seems to me that your relation to the "theory of exchanges" is not that of ignorance, for you have the very best kind of knowledge of it, namely, that got by making it out for yourself. But I do not find any mention in your letter of those who established the theory, or any reference to the theorems as established, although most if not all of what you say is consistent with those theorems. Now whatever may be the case ⟨with⟩ on your side of the water, the scientific public here always require to be reminded of the author of any important statement, or else they put it down to the speaker as his own notion.

The first eminent exponent of the theory of the exchange of radiations was Pierre Prevost of Geneva, the pupil and follower of Le Sage.[b]

After being held in no great esteem for a long time* it was taken up with

---

*I forgot to say that Fourier makes an important application of it to determine the amount of radiation from a surface at different degrees of obliquity.[c]

---

great ability and ingenuity by Balfour Stewart and by Kirchhoff.[d] Afterwards Clausius investigated some very interesting cases of it (See his Theory of Heat) and Lord Rayleigh has shown how to deduce it from purely dynamical principles.

B. Stewart has written on the history of the theory in the Transactions of the Royal Society of Edinburgh and in the Reports of the British Association.

The principal theorem may be stated as follows. Let there be two bodies *A* and *B* at the same temperature. Between these bodies let there be any arrangement whatever of transparent media, mirrors, and screens including a closed curve *C* which defines the rays which we have to consider.

Then if of radiations of a given kind emitted by *A* and passing through the aperture *C* and falling on *B*, a certain proportion are absorbed by *B*, then of radiations from *B* of the same kind passing through *C* and falling on *A* the same proportion will be absorbed.

Or leaving out *B*.

If at a certain temperature, *A* radiates through the aperture *C* rays of a certain kind in a certain quantity, then, at the same temperature, *A* will absorb rays of that kind falling on it after passing through *C*, and the coefficient of absorption will be equal to the ratio of the emissive power of *A* to that of a perfectly absorbing body.

In these statements rays differing in wave length and in polarization are understood to be of different kinds.

You have yourself observed, that if the sun's disk were broken up and strewed all over the sky, the result would not be a shell of uniform temperature, and therefore the reasoning about such a shell would not apply.

With respect to the radiometer Mr Crookes has shown that it will turn under the action of radiations from all parts of a hollow sphere, provided the temperature of the sphere is greater than that of the fly. As soon as the fly gets warmed up to the temperature of the shell, it is no longer driven, and soon stops.[e]

But if the spherical surface is only a number of radiant points with nothing between them, a large part of the radiation from the radiometer goes off into space through the intervals.

The radiometer therefore never arrives at the temperature of the radiant points, and therefore emits radiations of a different kind from those it receives. It will therefore make a great difference to it, whether it is surrounded with a glass globe which is more transparent to radiations from the sun than to those from a cooler body. The late Mr Hopkins[f] of Cambridge wrote several papers on the effect of an atmosphere on the mean temperature of a planet. I think they are in the Cambridge Transactions but at present I have no opportunity of giving references.

Yours faithfully

J. Clerk Maxwell

P. S. June 2nd 1879

I have just received your letter of 20th May in which you state in a very clear and correct manner the nature of the whole radiation, (emitted, reflected, transmitted, dispersed, fluorescent or phosphorescent) which leaves the surface of a body when entirely surrounded with an opaque shell at the same temperature.

Your views on this subject, as those of an original and independent investigator, have a scientific value which even to strangers is greater, to those whom you teach very much greater, and to yourself infinitely greater than that of any statements derived from other sources.

Still I think that the value of a printed paper, considered in itself apart from its author, is greatly increased if it contains an historical statement of what has already been done by others, and if it states explicitly the nature of the contribution to science made by the author, whether new facts or theories or verification or coordination of known ones.

If as often must be the case, the author has himself made considerable progress in an investigation before he learns that something of the same kind has been already done, he ought, of course to state as much as he can ascertain of the external history of the subject, but it still remains for him to judge whether the process by which he himself learned what he knows may not be worthy to be recorded elsewhere than in his own memory.

Of course new methods of proving old theorems and new methods of illustrating them rank among contributions to that storehouse of printed matter which is so useful to us all and which we often think of when we speak of Science. Although we known that Science does not really eixst in libraries, but in the minds of men, and the outward and visible thing which comes nearest to real science is a living man of science actually discovering something in the presence of his pupils, or going through the process by which he has already made the discovery or even giving an account of it in print. To have learned Euclid I.49 is one thing. To have assisted Pythagoras in sacrificing his hecatomb is another.

In the end of your letter you indicate that the law of exchanges is connected with a law of molecular motion common to all bodies. Here you touch upon the leading idea of contemporary science. As the conservation of energy was the leading idea of the middle of the century, so the dissipation, communication, and equilibrium of energy is the subject which at present best repays the labor of research.

Boltzmann, of Graz, has contributed greatly to the ⟨mathematical⟩ dynamical, theory of the subject. See his papers in the Sitzungsberichte of the Vienna Academy.[g] Willard Gibbs, of Yale, has attacked the same subject by pure thermodynamics, and has laid a massive foundation for a splendid structure in his papers "On the Equilibrium of Heterogeneous Substances," in the Trans. Acad. Connecticut.[h]

I wish you all success in your scientific speculations, and I also hope that your experimental quest of the velocity of light by means of the revolving mirror may turn out successful.

Yours very truly

J. Clerk Maxwell

a. Simon Newcomb (1835–1909), Director of the U. S. Naval Observatory, was "the most honored American scientist of his time" (Brian G. Marsden, "Newcomb, Simon," *Dictionary of Scientific Biography*, 10 (1974), 33–36).

b. Pierre Prevost (1751–1839), Swiss physicist; see "Sur l'equilibre du Feu," *Journal de Physique* 38 (1791), 314–322; translated in *The Laws of Radiation and Absorption*, D. B. Brace, ed. (New York: American Book Co., 1901), 1–13.

c. J. B. J. Fourier, "Questions sur la Theorie Physique de la Chaleur Rayonnante," *Ann. Chim. Phys.* 6 (1817), 259–303.

d. Balfour Stewart, "An Account of Some Experiments on Radiant Heat, Involving an Extension of Prevost's Theory of Exchanges," *Rep. 28th Meeting BAAS* (1858), 23–24; *Proc. R. Soc. Edinburgh* 22 (1861), 1–20; reprinted in *The Laws of Radiation* (op. cit., note b), 21–50. Gustav Kirchhoff, "Ueber den Zusammenhang zwischen Emission und Absorption von Licht und Wärme," *Monats. Akad. Wiss. Berlin* (1860), 783–787. See Daniel M. Siegel, "Balfour Stewart and Gustav Robert Kirchhoff: Two Independent Approaches to 'Kirchhoff's Ratiation Law,'" *Isis* 67 (1976), 565–600.

e. The radiometer is discussed in our next volume of Maxwell's papers.

f. Wiliam Hopkins (1793–1866), the great Cambridge private tutor who coached Stokes, Cayley, William Thomson, Tait, Maxwell, and many others for the Cambridge Mathematical Tropos, was also the founder of mathematical geophysics in Britain. Maxwell may have been thinking of Hopkins's paper "On the External Temperature of the Earth" *Monthly Notices of the Royal Astronomical Society* 17 (1856), 190–195; see other papers listed in the Royal Society *Catalogue*.

g. Ludwig Boltzmann, *Wissenschaftliche Abhandlungen* (New York: Chelsea, 1968, reprint of the 1909 edition), 3 vols.; *Ludwig Boltzmann Gesamtausgabe*, Roman U. Sexl, ed. (Graz: Akademische Druck-u. Verlagsanstalt, 1981–, in progress). English translations of two papers that Maxwell may have had in mind (1872 and 1877) are in Brush, *Kinetic Theory*, vol. 2.

h. See *The Collected Works of J. Willard Gibbs* (New Haven: Yale University Press, 1948), vol. 1.

III Documents on the Kinetic Theory of Gases

1. Letter from Maxwell to George Gabriel Stokes, May 30, 1859

Memoir and Scientific Correspondence of the Late Sir George Gabriel Stokes, Joseph Larmor, ed. (Cambridge: At the University Press, 1907), vol. 2, 8–11.

Glenlair, Springholm, Dumfries,[a]
30 May 1859

Dear Stokes,[b]

I saw in the *Philosophical Magazine* of February, '59, a paper by Clausius on the "mean length of path of a particle of air or gas between consecutive collisions,"[c] on the hypothesis of the elasticity of a gas being due to the velocity of its particles and of their paths being rectilinear except when they come into close proximity to each other, which event may be called a collision.

The result arrived at by Clausius is that[d] of N particles Ne^l reach a distance greater than nl where l is the "mean path" and that

$$\frac{1}{l} = \frac{4}{3}\pi s^2 N$$

where s is the radius of the sphere of action of a particle, and N the number of particles in unit of volume. Note.—I find $1/l = \sqrt{2}\pi s^2 N$

As we know nothing about either s or N, I thought that it might be worth while examining the hypothesis of free particles acting by impact and comparing it with phenomena which seem to depend on this "mean path." I have therefore begun at the beginning and drawn up the theory of the motions and collisions of free particles acting only by impact, applying it to internal friction of gases, diffusion of gases, and conduction of heat through a gas (without radiation). Here is the theory of gaseous friction with its results [Diagram 1].

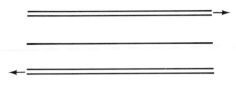

[Diagram 1]

Divide the gas into layers on each side of the plane in which you measure the friction, and suppose the tangential motion uniform in each layer, but varying from one to another. Then particles from a layer on one side of the plane will always be darting about, and some of them will strike the particles belonging to a layer on the other side of the plane, moving with a different mean velocity of translation.

Now though the velocities of these particles are very great and in all directions, their mean velocity, that of their centre of gravity, is that of the layer from which they started, so that the other layer will receive so many particles per second, having a different velocity from its own.

Taking the action of all the layers on the one side and all the layers on the other side I find for the force on unit of area

$$F = \frac{1}{3} M N l v \frac{du}{dr}$$

where M = mass of a particle, N no. of particles in unit of volume, l = mean path, and v = mean molecular velocity and $\frac{du}{dr}$ velocity of slipping. Now you put[e,f]

$$F = \mu \frac{du}{dr}$$

$$\therefore \mu = M N l v = \rho l v$$

Now[g]

$$v = \sqrt{\frac{8k}{\pi}}$$

and if we take

$$\sqrt{k} = 930 \text{ feet per second}, \ v = 1505 \text{ feet per second},$$

and if

$$\sqrt{\frac{\mu}{\rho}} = .116 \text{ in inches \& seconds},$$

we find

$$l = \frac{1}{447000} \text{ of an inch,}$$

and the number of collisions per second

8,077,000,000,

for each particle.

The rate of diffusion of gases depends on more particles passing a given plane in one direction than another, owing to one gas being denser towards the one side, and the other towards the other.

It appears that l should be the same [?] for all pure gases at the same pressure and temperature, but I have found only a very few experiments by Prof Graham on diffusion through measurable apertures, and these seem to give values of l much larger than that derived from friction.

I should think it would be very difficult to make experiments on the conductivity of a gas so as to eliminate the effect of radiation from the sides of the containing vessel, so that this would not give so good a method of determining l.

I do not know how far such speculations may be found to agree with facts, even if they do not it is well to know that Clausius' (or rather Herapath's)[h] theory is wrong,[i] and at any rate as I found myself able and willing to deduce the laws of motion of systems of particles acting on each other only by impact, I have done so as an exercise in mechanics. Now do you think there is any so complete a refutation of this theory of gases as would make it absurd to investigate it further so as to found arguments upon measurements of strictly "molecular" quantities before we know whether there be any molecules? One curious result is that μ is independent of the density, for

$$\mu = MNlv = \frac{Mv}{\sqrt{2\pi s^2}}$$

This is certainly very unexpected, that the friction should be as great in a rare as in a dense gas. The reason is, that in the rare gas the mean path is greater, so that the frictional action extends to greater distances.

Have you any means of refuting this result of the hypothesis?

Of course my particles have not all the same velocity, but the velocities are distributed according to the same formula as the errors are distributed in the theory of least squares.

If two sets of particles act on each other the mean *vis viva* of a particle will become the same in both, which implies, that equal volumes of gases at same press. and temp. have the same number of particles, that is, are chemical equivalents. This is one satisfactory result at least.

I have been rather diffuse on gases but I have taken to the subject for mathematical work lately and I am getting fond of it and require to be snubbed a little by experiments, and I have only a few of Prof. Graham's,[j] quoted by Herapath, on diffusion so that I am tolerably high-minded still.

I had a colour-blind student last season who was a good student and a prizeman so I got pretty good observations out of him.[k] Any four colours can be arranged, three against one, or two against two, so as to form an equation to a colour-blind eye. I took 6 colours, Red, Blue, Green, Yellow, White, Black and formed the 15 sets of four of them and tried 14 of these by the help of my student, making no suggestions of course to him, after I had instructed him in the things to be observed.

I then found the most probable values of the 3 equations, which ought to contain the whole 15, and then deducing the 15 equations from the 3 I found them very near the observed ones, the average error being 2 hundredths of the circle on each colour observed, and much less where the colours were very decided to a colour-blind eye. I hope to see him again next year.

I suppose we shall hear something of the theory of absorption of light in crystals from you in due time. I hope Cambridge takes an interest in light in May.

Yours truly,
James Clerk Maxwell

a. Glenlair was Maxwell's family estate in Scotland to which he retired every summer. As one can see from his correspondence he lacked data and other sources and had to appeal to colleagues.

b. George Gabriel Stokes (1819–1903) was made a baronet in 1889. He was Lucasian Professor at Cambridge University from 1849 until his death, Secretary of the Royal Society of London from 1854 to 1885, and its President from 1889 to 1890. His early work on the hydrodynamics of viscous fluids furnished the basic macroscopic equations for Maxwell's kinetic theory (*Scientific Papers*, vol. 2, 69 (document III-30) and elsewhere).

c. Rudolph Julius Emmanuel Clausius (1822–1888); see note 19 of chapter I. The contributions of Clausius to kinetic theory have been discussed by S. G. Brush, *The Kind of Motion We Call Heat* (New York: North-Holland Pub. Co., 1976), chap. 4, and by E. W. Garber, *Maxwell, Clausius and Gibbs: Aspects of the Development of Kinetic Theory and Thermodynamics* (Ph.D. Dissertation, Case Institute of Technology, 1966); "Clausius and Maxwell's Kinetic Theory of Gases," *Historical Studies in the Physical Sciences* 2 (1970), 299–319.

d. This expression is quoted incorrectly. If l is the mean free path, then of N particles Ne^{-nl} reach a distance greater than nl. For the discrepancy between Clausius's factor 4/3 and Maxwell's factor $\sqrt{2}$, see Niven's note to Prop. X of "Illustrations of the

Dynamical Theory of Gases," in Maxwell's *Scientific Papers*, vol. 1, 377–409 (document III-6).

e. μ is the coefficient of viscosity for the gas. The word "you" refers to Stokes's papers on viscosity of fluids.

f. ρ is the density of the gas. The factor $1/3$ derivable from the two previous equations is missing in the original.

g. The expression $v = \sqrt{(8k/\pi)}$ is derived from Maxwell's velocity distribution function, where $k \propto \alpha^2$ and u^2 is the mean square velocity, $\overline{u^2} = (3/2)\alpha^2$. In the published version of his paper Maxwell wrote $v = 2\alpha/\sqrt{\pi} = 2/(2k/\pi)$. See *Scientific Papers*, vol. 1, 391 (document III-6).

h. John Herapath (1790–1868) developed the theory that heat was the manifestation of "intestine motion." He derived the relation between pressure, volume, mass, and the speed of the particles of a gas. See notes 15 and 17, chapter I, and document II-3.

i. Larmor, evidently disturbed by Maxwell's bold use of "wrong" in connection with Herapath's kinetic theory, added a footnote "i.e. inadequate."

j. Thomas Graham (1805–1869). The paper to which Maxwell refers is cited, along with others, in note 57, chapter I; cf. note c to document III-2.

k. James Simpson; see *Scientific Papers*, vol. 1, 438f.

2. "On the Dynamical Theory of Gases"

Report of the 29th Meeting of the British Association for the Advancement of Science, Notices and Abstracts (Aberdeen, September 1859), 9.

The phenomena of the expansion of gases by heat, and their compression by pressure, have been explained by Joule,[a] Claussens,[b] Herapath, &c., by the theory of their particles being in a state of rapid motion, the velocity depending on the temperature. These particles must not only strike against the sides of the vessel, but against each other, and the calculation of their motions is therefore complicated. The author has established the following results:—1. The velocities of the particles are not uniform, but vary so that they deviate from the mean value by a law well known in the "method of least squares." 2. Two different sets of particles will distribute their velocities, so that their *vires vivae* will be equal; and this leads to the chemical law, that the equivalents of gases are proportional to their specific gravities. 3. From Prof. Stokes's experiments on friction in air, it appears that the distance travelled by a particle between consecutive collisions is about $1/447,000$th of an inch, the mean velocity being about 1505 feet per second; and therefore each particle makes 8,077,200,000 collisions per second. 4. The laws of the diffusion of gases, as established by the Master of the Mint,[c] are deduced from this theory, and the absolute rate of diffusion through an opening can be calculated. The author intends to apply his mathematical methods to the explanation on this hypothesis of the propagation of sound,[d] and expects some light on the mysterious question of the absolute number of such particles in a given mass.[e]

a. James Prescott Joule (1818–1889). Joule's first theory of gases, first presented at a meeting of the Manchester Literary and Philosophical Society in 1848, was heavily influenced by Herapath; for references see note 18 to chapter I.

b. Claussens, i.e., Clausius, evidently a printer's error.

c. Thomas Graham was appointed Master of the Mint in 1854, succeeding John Herschel. His papers on diffusion are cited in note 57, chapter I.

d. Maxwell did not include any estimate of the speed of sound in his 1860 paper. Waterston's work on this problem was brought to his attention by S. Tolver Preston (document II-29), and this led Maxwell to give it brief consideration (note c to document II-29), but his main concern was with the related problem of the ratio of specific heats (see the end of document III-8).

e. This came within a decade, when Loschmidt and others used Maxwell's theory to estimate the size of a molecule and thus the number of them in a given mass or volume; see chapter I, notes 138–142, and documents III-35, 36, and 42.

3. Letter[a] from Maxwell to George Gabriel Stokes, October 8, 1859

Memoir and Scientific Correspondence of the Late Sir George Gabriel Stokes, Joseph Larmor, ed. (Cambridge: At the University Press, 1907), vol. 2, 11.

Glenlair, Springholm, Dumfries,
1859, Oct. 8.

Dear Stokes,

I received your letter some days ago. I intend to arrange my propositions about the motions of elastic spheres in a manner independent of the speculations about gases, and I shall probably send them to the *Phil. Magazine,* which publishes a good deal about the dynamical theories of matter & heat.[b] I have not done much to it since I wrote to you, as all my optical observations must be done when I have sun and that is getting lower every day. But I think I have made as much optical hay as will make a small bundle for you as Secretary R. S., some time in winter if you can tell me under what forms it should be sent in.[c] I would call it On the Relations of the Colours of the Spectrum as seen by the Human Eye.[d]

a. Only the first paragraph of this long letter has been reproduced here. The rest of the letter contains Maxwell's experimental results on color.

b. For the published version, see document III-6.

c. Stokes became Secretary of the Royal Society of London in 1854 and remained in that office until 1885.

d. Maxwell sent Stokes the completed manuscript the following January; it was read as the Bakerian Lecture for 1860 on March 22, 1860, and was published under the title "On the Theory of Compound Colours, and the Relations of the Colours of the Spectrum," *Phil. Trans. R. Soc. London* (1860), 57–84. See *Memoir and Scientific Correspondence of the Late Sir George Gabriel Stokes*, vol. 2, 14–16, and Maxwell's *Scientific Papers*, vol. 1, 410–444.

4. [Notes for "Illustrations of the Dynamical Theory of Gases" [a]

Cambridge University Library, Maxwell Collection, Scientific Papers 6.

To find the number of collisions in unit of time between n particles having a given velocity v and N particles having velocities distributed ⟨according⟩ so that the number having velocities between u and $u + du$ is $\dfrac{4N}{\alpha^3 \sqrt{\pi}} u^2 e^{-u^2/\alpha^2}\, du$,

the motion being in one unit of volume. Let the distance between the centres at collision be s and the relative velocity of a pair of particles r, then the number of collisions in unit of time between that pair of particles will be $\pi s^2 r$.

Let lines representing the velocities of the N particles be drawn from a point, then the number of such lines having their extremities in a space $=1$ at a distance u will be [b]

$$\frac{N}{\alpha^3 \pi^{7/2}} e^{-u^2/\alpha^2}$$

Let us call this the "density" of the distribution of velocities corresponding to a given direction and magnitude of u. In general it will depend on the direction as well as the magnitude of u.

Now let v be the velocity of the n particles, let r be their velocity relative to a particle of the other system and let θ be the angle between v and r, then

$$u^2 = v^2 + r^2 - 2vr \cos\theta$$

and the number of particles of the second system for which r lies between r and $r + dr$ and θ between θ and $\theta + d\theta$ will be

$$\frac{N}{\alpha^3 \pi^{3/2}} e^{-u^2/\alpha^2} 2\pi r^2 \sin\theta\, dr\, d\theta = \frac{2N}{\alpha^3 \pi^{1/2}} \frac{ru}{v} e^{-u^2/\alpha^2}\, dr\, du$$

if we make r & u the variables instead of v & θ. The number of collisions between these and the n particles whose relative velocity is r, is

$$\frac{2\sqrt{\pi}\, s^2 Nn}{\alpha^3 v} r^2 u\, e^{-u^2/\alpha^2}\, dr\, du$$

Integrating with respect to u from $u^2 = (v - r)^2$ to $u^2 = (v + r)^2$

$$\frac{\sqrt{\pi}\, s^2 Nn}{\alpha v} r^2 \left[e^{-(r-v)^2/\alpha^2} - e^{-(r+v)^2/\alpha^2} \right] dr$$

Integrating with respect to r from $r = 0$ to $r = \infty$,

$$\pi s^2 Nn\alpha \left[e^{-v^2/\alpha^2} + \left(\frac{\alpha}{v} + 2\frac{v}{\alpha} \right) \int_0^v e^{-v^2/\alpha^2}\, dv \right]$$

This is the number of collisions in unit of time between n particles moving with velocity v and N particles whose modulus of velocity is α.

a. This calculation probably formed part of the basis for Prop. VIII of Maxwell's 1860 paper; see *Scientific Papers*, vol. 1, 384–385. The following set of notes clusters around problems associated with the relative motion of two sets of particles. Maxwell wanted to know the distribution function for the relative velocity for two sets of particles whose distribution followed the mean square law. Probably this interest sprang from the only phenomenon for which he had any data, diffusion. In the diffusion problem two sets of gas molecules move with respect to one another, and the distribution of their respective velocities is not spherical. Maxwell even tried to take this complication into account in his notes.

b. The exponent for π should be $\frac{1}{2}$. Maxwell corrected this in his later expressions.

5. [Notes on Graham's Diffusion Data]

Cambridge University Library, Maxwell Collection, Scientific Papers 6.

Thomas Graham on Diffusion of Gases, Brandes Journal,[a] 1829 pt. 2 p. 74 Tube 9 inches long and 0.9 internal diameter diffusion tube 2 inches long & 0.12 diam. [Diagram 1]

Gas	sp.g.[b]	by diffusion in 10 hours out of 150 parts there were left	after 4 hours out of 152 there were left
Hydrogen	0.0694	8.3	28.1
Marsh Gas[c]	0.5555	56	86
Ammonia	0.59027	61	89
Olefiant[d]	0.9722	77.5	99
Carbonic acid	1.5277	79.5	104
Sulphurous acid	2.2222	81.	110
Chlorine	2.5	91.	116

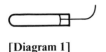

[Diagram 1]

	Time	Left		
75H + 75 olefiant gas	10 hours	H 3.5 Olefiant gas 56.6 Air 89.9		
75CO₂ + 75H	10	CO₂ 45	H 4.65	Air 100.35
102 + 50	4 hours	76	10.3	65.7
74CO₂ + 76 Marsh Gas	4	57 Marsh Gas 35.3		59.7
52 100	4	39	51.6	61.4
31 121	4	23	71	62.85
Olefiant gas 76 Marsh gas 76	4	Olefiant 47.75 41.40		62.85

$A = 5.2$ cub[ic] inches $B = 37$ cubic inches tube 0.12 bore length not given. A equal parts of olefiant gas and H. B, CO_2. After 10 hours A contained olefiant gas 12 Hydrogen 3.1 [Diagram 2].

[Diagram 2]

a. Thomas Graham, "A Short Account of Experimental Researches on the Diffusion of Gases through Each Other, and their Separation by Mechanical Means," *Quarterly Journal of Science and the Arts*, 27 (1829), 74–83. Maxwell cites this paper of Graham in "Illustrations of the Dynamical Theory of Gases" (document III-6) and from the data got an estimate of the mean free path (see *Scientific Papers*, vol. 1, 403). He also cited these data in his later paper "On the Dynamical Theory of Gases" (document III-30; see *Scientific Papers*, vol. 2, 61).

b. Specific gravity.

c. A mixture of gases evolved during the breakdown of organic material, mainly methane.

d. Ethylene.

6. "Illustrations of the Dynamical Theory of Gases"

Phil. Mag. [4] 19 (1860), 19–32; 20 (1860), 21–37. *Scientific Papers*, vol. 1, 377–409. Footnotes by W. D. Niven.

[From the *Philosophical Magazine* for January and July, 1860.]

XX.   *Illustrations of the Dynamical Theory of Gases**.

PART I.

On the Motions and Collisions of Perfectly Elastic Spheres.

So many of the properties of matter, especially when in the gaseous form, can be deduced from the hypothesis that their minute parts are in rapid motion, the velocity increasing with the temperature, that the precise nature of this motion becomes a subject of rational curiosity.   Daniel Bernouilli, Herapath, Joule, Krönig, Clausius, &c. have shewn that the relations between pressure, temperature, and density in a perfect gas can be explained by supposing the particles to move with uniform velocity in straight lines, striking against the sides of the containing vessel and thus producing pressure.   It is not necessary to suppose each particle to travel to any great distance in the same straight line; for the effect in producing pressure will be the same if the particles strike against each other; so that the straight line described may be very short. M. Clausius has determined the mean length of path in terms of the average distance of the particles, and the distance between the centres of two particles when collision takes place.   We have at present no means of ascertaining either of these distances; but certain phenomena, such as the internal friction of gases, the conduction of heat through a gas, and the diffusion of one gas through another, seem to indicate the possibility of determining accurately the mean length of path which a particle describes between two successive collisions.   In order to lay the foundation of such investigations on strict mechanical principles, I shall demonstrate the laws of motion of an indefinite number of small, hard, and perfectly elastic spheres acting on one another only during impact.

* Read at the Meeting of the British Association at Aberdeen, September 21, 1859.

48

378   ILLUSTRATIONS OF THE DYNAMICAL THEORY OF GASES.

If the properties of such a system of bodies are found to correspond to those of gases, an important physical analogy will be established, which may lead to more accurate knowledge of the properties of matter. If experiments on gases are inconsistent with the hypothesis of these propositions, then our theory, though consistent with itself, is proved to be incapable of explaining the phenomena of gases. In either case it is necessary to follow out the consequences of the hypothesis.

Instead of saying that the particles are hard, spherical, and elastic, we may if we please say that the particles are centres of force, of which the action is insensible except at a certain small distance, when it suddenly appears as a repulsive force of very great intensity. It is evident that either assumption will lead to the same results. For the sake of avoiding the repetition of a long phrase about these repulsive forces, I shall proceed upon the assumption of perfectly elastic spherical bodies. If we suppose those aggregate molecules which move together to have a bounding surface which is not spherical, then the rotatory motion of the system will store up a certain proportion of the whole *vis viva*, as has been shewn by Clausius, and in this way we may account for the value of the specific heat being greater than on the more simple hypothesis.

On the Motion and Collision of Perfectly Elastic Spheres.

Prop. I. Two spheres moving in opposite directions with velocities·inversely as their masses strike one another; to determine their motions after impact.

Let P and Q be the position of the centres at impact; AP, BQ the directions and magnitudes of the velocities before impact; Pa, Qb the same after impact; then, resolving the velocities parallel and perpendicular to PQ the line of centres, we find that the velocities parallel to the line of centres are exactly reversed, while those perpendicular to that line are

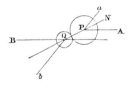

unchanged. Compounding these velocities again, we find that the velocity of each ball is the same before and after impact, and that the directions before and after impact lie in the same plane with the line of centres, and make equal angles with it.

Prop. II.   To find the probability of the direction of the velocity after impact lying between given limits.

In order that a collision may take place, the line of motion of one of the balls must pass the centre of the other at a distance less than the sum of their radii; that is, it must pass through a circle whose centre is that of the other ball, and radius (s) the sum of the radii of the balls. Within this circle every position is equally probable, and therefore the probability of the distance from the centre being between r and $r + dr$ is

$$\frac{2r\,dr}{s^2}.$$

Now let ϕ be the angle APa between the original direction and the direction after impact, then $APN = \frac{1}{2}\phi$, and $r = s \sin \frac{1}{2}\phi$, and the probability becomes

$$\tfrac{1}{2} \sin \phi\, d\phi.$$

The area of a spherical zone between the angles of polar distance ϕ and $\phi + d\phi$ is

$$2\pi \sin \phi\, d\phi\,;$$

therefore if ω be any small area on the surface of a sphere, radius unity, the probability of the direction of rebound passing through this area is

$$\frac{\omega}{4\pi}\,;$$

so that the probability is independent of ϕ, that is, all directions of rebound are equally likely.

Prop. III.   Given the direction and magnitude of the velocities of two spheres before impact, and the line of centres at impact; to find the velocities after impact.

Let OA, OB represent the velocities before impact, so that if there had been no action between the bodies they would have been at A and B at the end of a second. Join AB, and let G be their centre of gravity, the position of which is not affected by their mutual action. Draw GN parallel to the line of centres at impact (not necessarily in the plane AOB). Draw aGb in the plane AGN, making $NGa = NGA$, and $Ga = GA$ and $Gb = GB$; then by

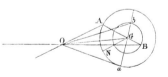

Prop. I. Ga and Gb will be the velocities relative to G; and compounding these with OG, we have Oa and Ob for the true velocities after impact.

By Prop. II. all directions of the line aGb are equally probable. It appears therefore that the velocity after impact is compounded of the velocity of the centre of gravity, and of a velocity equal to the velocity of the sphere relative to the centre of gravity, which may with equal probability be in any direction whatever.

If a great many equal spherical particles were in motion in a perfectly elastic vessel, collisions would take place among the particles, and their velocities would be altered at every collision; so that after a certain time the *vis viva* will be divided among the particles according to some regular law, the average number of particles whose velocity lies between certain limits being ascertainable, though the velocity of each particle changes at every collision.

Prop. IV. To find the average number of particles whose velocities lie between given limits, after a great number of collisions among a great number of equal particles.

Let N be the whole number of particles. Let x, y, z be the components of the velocity of each particle in three rectangular directions, and let the number of particles for which x lies between x and $x+dx$, be $Nf(x)dx$, where $f(x)$ is a function of x to be determined.

The number of particles for which y lies between y and $y+dy$ will be $Nf(y)dy$; and the number for which z lies between z and $z+dz$ will be $Nf(z)dz$, where f always stands for the same function.

Now the existence of the velocity x does not in any way affect that of the velocities y or z, since these are all at right angles to each other and independent, so that the number of particles whose velocity lies between x and $x+dx$, and also between y and $y+dy$, and also between z and $z+dz$, is

$$Nf(x)f(y)f(z)\,dx\,dy\,dz.$$

If we suppose the N particles to start from the origin at the same instant, then this will be the number in the element of volume $(dx\,dy\,dz)$ after unit of time, and the number referred to unit of volume will be

$$Nf(x)f(y)f(z).$$

But the directions of the coordinates are perfectly arbitrary, and therefore this number must depend on the distance from the origin alone, that is

$$f(x)f(y)f(z) = \phi(x^2 + y^2 + z^2).$$

Solving this functional equation, we find

$$f(x) = Ce^{Ax^2}, \qquad \phi(r^2) = C^3 e^{Ar^2}.$$

If we make A positive, the number of particles will increase with the velocity, and we should find the whole number of particles infinite. We therefore make A negative and equal to $-\dfrac{1}{a^2}$, so that the number between x and $x + dx$ is

$$NCe^{-\frac{x^2}{a^2}} dx.$$

Integrating from $x = -\infty$ to $x = +\infty$, we find the whole number of particles,

$$NC\sqrt{\pi}a = N, \quad \therefore \ C = \frac{1}{a\sqrt{\pi}},$$

$f(x)$ is therefore

$$\frac{1}{a\sqrt{\pi}} e^{-\frac{x^2}{a^2}}.$$

Whence we may draw the following conclusions :—

1st. The number of particles whose velocity, resolved in a certain direction, lies between x and $x + dx$ is

$$N \frac{1}{a\sqrt{\pi}} e^{-\frac{x^2}{a^2}} dx \ \dotfill \ (1).$$

2nd. The number whose actual velocity lies between v and $v + dv$ is

$$N \frac{4}{a^3 \sqrt{\pi}} v^2 e^{-\frac{v^2}{a^2}} dv \dotfill (2).$$

3rd. To find the mean value of v, add the velocities of all the particles together and divide by the number of particles ; the result is

$$\text{mean velocity} = \frac{2a}{\sqrt{\pi}} \dotfill (3).$$

4th. To find the mean value of v^2, add all the values together and divide by N,

$$\text{mean value of } v^2 = \tfrac{3}{2}a^2 \dotfill (4).$$

This is greater than the square of the mean velocity, as it ought to be.

It appears from this proposition that the velocities are distributed among the particles according to the same law as the errors are distributed among the observations in the theory of the "method of least squares." The velocities range from 0 to ∞, but the number of those having great velocities is comparatively small. In addition to these velocities, which are in all directions equally, there may be a general motion of translation of the entire system of particles which must be compounded with the motion of the particles relatively to one another. We may call the one the motion of translation, and the other the motion of agitation.

PROP. V. Two systems of particles move each according to the law stated in Prop. IV.; to find the number of pairs of particles, one of each system, whose relative velocity lies between given limits.

Let there be N particles of the first system, and N' of the second, then NN' is the whole number of such pairs. Let us consider the velocities in the direction of x only; then by Prop. IV. the number of the first kind, whose velocities are between x and $x+dx$, is

$$N \frac{1}{a\sqrt{\pi}} e^{-\frac{x^2}{a^2}} dx.$$

The number of the second kind, whose velocity is between $x+y$ and $x+y+dy$, is

$$N' \frac{1}{\beta\sqrt{\pi}} e^{-\frac{(x+y)^2}{\beta^2}} dy,$$

where β is the value of a for the second system.

The number of pairs which fulfil both conditions is

$$NN' \frac{1}{a\beta\pi} e^{-\left(\frac{x^2}{a^2}+\frac{(x+y)^2}{\beta^2}\right)} dx\, dy.$$

Now x may have any value from $-\infty$ to $+\infty$ consistently with the difference of velocities being between y and $y+dy$; therefore integrating between these limits, we find

$$NN' \frac{1}{\sqrt{a^2+\beta^2}\sqrt{\pi}} e^{-\frac{y^2}{a^2+\beta^2}} dy \dots\dots\dots\dots\dots\dots (5)$$

for the whole number of pairs whose difference of velocity lies between y and $y+dy$.

This expression, which is of the same form with (1) if we put NN' for N, $a^2 + \beta^2$ for a^2, and y for x, shews that the distribution of relative velocities is regulated by the same law as that of the velocities themselves, and that the mean relative velocity is the square root of the sum of the squares of the mean velocities of the two systems.

Since the direction of motion of every particle in one of the systems may be reversed without changing the distribution of velocities, it follows that the velocities compounded of the velocities of two particles, one in each system, are distributed according to the same formula (5) as the relative velocities.

PROP. VI.   Two systems of particles move in the same vessel ; to prove that the mean *vis viva* of each particle will become the same in the two systems.

Let P be the mass of each particle of the first system, Q that of each particle of the second. Let p, q be the mean veloci- ties in the two systems before impact, and let p', q' be the mean velocities after one impact. Let $OA = p$ and $OB = q$, and let AOB be a right angle; then, by Prop. V., AB will be the mean relative velocity, OG will be the mean velocity of the centre of gravity ; and drawing aGb at right angles to OG, and making $aG = AG$ and $bG = BG$, then Oa will be the mean velocity of P after impact, compounded of OG and Ga, and Ob will be that of Q after impact.

Now $\qquad AB = \sqrt{p^2 + q^2}, \ AG = \dfrac{Q}{P+Q} \sqrt{p^2 + q^2}, \ BG = \dfrac{P}{P+Q} \sqrt{p^2 + q^2},$

$$OG = \frac{\sqrt{P^2 p^2 + Q^2 q^2}}{P+Q},$$

therefore $\qquad p' = Oa = \dfrac{\sqrt{Q^2 (p^2 + q^2) + P^2 p^2 + Q^2 q^2}}{P+Q},$

and $\qquad q' = Ob = \dfrac{\sqrt{P^2 (p^2 + q^2) + P^2 p^2 + Q^2 q^2}}{P+Q},$

and $\qquad Pp'^2 - Qq'^2 = \left(\dfrac{P-Q}{P+Q}\right)^2 (Pp^2 - Qq^2) \ \dotfill \ (6).$

It appears therefore that the quantity $Pp^2 - Qq^2$ is diminished at every impact in the same ratio, so that after many impacts it will vanish, and then

$$Pp^2 = Qq^2.$$

Now the mean *vis viva* is $\frac{3}{2}Pa^2 = \frac{3\pi}{8}Pp^2$ for P, and $\frac{3\pi}{8}Qq^2$ for Q; and it is manifest that these quantities will be equal when $Pp^2 = Qq^2$.

If any number of different kinds of particles, having masses P, Q, R and velocities p, q, r respectively, move in the same vessel, then after many impacts

$$Pp^2 = Qq^2 = Rr^2, \ \&c. \ \dots\dots\dots\dots\dots\dots\dots(7).$$

PROP. VII.   A particle moves with velocity r relatively to a number of particles of which there are N in unit of volume; to find the number of these which it approaches within a distance s in unit of time.

If we describe a tubular surface of which the axis is the path of the particle, and the radius the distance s, the content of this surface generated in unit of time will be πrs^2, and the number of particles included in it will be

$$N\pi rs^2\dots\dots\dots\dots\dots\dots\dots\dots\dots\dots\dots(8),$$

which is the number of particles to which the moving particle approaches within a distance s.

PROP. VIII.   A particle moves with velocity v in a system moving according to the law of Prop. IV.; to find the number of particles which have a velocity relative to the moving particle between r and $r+dr$.

Let u be the actual velocity of a particle of the system, v that of the original particle, and r their relative velocity, and θ the angle between v and r, then

$$u^2 = v^2 + r^2 - 2vr \cos \theta.$$

If we suppose, as in Prop. IV., all the particles to start from the origin at once, then after unit of time the "density" or number of particles to unit of volume at distance u will be

$$N\frac{1}{a^3\pi^{\frac{3}{2}}}e^{-\frac{u^2}{a^2}}.$$

From this we have to deduce the number of particles in a shell whose centre is at distance v, radius $= r$, and thickness $= dr$,

$$N \frac{1}{a\sqrt{\pi}} \frac{r}{v} \{e^{-\frac{(r-v)^2}{a^2}} - e^{-\frac{(r+v)^2}{a^2}}\} dr \dots\dots\dots\dots\dots (9),$$

which is the number required.

COR. It is evident that if we integrate this expression from $r = 0$ to $r = \infty$, we ought to get the whole number of particles $= N$, whence the following mathematical result,

$$\int_0^\infty dx \cdot x \left(e^{-\frac{(x-a)^2}{a^2}} - e^{-\frac{(x+a)^2}{a^2}}\right) = \sqrt{\pi} a a \dots\dots\dots\dots (10).$$

PROP. IX. Two sets of particles move as in Prop. V.; to find the number of pairs which approach within a distance s in unit of time.

The number of the second kind which have a velocity between v and $v + dv$ is

$$N' \frac{4}{\beta^3 \sqrt{\pi}} v^2 e^{-\frac{v^2}{\beta^2}} dv = n'.$$

The number of the first kind whose velocity relative to these is between r and $r + dr$ is

$$N \frac{1}{a\sqrt{\pi}} \frac{r}{v} \left(e^{-\frac{(r-v)^2}{a^2}} - e^{-\frac{(r+v)^2}{a^2}}\right) dr = n,$$

and the number of pairs which approach within distance s in unit of time is

$$nn'\pi rs^2,$$

$$= NN' \frac{4}{a\beta^3} s^2 r^2 v e^{-\frac{v^2}{\beta}} \{e^{-\frac{(v-r)^2}{a^2}} - e^{-\frac{(v+r)^2}{a^2}}\} dr \, dv.$$

By the last proposition we are able to integrate with respect to v, and get

$$NN' \frac{4\sqrt{\pi}}{(a^2+\beta^2)^{\frac{3}{2}}} s^2 r^3 e^{-\frac{r^2}{a^2+\beta^2}} dr.$$

Integrating this again from $r = 0$ to $r = \infty$,

$$2NN'\sqrt{\pi} \sqrt{a^2+\beta^2} s^2 \dots\dots\dots\dots\dots\dots (11)$$

is the number of collisions in unit of time which take place in unit of volume between particles of different kinds, s being the distance of centres at collision.

The number of collisions between two particles of the first kind, s_1 being the striking distance, is

$$2N^2 \sqrt{\pi} \sqrt{2a^2 s_1^2} \, ;$$

and for the second system it is

$$2N'^2 \sqrt{\pi} \sqrt{2\beta^2 s_2^2}.$$

The mean velocities in the two systems are $\dfrac{2a}{\sqrt{\pi}}$ and $\dfrac{2\beta}{\sqrt{\pi}}$; so that if l_1 and l_2 be the mean distances travelled by particles of the first and second systems between each collision, then

$$\frac{1}{l_1} = \pi N_1 \sqrt{2} s_1^2 + \pi N_2 \frac{\sqrt{a^2 + \beta^2}}{a} s^2,$$

$$\frac{1}{l_2} = \pi N_1 \frac{\sqrt{a^2 + \beta^2}}{\beta} s^2 + \pi N_2 \sqrt{2} s_2^2.$$

Prop. X. To find the probability of a particle reaching a given distance before striking any other.

Let us suppose that the probability of a particle being stopped while passing through a distance dx, is $a\,dx$; that is, if N particles arrived at a distance x, $Na\,dx$ of them would be stopped before getting to a distance $x + dx$. Putting this mathematically,

$$\frac{dN}{dx} = -Na, \text{ or } N = Ce^{-ax}.$$

Putting $N = 1$ when $x = 0$, we find e^{-ax} for the probability of a particle not striking another before it reaches a distance x.

The *mean distance* travelled by each particle before striking is $\dfrac{1}{a} = l$. The probability of a particle reaching a distance $= nl$ without being struck is e^{-n}. (See a paper by M. Clausius, *Philosophical Magazine*, February 1859.)

If all the particles are at rest but one, then the value of a is

$$a = \pi s^2 N,$$

where s is the distance between the centres at collision, and N is the number of particles in unit of volume. If v be the velocity of the moving particle relatively to the rest, then the number of collisions in unit of time will be

$$v\pi s^2 N \, ;$$

and if v_1 be the actual velocity, then the number will be $v_1 a$; therefore

$$a = \frac{v}{v_1} \pi s^2 N,$$

where v_1 is the actual velocity of the striking particle, and v its velocity relatively to those it strikes. If v_2 be the actual velocity of the other particles, then $v = \sqrt{v_1^2 + v_2^2}$. If $v_1 = v_2$, then $v = \sqrt{2} v_1$, and

$$a = \sqrt{2} \pi s^2 N.$$

Note*. M. Clausius makes $a = \frac{4}{3} \pi s^2 N$.

PROP. XI. In a mixture of particles of two different kinds, to find the mean path of each particle.

Let there be N_1 of the first, and N_2 of the second in unit of volume. Let s_1 be the distance of centres for a collision between two particles of the first set, s_2 for the second set, and s' for collision between one of each kind. Let v_1 and v_2 be the coefficients of velocity, M_1, M_2 the mass of each particle.

The probability of a particle M_1 not being struck till after reaching a distance x_1 by another particle of the same kind is

$$e^{-\sqrt{2} \pi s_1^2 N_1 x}.$$

---

* [In the *Philosophical Magazine* of 1860, Vol. I. pp. 434—6 Clausius explains the method by which he found his value of the mean relative velocity. It is briefly as follows: If u, v be the velocities of two particles their relative velocity is $\sqrt{u^2 + v^2 - 2uv \cos \theta}$ and the mean of this as regards direction only, all directions of v being equally probable, is shewn to be

$$v + \frac{1}{3} \frac{u^2}{v} \quad \text{when } u < v, \text{ and } u + \frac{1}{3} \frac{v^2}{u} \quad \text{when } u > v.$$

If $v = u$ these expressions coincide. Clausius in applying this result and putting u, v for the mean velocities assumes that the mean relative velocity is given by expressions of the same form, so that when the mean velocities are each equal to u the mean relative velocity would be $\frac{4}{3} u$. This step is, however, open to objection, and in fact if we take the expressions given above for the mean velocity, treating u and v as the velocities of two particles which may have any values between 0 and ∞, to calculate the mean relative velocity we should proceed as follows: Since the number of particles with velocities between u and $u + du$ is $N \dfrac{4}{a^3 \sqrt{\pi}} u^2 e^{-\frac{u^2}{a^2}} du$, the mean relative velocity is

$$\frac{16}{a^3 \beta^3 \pi} \int_0^\infty \int_v^\infty u^2 v^2 e^{-\left(\frac{u^2}{a^2} + \frac{v^2}{\beta^2}\right)} \left(u + \frac{1}{3} \frac{v^2}{u}\right) du\, dv + \frac{16}{a^3 \beta^3 \pi} \int_0^\infty \int_0^v u^2 v^2 e^{-\left(\frac{u^2}{a^2} + \frac{v^2}{\beta^2}\right)} \left(v + \frac{1}{3} \frac{u^2}{v}\right) du\, dv.$$

This expression, when reduced, leads to $\dfrac{2}{\sqrt{\pi}} \sqrt{a^2 + \beta^2}$, which is the result in the text. Ed.]

The probability of not being struck by a particle of the other kind in the same distance is

$$e^{-\sqrt{1+\frac{v_2^2}{v_1^2}}\pi s'^2 N_2 x}$$

Therefore the probability of not being struck by any particle before reaching a distance x is

$$e^{-\pi\left(\sqrt{2}s_1^2 N_1 + \sqrt{1+\frac{v_2^2}{v_1^2}}s'^2 N_2\right)x};$$

and if l_1 be the *mean distance* for a particle of the first kind,

$$\frac{1}{l_1} = \sqrt{2}\pi s_1^2 N_1 + \pi\sqrt{1+\frac{v_2^2}{v_1^2}}\,s'^2 N_2 \quad\dots\dots\dots\dots\dots (12).$$

Similarly, if l_2 be the mean distance for a particle of the second kind,

$$\frac{1}{l_2} = \sqrt{2}\pi s_2^2 N_2 + \pi\sqrt{1+\frac{v_1^2}{v_2^2}}\,s'^2 N_1 \dots\dots\dots\dots\dots\dots(13).$$

The mean density of the particles of the first kind is $N_1 M_1 = \rho_1$, and that of the second $N_2 M_2 = \rho_2$. If we put

$$A = \sqrt{2}\,\frac{\pi s_1^2}{M_1}, \quad B = \pi\sqrt{1+\frac{v_2^2}{v_1^2}}\,\frac{s'^2}{M_2}, \quad C = \pi\sqrt{1+\frac{v_1^2}{v_2^2}}\,\frac{s'^2}{M_1}, \quad D = \sqrt{2}\,\frac{\pi s_2^2}{M_2}\dots\dots(14),$$

$$\frac{1}{l_1} = A\rho_1 + B\rho_2, \quad \frac{1}{l_2} = C\rho_1 + D\rho_2\dots\dots\dots\dots\dots\dots(15),$$

and

$$\frac{B}{C} = \frac{M_1 v_2}{M_2 v_1} = \frac{v_2^3}{v_1^3}\dots\dots\dots\dots\dots\dots\dots\dots\dots(16).$$

PROP. XII. To find the pressure on unit of area of the side of the vessel due to the impact of the particles upon it.

Let N = number of particles in unit of volume;

M = mass of each particle;

v = velocity of each particle;

l = mean path of each particle;

then the number of particles in unit of area of a stratum dz thick is

$$N dz \dots\dots\dots\dots\dots\dots\dots\dots\dots\dots\dots (17).$$

The number of collisions of these particles in unit of time is

$$N dz\,\frac{v}{l}\dots\dots\dots\dots\dots\dots\dots\dots\dots\dots(18).$$

The number of particles which after collision reach a distance between nl and $(n+dn)\,l$ is

$$N\,\frac{v}{l}\,e^{-n}\,dz\,dn \dots\dots\dots\dots\dots\dots (19).$$

The proportion of these which strike on unit of area at distance z is

$$\frac{nl-z}{2nl} \dots\dots\dots\dots\dots\dots\dots\dots\dots (20)\,;$$

the mean velocity of these in the direction of z is

$$v\,\frac{nl+z}{2nl} \dots\dots\dots\dots\dots\dots\dots\dots (21).$$

Multiplying together (19), (20), and (21), and M, we find the momentum at impact

$$MN\,\frac{v^2}{4n^2l^3}\left(n^2l^2-z^2\right)e^{-n}\,dz\,dn.$$

Integrating with respect to z from 0 to nl, we get

$$\tfrac{1}{6}MNv^2\,ne^{-n}\,dn.$$

Integrating with respect to n from 0 to ∞, we get

$$\tfrac{1}{6}MNv^2$$

for the momentum in the direction of z of the striking particles; for the momentum of the particles after impact is the same, but in the opposite direction; so that the whole pressure on unit of area is twice this quantity, or

$$p=\tfrac{1}{3}MNv^2.$$

This value of p is independent of l the length of path. In applying this result to the theory of gases, we put $MN=\rho$, and $v^2=3k$, and then

$$p=k\rho,$$

which is Boyle and Mariotte's law. By (4) we have

$$v^2=\tfrac{3}{2}a^2,\quad \therefore\ a^2=2k \dots\dots\dots\dots\dots (23).$$

We have seen that, on the hypothesis of elastic particles moving in straight lines, the pressure of a gas can be explained by the assumption that the square of the velocity is proportional directly to the absolute temperature, and inversely to the specific gravity of the gas at constant temperature, so that at the same pressure and temperature the value of NMv^2 is the same for all gases. But we found in Prop. VI. that when two sets of particles communicate agitation to one another, the value of Mv^2 is the same in each. From this it appears that N, the number of particles in unit of volume, is the same for all gases at the same pressure and temperature. This result agrees with the chemical law, that equal volumes of gases are chemically equivalent.

We have next to determine the value of l, the mean length of the path of a particle between consecutive collisions. The most direct method of doing this depends upon the fact, that when different strata of a gas slide upon one another with different velocities, they act upon one another with a tangential force tending to prevent this sliding, and similar in its results to the friction between two solid surfaces sliding over each other in the same way. The explanation of gaseous friction, according to our hypothesis, is, that particles having the mean velocity of translation belonging to one layer of the gas, pass out of it into another layer having a different velocity of translation; and by striking against the particles of the second layer, exert upon it a tangential force which constitutes the internal friction of the gas. The whole friction between two portions of gas separated by a plane surface, depends upon the total action between all the layers on the one side of that surface upon all the layers on the other side.

PROP. XIII. To find the internal friction in a system of moving particles.

Let the system be divided into layers parallel to the plane of xy, and let the motion of translation of each layer be u in the direction of x, and let $u = A + Bz$. We have to consider the mutual action between the layers on the positive and negative sides of the plane xy. Let us first determine the action between two layers dz and dz', at distances z and $-z'$ on opposite sides of this plane, each unit of area. The number of particles which, starting from dz in unit of time, reach a distance between nl and $(n+dn)\,l$ is by (19),

$$N \frac{v}{l}\, e^{-n}\, dz\, dn.$$

The number of these which have the ends of their paths in the layer dz' is

$$N \frac{v}{2nl^2}\, e^{-n}\, dz\, dz'\, dn.$$

The mean velocity in the direction of x which each of these has before impact is $A + Bz$, and after impact $A + Bz'$; and its mass is M, so that a mean momentum $= MB\,(z-z')$ is communicated by each particle. The whole action due to these collisions is therefore

$$NMB\,\frac{v}{2nl^2}\,(z-z')\,e^{-n}\,dz\,dz'\,dn.$$

We must first integrate with respect to z' between $z'=0$ and $z'=z-nl$; this gives

$$\tfrac{1}{2}NMB\,\frac{v}{2nl^2}\,(n^2l^2-z^2)\,e^{-n}\,dz\,dn$$

for the action between the layer dz and all the layers below the plane xy. Then integrate from $z=0$ to $z=nl$,

$$\tfrac{1}{6}MNBlvn^2e^{-n}\,dn.$$

Integrate from $n=0$ to $n=\infty$, and we find the whole friction between unit of area above and below the plane to be

$$F=\tfrac{1}{3}MNlvB=\tfrac{1}{3}\rho lv\,\frac{du}{dz}=\mu\,\frac{du}{dz},$$

where μ is the ordinary coefficient of internal friction,

$$\mu=\tfrac{1}{3}\rho lv=\frac{1}{3\sqrt{2}}\,\frac{Mv}{\pi s^3}\,\dots\dots\dots\dots\dots\dots\dots (24),$$

where ρ is the density, l the mean length of path of a particle, and v the mean velocity $v=\dfrac{2a}{\sqrt{\pi}}=2\sqrt{\dfrac{2k}{\pi}}$,

$$l=\tfrac{3}{2}\,\frac{\mu}{\rho}\sqrt{\frac{\pi}{2k}}\,\dots\dots\dots\dots\dots\dots\dots\dots (25).$$

Now Professor Stokes finds by experiments on air,

$$\sqrt{\frac{\mu}{\rho}}=\cdot116.$$

If we suppose $\sqrt{k}=930$ feet per second for air at 60°, and therefore the mean velocity $v=1505$ feet per second, then the value of l, the mean distance travelled over by a particle between consecutive collisions, $=\frac{1}{447000}$th of an inch, and each particle makes 8,077,200,000 collisions per second.

A remarkable result here presented to us in equation (24), is that if this explanation of gaseous friction be true, the coefficient of friction is independent of the density. Such a consequence of a mathematical theory is very startling, and the only experiment I have met with on the subject does not seem to confirm it. We must next compare our theory with what is known of the diffusion of gases, and the conduction of heat through a gas.

PART II.

* ON THE PROCESS OF DIFFUSION OF TWO OR MORE KINDS OF MOVING PARTICLES AMONG ONE ANOTHER.

We have shewn, in the first part of this paper, that the motions of a system of many small elastic particles are of two kinds: one, a general motion of translation of the whole system, which may be called the motion in mass; and the other a motion of agitation, or molecular motion, in virtue of which velocities in all directions are distributed among the particles according to a certain law. In the cases we are considering, the collisions are so frequent that the law of distribution of the molecular velocities, if disturbed in any way, will be re-established in an inappreciably short time; so that the motion will always consist of this definite motion of agitation, combined with the general motion of translation.

When two gases are in communication, streams of the two gases might run freely in opposite directions, if it were not for the collisions which take place between the particles. The rate at which they actually interpenetrate each other must be investigated. The diffusion is due partly to the spreading of the particles by the molecular agitation, and partly to the actual motion of the two opposite currents in mass, produced by the pressure behind, and resisted

---

* [The methods and results of this paper have been criticised by Clausius in a memoir published in Poggendorff's *Annalen*, Vol. cxv., and in the *Philosophical Magazine*, Vol. xxiii. His main objection is that the various circumstances of the strata, discussed in the paper, have not been sufficiently represented in the equations. In particular, if there be a series of strata at different temperatures perpendicular to the axis of x, then the proportion of molecules whose directions form with the axis of x angles whose cosines lie between μ and $\mu + d\mu$ is not $\frac{1}{2}d\mu$ as has been assumed by Maxwell throughout his work, but $\frac{1}{2}Hd\mu$ where H is a factor to be determined. In discussing the steady conduction of heat through a gas Clausius assumes that, in addition to the velocity attributed to the molecule according to Maxwell's theory, we must also suppose a velocity normal to the stratum and depending on the temperature of the stratum. On this assumption the factor H is investigated along with other modifications, and an expression for the assumed velocity is determined from the consideration that when the flow of heat is steady there is no movement of the mass. Clausius combining his own results with those of Maxwell points out that the expression contained in (28) of the paper involves as a result the motion of the gas. He also disputes the accuracy of expression (59) for the Conduction of Heat. In the introduction to the memoir published in the *Phil. Trans.*, 1866, it will be found that Maxwell expresses dissatisfaction with his former theory of the Diffusion of Gases, and admits the force of the objections made by Clausius to his expression for the Conduction of Heat. Ed.]

by the collisions of the opposite stream.   When the densities are equal, the diffusions due to these two causes respectively are as 2 to 3.

PROP. XIV. *In a system of particles whose density, velocity, &c. are functions of* x, *to find the quantity of matter transferred across the plane of* yz, *due to the motion of agitation alone.*

If the number of particles, their velocity, or their length of path is greater on one side of this plane than on the other, then more particles will cross the plane in one direction than in the other ; and there will be a transference of matter across the plane, the amount of which may be calculated.

Let there be taken a stratum whose thickness is dx, and area unity, at a distance x from the origin.   The number of collisions taking place ·in this stratum in unit of time will be

$$N \frac{v}{l} \, dx.$$

The proportion of these which reach a distance between nl and $(n + dn)l$ before they strike another particle is

$$e^{-n} \, dn.$$

The proportion of these which pass through the plane yz is

$$\frac{nl + x}{2nl} \text{ when } x \text{ is between } -nl \text{ and } 0,$$

and $$-\frac{nl - x}{2nl} \text{ when } x \text{ is between } 0 \text{ and } +nl ;$$

the sign being negative in the latter case, because the particles cross the plane in the negative direction.   The mass of each particle is M ; so that the quantity of matter which is projected from the stratum dx, crosses the plane yz in a positive direction, and strikes other particles at distances between nl and $(n + dn)l$ is

$$\frac{MNv \, (x \mp nl)}{2nl^2} \, dx \, e^{-n} dn \dots\dots\dots\dots\dots\dots(26),$$

where x must be between $\pm nl$, and the upper or lower sign is to be taken according as x is positive or negative.

In integrating this expression, we must remember that N, v, and l are functions of x, not vanishing with x, and of which the variations are very small between the limits $x = -nl$ and $x = +nl$.

50

As we may have occasion to perform similar integrations, we may state here, to save trouble, that if U and r are functions of x not vanishing with x, whose variations are very small between the limits $x = +r$ and $x = -r$,

$$\int_{-r}^{+r} \pm U x^m dx = \frac{2}{m+2} \frac{d}{dx} (U r^{m+2}) \dots\dots\dots\dots\dots (27).$$

When m is an odd number, the upper sign only is to be considered; when m is even or zero, the upper sign is to be taken with positive values of x, and the lower with negative values. Applying this to the case before us,

$$\int_{-nl}^{+nl} \frac{MNvx}{2nl^2} dx = \tfrac{1}{3} \frac{d}{dx} (MNvn^2l),$$

$$\int_{-nl}^{+nl} \mp \frac{MNv}{2l} dx = -\tfrac{1}{2} \frac{d}{dx} (MNvn^2l).$$

We have now to integrate

$$\int_{0}^{\infty} -\tfrac{1}{6} \frac{d}{dx} (MNvl) \, n^2 e^{-n} dn,$$

n being taken from 0 to ∞. We thus find for the quantity of matter transferred across unit of area by the motion of agitation in unit of time,

$$q = -\tfrac{1}{3} \frac{d}{dx} (\rho vl) \dots\dots\dots\dots\dots\dots\dots (28),$$

where $\rho = MN$ is the density, v the mean velocity of agitation, and l the mean length of path.

PROP. XV. The quantity transferred, in consequence of a mean motion of translation V, would obviously be

$$Q = V\rho \dots\dots\dots\dots\dots\dots\dots\dots\dots (29).$$

PROP. XVI. *To find the resultant dynamical effect of all the collisions which take place in a given stratum.*

Suppose the density and velocity of the particles to be functions of x, then more particles will be thrown into the given stratum from that side on which the density is greatest; and those particles which have greatest velocity will have the greatest effect, so that the stratum will not be generally in equilibrium, and the dynamical measure of the force exerted on the stratum will be the resultant momentum of all the particles which lodge in it during unit of time. We shall first take the case in which there is no mean motion of translation, and then consider the effect of such motion separately.

Let a stratum whose thickness is a (a small quantity compared with l), and area unity, be taken at the origin, perpendicular to the axis of x; and let another stratum, of thickness dx, and area unity, be taken at a distance x from the first.

If M_1 be the mass of a particle, N the number in unit of volume, v the velocity of agitation, l the mean length of path, then the number of collisions which take place in the stratum dx is

$$N \frac{v}{l} dx.$$

The proportion of these which reach a distance between nl and $(n+dn)l$ is

$$e^{-n} dn.$$

The proportion of these which have the extremities of their paths in the stratum a is

$$\frac{a}{2nl}.$$

The velocity of these particles, resolved in the direction of x, is

$$-\frac{vx}{nl},$$

and the mass is M; so that multiplying all these terms together, we get

$$\frac{NMv^2ax}{2n^2l^3} e^{-n} dx \, dn \dots\dots\dots\dots\dots\dots\dots\dots\dots (30)$$

for the momentum of the particles fulfilling the above conditions.

To get the whole momentum, we must first integrate with respect to x from $x = -nl$ to $x = +nl$, remembering that l may be a function of x, and is a very small quantity. The result is

$$\frac{d}{dx}\left(\frac{NMv^2}{3}\right) ane^{-n} dn.$$

Integrating with respect to n from $n = 0$ to $n = \infty$, the result is

$$-a \frac{d}{dx}\left(\frac{NMv^2}{3}\right) = aX\rho \dots\dots\dots\dots\dots (31)$$

as the whole resultant force on the stratum a arising from these collisions. Now $\frac{NMv^2}{3} = p$ by Prop. XII., and therefore we may write the equation

$$-\frac{dp}{dx} = X\rho \dots\dots\dots\dots\dots\dots (32),$$

the ordinary hydrodynamical equation.

PROP. XVII. *To find the resultant effect of the collisions upon each of several different systems of particles mixed together.*

Let M_1, M_2, &c. be the masses of the different kinds of particles, N_1, N_2, &c. the number of each kind in unit of volume, v_1, v_2, &c. their velocities of agitation, l_1, l_2 their mean paths, p_1, p_2, &c. the pressures due to each system of particles; then

$$\left.\begin{aligned}\frac{1}{l_1} &= A\rho_1 + B\rho_2 + \&c. \\ \frac{1}{l_2} &= C\rho_1 + D\rho_2 + \&c.\end{aligned}\right\} \dots\dots\dots\dots\dots (33).$$

The number of collisions of the first kind of particles with each other in unit of time will be

$$N_1 v_1 A\rho_1.$$

The number of collisions between particles of the first and second kinds will be

$$N_1 v_1 B\rho_2, \text{ or } N_2 v_2 C\rho_1, \text{ because } v_1^2 B = v_2^2 C.$$

The number of collisions between particles of the second kind will be $N_2 v_2 D\rho_2$, and so on, if there are more kinds of particles.

Let us now consider a thin stratum of the mixture whose volume is unity.

The resultant momentum of the particles of the first kind which lodge in it during unit of time is

$$-\frac{dp_1}{dx}.$$

The proportion of these which strike particles of the first kind is

$$A\rho_1 l_1.$$

The whole momentum of these remains among the particles of the first kind. The proportion which strike particles of the second kind is

$$B\rho_2 l_1.$$

The momentum of these is divided between the striking particles in the ratio of their masses; so that $\dfrac{M_1}{M_1+M_2}$ of the whole goes to particles of the first kind, and $\dfrac{M_2}{M_1+M_2}$ to particles of the second kind.

The effect of these collisions is therefore to produce a force

$$-\frac{dp_1}{dx}\left(A\rho_1 l_1 + B\rho_2 l_1 \,\frac{M_1}{M_1+M_2}\right)$$

on particles of the first system, and

$$-\frac{dp_1}{dx}\, B\rho_2 l_1 \,\frac{M_2}{M_1+M_2}$$

on particles of the second system.

The effect of the collisions of those particles of the second system which strike into the stratum, is to produce a force

$$-\frac{dp_2}{dx}\, C\rho_1 l_2 \,\frac{M_1}{M_1+M_2}$$

on the first system, and

$$-\frac{dp_2}{dx}\left(C\rho_1 l_2 \,\frac{M_2}{M_1+M_2} + D\rho_2 l_2\right)$$

on the second.

The whole effect of these collisions is therefore to produce a resultant force

$$-\frac{dp_1}{dx}\left(A\rho_1 l_1 + B\rho_2 l_1 \,\frac{M_1}{M_1+M_2}\right) - \frac{dp_2}{dx}\, C\rho_1 l_2 \,\frac{M_1}{M_1+M_2} + \&\text{c}.\ldots\ldots(34)$$

on the first system,

$$-\frac{dp_1}{dx}\, B\rho_2 l_1 \,\frac{M_2}{M_1+M_2} - \frac{dp_2}{dx}\left(C\rho_1 l_2 \,\frac{M_2}{M_1+M_2} + D\rho_2 l_2\right) + \&\text{c}.\ldots\ldots(35)$$

on the second, and so on.

Prop. XVIII. *To find the mechanical effect of a difference in the mean velocity of translation of two systems of moving particles.*

Let V_1, V_2 be the mean velocities of translation of the two systems respectively, then $\dfrac{M_1 M_2}{M_1 + M_2}(V_1 - V_2)$ is the mean momentum lost by a particle of the first, and gained by a particle of the second at collision. The number of such collisions in unit of volume is

$$N_1 B \rho_2 v_1, \text{ or } N_2 C \rho_1 v_2;$$

therefore the whole effect of the collisions is to produce a force

$$= - N_1 B \rho_2 v_1 \frac{M_1 M_2}{M_1 + M_2}(V_1 - V_2) \dots\dots\dots\dots\dots(36)$$

on the first system, and an equal and opposite force

$$= + N_2 C \rho_1 v_2 \frac{M_1 M_2}{M_1 + M_2}(V_1 - V_2) \dots\dots\dots\dots\dots (37)$$

on unit of volume of the second system.

Prop. XIX. *To find the law of diffusion in the case of two gases diffusing into each other through a plug made of a porous material, as in the case of the experiments of Graham.*

The pressure on each side of the plug being equal, it was found by Graham that the quantities of the gases which passed in opposite directions through the plug in the same time were directly as the square roots of their specific gravities.

We may suppose the action of the porous material to be similar to that of a number of particles fixed in space, and obstructing the motion of the particles of the moving systems. If L_1 is the mean distance a particle of the first kind would have to go before striking a fixed particle, and L_2 the distance for a particle of the second kind, then the mean paths of particles of each kind will be given by the equations

$$\frac{1}{l_1} = A\rho_1 + B\rho_2 + \frac{1}{L_1}, \quad \frac{1}{l_2} = C\rho_1 + D\rho_2 + \frac{1}{L_2} \dots\dots\dots\dots (38).$$

The mechanical effect upon the plug of the pressures of the gases on each side, and of the percolation of the gases through it, may be found by Props. XVII. and XVIII. to be

$$\frac{M_1 N_1 v_1 V_1}{L_1} + \frac{M_2 N_2 v_2 V_2}{L_2} - \frac{dp_1}{dx}\frac{l_1}{L_1} - \frac{dp_2}{dx}\frac{l_2}{L_2} = 0 \dots\dots\dots\dots(39);$$

and this must be zero, if the pressures are equal on each side of the plug. Now if Q_1, Q_2 be the quantities transferred through the plug by the mean motion of translation, $Q_1 = \rho_1 V_1 = M_1 N_1 V_1$; and since by Graham's law

$$\frac{Q_1}{Q_2} = -\sqrt{\frac{M_1}{M_2}} = -\frac{v_2}{v_1},$$

we shall have

$$M_1 N_1 v_1 V_1 = -M_2 N_2 v_2 V_2 = U \text{ suppose};$$

and since the pressures on the two sides are equal, $\dfrac{dp_2}{dx} = -\dfrac{dp_1}{dx}$, and the only way in which the equation of equilibrium of the plug can generally subsist is when $L_1 = L_2$ and $l_1 = l_2$. This implies that $A = C$ and $B = D$. Now we know that $v_1^3 B = v_2^3 C$. Let $K = 3\dfrac{A}{v_1^3}$, then we shall have

$$A = C = \tfrac{1}{3} K v_1^3, \quad B = D = \tfrac{1}{3} K v_2^3 \dots\dots\dots\dots\dots (40),$$

and

$$\frac{1}{l_1} = \frac{1}{l_2} = K(v_1 p_1 + v_2 p_2) + \frac{1}{L} \dots\dots\dots\dots\dots\dots(41).$$

The diffusion is due partly to the motion of translation, and partly to that of agitation. Let us find the part due to the motion of translation.

The equation of motion of one of the gases through the plug is found by adding the forces due to pressures to those due to resistances, and equating these to the moving force, which in the case of slow motions may be neglected altogether. The result for the first is

$$\frac{dp_1}{dx}\left(A\rho_1 l_1 + B\rho_2 l_1 \frac{M_1}{M_1 + M_2}\right) + \frac{dp_2}{dx} C\rho_1 l_2 \frac{M_1}{M_1 + M_2}$$

$$+ N_1 B \rho_2 v_1 \frac{M_1 M_2}{M_1 + M_2}(V_1 - V_2) + \frac{\rho_1 v_1 V_1}{L} = 0 \dots\dots(42).$$

Making use of the simplifications we have just discovered, this becomes

$$\frac{dp}{dx}\frac{Kl}{v_1^2 + v_2^2}(v_1^3 p_1 + v_2^3 p_2) + K\frac{v_1 v_2}{v_1^2 + v_2^2}(p_1 v_2 + p_2 v_1) U + \frac{1}{L} U \dots\dots(43),$$

whence

$$U = -\frac{dp}{dx}\frac{Kl(v_1^3 p_1 + v_2^3 p_2)}{Kv_1 v_2 (p_1 v_2 + p_2 v_1) + \dfrac{v_1^2 + v_2^2}{L}} \dots\dots\dots\dots (44);$$

whence the rate of diffusion due to the motion of translation may be found; for

$$Q_1 = \frac{U}{v_1}, \text{ and } Q_2 = -\frac{U}{v_2} \quad \text{.........................} (45).$$

To find the diffusion due to the motion of agitation, we must find the value of q_1.

$$q_1 = -\tfrac{1}{3} \frac{d}{dx} (\rho_1 v_1 l_1),$$

$$= -\frac{L}{v_1} \frac{d}{dx} \frac{p_1}{1 + KL\,(v_1 p_1 + v_2 p_2)},$$

$$q_1 = -\frac{l^2}{v_1 L} \frac{dp}{dx} \{1 + KLv_2\,(p_1 + p_2)\} \quad \text{.................} (46).$$

Similarly,

$$q_2 = +\frac{l^2}{v_2 L} \frac{dp}{dx} \{1 + KLv_1\,(p_1 + p_2)\} \quad \text{.................} (47).$$

The whole diffusions are $Q_1 + q_1$ and $Q_2 + q_2$. The values of q_1 and q_2 have a term not following Graham's law of the square roots of the specific gravities, but following the law of equal volumes. The closer the material of the plug, the less will this term affect the result.

Our assumptions that the porous plug acts like a system of fixed particles, and that Graham's law is fulfilled more accurately the more compact the material of the plug, are scarcely sufficiently well verified for the foundation of a theory of gases; and even if we admit the original assumption that they are systems of moving elastic particles, we have not very good evidence as yet for the relation among the quantities A, B, C, and D.

Prop. XX.   *To find the rate of diffusion between two vessels connected by a tube.*

When diffusion takes place through a large opening, such as a tube connecting two vessels, the question is simplified by the absence· of the porous diffusion plug; and since the pressure is constant throughout the apparatus, the volumes of the two gases passing opposite ways through the tube at the same time must be equal. Now the quantity of gas which passes through the tube is due partly to the motion of agitation as in Prop. XIV., and partly to the mean motion of translation as in Prop. XV.

Let us suppose the volumes of the two vessels to be a and b, and the length of the tube between them c, and its transverse section s. Let a be filled with the first gas, and b with the second at the commencement of the experiment, and let the pressure throughout the apparatus be P.

Let a volume y of the first gas pass from a to b, and a volume y' of the second pass from b to a; then if p_1 and p_2 represent the pressures in a due to the first and second kinds of gas, and p'_1 and p'_2 the same in the vessel b,

$$p_1 = \frac{a-y}{a}\,P, \quad p_2 = \frac{y'}{a}\,P, \quad p'_1 = \frac{y}{b}\,P, \quad p'_2 = \frac{b-y'}{b}\,P \dots\dots\dots\dots(48).$$

Since there is still equilibrium,

$$p_1 + p_2 = p'_1 + p'_2,$$

which gives

$$y = y' \text{ and } p_1 + p_2 = P = p'_1 + p'_2 \dots\dots\dots\dots\dots(49).$$

The rate of diffusion will be $+\dfrac{dy}{dt}$ for the one gas, and $-\dfrac{dy}{dt}$ for the other, measured in volume of gas at pressure P.

Now the rate of diffusion of the first gas will be

$$\frac{dy}{dt} = s\,\frac{k_1 q_1 + p_1 V_1}{P} = s\,\frac{-\frac{1}{3}v_1\dfrac{d}{dx}(p_1 l_1) + p_1 V_1}{P} \dots\dots\dots\dots(50);$$

and that of the second,

$$-\frac{dy}{dt} = s\,\frac{-\frac{1}{3}v_2\dfrac{d}{dx}(p_2 l_2) + p_2 V_2}{P} \dots\dots\dots\dots(51).$$

We have also the equation, derived from Props. XVI. and XVII.,

$$\frac{dp_1}{dx}\{A\rho_1 l_1(M_1 + M_2) + B\rho_2 l_2 M_1 - C\rho_1 l_2 M_1\} + B\rho_1 \rho_2 v_1 M_2(V_1 - V_2) = 0 \dots(52).$$

From these three equations we can eliminate V_1 and V_2, and find $\dfrac{dy}{dt}$ in terms of p and $\dfrac{dp}{dx}$, so that we may write

$$\frac{dy}{dt} = f\left(p_1, \frac{dp_1}{dx}\right) \dots\dots\dots\dots\dots(53).$$

Since the capacity of the tube is small compared with that of the vessels, we may consider $\dfrac{dy}{dt}$ constant through the whole length of the tube. We may then solve the differential equation in p and x; and then making $p = p_1$ when $x = 0$, and $p = p'_1$ when $x = c$, and substituting for p_1 and p'_1 their values in terms of y, we shall have a differential equation in y and t, which being solved, will give the amount of gas diffused in a given time.

The solution of these equations would be difficult unless we assume relations among the quantities A, B, C, D, which are not yet sufficiently established in the case of gases of different density. Let us suppose that in a particular case the two gases have the same density, and that the four quantities A, B, C, D are all equal.

The volume diffused, owing to the motion of agitation of the particles, is then

$$-\tfrac{1}{3}\frac{s}{P}\frac{dp}{dx}\,vl,$$

and that due to the motion of translation, or the interpenetration of the two gases in opposite streams, is

$$-\frac{s}{P}\frac{dp}{dx}\frac{kl}{v}.$$

The values of v are distributed according to the law of Prop. IV., so that the mean value of v is $\dfrac{2a}{\sqrt{\pi}}$, and that of $\dfrac{1}{v}$ is $\dfrac{2}{\sqrt{\pi a}}$, that of k being $\tfrac{1}{2}a^2$. The diffusions due to these two causes are therefore in the ratio of 2 to 3, and their sum is

$$\frac{dy}{dt} = -\tfrac{4}{3}\sqrt{\frac{2k}{\pi}}\,\frac{sl}{P}\frac{dp}{dx} \quad\text{.........................(54).}$$

If we suppose $\dfrac{dy}{dt}$ constant throughout the tube, or, in other words, if we regard the motion as *steady* for a short time, then $\dfrac{dp}{dx}$ will be constant and equal to $\dfrac{p'_1 - p_1}{c}$; or substituting from (48),

$$\frac{dy}{dt} = -\tfrac{4}{3}\sqrt{\frac{2k}{\pi}}\,\frac{sl}{abc}\{(a+b)\,y - ab\} \quad\text{.................(55);}$$

whence
$$y = \frac{ab}{a+b}\left(1 - e^{-\tfrac{4}{3}\sqrt{\frac{2k}{\pi}}\frac{sl}{abc}(a+b)t}\right) \quad\text{.........................(56).}$$

By choosing pairs of gases of equal density, and ascertaining the amount of diffusion in a given time, we might determine the value of l in this expression. The diffusion of nitrogen into carbonic oxide or of deutoxide of nitrogen into carbonic acid, would be suitable cases for experiment. The only existing experiment which approximately fulfils the conditions is one by Graham, quoted by Herapath from Brande's *Quarterly Journal of Science*, Vol. XVIII. p. 76.

A tube 9 inches long and 0·9 inch diameter, communicated with the atmosphere by a tube 2 inches long and 0·12 inch diameter; 152 parts of olefiant gas being placed in the tube, the quantity remaining after four hours was 99 parts.

In this case there is not much difference of specific gravity between the gases, and we have $a = 9 \times (0\cdot9)^2 \frac{\pi}{4}$ cubic inches, $b = \infty$, $c = 2$ inches, and $s = (0\cdot12)^2 \frac{\pi}{4}$ square inches;

$$l = \sqrt{\frac{\pi}{2k}} \frac{3}{4} \frac{ac}{s} \log_e 10 \cdot \frac{1}{t} \cdot \log_{10}\left(\frac{a}{a-y}\right) \dots\dots\dots\dots(57);$$

$$\therefore\ l = 0\cdot00000256 \text{ inch} = \tfrac{1}{389000} \text{ inch} \dots\dots\dots\dots(58).$$

PROP. XXI. *To find the amount of energy which crosses unit of area in unit of time when the velocity of agitation is greater on one side of the area than on the other.*

The energy of a single particle is composed of two parts,—the *vis viva* of the centre of gravity, and the *vis viva* of the various motions of rotation round that centre, or, if the particle be capable of internal motions, the *vis viva* of these. We shall suppose that the whole *vis viva* bears a constant proportion to that due to the motion of the centre of gravity, or

$$E = \tfrac{1}{2}\beta M v^2,$$

where β is a coefficient, the experimental value of which is 1·634. Substituting E for M in Prop. XIV., we get for the transference of energy across unit of area in unit of time,

$$Jq = -\tfrac{1}{3}\frac{d}{dx}\left(\tfrac{1}{2}\beta M v^2 N v l\right),$$

404   ILLUSTRATIONS OF THE DYNAMICAL THEORY OF GASES.

where J is the mechanical equivalent of heat in foot-pounds, and q is the transfer of heat in thermal units.

Now $MN = \rho$, and $l = \dfrac{1}{A\rho}$, so that $MNl = \dfrac{1}{A}$;

$$\therefore\ Jq = -\tfrac{1}{2}\frac{\beta v^2}{A}\frac{dv}{dx} \quad\dotfill (59).$$

Also, if T is the absolute temperature,

$$\frac{1}{T}\frac{dT}{dx} = \frac{2}{v}\frac{dv}{dx};$$

$$\therefore\ Jq = -\tfrac{3}{4}\beta plv\,\frac{1}{T}\frac{dT}{dx}\quad\dotfill (60),$$

where p must be measured in dynamical units of force.

Let $J = 772$ foot-pounds, $p = 2116$ pounds to square foot, $l = \frac{1}{400000}$ inch, $v = 1505$ feet per second, $T = 522$ or $62°$ Fahrenheit ; then

$$q = \frac{T' - T}{40000x}\quad\dotfill (61),$$

where q is the flow of heat in thermal units per square foot of area ; and T' and T are the temperatures at the two sides of a stratum of air x *inches* thick.

In Prof. Rankine's work on the Steam-engine, p. 259, values of the *thermal resistance*, or the reciprocal of the *conductivity*, are given for various substances as computed from a Table of conductivities deduced by M. Peclet from experiments by M. Despretz :—

	Resistance.
Gold, Platinum, Silver	0·0036
Copper	0·0040
Iron	0·0096
Lead	0·0198
Brick	0·3306
Air by our calculation	40000

It appears, therefore, that the resistance of a stratum of air to the conduction of heat is about 10,000,000 times greater than that of a stratum of copper of equal thickness. It would be almost impossible to establish the value of the conductivity of a gas by direct experiment, as the heat radiated from the sides of the vessel would be far greater than the heat conducted through the air, even if currents could be entirely prevented[*].

PART III.

ON THE COLLISION OF PERFECTLY ELASTIC BODIES OF ANY FORM.

When two perfectly smooth spheres strike each other, the force which acts between them always passes through their centres of gravity; and therefore their motions of rotation, if they have any, are not affected by the collision, and do not enter into our calculations. But, when the bodies are not spherical, the force of compact will not, in general, be in the line joining their centres of gravity; and therefore the force of impact will depend both on the motion of the centres and the motions of rotation before impact, and it will affect both these motions after impact.

In this way the velocities of the centres and the velocities of rotation will act and react on each other, so that finally there will be some relation established between them; and since the rotations of the particles about their three axes are quantities related to each other in the same way as the three velocities of their centres, the reasoning of Prop. IV. will apply to rotation as well as velocity, and both will be distributed according to the law

$$\frac{dN}{dx} = N \, \frac{1}{a\sqrt{\pi}} \, e^{-\frac{x}{a^2}}.$$

* [Clausius, in the memoir cited in the last foot-note, has pointed out two oversights in this calculation. In the first place the numbers have not been properly reduced to English measure, and have still to be multiplied by ·4356, the ratio of the English pound to the kilogramme. The numbers have, further, been calculated with one hour as the unit of time, whereas Maxwell has used them as if a second had been the unit. Taking account of these circumstances and using his own expression for the conduction which differs from (59) only in having $\frac{5}{12}$ in place of $\frac{1}{2}$ on the right-hand side, Clausius finds that the resistance of a stratum of air to the conduction of heat is 1400 times greater than that of a stratum of lead of the same thickness, or about 7000 times greater than that of copper.   Ed.]

Also, by Prop. V., if x be the average velocity of one set of particles, and y that of another, then the average value of the sum or difference of the velocities is

$$\sqrt{x^2 + y^2} \, ;$$

from which it is easy to see that, if in each individual case

$$u = ax + by + cz,$$

where x, y, z are independent quantities distributed according to the law above stated, then the *average values* of these quantities will be connected by the equation

$$u^2 = a^2 x^2 + b^2 y^2 + c^2 z^2.$$

PROP. XXII. *Two perfectly elastic bodies of any form strike each other: given their motions before impact, and the line of impact, to find their motions after impact.*

Let M_1 and M_2 be the centres of gravity of the two bodies. $M_1 X_1$, $M_1 Y_1$, and $M_1 Z_1$ the principal axes of the first; and $M_2 X_2$, $M_2 Y_2$ and $M_2 Z_2$ those of the second. Let I be the point of impact, and $R_1 I R_2$ the line of impact.

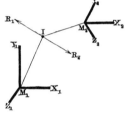

Let the co-ordinates of I with respect to M_1 be $x_1 y_1 z_1$, and with respect to M_2 let them be $x_2 y_2 z_2$.

Let the direction-cosines of the line of impact $R_1 I R_2$ be $l_1 m_1 n_1$ with respect to M_1, and $l_2 m_2 n_2$ with respect to M_2.

Let M_1 and M_2 be the masses, and $A_1 B_1 C_1$ and $A_2 B_2 C_2$ the moments of inertia of the bodies about their principal axes.

Let the velocities of the centres of gravity, resolved in the direction of the principal axes of each body, be

$$U_1, \ V_1, \ W_1, \ \text{and} \ U_2, \ V_2, \ W_2, \ \text{before impact,}$$

and

$$U'_1, \ V'_1, \ W'_1, \ \text{and} \ U'_2, \ V'_2, \ W'_2, \ \text{after impact.}$$

Let the angular velocities round the same axes be

$$p_1, \ q_1, \ r_1, \ \text{and} \ p_2, \ q_2, \ r_2, \ \text{before impact,}$$

and

$$p'_1, \ q'_1, \ r'_1, \ \text{and} \ p'_2, \ q'_2, \ r'_2, \ \text{after impact.}$$

Let R be the impulsive force between the bodies, measured by the momentum it produces in each.

Then, for the velocities of the centres of gravity, we have the following equations :

$$U'_1 = U_1 + \frac{Rl_1}{M_1}, \quad U'_2 = U_2 - \frac{Rl_2}{M_2} \dots\dots\dots\dots\dots (62),$$

with two other pairs of equations in V and W.

The equations for the angular velocities are

$$p'_1 = p_1 + \frac{R}{A_1}(y_1 n_1 - z_1 m_1), \quad p'_2 = p_2 - \frac{R}{A_2}(y_2 n_2 - z_2 m_2) \dots\dots\dots (63),$$

with two other pairs of equations for q and r.

The condition of perfect elasticity is that the whole *vis viva* shall be the same after impact as before, which gives the equation

$$M_1(U'^2_1 - U^2_1) + M_2(U'^2_2 - U^2_2) + A_1(p'^2_1 - p^2_1) + A_2(p'^2_2 - p^2_2) + \&c. = 0 \dots (64).$$

The terms relating to the axis of x are here given; those relating to y and z may be easily written down.

Substituting the values of these terms, as given by equations (62) and (63), and dividing by R, we find

$$l_1(U'_1 + U_1) - l_2(U'_2 + U_2) + (y_1 n_1 - z_1 m_1)(p'_1 + p_1) - (y_2 n_2 - z_2 m_2)(p'_2 + p_2) + \&c. = 0 \dots (65).$$

Now if v_1 be the velocity of the striking-point of the first body before impact, resolved along the line of impact,

$$v_1 = l_1 U_1 + (y_1 n_1 - z_1 m_1) p_1 + \&c. ;$$

and if we put v_2 for the velocity of the other striking-point resolved along the same line, and v'_1 and v'_2 the same quantities after impact, we may write, equation (65),

$$v_1 + v'_1 - v_2 - v'_2 = 0 \dots\dots\dots\dots\dots\dots (66),$$

or

$$v_1 - v_2 = v'_2 - v'_1 \dots\dots\dots\dots\dots\dots (67),$$

which shows that the velocity of separation of the striking-points resolved in the line of impact is equal to that of approach.

Substituting the values of the accented quantities in equation (65) by means of equations (63) and (64), and transposing terms in R, we find

$$2\left\{U_1 l_1 - U_2 l_2 + p_1\left(y_1 n_1 - z_1 m_1\right) - p_2\left(y_2 n_2 - z_2 m_2\right)\right\} + \&\text{c.}$$

$$= -R\left\{\frac{l_1^2}{M_1} + \frac{l_2^2}{M_2} + \frac{\left(y_1 n_1 - z_1 m_1\right)^2}{A_1} + \frac{\left(y_2 n_2 - z_2 m_2\right)^2}{A_2} + \&\text{c.}\right.\quad\text{.........(68)},$$

the other terms being related to y and z as these are to x. From this equation we may find the value of R; and by substituting this in equations (63), (64), we may obtain the values of all the velocities after impact.

We may, for example, find the value of U'_1 from the equation

$$\left.\begin{aligned}
U'_1&\left\{\frac{l_1^2}{M_1} + \frac{l_2^2}{M_2} + \frac{\left(y_1 n_1 - z_1 m_1\right)^2}{A_1} + \frac{\left(y_2 n_2 - z_2 m_2\right)^2}{A_2} + \&\text{c.}\right\}\frac{M_1}{l_1}\\
&= U_1\left\{-\frac{l_1^2}{M_1} + \frac{l_2^2}{M_2} + \frac{\left(y_1 n_1 - z_1 m_1\right)^2}{A_1} + \frac{\left(y_2 n_2 - z_2 m_2\right)^2}{A_2} + \&\text{c.}\right\}\frac{M_1}{l_1}\\
&\quad + 2U_2 l_2 - 2p_1\left(y_1 n_1 - z_1 m_1\right) + 2p_2\left(y_2 n_2 - z_2 m_2\right) - \&\text{c.}
\end{aligned}\right\}\quad\text{..........(69)}.$$

PROP. XXIII. *To find the relations between the average velocities of trans-lation and rotation after many collisions among many bodies.*

Taking equation (69), which applies to an individual collision, we see that U'_1 is expressed as a linear function of U_1, U_2, p_1, p_2, &c., all of which are quantities of which the values are distributed among the different particles according to the law of Prop. IV. It follows from Prop. V., that if we square every term of the equation, we shall have a new equation between the *average values* of the different quantities. It is plain that, as soon as the required relations have been established, they will remain the same after collision, so that we may put $U'^2_1 = U^2_1$ in the equation of averages. The equation between the average values may then be written

$$\left(M_1 U_1^2 - M_2 U_2^2\right)\frac{l_2^2}{M_2} + \left(M_1 U_1^2 - A_1 p_1^2\right)\frac{\left(y_1 n_1 - z_1 m_1\right)^2}{A_1} + \left(M_1 U_1^2 - A_2 p_2^2\right)\frac{\left(y_2 n_2 - z_2 m_2\right)^2}{A_2} + \&\text{c.} = 0.$$

Now since there are collisions in every possible way, so that the values of l, m, n, &c. and x, y, z, &c. are infinitely varied, this equation cannot subsist unless

$$M_1 U_1^2 = M_2 U_2^2 = A_1 p_1^2 = A_2 p_2^2 = \&\text{c.}$$

The final state, therefore, of any number of systems of moving particles of any form is that in which the average *vis viva* of translation along each of the three axes is the same in all the systems, and equal to the average *vis viva* of rotation about each of the three principal axes of each particle.

Adding the *vires vivæ* with respect to the other axes, we find that the whole *vis viva* of translation is equal to that of rotation in each system of particles, and is also the same for different systems, as was proved in Prop. VI.

This result (which is true, however nearly the bodies approach the spherical form, provided the motion of rotation is at all affected by the collisions) seems decisive against the unqualified acceptation of the hypothesis that gases are such systems of hard elastic particles. For the ascertained fact that γ, the ratio of the specific heat at constant pressure to that at constant volume, is equal to 1·408, requires that the ratio of the whole *vis viva* to the *vis viva* of translation should be

$$\beta = \frac{2}{3\,(\gamma - 1)} = 1\cdot634 \; ;$$

whereas, according to our hypothesis, $\beta = 2$.

We have now followed the mathematical theory of the collisions of hard elastic particles through various cases, in which there seems to be an analogy with the phenomena of gases. We have deduced, as others have done already, the relations of pressure, temperature, and density of a single gas. We have also proved that when two different gases act freely on each other (that is, when at the same temperature), the mass of the single particles of each is inversely proportional to the square of the molecular velocity ; and therefore, at equal temperature and pressure, *the number of particles in unit of volume is the same.*

We then offered an explanation of the internal friction of gases, and deduced from experiments a value of the mean length of path of a particle between successive collisions.

We have applied the theory to the law of diffusion of gases, and, from an experiment on olefiant gas, we have deduced a value of the length of path not very different from that deduced from experiments on friction.

Using this value of the length of path between collisions, we found that the resistance of air to the conduction of heat is 10,000,000 times that of copper, a result in accordance with experience.

Finally, by establishing a necessary relation between the motions of translation and rotation of all particles not spherical, we proved that a system of such particles could not possibly satisfy the known relation between the two specific heats of all gases.

7. Letter from Maxwell to Lewis Campbell,[a] January 5, 1860

Lewis Campbell and William Garnett, *The Life of James Clerk Maxwell* (London, 1882), 328.

Marischal College
Aberdeen, 5th January 1860

... I have been publishing my views about Elastic Spheres in the *Phil Mag* for Jany.,[b] and am going to go on with it as I get the prop^{ns} [propositions] written out. I have also sent my experiments on Colours to the Royal Society of London, so I have two sets of irons in the fire, besides class work. I hope you get on with Plato, and that your pupils are all Theaetetuses, and that wisdom soaks like oil into their inwards. There is a man here striving after a general theory of things, but he has great difficulty in so churning his thoughts as to coagulate and solidify the vague and nebulous notions which wander in his head. He has been applying to me very steadily whenever he can pounce on me, and I have prescribed for him as I best could, and I hope his abstract of his general theory of things will be palatable to the readers of the *British Ass. Reports* for 1859.[c]

a. Lewis Campbell (1830–1908) was a close friend of Maxwell from 1841, when they both entered Edinburgh Academy, until Maxwell's death. Campbell was a distinguished classical scholar, professor of Greek at St. Andrews for many years, and author of about twenty books on classical, educational, and religious subjects, including the first biography of Maxwell. His edition of Plato's Theaetatus (first edition, 1871) became a standard work and remained in print until 1930 or later. There is a biographical notice of Campbell in the British *Dictionary of National Biography*. See also Robert Kargon's preface to the 1969 reprint of Campbell and Garnett's *Life of James Clerk Maxwell* (New York: Johnson Reprint Corp., 1969), ix–xii, and C. W. F. Everitt, *James Clerk Maxwell* (New York: Scribner's, 1975), 45.

b. See document III-6.

c. We have no positive information about the identity of the person mentioned here—indeed nothing to prove that it is not an ironical reference to himself. The "Notices and Abstracts" section of the *Report* of the 1859 meeting includes two papers that might fit Maxwell's description: "A Proposal of a General Mechanical Theory of Physics" by J. S. Stuart Glennie, pp. 58–59, and "On the Philosophy of Physics" by John G. Macvicar, pp. 59–60. Neither gives an address.

8. "On the Results of Bernoulli's Theory of Gases as Applied to Their Internal Friction, Their Diffusion, and Their Conductivity for Heat"

Report of the 30th Meeting of the British Association for the Advancement of Science, Notices and Abtracts (Oxford, June and July 1860), 15–16.

The substance of this paper is to be found in the 'Philosophical Magazine' for January and July 1860.[a] Assuming that the elasticity of gases can be accounted for by the impact of their particles against the sides of the containing vessel, the laws of motion of an immense number of very small elastic particles impinging on each other, are deduced from mathematical principles; and it is shown,— 1st, that the velocities of the particles vary from 0 to ∞, but that the number at any instant having velocities between given limits follows a law similar in its expression to that of the distribution of errors according to the theory of the "Method of least squares." 2nd. That the relative velocities of particles of two different systems are distributed according to a similar law, and that the mean relative velocity is the square root of the sum of the squares of the two mean velocities. 3rd. That the pressure is one-third of the density multiplied by the mean square of the velocity. 4th. That the mean *vis viva* of a particle is the same in each of two systems in contact, and that temperature may be represented by the *vis viva* of a particle, so that at equal temperatures and pressures, equal volumes of different gases must contain equal numbers of particles. 5th. That when layers of gas have a motion of sliding over each other, particles will be projected from one layer into another, and thus tend to resist the sliding motion. The amount of this will depend on the average distance described by a particle between successive collisions. From the coefficient of friction in air, as given by Professor Stokes, it would appear that this distance is 1/447,000 inch; the mean velocity being 1505 feet per second, so that each particle makes 8,077,200,000 collisions per second. 6th. That diffusion of gases is due partly to the agitation of the particles tending to mix them, and partly to the existence of opposing currents of the two gases through each other. From experiments of Graham on the diffusion of olefiant gas into air, the value of the distance described by a particle between successive collisions is found to be 1/389,000 of an inch, agreeing with the value derived from friction as closely as rough experiments of this kind will permit. 7th. That conduction of heat consists in the propagation of the motion of agitation from one part of the system to another, and may be calculated when we know the nature of the motion. Taking 1/400,000 of an inch as a probable value of the distance that a particle moves between successive collisions, it appears that the quantity of heat transmitted through a stratum of air by conduction would be 1/10,000,000 of that transmitted by a stratum of copper of equal thickness, the difference of the temperatures of the two sides being the same in both cases. This shows that the observed low conductivity of air is no objection to the theory, but a result of it. 8th. That if the collisions produce rotation of the particles at all, the *vis viva* of rotation will be equal to that of translation. This relation would make the ratio of specific heat at constant pressure to that at constant volume to be 1.33, whereas we know that for air it is 1.408. This result of the dynamical theory, being at variance with experiment, overturns the whole hypothesis, however satisfactory the other results may be.

a. See document III-6.

9. [Miscellaneous Problems]

Cambridge University Library, Maxwell Collection, Scientific Papers 6.

[The first three sheets refer to Maxwell's attempts to deduce the distribution of the relative velocities of the molecules of a gas. His concern sprang from Clausius's expression for the mean free path (noted in his letter to Stokes (document III-1)), which depended on knowing which molecule was the faster. This explains his attempt to establish the mean relative velocity, taking the velocity of one of two molecules to be the faster in separate cases. He deduced the same result in each case, demonstrating the independence of his own expression for the mean velocity from which molecule was the faster of the two in the collision. The problem of which molecule is faster and how to take this into account surfaces again in the other sheets. The physical problem around which these notes are grouped is that of finding the probability that the velocity of a molecule after a collision lies within certain velocity and space limits. The concerns that are discussed in these notes do not appear in his published papers; the problem of knowing which molecule is faster is not even mentioned.]

To find the mean relative velocity of particles whose direction makes an angle $\cos^{-1} \lambda$ with the axis with respect to particles the directions of which are

[Diagram 1]

distributed according to the law $(1 + F\mu)$ where $\cos^{-1}\mu$ is the angle with the axis [Diagram 1]. Let OA be the axis, OB the velocity of the first particle $= v$, $u = OC$ that of any other and BC the relative velocity $= r$. Let $\cos AOB = \lambda$, $\cos AOC = \mu$. Let $BOC = \theta$ and the angle between planes AOB, $AOC = \phi$.

Density of particles at $C = \dfrac{N}{\alpha^3\pi^{3/2}}e^{-u^2/\alpha^2}(1 + F\mu)$.

Differential element at $C = u^2 \sin\theta\, du\, d\theta\, d\phi$

relative velocity $r = \sqrt{u^2 + v^2 - 2uv\cos\theta}$, $r\,dr = 2uv\sin\theta\, d\theta$

Sum of relative velocities $= \displaystyle\iiint \dfrac{N}{\alpha^3\pi^{3/2}}e^{-u^2/\alpha^2}(1 + F[\lambda\cos\theta$

$+ \sqrt{1 - \lambda^2}\sin\theta\cos\phi\sqrt{u^2 + v^2 - 2uv\cos\theta}\,u^2\sin\theta\, du\, d\theta\, d\phi$
Integrate from $\phi = 0$ to $\phi = 2\pi$.

$$\iint \dfrac{2N}{\alpha^3\pi^{1/2}}e^{-u^2/\alpha^2}(1 + F\lambda\cos\theta)\sqrt{u^2 + v^2 - 2uv\cos\theta}\,u^2\sin\theta\, du\, d\theta$$

$\langle$from $\theta = 0$ to $\theta = \pi\rangle^a$

$$= \iint \dfrac{2N}{\alpha^3\pi^{1/2}}e^{-u^2/\alpha^2}u^2\, du\left(1 + F\lambda\dfrac{u^2 + v^2 - r^2}{2uv}\right)\dfrac{r^2}{2uv}\,dr$$

$$= \int \dfrac{2N}{\alpha^3\pi^{1/2}}e^{-u^2/\alpha^2}u^2\, du\,\dfrac{1}{4u^2v^2}\left[\dfrac{2uv}{3}r^3 + \dfrac{1}{3}F\lambda(u^2 + v^2)r^3 - \dfrac{1}{5}F\lambda r^5\right]$$

The superior limit of r is $u + v$ and the inferior, $u - v$ or $v - u$ according as u or v is greater. The quantity within brackets becomes

$$\dfrac{4uv^2}{3}(3u^2 + v^2) + \dfrac{4}{15}F\lambda(v^5 - 5u^2v^3)\quad\text{when } u \text{ is greater than } v$$

and $\dfrac{4u^2v}{3}(u^2 + 3v^2) + \dfrac{4}{15}F\lambda(u^5 - 5u^3v^2)\quad$ when v is greater than u

If N' is the number of particles projected at an angle λ with the axes & if β is their coeff[ᴸ] [coefficient] of velocity the number between v & $v + dv$ is

$N\dfrac{4}{\beta^3\sqrt{\pi}}v^3 e^{-v^2/\beta^2}\, dv$. The sum of relative velocities is therefore,

$$\iint \dfrac{8NN'}{\alpha^3\beta^3\pi}e^{-u^2/\alpha^2}e^{-v^2/\beta^2}\, du\, dv\left[\left(u^3v^2 + \dfrac{1}{3}uv^4\right) + F\lambda\left(\dfrac{v^5}{15} - \dfrac{u^2v^3}{3}\right)\right]\quad\text{when}$$

$u > v$

Integrate the first term with respect to u from $u = v$ to $u = \infty$ and the second with respect to v from $v = 0$ to $v = u$.·

$$\int \frac{8NN'}{\alpha^3 \beta^3 \pi} e^{-u^2/\alpha^2} e^{-v^2/\beta^2} \left(\frac{2}{3} \alpha^2 v^4 + \frac{1}{2} \alpha^4 v^2 \right) dv$$

$$+ F\lambda \int \frac{8NN'}{\alpha^3 \beta^3 \pi} e^{-u^2/\alpha^2} e^{-v^2/\beta^2} \left(\frac{2}{15} \beta^2 u^4 + \frac{1}{10} \beta^4 u^2 + \frac{1}{15} \beta^6 \right) dv$$

Integrating from 0 to ∞

$$\frac{8NN'}{\alpha^3 \beta^3} \frac{1}{\sqrt{\pi}} \left[\left(\frac{1}{2} \alpha^2 \beta^5 + \frac{1}{4} \alpha^4 \beta^3 \right) + F\lambda \left(\frac{1}{10} \beta^2 \alpha^5 + \frac{1}{20} \beta^4 \alpha^3 + \frac{1}{15} \beta^6 \alpha \right) \right]$$

for all cases where $u > v$.
Similarily for $v > u$,

$$\frac{8NN'}{\alpha^3 \beta^3} \frac{1}{\sqrt{\pi}} \left[\frac{1}{2} \beta^2 \alpha^5 + \frac{1}{4} \beta^4 \alpha^3 + F\lambda \left(\frac{1}{10} \alpha^2 \beta^5 + \frac{1}{20} \alpha^4 \beta^3 + \frac{1}{15} \alpha^6 \beta \right) \right]$$

[The sheet ends with this equation. The following material, beginning on a new sheet, may not have been intended as part of the same set of notes, but we adjoin it here because of similarity in content.]

Let there be n_1 molecules in unit of volume having a relative velocity r to n_2 other molecules in the same unit of volume. Then the number of encounters between molecules of the two kinds in which the alteration of the direction of the relative velocity lies between γ and $\gamma + d\gamma$ will be

$$n_1 n_2 f(r, \gamma) \sin \gamma \, d\gamma$$

where $f(r, \gamma)$ is some function of r and γ depending on the law of action between the molecules.

Now let us assume that there are N_1 molecules of the first kind in unit of volume and that the number of these which have component velocities between ξ and $\xi + d\xi$, η and $\eta + d\eta$ and ζ and $\zeta + d\zeta$ is

$$\frac{N_1}{\alpha^3 \pi^{3/2}} e^{-(\xi^2 + \eta^2 + \zeta^2)/\alpha^2} \, d\xi \, d\eta \, d\zeta$$

if we put

$$\frac{N_1}{\alpha^3 \pi^{3/2}} e^{-(\xi^2 + \eta^2 + \zeta^2)/\alpha^2} = \rho_1$$

we may call ρ_1 the *density* of the velocities which are nearly equal to the given velocity and we may say that $\rho_1 = [N_1/\alpha^3 \pi^{3/2}] e^{-v^2/\alpha^2}$ is the density of the velocities near to v, v being given both in direction and magnitude and if dV_1 be a small element of volume on the diagram of velocities then $\rho \, dV$ will be the number of molecules such that the lines representing their velocities terminate within dV.

Let there be N_2 molecules of another kind in unit of volume and let their modulus of velocity be β. Let us find how many of these have a velocity relative to v lying between r and $r + dr$. Let θ be the angle between the directions r and v and ϕ the angle which the plane of this angle makes with a fixed plane through v, then we have to integrate

$$\frac{N_2}{\beta^3 \pi^{3/2}} e^{-(v^2 - 2vr \cos\theta + r^2)/\beta^2} r^2 \sin\theta \, dr \, d\theta \, d\phi$$

from $\phi = 0$ to $\phi = 2\pi$ and from $\theta = 0$ to $\theta = \pi$. This gives

$$\frac{N_2}{\beta \pi^{1/2}} \frac{r}{v} [e^{-(v-r)^2/\beta^2} - e^{-(v+r)^2/\beta^2} \, dr]$$

for the number required.[b] The number of molecules of velocity between v and $v + dv$ which in unit of time have their velocity altered so that γ lies between γ and $\gamma + d\gamma$ is

$$4v \, dV \frac{N_1}{\alpha^3 \pi^{1/2}} e^{-v^2/\alpha^2} \frac{N_2}{\beta^3 \pi^{1/2}} \int_0^\infty [e^{-(v-r)^2/\beta^2} - e^{-(v+r)^2/\beta^2}] r f(r, \gamma) \, dr.$$

[The sheet ends with this equation. The following material, beginning on a new sheet, may not belong with it.]
We have next to determine the number of molecules of the first kind which after an encounter with ⟨part⟩ molecules of the second kind in which this relative velocity is altered in direction through an angle between γ and $\gamma + d\gamma$ ⟨have⟩ acquire a velocity between v and $v + dv$. Let OA, OB [Diagram 2] be the original velocities of the molecules of the first and second kinds, $AB = r$ their relative velocity, G their centre of gravity.

$$AG = Ga = r\frac{\alpha^2}{\alpha^2 + \beta^2}, \quad AGa = \gamma, \; GaO = \theta, \quad Oa = [\text{left blank}]$$

Let ϕ be the angle between the plane GaO and a fixed plane through Oa and ψ the angle between the planes aGA and aGO. Let ρ_1 be the density of the distribution of the molecules of the first kind corresponding to the velocity OA

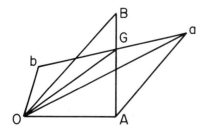

[Diagram 2]

and ρ_2 the density of distribution of the second kind corresponding to OB. To find the number of molecules of the first kind which can acquire a velocity Oa, let AB describe a conical surface about G whose solid angle is

$$\omega = \sin \gamma \, d\gamma \, d\psi$$

Next let r increase from r to $r + dr$, then the solid element described by A in all its positions will be

$$\omega G A^2 (1 - \cos \gamma) \, dGA = \frac{\alpha^6 r^2}{(\alpha^2 + \beta^2)^3} (1 - \cos \gamma) \sin \gamma \, d\gamma \, d\psi \, dr$$

The number of ⟨particles⟩ molecules of the first kind is therefore

$$\frac{N_1}{\pi^{3/2}} \frac{\alpha^3 r^2}{(\alpha^2 + \beta^2)^3} e^{-OA^2/\alpha^2} (1 - \cos \gamma) \sin \gamma \, d\gamma \, d\psi \, dr$$

Let us next find the number of molecules of the second kind with which each of these molecules may have an encounter so as to acquire such a velocity that a may fall within an element of volume dV described about a. Let A be fixed and let AB describe a solid angle ω' about A, Ga continuing parallel to itself, then B will describe an area $r^2\omega'$, and a an area $\dfrac{\alpha^2 r^2 \omega'}{\alpha^2 + \beta^2}$. Now let r increase from r to $r + dr$. B will advance dr and a will move to a distance $(1 - \cos \gamma) \dfrac{\alpha^2}{\alpha^2 + \beta^2} dr$ from the plane of the small area. The number of particles which A may meet will therefore be $\rho_2 r^2 \omega' \, dr$ and the space dV within which a will be found will be

$$dV = \frac{\alpha^4}{(\alpha^2 + \beta^2)^2} [1 - \cos \gamma] r^2 \omega' \, dr$$

[The sheet ends with this equation. The following material, beginning on a new sheet, may not belong with it.]

Two systems of particles moving in the same space have respectively N_1 and N_2 particles in unit of volume and α_1, α_2 as the modules [sic] of velocity. Required the number of collisions in unit of time after which a particle of the first system has a velocity between v and $v + dv$. Let OA [Diagram 3] represent the velocity of the first particle, OB that of the second and OG that of the centre of gravity, then if we draw $Ga = GA = r_1$ in any direction, the probability that Oa will represent the velocity of the first particle after collision will be the same in whatever direction we draw Ga. We have to determine the number of cases in which the point a will be within certain limits. If we suppose a volume V described about a then if the number of cases in which a lies within this volume be ρV we may call ρ the "density" corresponding to the velocity Oa. Let aG be fixed and let AB describe a conical surface whose

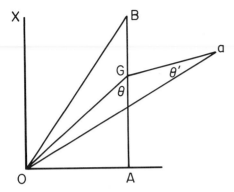

[Diagram 3]

solid angle is ω, then A will describe an area $= \omega r_1^2$.[c] Now let aG and GA be increased simultaneously by dr_1[d] $\langle$and A will move the$\rangle$ the displacement of A $\langle$(perpendi$\rangle$ in the direction of GA will be $(1 + \cos BGa)dr_1$[e] so that the volume at A corresponding to the solid angle ω and the variation dr_1 is $(1 + \cos BGA)r_1^2 \omega \, dr_1$.[f] If ρ_1 is the "density" of distribution of particles at A, the number between these limits will be

$$(1 + \cos BGA)\rho_1 r_1^2 \omega \, dr_1 \tag{1f}$$

Each of these particles, after collision, may have a velocity Oa. Let us find the number of particles with which the collision may take place, so that a may lie within the sphere V. To do this let A be fixed and let AB describe a conical surface whose solid angle is ω' while Ga is always drawn parallel to itself and equal to GA.[g] Let $GA = r_1$, $GB = r_2$, then B describes an area $= (r_1 + r_2)^2 \omega'$ $\langle$and a an area $(1 + \cos BGA)r_1^2 \omega\rangle$ and a describes an area $= r_1^2 \omega'$. Now let r_1 increase by dr_1 and r_2 by $(r_2/r_1)\,dr_1$, then B will advance $[(r_1 + r_2)/r_1]\,dr_1$ and a will advance $(1 + \cos BGA)\,dr_1$ so that if ρ_2 be the "density" at B the number of particles with which A may meet will be

$$\langle \rho_2[1 + \cos BGA][r_1 + r_2]^2 \omega' \, dr'\rangle$$

$$\rho_2[(r_1 + r_2)^3/r_1]\omega' \, dr_1' \tag{2}$$

and the volume described about A will be

$$V = (1 + \cos BGA)r_1^2 \omega' \, dr_1' \tag{3}$$

We have hitherto supposed the direction of Ga fixed but if we describe about a a solid angle ω_2 the probability that Ga will be within this angle is

$$(1/4\pi)\omega_2 \tag{4}$$

By multiplying (1), (2) and (4) we get the number of particles of the first kind which, after collision with particles of the second kind, may have velocities comprized within the given limits under the conditions that r_1 is between r_1 & $r_1 + dr_1$, that the direction Ga lies within a solid angle ω_2 and AB within a solid angle ω. The number of collisions of each of these particles in unit of time is

$$\pi s^2 (r_1 + r_2) \tag{5}$$

We have therefore

$$\delta \rho V = \delta \rho (1 + \cos BGA) r_1^2 \omega' \, dr_1'$$

$$= (1 + \cos BGA) \rho_1 r_1^2 \omega \, dr_1 \rho_2 \frac{(r_1 + r_2)^3}{r_1} \omega' \, dr_1' \frac{1}{4\pi} \omega_2 \pi s^2 (r_1 + r_2)$$

$$\rho = \sum \frac{1}{4} \rho_1 \rho_2 s^2 \frac{(r_1 + r_2)^4}{r_1} \omega \omega_2 \, dr_1$$

where ρ_1 is the "density" of distribution of velocities in the first system corresponding to the velocity OA, ρ_2 that in the second system corresponding to OB, s the distance of centres at collision, r_1 the velocity of the 1st particle & r_2 that of the second relative to their centre of gravity, ω and ω_2 small solid angles. Taking the integral of this expression with respect to ω & ω_2 round the whole sphere and with respect to r_1 from 0 to ∞, we obtain ρ the number of particles of the first kind propelled in unit of time from unit of volume, after collision with particles of the second kind, with velocities whose extremities be within unit of volume described about a.

[The rest of this sheet is blank. The following material is on the back of this sheet.]

To integrate $\rho = \sum \frac{1}{4} \rho_1 \rho_2 s^2 \dfrac{(r_1 + r_2)^4}{r_1} \omega \omega_2 \, dr_1$ when $\rho_1 = \dfrac{N_1}{\alpha_1^3 \pi^{3/2}} e^{-x^2/\alpha_1^2}$,

$\rho_2 = \dfrac{N_2}{\alpha_2^3 \pi^{3/2}} e^{-y^2/\alpha_2^2}$ and $r_1 \alpha_2^2 = r_2 \alpha_1^2$

Let $OA = x$, $OB = y$, $OG = z$, $Oa = v$, $OGA = \theta$, $OaG = \theta$:
Let the angle between the planes XOG and $GOA = \phi$, and that between XOa and $aOG = \phi'$, then[h]

$$z^2 = v^2 + r_1^2 - 2vr_1 \cos \theta'$$

$$x^2 = z^2 + r_1^2 - 2zr_1 \cos \theta$$

$$y^2 = z^2 + r_2^2 + 2zr_2 \cos \theta$$

$$\frac{x^2}{\alpha_1^2} + \frac{y^2}{\alpha_2^2} = z^2 \left(\frac{1}{\alpha_1^2} + \frac{1}{\alpha_2^2} \right) + r_1^2 \left(\frac{1}{\alpha_1^2} + \frac{\alpha_2^2}{\alpha_1^4} \right)$$

$$\omega = \sin \theta \, d\theta \, d\phi \quad \omega_2 = \sin \theta' \, d\theta' \, d\phi'$$

$$\rho = \sum \frac{1}{4} \frac{N_1 N_2 s^2}{\alpha_1^3 \alpha_2^3 \pi^3} e^{-z^2(\alpha_1^2 + \alpha_2^2)/\alpha_1^2 \alpha_2^2} e^{-r_1^2(\alpha_1^2 + \alpha_2^2)/\alpha_1^4}$$

$$\times \left[\frac{\alpha_1^2 + \alpha_2^2}{\alpha_1^2} \right]^4 r_1^3 \sin\theta \sin\theta' \, d\theta \, d\theta' \, d\phi \, d\phi' \, dr_1$$

Integrating with respect to ϕ and ϕ' from 0 to 2π we get

$$\rho = \sum \frac{N_1 N_2 s^2}{\alpha_1^3 \alpha_2^3 \pi} \left[\frac{\alpha_1^2 + \alpha_2^2}{\alpha_1^2} \right]^4 e^{-z^2(\alpha_1^2 + \alpha_2^2)/\alpha_1^2 \alpha_2^2} \times e^{-r_1^2(\alpha_1^2 + \alpha_2^2)/\alpha_1^4} r_1^3 \sin\theta \sin\theta' \, d\theta \, d\theta' \, dr_1$$

Now let z and x be made independent variables instead of θ and θ'.

$$z \, dz = v r_1 \sin\theta' \, d\theta', \quad x \, dx = z r_1 \sin\theta \, d\theta$$

Therefore

$$\rho = \sum \frac{N_1 N_2 s^2}{\alpha_1^{11} \alpha_2^3 \pi} (\alpha_1^2 + \alpha_2^2)^4 e^{-z^2(\alpha_1^2 + \alpha_2^2)/\alpha_1^2 \alpha_2^2} \times e^{-r_1^2(\alpha_1^2 + \alpha_2^2)/\alpha_1^4} \frac{x r_1}{v} \, dr_1 \, dx \, dz$$

Integrating with respect to x from $x^2 = (z - r_1)^2$ to $x^2 = (z + r_1)^2$

$$\rho = \sum \frac{N_1 N_2 s^2}{\alpha_1^{11} \alpha_2^3 \pi} (\alpha_1^2 + \alpha_2^2)^4 e^{-z^2(\alpha_1^2 + \alpha_2^2)/\alpha_1^2 \alpha_2^2} \times e^{-r_1^2(\alpha_1^2 + \alpha_2^2)/\alpha_1^4} \frac{2 z r_1^2}{v} \, dr_1 \, dz$$

Integrating with respect to z from $z^2 = (v - r_1)^2$ to $z^2 = (v + r_1)^2$

$$\rho = \sum \frac{N_1 N_2 s^2}{\alpha_1^9 \alpha_2 \pi} (\alpha_1^2 + \alpha_2^2)^3 \frac{r_1^2}{v} e^{-v^2(\alpha_1^2 + \alpha_2^2)/\alpha_1^4}$$

$$\times e^{-r_1^2(\alpha_1^2 + \alpha_2^2)^2/\alpha_1^4 \alpha_2^2} \left[e^{-2 v r_1(\alpha_1^2 + \alpha_2^2)/\alpha_1^2 \alpha_2^2} - e^{-2 v r_1(\alpha_1^2 + \alpha_2^2)/\alpha_1^2 \alpha_2^2} \right]$$

Integrating with respect to r_1 from 0 to ∞

$$\rho = -\frac{N_1 N_2 s^2}{\alpha_1^3 \pi} \alpha_2 e^{-v^2/\alpha_1^2} \left[e^{-v^2/\alpha_1^2} + \left(\frac{\alpha_2}{v} + 2\frac{v}{\alpha_2} \right) - \int_0^v e^{-v^2/\alpha_2^2} \, dv \right]$$

This is the number of particles of the first system projected under the required conditions after striking those of the second system. By putting $(N_1/\alpha_1^3 \pi^{3/2}) e^{-v^2/\alpha_1^2}$ for n in the expression for the number of collisions made by such particles we find the same expression which shows that as many particles of the given velocity are projected as lose their velocity by striking.

[End of material on back of sheet. The following material begins a new sheet.]

Let a great number of equal particles be in motion in a vessel having two of its sides equal and parallel and let the motion go on till the velocities are distributed according to the law already found. Now let it be supposed that whenever a ball strikes the first of the parallel sides it is turned white and whenever a ball strikes the other side it is turned black, and let the motion go on in this way for a sufficient time. ⟨The mechanical⟩ The distribution of velocities among the balls will be as before if we pay no regard to colour but if we distinguish the colours we shall find that more white balls are moving in the positive direction than in the negative and that the contrary is the case with the black balls. Let us investigate the number of white balls which in unit of area and at a given part of the vessel have velocities within given limits of direction and magnitude.

Let OX be the positive direction and let us assume that the ⟨number of white balls⟩ "density" of the velocities of the white balls corresponding to OA is $(N/\alpha^3\pi^{3/2})e^{-x^2/\alpha^2}\xi \cos AOX = \rho$, where ξ is a function of OA or x and must lie between 0 and 1. Let these white balls strike other balls (either white or black) whose velocity $OB = y$ and whose "density" is $(N/\alpha^3\pi^{3/2})e^{-y^2/\alpha^2}$ and let the white balls after collision have the velocity $Oa = v$. Required the density of the projected balls whose velocity is Oa. We have to sum[i]

$$\rho = \sum \frac{1}{4}\frac{N^2 s^2}{\alpha^6 \pi^3} e^{-(x^2+y^2)/\alpha^2}\xi \cos AOX\, r_1^3\, dr_1 \sin\theta \sin\theta'\, d\theta\, d\theta'\, d\phi\, d\phi'$$

Let $XOa = \gamma$, $XOG = \beta$, then
$\cos AOX = \cos\beta \cos GOA + \sin\beta \sin GOA \cos\phi$
Integrating with respect to ϕ from 0 to 2π

$$\sum \cos AOX\, d\phi = 2\pi \cos\beta \cos GOA$$

$$= 2\pi \cos\beta \frac{x^2 + z^2 - r^2}{2xz}$$

Again $\cos\beta = \cos\gamma \cos GOa + \sin\gamma \sin GOa \cos\phi'$
Integrating with respect to ϕ' from 0 to 2π[j]

$$\sum \cos\beta\, d\phi' = 2\pi \cos\gamma \frac{v^2 + z^2 - r^2}{2vz}$$

Putting θ & θ' in terms of x and z we find[k]

$$\rho = \sum \frac{1}{4}\frac{N^2 s^2 \cos\gamma}{\alpha^6 \pi} e^{-2(r_1^2+z^2)/\alpha^2}\xi \frac{r}{v^2 z^2}(x^2 + z^2 - r_1^2)(v^2 + z^2 - r_1^2)\, dx\, dz\, dr$$

In this expression we do not know the form of the function ξ and we therefore cannot integrate with respect to x. We must therefore integrate first with respect to r, but we must remember that the limits of r_1 will be different according to the relations of x and z to v. The conditions are

(1) $r < z + v$ (2) $r > z - v$ (3) $r > v - z$

(4) $r < z + x$ (5) $r > z - x$ (6) $r > x - z$

By (1) & (6) $z > \dfrac{x - v}{2}$ By (4) and (3) $z > \dfrac{v - x}{2}$

When $x < v$ the upper limit of r is $z + x$, and the lower limit is

$r = v - z$ when z is between $(v - x)/2$ and $(v + x)/2$

or $r = z - x$ when z is between $(v + x)/2$ and

When $x > v$ the upper limit of r is $z + v$, and the lower limit is

$r = x - z$ when z is between $(x - v)/2$ and $(x + v)/2$

or $r = z - v$ when z is between $(x + v)/2$ and ∞

The part depending on r is

$$\int e^{-2r_1^2/\alpha^2}\left[r(x^2 + z^2)(v^2 + z^2) - r^3(x^2 + v^2 + 2z^2) + r^5\right]dr$$

The indefinite integral is

$$\left[\frac{\alpha^2}{4}(x^2 + z^2)(v^2 + z^2) - \left(\frac{\alpha^2}{4}r^2 + \frac{\alpha^4}{8}\right)(x^2 + v^2 + 2z^2)\right.$$

$$\left. + \frac{1}{4}\alpha^2 r^4 + \frac{1}{8}\alpha^6\right]e^{-2r^2/\alpha^2}$$

$\langle$When $r = z + v\rangle$

$$= \left[\frac{1}{4}\alpha^2\left(x^2 + z^2 - r^2 - \frac{1}{2}\alpha^2\right)\left(v^2 + z^2 - r^2 - \frac{1}{2}\alpha^2\right) + \frac{1}{4}\alpha^6\right]e^{-2r^2/\alpha^2}$$

$\Big\langle$When $r = z + v$ $\dfrac{N^2 s^2 \cos\gamma}{\alpha^4 \pi v^2 z^2}\xi\left[-\left(x^2 - v^2 - 2vz - \frac{1}{2}\alpha^2\right)\left(2vz + \frac{1}{2}\alpha^2\right)\right.$

$$\left. + \frac{1}{4}\alpha^6\right]e^{-2(v^2 + 2z^2 + 2vz)/\alpha^2}\,dz\,dx \qquad \text{A}$$

$r = z \sim v$ $\dfrac{N^2 s^2 \cos\gamma}{\alpha^4 \pi v^2 z^2}\xi\left[+\left(x^2 - v^2 + 2vz - \frac{1}{2}\alpha^2\right)\left(2vz - \frac{1}{2}\alpha^2\right)\right.$

$$\left. + \frac{1}{4}\alpha^6\right]e^{-2(v^2 + 2z^2 - 2vz)/\alpha^2}\,dz\,dx \qquad \text{B}$$

$r = z + x$ $\dfrac{N^2 s^2 \cos\gamma}{\alpha^4 \pi v^2 z^2}\xi\left[-\left(v^2 - x^2 - 2xz - \frac{1}{2}\alpha^2\right)\left(2xz + \frac{1}{2}\alpha^2\right)\right.$

$$\left. + \frac{1}{4}\alpha^6\right]e^{-2(x^2 + 2z^2 + 2xz)/\alpha^2}\,dz\,dx \qquad \text{C}$$

$$r = z \sim x \quad \frac{N^2 s^2 \cos \gamma}{\alpha^4 \pi v^2 z^2} \zeta \left[\left(v^2 - x^2 + 2xz - \frac{1}{2}\alpha^2 \right) \left(2xz - \frac{1}{2}\alpha^2 \right) \right.$$

$$\left. + \frac{1}{4}\alpha^6 \right] e^{-2(x^2 + 2z^2 - 2xz)/\alpha^2} \, dz \, dx \qquad D \Big\rangle$$

When $r = x \pm v$ $\quad \dfrac{N^2 s^2 \cos \gamma}{\alpha^4 \pi v^2} \zeta \left[4v^2 + (\alpha^2 + v^2 - x^2) \left(\dfrac{1}{2}\dfrac{\alpha^2}{z^2} \pm 2\dfrac{v}{z} \right) \right]$

$$\times \, e^{-2v^2/\alpha^2} e^{-4(z^2 \pm vz)/\alpha^2} \, dx \, dz \qquad \binom{A}{B}$$

When $r = z \pm x$ $\quad \dfrac{N^2 s^2 \cos \gamma}{\alpha^4 \pi v^2} \zeta \left[4x^2 + (\alpha^2 - v^2 + x^2) \left(\dfrac{1}{2}\dfrac{\alpha^2}{z^2} \pm 2\dfrac{x}{z} \right) \right]$

$$\times \, e^{-2x^2/\alpha^2} e^{-4(z^2 \pm xz)/\alpha^2} \, dx \, dz \qquad \binom{C}{D}$$

When $x < v$ the upper limit is C and the lower limit is B from $z = (v - x)/2$ to $z = (v + x)/2$ and D above this.

When $x > v$

B^I A   D $(x - v)/2$

Integrating these expressions with regard to z and writing $\Phi(x/\alpha)$ for $\int_0^{x/\alpha} e^{-t^2} \, dt$ they become

$$-\frac{N^2 s^2 \cos \gamma}{\alpha^3 \pi v^2} \zeta e^{-2v^2/\alpha^2} \left[2(\alpha^2 - x^2)\Phi \frac{(2z \pm v)}{\alpha} e^{v^2/\alpha^2} \right.$$

$$\left. + \frac{1}{2}\frac{\alpha}{z}(\alpha^2 + v^2 - x^2)e^{-4(z^2 \pm vz)/\alpha^2} \right] dx \qquad \binom{A}{B}$$

and

$$-\frac{N^2 s^2 \cos \gamma}{\alpha^3 \pi v^2} \zeta e^{-2v^2/\alpha^2} \left[2(\alpha^2 - v^2)\frac{(2z \pm x)}{\alpha} e^{x^2/\alpha^2} \right.$$

$$\left. + \frac{1}{2}\frac{\alpha}{z}(\alpha^2 - v^2 + x^2)e^{-4(z^2 \pm zx)/\alpha^2} \right] dx \qquad \binom{C}{D}$$

When $x < v$, C is the upper limit from $z = (v - x)/2$ to $z = \infty$. The integral within these limits is,

$$\frac{N^2 s^2 \cos \gamma}{\alpha^3 \pi v^2} \zeta e^{-x^2/\alpha^2} \left[2(\alpha^2 - v^2) \left(\frac{\sqrt{\pi}}{2} - \Phi(v/\alpha) \right) \right.$$

$$\left. - \left(\frac{\alpha^2}{v - x} - v - x \right) \alpha e^{-v^2/\alpha^2} \right] dx$$

The lower limit is B from $z = (v - x)/2$ to $z = (v + x)/2$.
Taking B within these limits

$$\frac{N^2s^2\cos\gamma}{\alpha^3\pi v^2}\xi e^{-v^2/\alpha^2}\left[2(\alpha^2 - x^2)(2\Phi(x/\alpha) - (\alpha^2 + v^2 - x^2)\frac{2\alpha x}{v^2 - x^2}e^{-x^2/\alpha^2}dx\right.$$

Between $x = (v + x)/2$ and $z = \infty$ the lower limit is D, therefore take D between these limits,

$$\frac{N^2s^2\cos\gamma}{\alpha^3\pi v^2}\xi e^{-v^2/\alpha^2}\left[2(\alpha^2 - v^2)\left(\frac{\sqrt{\pi}}{2} - \Phi(v/\alpha)\right) - \left(\frac{\alpha^2}{v + x} - v + x\right)\alpha e^{-v^2/\alpha^2}\right]dx$$

Subtracting these integrals at the lower limits from that at the upper one we get,

$$\int_0^v \frac{N^2s^2\cos\gamma}{\alpha^3\pi v^2}\xi e^{-v^2/\alpha^2}[4(\alpha^2 - x^2)\Phi(x/\alpha) + 4\alpha x e^{-x^2/\alpha^2}]dx$$

for the complete expression to be integrated between $x = 0$ and $x = v$. Similarly, for the expression between $x = v$ and $x = \infty$ we get,

$$\int_v^\infty \frac{N^2s^2\cos\gamma}{\alpha^3\pi v^2}\xi e^{-x^2/\alpha^2}[4(\alpha^2 - v^2)\Phi(v/\alpha) + 4\alpha v e^{-v^2/\alpha^2}]dx$$

The sum of these expressions gives the value of ρ, but we cannot integrate them as we do not know the form of the function ξ.
[End of sheet. The following material is on a separate sheet.]

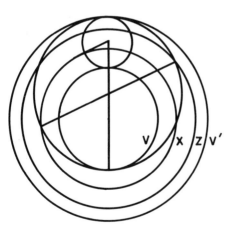

[Diagram 4]

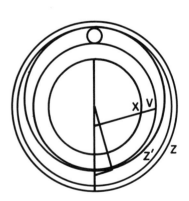

[Diagram 5]

$$x \text{ and } z \text{ both } > v \quad r\max = \sqrt{x^2 + y^2 - v^2} + v \quad y = \tfrac{1}{2}\left(\sqrt{x^2 + y^2 - v^2} - v\right) \quad x < v < z \quad \max r = x + z \quad y = \tfrac{1}{2}(z - x)$$

$$r\min = \sqrt{x^2 + y^2 - v^2} - v \quad y = \tfrac{1}{2}\left(\sqrt{x^2 + y^2 - v^2} - v\right) \qquad \min r = z - x \quad y = \tfrac{1}{2}(z + x)$$

$$x \text{ and } z \text{ both } < v \quad r\max = \sqrt{x^2 + z^2 - v^2} + v \quad y = \tfrac{1}{2}\left(v - \sqrt{x^2 + z^2 - v^2}\right) \quad x > v > z \quad \max r = x + z \quad y = \tfrac{1}{2}x - z$$

$$r\min = v - \sqrt{x^2 + z^2 - v^2} \quad y = \tfrac{1}{2}\left(v + \sqrt{x^2 + z^2 - v^2}\right) \qquad \min r = x - z \quad y = \tfrac{1}{2}x + z$$

$$\iint \rho_1 x \rho_2 zy \quad z < v \quad \sqrt{v^2 - z^2} < x < v - \iint \rho_1 x \rho_2 zy \sqrt{x^2 + z^2 - v^2}$$

$$x > v - \iint \rho_1 x \rho_2 zz$$

$$z + v \quad x - z$$

$$z > v$$

$$x < v - \iint \rho_1 x \rho_2 zx$$

$$x > v - \iint \rho_1 x \rho_2 zv$$

1 $r^2 < (v + z)^2$ therefore $z > r - v$ when $x > v$

2 $r^2 < (z + x)^2$ therefore $z > r - x$ when $x < v$

3 $r^2 > (v - z)^2$ $z < v + r$ when $z > v$

   $z > v - r$ when $z < v$

4 $r^2 > (z - x)^2$ $z < x + r$ when $z > x$

   $z > x - r$ when $z < x$

$$x > v > r, \quad z < v + r \quad \frac{x+v}{2} > r > v \quad z < v + r \quad x > r > \frac{x+y}{2} \quad z < v + r \quad rxv \quad \frac{4r}{\alpha^2}$$

$$z > x - r \qquad z > x - r \qquad z > r - v \qquad r^5 + 2r^3 + \frac{\alpha^4 r}{2}$$

$$v \qquad r \qquad\qquad v \qquad rx$$

$x > v, \ r > \dfrac{x+v}{2}$	$x > v \ \ r < \dfrac{x+v}{2}$	$x < v \ \ r > \dfrac{x+v}{2}$	$x < v \ \ r < \dfrac{x+v}{2}$
$z < v + r$ $z > r - v$	$z < v + r$ $z > x - r$	$z < x + r$ $z > r - x$	$z < x + r$ $z > v - r$

$$x \qquad z \qquad v$$

$$\text{minimum of } r = z - v \text{ when } z > \frac{x+v}{2}, \quad z = \infty \text{ to } z = \frac{x+v}{2}$$

$$\text{minimum of } r = z + v \text{ when } x > v = x - z \text{ when } z \ \frac{x+v}{2} \quad z = \frac{x+v}{2} \text{ to } z = \frac{x-v}{2}$$

$$= z + x \text{ when } v \ x \qquad z - x \qquad z > \frac{x+v}{2}$$

$$v - z \qquad z < \frac{x+v}{2}$$

$$\frac{\alpha^2}{2}\left[r(x^2 + z^2)(v^2 + z^2) - r^3(x^2 + v^2 + 2z^2) + r^5\right]$$

$$\frac{\alpha^2}{4} \qquad \frac{\alpha^2}{4}r^2 + \frac{\alpha^4}{8} \qquad \frac{\alpha^2}{4}r^2 + \frac{\alpha^4}{8} \qquad \left(\frac{\alpha^4}{4}r^4 + \frac{\alpha^4}{4}r^2 + \frac{\alpha^6}{8}\right)e^{2r^2/\alpha^2}$$

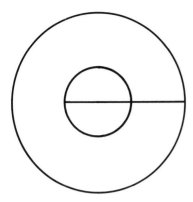

[Diagram 6]

a. Although the limits of integration over θ are scored through, the next line shows that Maxwell did the integration as indicated between the limits $\theta = 0$ to $\theta = \pi$.

b. The number to which Maxwell refers is the number of particles which have velocity relative to the moving particle of r. This result is deduced in the published 1860 paper (document III-6); see *Scientific Papers*, vol. 1, 385.

c. Written as $\omega r_1^2 GA^2$ with GA^2 scored through.

d. Maxwell does not allow GA and Ga to both increase by dr_1. Ga represents the final velocity of the particle whose original velocity was OA. Ga is allowed to increase by dr_1. Maxwell then asks: what is the change in the original velocity Oa, and then what is the change in its component along GA that produce the change in the final velocity. The velocity change is reduced to a "displacement" in his diagram.

e. This expression was originally written as $(1 + \cos BGa) dr_1$.

f. Both these expressions should be written as $(1 + \cos BGa)$.

g. The apex of the cone is A.

h. Therefore, $z = OG$, $x = OA$, and $y = OB$.

i. To understand Maxwell's notation refer to diagram 3. He also is using the same expressions for the velocities as in the previous problem. $Ox = x$ is the original velocity of the white balls; $OY = y$ is the original velocity of all the other balls.

j. Maxwell drops the suffix on r.

k. $z = OG$.

l. From the spacing of the letters on the manuscript page (not easily reproducible here) it appears that this sentence should be interpreted as: "When $x > v$ the upper limit is A and the lower is D from $r = (x - v)/2$ to $\dfrac{v + x}{2}$ and B above this."

10. Letter from Maxwell to H. R. Droop,[a] January 28, 1862

Lewis Campbell and William Garnett, *The Life of James Clerk Maxwell*, 332.

8, Palace Garden Terrace[b]
Kensington, London, W.
28th, January, 1862.

Some time ago, when investigating Bernoulli's theory of gases, I was surprised to find that the internal friction of a gas (if it depends on the collisions of particles) should be independent of the density.

Stokes has been examining Graham's experiments on the rate of flow of gases through fine tubes, and he finds that the friction, if independent of density, accounts for Graham's results, but, if taken proportional to density, differs from those results very much. This seems rather a curious result, and an additional phenomenon explained by the "collisions of particles" theory of gases. Still one phenomenon goes against that theory--the relation between specific heat at constant pressure and at constant volume, which is in air = 1.408, while it ought to be 1.333.

My brother-in-law,[c] who is still with us, is getting better, and had his first walk on crutches today across the room.

a. H. R. Droop (1832–1884) was an undergraduate at Trinity College Cambridge during the same years as Maxwell and a Third Wrangler in 1854. Maxwell referred to him as "Droop the ingenuous" in his correspondence. Droop later became a member of the Equity Bar. The surviving correspondence opens with Maxwell's appeal for funds to endow a chapel close to his estate in Scotland. The rest is on scientific matters.

b. Maxwell's home address during his tenure as Professor of Natural Philosophy, Kings College, London, 1860–1865.

c. Donald Dewar—see Campbell and Garnett, *Life*, 318.

11. Letter from Maxwell to Lewis Campbell, April 21, 1862

Lewis Campbell and William Garnett, *Life of James Clerk Maxwell*, 335–336.

8, Palace Gardens Terrace,
Kensington, W., 21st April 1862

It is now a long time since I wrote half a letter to you, but I never since had time to write or to find the scrap. I suppose, as it was more than a good intention, but less than a perfect act, it may be regarded as destined to paper purgatory. This is the season of work to you, when folk visit shrines in April and May, but I get holiday this week. I have been putting together a large optical box, 10 feet long containing two prisms of bisulphuret of carbon, the largest yet made in London, five lenses and two mirrors, and a set of movable slits. Everything requires to be adjusted over and over again if one thing is not quite right placed, so I have plenty of trial work to do before it is perfect, but the colours are most splendid.[a]

I think you asked me once about Helmholtz and his philosophy. He is not a philosopher in the exclusive sense, as Kant, Hegel, Mansel[b] are philosophers, but one who prosecutes physics and physiology, and acquires therein not only skill in discovering any desideratum, but wisdom to know what are the desiderata, *e.g.*, he was one of the first, and is one of the most active, preachers of the doctrine that since all kinds of energy are convertible, the first aim of science at this time should be to ascertain in what way particular forms of energy can be converted into each other, and what are the equivalent quantities of the two forms of energy.

The notion is as old as Descartes (if not Solomon), and one statement of it was familiar to Leibnitz. It was wholly unknown to Comte, but all sorts of people have worked at it, of late,—Joule and Thomson for heat and electricals, Andrews[c] for chemical combinations, Dr. E. Smith[d] for human food and labour. We can now assert that the power of our bodies is generated in the muscles, and is not conveyed to them by the nerves but produced during the transformation of substances in the muscle, which are supplied fresh by the blood.

We can also form a rough estimate of the efficiency of a man as a mere machine, and find that neither a perfect heat engine nor an electric engine could produce so much work and waste so little in heat. We therefore save our pains in investigating any theories of animal power based on heat and electricity. We see also that the soul is not the direct moving force of the body. If it were, it would only last till it had done a certain amount of work, like the spring of a watch, which works till it is run down. The soul is not the mere mover. Food is the mover, and perishes in the using, which the soul does not. There is action and reaction between body and soul, but it is not of a kind in which energy passes from the one to the other,—as when a man pulls a trigger it is the gunpowder that projects the bullet, or when a pointsman shunts a train it is the rails that bear the thrust. But the constitution of our nature is not explained by finding out what it is not. It is well that it will go, and that we remain in possession, though we do not understand it.

Hr. Clausius of Zurich, one of the heat philosophers, has been working at the theory of gases being little bodies flying about, and has found some cases in which he and I don't tally, so I am working it out again. Several experimental results have turned up lately, rather confirmatory than otherwise of that theory.

I hope you enjoy the absence of pupils. I find that the division of them into smaller classes is a great help to me and to them; but the total oblivion of them for definite intervals is a necessary condition of doing them justice at the proper time.

a. This was Maxwell's third and most advanced colorimeter, employing what afterward became known as the "double monochromator" principle. See *Scientific Papers*, vol. 2, 230–231, and C. W. F. Everitt, *James Clerk Maxwell*, 68–69.

b. Henry Longueville Mansel (1820–1871), educated at St. John's College Oxford, where he graduated with firstclass honors simultaneously in classics and mathematics. He tried to develop Sir William Hamilton's philosophy and ran into much controversy over the supposed agnosticism of his conclusions. He later edited Hamilton's lectures.

c. Thomas Andrews (1813–1885) in a series of papers published between 1841 and 1848 gave the details of his experiments on the heats of neutralization, heats of formation of water, oxides, metal halides, and the heat evolved when one metal replaces another in solution.

d. Edward Smith (1818–1874). His interests were in physiology, nutrition, and public health. Much of his work was motivated by social reform. From 1855 to 1865 he was assistant physician at Brompton Hospital for Consumption, where he studied respiratory function and physiology. His studies were made primarily on prisoners at hard labor who were required to spend hours every day on a treadmill. He measured respiration at rest and during exercise; this led to studies in the metabolism of foodstuffs and energy sources. His findings were published as, "Experimental Inquiries into Chemical and Other Phenomena of Respiration, and Their Modification by Various Physical Agencies," *Phil. Trans. R. Soc. London* 149 (1859), 681–714.

12. "On the Conduction of Heat in Gases"[a]

Cambridge University Library, Maxwell Manuscript Collection.

In the *Philosophical Magazine* for January and July 1860[b] I applied the theory of the motions and collisions of small elastic particles to the explanation of various properties of gases, according to the analogies already pointed out by Daniel Bernoulli and others and more recently by M. Clausius.[c] The theory supposes that the particles of gases are in rapid motion, ⟨but⟩ that they do not act on one another except within a very small distance, but that an exceedingly intense repulsive force comes into action when two particles come within this distance from each other. The particles therefore move in straight lines except when they come ⟨into each other's⟩ within the reach of the repulsive action of other particles which alters their motion with a suddenness like that of the impact of elastic bodies. The path of each particle is thus made up of a succession of straight lines and very sharp ⟨turnings⟩ curves, and the mode in which the motion of one set of particles influences that of another will depend upon the average length of the straight part of the path as well as on the mass and velocity of the particles. M. Clausius has recently published an investigation of the particular case of the conduction of heat through a gas which was very imperfectly treated by me in the paper referred to.[d] ⟨and as⟩ He has pointed out several oversights in my calculations, I have reexamined it and found some others ⟨so that⟩ the influence of which extends to other parts of my investigation.[e] I shall therefore state here as much of my former results as will make the ⟨corre⟩ requisite corrections intelligible, and I shall retain the methods used in my former paper except when obliged to compare them with those of M. Clausius.

(1) In my former paper I investigated the collision of two elastic spheres and found that the ⟨motion⟩ velocity of each after collision is resoluble into two parts, one of which is equal to the velocity of the centre of gravity of the two spheres before impact, and in the same direction while the other is equal to the relative velocity of the sphere with respect to the centre of gravity and may with equal probability be in any direction.

(2) If a great many particles are in motion in the same vessel they will not all have the same velocity, but the average number of particles whose velocity lies within the limits v and $v + dv$ will be

$$N \frac{4}{\alpha^3 \sqrt{\pi}} v^2 e^{-v^2/\alpha^2} \, dv \tag{1}$$

where N is the whole number of particles, and α a constant depending on the velocity. The velocities range through all possible values but more particles have a velocity $= \alpha$ than any other given velocity. The mean values of the different powers of v are ⟨given in⟩ found by integration to be

$$\frac{1}{v} = \frac{2}{\alpha\sqrt{\pi}}, \quad v = \frac{2\alpha}{\sqrt{\pi}}, \quad v^2 = (3/2)\alpha^2, \quad v^3 = \frac{4\alpha^3}{\sqrt{\pi}} \tag{2}$$

Wherever we have to take the mean value of any power of v we must use the value here given, and not that got by raising the mean value of v to the given power.

(3) The motion of the particles ⟨of any⟩ after a sufficient number of collisions will be compounded of a velocity V in a given direction, the same for all the particles, and a velocity v which may be in any direction and of which the values are distributed according to the law of eqⁿ (1). We shall call V the motion of translation and v the motion of agitation.

(4) If the velocities of agitation of two systems are distributed according to the law of eqⁿ (1) then the relative velocities of the particles, one in each system, will be distributed according to the same law, but the mean relative velocity of agitation will be the square root of the mean of the square of the mean velocity of agitation in the two systems.[f] By the same method of demonstration it may be shown that if any quantity u is a linear function of several independent quantities x, y, z of the form

$$u = ax + by + cz \tag{3}$$

then if x, y, z, are distributed according to the law of eqⁿ (1), u will also be distributed according to that law and the mean values of u, x, y and z will be connected by the equation

$$u^2 = a^2x^2 + b^2y^2 + c^2z^2 \tag{4}$$

(5) The pressure of a gas is one third of the density multiplied by the mean of the square of the velocity. Now if T be the absolute temperature[g] and p_0, ρ_0 the pressure and density ⟨at temperature⟩ when $T = T_0$ then we know by experience that[h]

$$p = \frac{p_0 \rho T}{\rho_0 T_0}$$

whence we find

$$\overline{v^2} = \frac{3p_0}{\rho_0 T_0} T$$

or the velocity varies as the square root of the temperature.

(6) When particles of different kinds are allowed to communicate their motion of agitation to each other the average *vis viva* of the particles tends to become

the same in both sets of particles, or if M_1 and M_2 are the ⟨atomic⟩ masses of ⟨the⟩ an atom in the two systems then when there is equilibrium of temperature,

$$M_1 v_1^2 = M_2 v_2^2$$

or $\quad \dfrac{M_1}{M_2} = \dfrac{v_2^2}{v_1^2} = \dfrac{p_2}{p_1}\dfrac{p_1}{p_2}$

so that when the pressure and temperature of the two gases are the same the atomic weights are proportional to the densities.

(7) By considering the effect of the collisions of bodies of any form not spherical it appears that the vis viva of rotation tends to become equal to that of translation so that the whole energy in unit of volume is not $\frac{1}{2}\rho v^2$ as in the case of perfect spheres, but ρv^2. In a medium consisting partly of perfect spheres and partly of other bodies the energy will be $\frac{1}{2}\beta\rho v^2$ where

$$\beta = \left\langle \frac{\frac{1}{2}\rho_1 + \rho_2}{\rho_1 + \rho_2} \right\rangle = \langle \rho_1 + 2\rho_2 \rangle = 1 + q$$

⟨ρ_1 being the weight of the spheres and ρ_2 of the other bodies in unit of volume⟩ where q is the ratio of the mass of the non-spherical particles to the whole mass. If γ is the ratio of the specific heat under constant pressure to that under constant volume

$$\gamma = 1 + \frac{2}{3\beta}$$

If the particles are all spherical with their centres of figure and mass coincident then $q = 0$, $\beta = 1$ and $\gamma = 1\frac{2}{3} = 1.\dot{6}$.
If none of the particles fulfil these conditions then
$\qquad q = 1$, $\beta = 2$ and $\gamma = 1\frac{1}{3} = 1.\dot{3}$.
These are the two extreme cases.
In the case of air $\gamma = 1.408$, $\beta = 1.634$, $q = 0.634$, $1 - q = 0.366$ so that ⟨on the theory of⟩ according to the theory we are treating of, we must suppose 0.634 of the weight of air to consist of non-spherical particles and 0.366 of its weight to consist of perfectly spherical particles having their centre of gravity at their centres.
In the case of Oxygen, Hydrogen and Nitrogen the proportion must be supposed nearly the same as in air, but in carbonic acid we must suppose the proportions of spherical particles to be smaller.
In the case of Steam, if we admit that the value of γ given at p. 320 of "Rankine on the Steam Engine"[i] is correct we find, $\gamma = 1.304$, $\beta = 2.19$, $q = 1.19$, $1 - q = -0.19$, that is, we must suppose a negative quantity of spherical particles

to exist, or in other words our theory fails to explain how the value of γ can be so low as 1.304.

(8) We come now to those properties of gases which depend on the distance which a particle travels between successive collisions. This distance depends on the number of particles in unit of volume and on the distance ⟨at which two particles comes into⟩ of the centres of two particles at the moment of collision. If l_1 be the *mean* length of path of a particle of a gas whose density is ρ_1, mixed with other gases whose densities in the mixture are ρ_2 &c.,

$$\frac{1}{l_1} = A\rho_1 + B\rho_2 + \text{etc.}$$

where $A = \sqrt{2}\,\dfrac{\pi s_1^2}{M_1}$ $B = \sqrt{\left(1 + \dfrac{v_2^2}{v_1^2}\right)} \cdot \dfrac{\pi s'^2}{M_2}$ &c.

where s_1 is the distance of centres at collision for two particles of the first kind and s' the same for a collision between one of the first and one of the second, M_1 and M_2 the masses of particles of each kind and v_1 and v_2 their velocities of agitation.

(9) The actual length of the path described by a particle between successive collisions is not always the same, but the values of the actual paths are distributed according to the following law, as M. Clausius has shown in his former paper.[j]

Let l be the mean length of path, then the proportion of the whole particles whose path exceeds nl is e^{-n}.

(10) The actual value of the length of path depends on the diameter of the particles and on the number in unit of volume, neither of which quantities are known, but if the internal friction of gases arises from the intermingling of particles from different layers of the moving gas then there will be a relation between l and the coefficient of internal friction, which may be determined by experiments on oscillating bodies and on ⟨fluids⟩ the passage of gases through long tubes. I have shown that if μ is the tangential force on unit of area when the velocity ⟨increases⟩ parallel to that area increases by unity for every unit of length ⟨measured⟩ normal to the area,

$$\mu = \tfrac{1}{3}\rho l v = \tfrac{1}{3}A$$

since $\rho l = A$. This shows that μ is proportional to the square root of the absolute temperature, and independent of the density. From the value of μ given by Professor Stokes[k] it appears that for air under the ordinary conditions, $\mu = 1/400{,}000$ inch nearly.

(11) I now come to the question which I neglected to consider in my former paper. When the density and temperature ⟨and composition⟩ of a gas or the composition of a system of mixed gases vary from one place to another, what is

the proportion of particles, which, starting from one given place, arrives at another given place without a collision.

Let us suppose the gas to be a mixture of gases whose densities are ρ_1, ρ_2 etc., their velocities of agitation v_1, v_2 etc. all these quantities being functions of x. Let s_1 be the distance of centres for collisions between two particles of the first kind, $\langle \& \rangle$ s' for a particle of the first with one of the second &c. Let λ_1 be the mean length of path which a particle of the first kind having a velocity v' differing slightly from v_1 would have if projected in a system for which ρ_1, ρ_2, v_1, v_2 are constants. We find by Prop. IX[1]

$$\frac{1}{\lambda_1} = \pi s_1^2 N_1 \sqrt{\left(1 + \frac{v_1^2}{v_2^2}\right)} + \pi s'^2 N_2 \sqrt{\left(1 + \frac{v_2^2}{v'^2}\right)}$$

Putting $\langle v' = v_1 + dv_1 \rangle$ $v_1 = v' + dv$ we get

$$\frac{1}{\lambda_1} = \sqrt{2} \pi s_1^2 N_1 \left[1 + \frac{1}{2}\frac{dv}{v_1}\right] + \pi s'^2 N_2 \sqrt{\left[1 + \frac{v_2^2}{v_1^2}\right]} \left[1 + \frac{v_2^2}{v_1^2 + v_2^2}\frac{dv}{v_1}\right]$$

or $\dfrac{1}{\lambda_1} = A\rho_1 \left(1 + \frac{1}{2}\frac{dv}{v_1}\right) + B\rho_2 \left[1 + \frac{v_2^2}{v_1^2 + v_2^2}\frac{dv}{v_1}\right]$

λ_1 is a function of x, and when $v' = v_1$, $\lambda_1 = l_1$.

If v be measured in any direction and if particles are projected with velocity v' in this direction, then if u represent the number of particles which arrive at distance r, the proportion of these which will be stopped in the succeeding portion dr will be dr/λ_1, or in symbols,

$du = u \, dr/\lambda_1$

or

$u = Ne^{-\int_0^r dr/\lambda_1}$

if N is the whole number projected and u the number which reach the distance r. Since r must be small because it is the path of a particle we may write $r = n\Lambda$, where $\Lambda = \frac{1}{2}(l_1 + \lambda_1)$ and then,

$u = Ne^{-n}$

$\langle$where $n = r/\lambda_1$ and $\Lambda_1 = (l_1 + \lambda_1)/2\rangle$

In all cases therefore, in which the properties of the gas vary from place to place, the number of particles which start from one place and pass through another, depends on the quantity Λ which is a mean between l_1, the length of path of the particles at the place where they started, and λ their length of path if they had been projected $\langle$among$\rangle$ with their actual velocity at the other place.

$\langle$The general value of Λ is

$$\frac{1}{\Lambda} = A\rho_1\left(1 + \frac{1}{4}\frac{dv}{v_1}\right) + B\rho_2\left(1 + \frac{1}{2}\frac{v_2^2}{v_1^2 + v_2^2}\frac{dv}{v_1}\right)\bigg\rangle$$

By overlooking the differences between l_1, λ_1 and Λ I have gone wrong in Props XIV and XVI^m of my former paper. I must therefore repeat the calculations of those propositions making the required corrections.

(12) Prop XIV (corrected) In a system of particles whose density, velocity etc., are functions of x, to find the quantity of matter transferred across the plane yz due to the motion of agitation alone. If there is a motion of translation we must suppose the plane of yz to move with the same velocity so as to reduce the relative motion of translation to zero.

Let N = number of particles in unit of volume

$\qquad M$ = mass of each particle

$\qquad v$ = velocity of agitation at distance x

$\qquad l$ = mean free path of particle at distance x

$\qquad v_0$ = velocity of agitation at origin

$\qquad \lambda$ = mean path of particles with velocity v among particles at origin

$\qquad \Lambda = \frac{1}{2}(l + \lambda)$

$\qquad q$ = quantity of matter transferred across unit of area in unit of time

Take a stratum whose thickness is dx and distance from origin x, the area being unity. The number of collisions taking place in the stratum is

$$N\frac{v}{l}dx.$$

These particles will move in all directions with velocity v till they strike other particles and their lengths of path will be different in different directions, because the properties of the system are different in different places. We have only, however, to ascertain what proportion of these particles pass through the plane yz and we have shown that this depends only on the value of Λ. We thus find for the number of particles projected from the stratum in unit of time whose paths are between $n\Lambda$ and $(n + dn)\Lambda$ and which pass through the plane yz in a positive direction,

$$\frac{Nv(x \mp n\Lambda)}{2nl\Lambda}e^{-n}dx\,dn$$

where x must be between $\pm n\Lambda$ and the upper or lower sign is to be taken according as x is positive or negative. We find by multiplying by M and integrating with respect to x from $x = -n\Lambda$ to $x = +n\Lambda$ and with respect to n from 0 to ∞,

$$q = -\frac{1}{3}\frac{d}{dx}\left(\rho v\frac{\Lambda^2}{l}\right)$$

Since Λ is a mean between l and λ, and these three quantities ⟨are very nearly equal⟩ differ by very small quantities we may write,

$$q = -\frac{1}{3}\frac{d}{dx}(\rho v \lambda)$$

as the most convenient expression for q. In differentiating λ we must remember that in the value of λ, ρ_1, ρ_2, v_1, v_2 are the values at the origin and therefore are to be regarded as constant, and that v' is the velocity at x, so that $v' = v_1 + (dv_1/dx)x$ and $dv = -(dv_1/dx)x$. We find

$$\frac{d\lambda}{dx} = l_1^2 \frac{1}{v_1}\frac{dv_1}{dx}\left(\frac{1}{2}A\rho_1 + \frac{v_2^2}{v_1^2 + v_2^2}B\rho_2\right)$$

and

$$q = -\frac{1}{3}\rho_1 v_1 l_1\left[\frac{1}{\rho_1}\frac{d\rho_1}{dx} + \frac{1}{v_1}\frac{dv_1}{dx}\left(1 + \frac{1}{2}Al_1\rho_1 + \frac{v_1^2}{v_1^2 + v_2^2}Bl_1\rho_2\right)\right]$$

which is the value of q for two mixed gases. For one gas, the expression becomes

$$q = -\frac{1}{3}\rho_1 v_1 l_1\left[\frac{1}{\rho_1}\frac{d\rho_1}{dx} + \frac{3}{2}\frac{1}{v_1}\frac{dv_1}{dx}\right]$$

This is the value to be employed in treating of the conduction of heat.
(13) Let us now apply similar corrections to Prop XVI. Prop XVI (corrected) To find the resultant dynamical effect of all the collisions which take place in a given stratum. We have to find the resultant momentum of all the particles which enter the stratum and strike there in unit of time. Using the same symbols we have to find the momentum of particles which starting from a stratum dx lodge in a stratum whose thickness is α at the origin.
The number of particles projected from dx is as before

$$N\frac{v}{l}dx.$$

The proportion of these whose directions make an angle with x whose cosine lies between μ and $\mu + d\mu$ is $\frac{1}{2}d\mu$. The proportion of these which reach a distance $r = n\Lambda - \frac{1}{\mu}x$ is

$$e^{-n}$$

The proportion of these which strike in the stratum α, that is between r and $r + dr$ where $\mu\,dr = \alpha$ is

$$\frac{\alpha}{\mu\lambda}$$

The velocity of these particles resolved along the axis of x is $-v\mu$ and the resolved momentum is $-Mv\mu$, Multiplying all these numbers together and remembering that $(d\mu/dn) = -(x/n^2\Lambda)$, we get for X the whole momentum,

$$\alpha\rho X = \int_0^\infty \int_{-n\Lambda}^{+n\Lambda} \frac{1}{2}\frac{\rho v^2\alpha}{l\lambda\Lambda n^2}\, xe^{-n}\, dx\, dn$$

$$= \frac{1}{3}\frac{d}{dx}\left[\frac{\rho v^2\Lambda^2}{l\lambda}\right]\alpha$$

and since Λ is a mean between l and λ this may be reduced to,

$$\rho X = \frac{1}{3}\frac{d}{dx}(\rho v^2) = \frac{dp}{dx} \quad \text{by Prop XI}$$

This result is the same as that which I obtained before, but the method of procedure is now rendered strict.

(14) In applying these results to the case of the conduction of heat through a stratum of air from a hot surface to a cold one we must ⟨make⟩ introduce the conditions that the transfer shall be of heat only and not of matter, and that every intermediate slice of air shall be in equilibrium. In my former paper I paid little attention to this subject as I had no experimental data to compare with the theory, but the errors of principle into which I fell ⟨should⟩ are worth correcting in order to compare the results of my method of calculation with those obtained by M. Clausius.

The whole quantity of matter transferred across unit of area in unit of time is

$$Q = q + pV$$

where q has the value given in Prop XIV and V is the velocity of translation. When there is no transfer of matter $Q = 0$. The resultant force of the collisions per unit of volume is

$$-\frac{dp}{dx}$$

When there is no resultant force, p must be constant. The quantity of energy in each particle depends partly on the velocity of its centre of gravity and partly on its velocity of rotation about that centre. As these velocities tend towards a constant ratio, we may assume that the energy of a particle is $\frac{1}{2}\beta Mv^2$. The amount of energy which is transferred across unit area in unit of time depends partly on the motion of translation and partly on that of agitation. That depending on the motion of translation is $\frac{1}{2}\beta V\rho v^2$, and that depending on the motion of agitation may be found ⟨from⟩ by the method of Prop XIV by substituting the energy instead of the mass of each particle. ⟨The energy transferred in unit of time acro⟩ The whole energy transferred being called E we find,

$$E = \frac{1}{2}\beta V\rho v^2 - \frac{1}{3}\frac{d}{dx}\left[\frac{1}{2}\beta\rho v^3\lambda\right]$$

with the conditions

$$p = \text{constant} = \frac{1}{3}\rho v^2$$

$$Q = V\rho - \frac{1}{3}\frac{d}{dx}(\rho v l) = 0$$

We find

$$E = -\frac{1}{12}\beta\rho l\frac{1}{v}\frac{dv}{dx}(v^2 \cdot v + 3v^3)$$

where the mean values of v, v^2 and v^3 must be taken as shown in section (2) so that $v^2 \cdot v + 3v^3 \neq 5v^2 \cdot v$.

If q is the quantity of heat transferred in unit of time measured in ordinary units, $E = Jgq$, where J is the mechanical equivalent of heat. If T be the absolute temperature,

$$\frac{1}{v}\frac{dv}{dx} = \frac{1}{2}\frac{1}{T}\frac{dT}{dx}$$

$$\rho v^2 = 3p$$

$$v = \sqrt{\frac{8p_0}{\pi\rho_0}\frac{T}{T_0}}$$

If l_0 be the value of l when $\rho = \rho_0$, then $l = l_0\rho_0/\rho$ and E becomes

$$Jgq = -\frac{5}{4}\sqrt{\frac{2}{\pi}}\beta l_0 p_0^{3/2} T_0^{-3/2}\rho_0^{-1/2} T^{1/2}\frac{dT}{dx}$$

where $\rho = MN$ is the density, v the mean velocity of agitation and l the mean length of path. This result differs from that obtained ⟨by⟩ as l does not come under the sign of differentiation."

a. This manuscript is the draft of a paper Maxwell never published. In it he tried to correct some of the omissions and errors of his 1860 paper.

b. Document III-6. The section on the conduction of heat is in *Scientific Papers*, vol. 1, 403–405.

c. Clausius's name is asterisked but Maxwell did not enter any reference. Presumably he intended to cite the 1857 paper which we cite in note 20 to chapter I.

d. Daggered, reference at bottom of page given as "Pogg. Annalen Jan. 1862." For the citation see chapter I, note 45. Clausius criticized Maxwell's original theory of conduction because it was based on a spherically symmetric velocity distribution function. In the case of conduction a small amount of vis viva (kinetic energy) is added to the particles, augmenting their velocity components in the direction of the temperature differential. The velocity distribution therefore cannot be spherical.

e. Double dagger, reference at the bottom of the page, "Prop. XXI." This refers to Proposition XXI of Maxwell's 1860 paper (document III-6), in the section on heat conduction; see *Scientific Papers*, vol. 1, 403–405.

f. Asterisk, footnote at the bottom of the page: "Note. In his original investigation of the length of path described by a particle M. Clausius has assumed the velocities of all the particles in each system to be equal, and in a communication to the Philosophical Magazine he has shown that on that supposition if v_1 and v_2 be the velocities in each system the mean relative velocity is not $\sqrt{v_1^2 + v_2^2}$ but $v_1 + (1/3)v_2^2/v_1$ where v_1 is the greater of the two velocities. If however the velocities in each system follow the law of eq^n (1) then the mean velocity is $\sqrt{v_1^2 + v_2^2}$." See Clausius, "On the Dynamical Theory of Gases," *Phil. Mag.* [4] 19 (1860), 434–436, and Niven's note to Prop. X in *Scientific Papers*, vol. 1, 387 (document III-6).

g. Word scored through and illegible.

h. The right-hand side of the equation has been rewritten. The first version was altered and ambiguous, but of the same form as the final one, though scored through.

i. William John Macquorn Rankine (1820–1872), one of the founders of thermodynamics. See *A Manual of the Steam Engine and Other Prime Movers* (London and Glasgow, 1859).

j. Asterisk with reference at the bottom of the page to *Phil. Mag.*, Feb. 1859; this is the English translation of Clausius's 1858 paper. The following section of the manuscript refers to Clausius's argument on pages 85–87.

k. Stokes's name is asterisked but no reference is given. The reference should be to "On the Effect of the Internal Friction of Fluids on the Motion of Pendulums," *Trans. Cambridge Phil. Soc.* 9 (1850, pub. 1856), [8]–[106], where he used the coefficient of friction but called it the index of friction.

l. *Scientific Papers*, vol. 1, 387–388 (Document III-6). In this proposition Maxwell first deduced the mean free path for a mixture of two kinds of particles.

m. Maxwell is referring to his sections on diffusion, proposition XIV, and a section in which he examined the effect of all collisions in a stratum from which he deduced the ordinary equations of hydrodynamics, proposition XVI. See *Scientific Papers*, vol. 1, 393–396.

n. In Maxwell's earlier derivation for heat conduction, in the same units,

$$Jq = \frac{1}{3}\frac{d}{dx}\left[\frac{1}{2}\beta M v^2 N v l\right]$$

reduced to

$$Jq = -(\beta v^2/2A)(dv/dx)$$

where $MN = \rho$ (density), $l = 1/A$ so that $MN = 1/A$. See *Scientific Papers*, vol. 1, 404–405.

13. Letter from Maxwell to C. H. Cay,[a] January 5, 1865

Lewis Campbell and William Garnett, *Life of James Clerk Maxwell*, 341–342.

Glenlair, 5th January, 1865

We are sorry to hear you cannot come and see us, but you seem better by your letter, and I hope you will be able for your travels, and be better able for your work afterwards, and not take it too severely, and avoid merimnosity and taking over too much thought, which greatly diminishes the efficiency of young teachers. We have been here since 22d ult., and are in the process of dining the valley in appropriate batches. We have had very rough weather this week, which, combined with the dining, has prevented our usual airings. The ordinary outing is to the Brig of Urr, Katherine on Charlie[b] and I on Darling. Charlie has got a fine band on his forehead, with his name in blue and white beads.

The Manse of Corsock is now finished; it is near the river, not far from the deep pool where we used to bathe.

I set Prof. W. Thomson a prop. [proposition] which I had been working with for a long time. He sent me 18 pages of letter of suggestions about it, none of which would work; but on Jan 3, in the railway from Largs, he got the way to it, which is all right; so we are jolly, having stormed the citadel, when we only hoped to sap it by approximations.

The prop. was to draw a set of lines like this [Diagram 1] so that the ultimate reticulations shall all be squares.

The solution is exact, but rather stiff. Now I have [Diagram 2] a disc A hung by wire D, between two discs B, C, the interval being occupied by air, hydrogen, carbonic acid, etc., the friction of which gradually brings A to rest. In order to calculate the thickdom or viscosity of the gas, I require to solve the problem above mentioned, which is now done, and I have the apparatus now ready to begin. We are also intent on electrical measure-

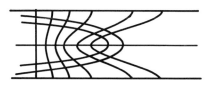

[Diagram 1]

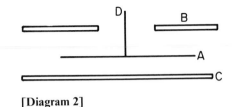

[Diagram 2]

ments, and are getting up apparatus, and have made sets of wires of alloy of platinum and silver, which are to be sent all abroad as standards of resistance. I have also a paper afloat, with an electro-magnetic theory of light, which, till I am convinced to the contrary, I hold to be great guns.[c]

Spice[d] is becoming first-rate: she is the principal patient under the opthalmoscope, and turns her eyes at command, so as to show the tapetum, the optic nerve, or any required part. Dr. Bowman,[e] the great oculist, came to see the sight, and when we were out of town he came again and brought Donders[f] of Utrecht with him to visit Spice.

a. Charles Hope Cay (1841–1869), a cousin of Maxwell's and mathematics master at Clifton College, who helped in experiments on color vision but died prematurely, partly from overwork. See Campbell and Garnett, *Life*, esp. 315, 324, and 343–344. Mrs. Maxwell's pony, Charlie, mentioned in the first paragraph of the letter was named after him. See note b.

b. Charlie: "He was a high-bred, spirited, light bay Galloway, with arched neck and flowing tail. Maxwell himself broke him in, riding side-saddle, with a piece of carpet to take the place of a habit" (*Life*, 319).

c. Maxwell is referring to "A Dynamical Theory of the Electromagnetic Field," *Phil. Trans. R. Soc. London* 155 (1865), 459–512; *Scientific Papers*, vol. 1, 526–597. "Great guns" is something of an understatement; this paper contains what is arguably the most brilliant achievement of nineteenth-century theoretical physics.

d. Spice was the current Scotch terrier that Maxwell used in his experiments on color.

e. William Bowman (1816–1892) was chief surgeon at King's College Hospital (attached to King's College, London, where Maxwell was professor). His treatise *Physiological Anatomy and Physiology of Man*, written jointly with Richard Bentley Todd, is described in the *Dictionary of Scientific Biography* as constituting "an epoch in physiology."

f. Francis Cornelis Donders (1818–1889), professor of ophthalmology and later physiology at Utrecht, author of more than 340 works in these fields. Bowman wrote a biographical notice of him for the Royal Society; see *Proc. R. Soc. London* 49 (1891), vii–xxiv. His 1865 work on the speed of mental processes is considered an important precursor for modern research in cognitive psychology; see *Acta Psychologica* 30 (1969).

14. Letter from Maxwell to Thomas Graham, May 1, 1865

Cambridge University Library.

8, Palace Terrace
London W
1865 May 1
Dear Sir,

I have now got a few results to compare with your experiments on the transpiration of gases but I find that my method of observation is more exact than some of my data derived from measurements so that at present I have taken the apparatus down to measure everything and so get results worthy of the trouble.

I have tried air at ⟨densities⟩ pressures from 30 inches to 0.7 inches and from 42° to 158° Fah [Fahrenheit.][a]

The friction is the same for all densities, but increases with the temperature, apparently in the same proportion as the air expands. I expected it would be as the square root of the absolute temperature but I am wrong.

These results agree with yours. I have also tried hydrogen and carbonic acid and find the velocity for hydrogen about 2.16 of that for air a little more than yours but my results are not fully reduced yet. Carbonic acid is also a good deal smoother than air.

Mixtures of air and hydrogen are more like air than like hydrogen a little air making it very rough. Damp air does not differ very much from dry.

Have you got any more results about the transpiration velocity of mixed gases, especially hydrogen and oxygen or ether vapour and oxygen. I see you have determined the transpiration of equal volumes of H and O. If you have also that of two volumes of H and 1 volume O it would serve as a test for the theory. I think the absolute value of the friction of a few gases may be best determined by my method but the comparison of gases can be best done by transpiration through tubes by your method.

Has anyone but you made such experiments on gases, of course Poiseuille and others have tried liquids. Have you any remaining copy of your papers on Transpiration in the Phil Trans.[b] I have your papers on Molecular Mobility[c] etc and Liquid Diffusion[d] and on Gaseous Diffusion[e] too and should like to put them all together if you have a copy of the Transpiration to spare.

I suppose the hydrogen particles must either be much bigger than the oxygen ones or else they must act on one another at a greater distance, though they are 16 times less in mass,

Yours truly

J. Clerk Maxwell

Master of the Mint.[f]

a. In a letter to P. G. Tait, March 7, 1865 (at Cambridge University), Maxwell reported: "I made an erroneous estimate by rule of thumb as to the strength of a glass plate 1/2 inch thick in consequence of which when exposed to a pressure of 3/4 atmosphere it succumbed with a stunning implosion and set me a month back with regard to the friction of gases."

b. Thomas Graham, "On the Motion of Gases," *Phil. Trans. R. Soc. London* 136 (1846), 573–632; 139 (1849), 349–392.

c. Thomas Graham, "On the Molecular Mobility of Gases," *Phil. Trans. R. Soc. London* 153 (1863), 385–405; *Ann. Phys.* [2] 120 (1863), 415–425.

d. Thomas Graham, "Liquid Diffusion Applied to Analysis," *Phil. Mag.* [4] 23 (1861), 204–223; *Phil. Trans. R. Soc. London* 151 (1861), 183–224.

e. Thomas Graham, "On the Laws of Diffusion of Gases," *Phil. Mag.* 2 (1833), 175– 190, 267–276, 351–358; *Trans. R. Soc. Edinburgh* 12 (1834), 222–258, translated in *Ann. Phys.* [2] 28 (1833), 331–358.

f. Thomas Graham (1805–1869) was Master of the Mint from 1854 to 1869. He succeeded Sir John Herschel in this position, who in turn had succeeded Maxwell's uncle, Sir George Clerk.

15. Letter from Maxwell to Peter Guthrie Tait, June 17, 1865

Cambridge University Library.

Craiglackie

Errol

1865 June 17

Dear Tait,

23 is founded on fact.[a] The problem like others about that date was anon, but was legitimized in the L J of M,[b] March 1858 at the end of a theory of pt I vols.

I am reducing friction of air to numerical values, but I get on slowly by reason of great heat and no logarithms at hand. However in foot, grains, second measure,

$$\mu = 0.0925$$

for air at 60.6.[c]

μ is independent of pressure and proportional to absolute temperature, not to square root thereof. Hydrogen about $\frac{1}{2}$ of air.

The value above given is nearly double what Stokes deduces from pendulum expts but agrees with Graham's expts on transpiration.

To reduce to French or other measure divide by Foot/grain second and multiply by corresponding foreign measure,

Yours truly,

J. C. Maxwell.

a. 23 probably refers to the number in the list of 68 mentioned in document III-16. Maxwell was preparing questions for the Cambridge Mathematical Tripos examination in January 1866, of which he was one of the two "moderators."

b. Possibly Liouville's *J. Math. Pures Appl.*

c. °F. Converting Maxwell's value for the viscosity to inches-grains-second measure this is 0.00771 compared with his published figure of 0.007802, in the same units, at 62° F in document III-18; see *Scientific Papers*, vol. 2, 12. He noted on page 21 that his published value for the viscosity was too large by a factor of 1.012.

16. Letter from Maxwell to H. R. Droop, July 19, 1865

Lewis Campbell and William Garnett, *Life of James Clerk Maxwell*, 342–343.

Glenlair, Dalbeattie, 19th July, 1865

There are so many different forms in which Societies may be cast, that I should like very much to hear something of what those who have been thinking about it propose as the plan of it.

There is the association for publishing each other's productions; for delivering lectures for the good of the public and the support of the Society; for keeping a reading room or club, frequented by men of a particular turn; for dining together once a month, etc.

I suppose W——'s[a] object is to increase the happiness of men in London who cultivate physical sciences, by their meeting together to read papers and discuss them, the publication of these papers being only one, and not the chief end of the Society, which fulfils its main purpose in the act of meeting and enjoying itself.

The Royal Society of Edinburgh used to be a very sociable body, but it had several advantages. Most of the fellows lived within a mile of the Society's rooms. They did not need to disturb their dinner arrangements in order to attend.

Many of them were good speakers as well as sensible men, whose mode of considering a subject was worth hearing, even if not correct.

The subjects were not limited to mathematics and physics, but included geology, physiology, and occasionally antiquities and even literary subjects. Biography of deceased fellows is still a subject of papers. Now those who cultivate the mathematical and physical sciences are sometimes unable to discuss a paper, because they would require to keep it some days by them to form an opinion on it, and physical men can get up a much better discussion about armour plates or the theory of glaciers than about the conduction of heat or capillary attraction.

The only man I know who can make everything the subject of discussion is Dr. Tyndall. Secure his attendance and that of somebody to differ from him, and you are all right for a meeting.

If we can take the field with a plan in our head, I dare say we could find a good many men who would cooperate.

We ride every day, sometimes both morning and evening, and so we consume the roads. I have made 68 problems, all stiff ones not counting riders.[b]

I am now getting the general equations for the motion of a gas considered as an assemblage of molecules flying about with great velocity. I find they must repel as inverse fifth power of distance.[c]

a. Maxwell is referring to the early attempts to establish what finally became the Physical Society of London, officially founded in 1874. See also the letter from Maxwell to his successor as Professor of Natural Philosophy at King's College London—W. G. Adams—quoted in Campbell and Garnett, *Life*, 384–385, comparing the behavior of human beings at scientific meetings to molecules in solid, gaseous, or the "intermediate plastic or colloidal condition." The identity of W—— is uncertain: none of the people mentioned by J. A. Ratcliffe in "The History of the Physical Society" (*Physics Bulletin* 25 (1974), 355–358) as founders of the Society have last names beginning with W. A likely guess is Charles Wheatstone (1802–1875) whose collected scientific papers were published by the Physical Society in 1879 and who was a friend of Adams's.

b. See note a to document III-15. There may be a deliberate play on words in the coupling of horseback riding to examination questions with "riders."

c. See document III-30; *Scientific Papers*, vol. 2, 41.

17. "On the Viscosity or Internal Friction of Air and other Gases"[a] [Abstract]

Proceedings of the Royal Society of London, 15 (1867), 14–17.

All bodies which are capable of having their form indefinitely altered, and which resist the change of form with a force depending on the rate of deformation, may be called Viscous bodies. Taking tar or treacle as an instance in which both the change of form and the resistance opposed to it are easily observed, we may pass in one direction through the series of soft solids up to the materials commonly supposed to be most unyielding, such as glass and steel, and in the other direction through the series of liquids of various degrees of mobility to the gases, of which oxygen is the most viscous, and hydrogen the least.

The viscosity of elastic solids has been investigated by M. F. Kohlrausch[b] and Professor W. Thomson[c]; that of gases by Professor Stokes[d], M. O. E. Meyer[e] and Mr. Graham.[f]

The author has investigated the laws of viscosity in air by causing three horizontal glass disks, 10.56 inches diameter, to perform rotatory oscillations about a vertical axis by means of the elasticity of a steel suspension wire about 4 feet long. The period of a complete oscillation was 72 seconds, and the maximum velocity of the edge of the disks was about 1/12 inch per second.

The three disks were placed at known intervals on the vertical axis, and four larger fixed disks were so adjusted above and below them and in the intervals between them, that strata of air of known thickness were intercepted between the surfaces of the moving disks and the fixed disks. During the oscillations of the moveable disks, the viscosity of the air in these six strata caused a gradual diminution of the amplitude of oscillation, which was measured by means of the reflexion of a circular scale in a mirror attached to the axis.

The whole apparatus was enclosed in an air-tight case, so that the air might be exhausted or exchanged for another gas, or heated by a current of steam round the receiver. The observed diminution in the arc of oscillation is in part due to the viscosity of the suspending wire. To eliminate the effect of the wire from that of the air, the arrangement of the disks was altered, and the three disks, placed in contact, were made to oscillate midway between two fixed glass disks, at distances sometimes of 1 inch, and sometimes of .5 inch.

From these experiments on two strata of air, combined with three sets of experiments on six strata of thicknesses .683, .425, and .1847 inches respectively, the value of the coefficient of viscosity or internal friction was determined.

Let two infinite planes be separated by a stratum of air whose thickness is unity. Let one of these planes be fixed, while the other moves in its own plane with a uniform velocity unity; then, if the air in immediate contact with either plane has the same velocity as the plane, every unit of surface of either plane will experience a tangential force μ, where μ is the coefficient of viscosity of the air between the planes.

The force μ is understood to be measured by the velocity which it would communicate in unit of time to unit of mass.

If L, M, T be the units of length, mass, and time, then the dimensions of μ are $L^{-1}MT^{-1}$.

In the actual experiment, the motion of the surfaces is rotatory instead of rectilinear, oscillatory instead of uniform, and the surfaces are bounded instead of infinite. These considerations introduce certain complications into the theory, which are separately considered.

The conclusions which are drawn from the experiments agree, as far as they go, with those of Mr. Graham on the Transpiration of Gases.[g] They are as follows:—

1. The coefficient of viscosity is independent of the density, the temperature being constant. No deviation from this law is observed between the atmospheric density and that corresponding to a pressure of half an inch of mercury.

This remarkable result was shown by the author in 1860[h] to be a consequence of the Dynamical Theory of Gases. It agrees with the conclusions of Mr. Graham, deduced from experiments on the transpiration of gases through capillary tubes. The considerable thickness of the strata of air in the present experiments shows that the property of air, to be equally viscous at all densities, is quite independent of any molecular action between its particles and those of solid surfaces, such as those of the capillary tubes employed by Graham.

2. The coefficient of viscosity increases with the temperature, and is proportional to $1 + \alpha\theta$, where θ is the temperature and α is the coefficient of expansion per degree for air.

This result cannot be considered so well established as the former, owing to the difficulty of maintaining a high temperature constant in so large an apparatus, and measuring it without interfering with the motion. Experiments, in which the temperature ranged from 50° to 185° F., agreed with the theory to within 0.8 per cent., so that it is exceedingly probable that this is the true relation to the temperature.

The experiments of Graham led him to this conclusion also.

3. The coefficient of viscosity of hydrogen is much less than that of air. I have never succeeded in filling my apparatus with perfectly pure hydrogen, for air leaks into the vacuum during the admission of so large a quantity of hydrogen as is required to fill it. The ratio of the viscosity of my hydrogen to that of air was .5156. That obtained by Graham was .4855.[i]

4. The ratio for carbonic acid was found to be .859. Graham makes it .807. It is probable that the comparative results of Graham are more exact than those of this paper, owing to the difficulty of introducing so large a volume of gas without letting in any air during the time of filling the receiver. I find also that a very small proportion of air causes a considerable increase in the viscosity of hydrogen. This result also agrees with those of Mr. Graham.

5. Forty experiments on dry air were investigated to determine whether any slipping takes place between the glass and the air in immediate contact with it.

The result was, that if there were any slipping, it is of exceedingly small amount; and that the evidence in favour of the indicated amount being real is very precarious.

The results of the hypothesis, that there is no slipping, agree decidedly better with the experiments.

6. The actual value of the coefficient of viscosity of dry air was determined, from forty experiments of five different kinds, to be[j]

$$\mu = .0000149(461° + \theta)$$

where the inch, the grain, and the second are the units, and the temperature is on Fahrenheit's scale.

At 62° this gives $\mu = .007802$.

Professor Stokes from the experiments of Baily on pendulums, has found

$$\sqrt{\frac{\mu}{\rho}} = .116,$$

which, with the average temperature and density of air, would give

$$\mu = .00417,$$

a much smaller value than that here found.

If the value of μ is expressed in feet instead of inches, so as to be uniform with the British measures of magnetic and electric phenomena, as recorded at the observatories,

$$\mu = .000179(461 + \theta)$$

$$= .08826 \quad \text{at } 32°$$

In metre-gramme-second measure and Centigrade temperature,

$$\mu = .01878(1 + .003660)$$

M. O. E. Meyer (Pogg. Ann. cxiii. (1861) p. 383)[k] makes μ at 18° C. = .000360 in centimetres, cubic centimetres of water, and seconds as units or in metrical units, $\mu = .0360$.

According to the experiments here described, μ at 18° C. = .02.

M. Meyer's value is therefore nearly twice as great as that of this paper, while that of Professor Stokes is only half as great.

In M. Meyer's experiments, which were with one disk at a time in an open space of air, the influence of the air near the edge of the disk is very considerable; but M. Meyer (Crelle, 59; Pogg. cxiii. 76)[k] seems to have arrived at the conclusion that the additional effect of the air at the edge is proportional to the thickness of the disk. If the additional force near the edge is underestimated, the resulting value of the viscosity will be in excess.

7. Each of the forty experiments on dry air was calculated from the concluded values of the viscosity of the air and of the wire, and the result compared with the observed result. In this way the error of mean square of each observation

was determined, and from this the "probable error" of μ was found to be .036 per cent of its value. These experiments, it must be remembered, were made with five different arrangements of the disks, at pressures ranging from 0.5 inch to 30 inches, and at temperatures from 51° to 74° F.; so that their agreement does not arise from a mere repetition of the same conditions, but from an agreement between the properties of air and the theory made use of in the calculations.

a. This is an abstract of Maxwell's longer paper of the same title (document III-18).

b. Asterisked, reference at the bottom of the page to *Pogg. Ann.* cxix (1863). The complete citation is F. W. G. Kohlrausch, "Ueber die elastische Nachwirkung bei der Torsion," *Ann. Phys.* [2] 119 (1863), 337–368. Friedrich Wilhelm Georg Kohlrausch (1840–1910) is best known for his work on the conductivity of electrolytic solutions. His father Rudolph Kohlrausch (1809–1858)—to whom the Royal Society *Catalogue of Scientific Papers* erroneously attributes this paper—performed with Wilhelm Weber an experiment showing that the ratio of electrostatic to electromagnetic units is equal to the speed of light (1856), thereby providing the empirical basis for Maxwell's electromagnetic theory of light.

c. Asterisked, reference at the bottom of the page to *Proceedings of the Royal Society*, May 18, 1865. The complete citation is William Thomson, "On the Elasticity and Viscosity of Metals," *Proc. R. Soc. London* 14 (1865), 289–297; also in *Phil. Mag.* [4] 30 (1865), 63–71.

d. Asterisked, reference at the bottom of the page to *Cambridge Philosophical Transactions*, 1850. The complete citation is given in note k to document III-12.

e. Asterisked, reference at the bottom of the page to Pogg. Ann. cxiii (1861); see note 113 to chapter I. Oskar Emil Meyer (1834–1909) used the same method as Maxwell but did not control the air currents around the torsion pendulum nearly as carefully. He later revised his results on the basis of further experiments, bringing them into fairly good agreement with Maxwell's (see our discussion in chapter I).

f. Asterisked, reference at the bottom of the page to *Phil. Trans.* [R. Soc. London] 1846 and 1849. See note 57 to chapter I.

g. Asterisked, reference at the bottom of the page to *Phil. Trans.* [R. Soc. London] 1846.

h. Asterisked, reference at the bottom of the page to *Phil. Mag.* Jan. 1860 (see document III-6).

i. These experiments were all done in the garret of Maxwell's house. The modern experimentalist is startled to read of an apparatus containing a mixture of hydrogen and air in a room with a large open fire (Campbell and Garnett, *Life*, 318). So might Maxwell's neighbors have been if they had known what was going on.

j. This value incorporates a correction which Maxwell explains in a note added to the published memoir, February 6, 1866 (document III-18; *Scientific Papers*, vol. 2, 21–22). A further correction, reducing all the numerical values of the viscosity, was noted by Leahy; see Niven's note in *Scientific Papers*, vol. 2, viii.

k. See note e.

l. See note e.

18. "On the Viscosity or Internal Friction of Air and other Gases," Bakerian Lecture, February 8, 1866

Phil. Trans. R. Soc. London 156 (1866), 249–268; *Scientific Papers*, vol. 2, 1–25.

[From the *Philosophical Transactions*, Vol. CLVI.]

XXVII. THE BAKERIAN LECTURE.—*On the Viscosity or Internal Friction of Air and other Gases.*

Received November 23, 1865,—Read February 8, 1866.

THE gaseous form of matter is distinguished by the great simplification which occurs in the expression of the properties of matter when it passes into that state from the solid or liquid form. The simplicity of the relations between density, pressure, and temperature, and between the volume and the number of molecules, seems to indicate that the molecules of bodies, when in the gaseous state, are less impeded by any complicated mechanism than when they subside into the liquid or solid states. The investigation of other properties of matter is therefore likely to be more simple if we begin our research with matter in the form of a gas.

The viscosity of a body is the resistance which it offers to a continuous change of form, depending on the rate at which that change is effected.

All bodies are capable of having their form altered by the action of sufficient forces during a sufficient time. M. Kohlrausch* has shewn that torsion applied to glass fibres produces a permanent set which increases with the time of action of the force, and that when the force of torsion is removed the fibre slowly untwists, so as to do away with part of the set it had acquired. Softer solids exhibit the phenomena of plasticity in a greater degree; but the investigation of the relations between the forces and their effects is extremely difficult, as in most cases the state of the solid depends not only on the forces actually impressed on it, but on all the strains to which it has been subjected during its previous existence.

---

* "Ueber die elastische Nachwerkung bei der Torsion," Pogg. *Ann.* CXIX. 1863.

2   ON THE VISCOSITY OR INTERNAL FRICTION

Professor W. Thomson* has shewn that something corresponding to internal friction takes place in the torsional vibrations of wires, but that it is much increased if the wire has been previously subjected to large vibrations. I have also found that, after heating a steel wire to a temperature below 120°, its elasticity was permanently diminished and its internal friction increased.

The viscosity of fluids has been investigated by passing them through capillary tubes†, by swinging pendulums in them‡, and by the torsional vibrations of an immersed disk§, and of a sphere filled with the fluid ‖.

The method of transpiration through tubes is very convenient, especially for comparative measurements, and in the hands of Graham and Poiseuille it has given good results, but the measurement of the diameter of the tube is difficult, and on account of the smallness of the bore we cannot be certain that the action between the molecules of the gas and those of the substance of the tubes does not affect the result. The pendulum method is capable of great accuracy, and I believe that experiments are in progress by which its merits as a means of determining the properties of the resisting medium will be tested. The method of swinging a disk in the fluid is simple and direct. The chief difficulty is the determination of the motion of the fluid near the edge of the disk, which introduces very serious mathematical difficulties into the calculation of the result. The method with the sphere is free from the mathematical difficulty, but the weight of a properly constructed spherical shell makes it un-suitable for experiments on gases.

In the experiments on the viscosity of air and other gases which I propose to describe, I have employed the method of the torsional vibrations of disks, but instead of placing them in an open space, I have placed them each between two parallel fixed disks at a small but easily measurable distance, in which case, when the period of vibration is long, the mathematical difficulties of deter-mining the motion of the fluid are greatly reduced. I have also used three

---

* *Proceedings of the Royal Society*, May 18, 1865.

† Liquids : Poiseuille, *Mém. de Savants Étrangers*, 1846.   Gases : Graham, *Philosophical Transactions*, 1846 and 1849.'

‡ Baily, *Phil. Trans.* 1832 ; Bessel, *Berlin Acad.* 1826 ; Dubuat, *Principes d'Hydraulique*, 1786.   All these are discussed in Professor Stokes's paper "On the Effect of the Internal Friction of Fluids on the Motion of Pendulums," *Cambridge Phil. Trans.* vol. IX. pt. 2 (1850).

§ Coulomb, *Mém. de l'Institut national*, III. p. 246 ; O. E. Meyer, Pogg. *Ann.* CXIII. (1861) p. 55, and *Crelle's Journal*, Bd. 59.

‖ Helmholtz and Pietrowski, *Sitzungsberichte der k. k. Akad.* April, 1860.

disks instead of one, so that there are six surfaces exposed to friction, which may be reduced to two by placing the three disks in contact, without altering the weight of the whole or the time of vibration. The apparatus was constructed by Mr Becker, of Messrs Elliott Brothers, Strand.

Description of the Apparatus.

Plate XXI. p. 30, fig. 1 represents the vacuum apparatus one-eighth of the actual size. *MQRS* is a strong three-legged stool supporting the whole. The top (*MM*) is in the form of a ring. *EE* is a brass plate supported by the ring *MM*. The under surface is ground truly plane, the upper surface is strengthened by ribs cast in the same piece with it. The suspension-tube *AC* is screwed into the plate *EE*, and is 4 feet in height. The glass receiver *N* rests on a wooden ring *PP* with three projecting pieces which rest on the three brackets *QQ*, of which two only are seen. The upper surfaces of the brackets and the under surfaces of the projections are so bevilled off, that by slightly turning the wooden ring in its own plane the receiver can be pressed up against the plate *EE*.

F, G, H, K are circular plates of glass of the form represented in fig. 2. Each has a hole in the centre 2 inches in diameter, and three holes near the circumference, by which it is supported on the screws *LL*.

Fig. 6 represents the mode of supporting and adjusting the glass plates. *LL* is one of the screws fixed under the plate *EE*. *S* is a nut, of which the upper part fits easily in the hole in the glass plate *F*, while the under part is of larger diameter, so as to support the glass plate and afford the means of turning the nut easily by hand. These nuts occupy little space, and enable the glass disks to be brought very accurately to their proper position.

ACB, fig. 1, is a siphon barometer, closed at *A* and communicating with the interior of the suspension-tube at *B*. The scale is divided on both sides, so that the difference of the readings gives the pressure within the apparatus. *T* is a thermometer, lying on the upper glass plate. *V* is a vessel containing pumice-stone soaked in sulphuric acid, to dry the air. Another vessel, containing caustic potash, is not shewn. *D* is a tube with a stopcock, leading to the air-pump or the gas generator. *C* is a glass window, giving a view of the suspended mirror *d*.

For high and low temperatures the tin vessel (fig. 10) was used. When the receiver was exhausted, the ring *P* was removed, and the tin vessel raised

4   ON THE VISCOSITY OR INTERNAL FRICTION

so as to envelope the receiver, which then rested on the wooden support *YY*. The tin vessel itself rested, by means of projections on the brackets *QQ*. The outside of the tin vessel was then well wrapped up in blankets, and the top of the brass plate *EE* covered with a feather cushion; and cold water, hot water, or steam was made to flow through the tin vessel till the thermometer *T*, seen through the window *W*, became stationary.

The moveable parts of the apparatus consist of—

The suspension-piece *a*, fitting air-tight into the top of the tube and holding the suspension-wire by a clip, represented in fig. 5.

The axis *cdek*, suspended to the wire by another clip at *C*.

The wire was a hard-drawn steel wire, one foot of which weighed 2·6 grains.

The axis carries the plane mirror *d*, by which its angular position is observed through the window *C*, and the three vibrating glass disks *f*, *g*, *h*, represented in fig. 3. Each disk is 10·56 inches diameter and about ·076 thick, and has a hole in the centre ·75 diameter. They are kept in position on the axis by means of short tubes of accurately known length, which support them on the axis and separate them from each other.

The whole suspended system weighs three pounds avoirdupoise.

In erecting the apparatus, the lower part of the axis *ek* is screwed off. The fixed disks are then screwed on, with a vibrating disk lying between each. Tubes of the proper lengths are then placed on the lower part of the axis and between the disks. The axis is then passed up from below through the disks and tubes, and is screwed to the upper part at *e*. The vibrating disks are now hanging by the wire and in their proper places, and the fixed disks are brought to their proper distances from them by means of the adjusting nuts.

ns is a small piece of magnetized steel wire attached to the axis.

When it is desired to set the disks in motion, a battery of magnets is placed under *N*, and so moved as to bring the initial arc of vibration to the proper value.

Fig. 4 is a brass ring whose moment of inertia is known. It is placed centrically on the vibrating disk by means of three radial wires, which keep it exactly in its place.

Fig. 7 is a tube containing two nearly equal weights, which slide inside it, and whose position can be read off by verniers.

The ring and the tube are used in finding the moment of inertia of the vibrating apparatus.

The extent and duration of the vibrations are observed in the ordinary way by means of a telescope, which shews the reflexion of a scale in the mirror d. The scale is on a circular arc of six feet radius, concentric with the axis of the instrument. The extremities of the scale correspond to an arc of vibration of 19° 36', and the divisions on the scale to 1·7. The readings are usually taken to tenths of a division.

Method of Observation.

When the instrument was properly adjusted, a battery of magnets was placed on a board below N, and reversed at proper intervals till the arc of vibration extended slightly beyond the limits of the scale. The magnets were then removed, and any accidental pendulous oscillations of the suspended disks were checked by applying the hand to the suspension-tube. The barometer and thermometer were then read off, and the observer took his seat at the telescope and wrote down the extreme limits of each vibration as shewn by the numbers on the scale. At intervals of five complete vibrations, the time of the transits of the middle point of the scale was observed (see Table I.). When the amplitude decreased rapidly, the observations were continued throughout the experiment; but when the decrement was small, the observer generally left the room for an hour, or till the amplitude was so far reduced as to furnish the most accurate results.

In observing a quantity which decreases in a geometrical ratio in equal times, the most accurate value of the rate of decrement will be deduced from a comparison of the initial values with values which are to these in the ratio of e to 1, where $e = 2·71828$, the base of the Napierian system of logarithms. In practice, however, it is best to stop the experiment somewhat before the vibrations are so much reduced, as the time required would be better spent in beginning a new experiment.

In reducing the observations, the sum of every five maxima and of the consecutive five minima was taken, and the differences of these were written as the terms of the series the decrement of which was to be found.

In experiments where the law of decrement is uncertain, this rough method is inapplicable, and Gauss's method must be applied; but the series of amplitudes

ON THE VISCOSITY OR INTERNAL FRICTION

in these experiments is so accurately geometrical, that no appreciable difference between the results of the two methods would occur.

The logarithm of each term of the series was then taken, and the mean logarithmic decrement ascertained by taking the difference of the first and last, of the second and last but one, and so on, multiplying each difference by the interval of the terms, and dividing the sum of the products by the sum of the squares of these intervals. Thus, if fifty observations were taken of the extreme limits of vibration, these were first combined by tens, so as to form five terms of a decreasing series. The logarithms of these terms were then taken. Twice the difference of the first and fifth of these logarithms was then added to the difference of the second and third, and the result divided by ten for the mean logarithmic decrement in five complete vibrations.

The times were then treated in the same way to get the mean time of five vibrations. The numbers representing the logarithmic decrement, and the time for five vibrations, were entered as the result of each experiment*.

The series found from ten different experiments were examined to discover any departure from uniformity in the logarithmic decrement depending on the amplitude of vibration. The logarithmic decrement was found to be constant in each experiment to within the limits of probable error; the deviations from uniformity were sometimes in one direction and sometimes in the opposite, and the ten experiments when combined gave no evidence of any law of increase or diminution of the logarithmic decrement as the amplitudes decrease. The forces which retard the disks are therefore as the first power of the velocity, and there is no evidence of any force varying with the square of the velocity, such as is produced when bodies move rapidly through the air. In these experiments the maximum velocity of the circumference of the moving disks was about $\frac{1}{12}$ inch per second. The changes of form in the air between the disks were therefore effected very slowly, and eddies were not produced†.

The retardation of the motion of the disks is, however, not due entirely to the action of the air, since the suspension wire has a viscosity of its own, which must be estimated separately. Professor W. Thomson has observed great changes in the viscosity of wires after being subjected to torsion and longitudinal strain. The wire used in these experiments had been hanging up for

---

* See Table II.

† The total moment of the resistances never exceeded that of the weight of $\frac{1}{30}$ grain acting at the edge of the disks.

some months before, and had been set into torsional vibrations with various weights attached to it, to determine its moment of torsion. Its moment of torsion and its viscosity seem to have remained afterwards nearly constant, till steam was employed to heat the lower part of the apparatus. Its viscosity then increased, and its moment of torsion diminished permanently, but when the apparatus was again heated, no further change seems to have taken place. During each course of experiments, care was taken not to set the disks vibrating beyond the limits of the scale, so that the viscosity of the wire may be supposed constant in each set of experiments.

In order to determine how much of the total retardation of the motion is due to the viscosity of the wire, the moving disks were placed in contact with each other, and fixed disks were placed at a measured distance above and below them. The weight and moment of inertia of the system remained as before, but the part of the retardation of the motion due to the viscosity of the air was less, as there were only two surfaces exposed to the action of the air instead of six. Supposing the effect of the viscosity of the wire to remain as before, the difference of retardation is that due to the action of the four additional strata of air, and is independent of the value of the viscosity of the wire.

In the experiments which were used in determining the viscosity of air, five different arrangements were adopted.

Arrangement 1. Three disks in contact, fixed disks at 1 inch above and below.
,,   2.   ,,   ,,   ,,   0·5 inch.
,,   3. Three disks, each between two fixed disks at distance 0·683.
,,   4.   ,,   ,,   ,,   ,,   ,,   0·425.
,,   5.   ,,   ,,   ,,   ,,   ,,   0·18475.

By comparing the results of these different arrangements, the coefficient of viscosity was obtained, and the theory at the same time subjected to a rigorous test.

Definition of the Coefficient of Viscosity.

The final result of each set of experiments was to determine the value of the coefficient of viscosity of the gas in the apparatus. This coefficient may be best defined by considering a stratum of air between two parallel horizontal planes of indefinite extent, at a distance a from one another. Suppose the upper plane to be set in motion in a horizontal direction with a velocity of v feet per second, and to continue in motion till the air in the different parts of the stratum has taken up its final velocity, then the velocity of the air will increase uniformly as we pass from the lower plane to the upper. If the air in contact with the planes has the same velocity as the planes themselves, then the velocity will increase $\dfrac{v}{a}$ feet per second for every foot we ascend.

The friction between any two contiguous strata of air will then be equal to that between either surface and the air in contact with it. Suppose that this friction is equal to a tangential force f on every square foot, then

$$f = \mu \frac{v}{a},$$

where μ is the coefficient of viscosity, v the velocity of the upper plane, and a the distance between them.

If the experiment could be made with the two infinite planes as described, we should find μ at once, for

$$\mu = \frac{fa}{v}.$$

In the actual case the motion of the planes is rotatory instead of rectilinear, oscillatory instead of constant, and the planes are bounded instead of infinite.

It will be shewn that the rotatory motion may be calculated on the same principles as rectilinear motion; but that the oscillatory character of the motion introduces the consideration of the inertia of the air in motion, which causes the middle portions of the stratum to lag behind, as is shewn in fig. 8, where the curves represent the successive positions of a line of particles of air, which, if there were no motion, would be a straight line perpendicular to the planes.

The fact that the moving planes are bounded by a circular edge introduces another difficulty, depending on the motion of the air near the edge being different from that of the rest of the air.

The lines of equal motion of the air are shewn in fig. 9.

The consideration of these two circumstances introduces certain corrections into the calculations, as will be shewn hereafter.

In expressing the viscosity of the gas in absolute measure, the measures of all velocities, forces, &c. must be taken according to some consistent system of measurement.

If L, M, T represent the units of length, mass, and time, then the dimensions of f (a pressure per unit of surface) are $L^{-1}MT^{-2}$; a is a length, and v is a velocity whose dimensions are LT^{-1}, so that the dimensions of μ are $L^{-1}MT^{-1}$.

Thus if μ be the viscosity of a gas expressed in inch-grain-second measure, and μ' the same expressed in foot-pound-minute measure, then

$$\frac{\mu}{\mu'} \cdot \frac{1 \text{ inch}}{1 \text{ foot}} \cdot \frac{1 \text{ pound}}{1 \text{ grain}} \cdot \frac{1 \text{ second}}{1 \text{ minute}} = 1.$$

According to the experiments of MM. Helmholtz and Pietrowski*, the velocity of a fluid in contact with a surface is not always equal to that of the surface itself, but a certain amount of actual slipping takes place in certain cases between the surface and the fluid in immediate contact with it. In the case which we have been considering, if v_0 is the velocity of the fluid in contact with the fixed plane, and f the tangential force per unit of surface, then

$$f = \sigma v_0,$$

where σ is the coefficient of superficial friction between the fluid and the particular surface over which it flows, and depends on the nature of the surface as well as on that of the fluid. The coefficient σ is of the dimensions $L^{-2}MT^{-1}$. If v_1 be the velocity of the fluid in contact with the plane which is moving with velocity v, and if σ' be the coefficient of superficial friction for that plane,

$$f = \sigma'(v - v_1).$$

The internal friction of the fluid itself is

$$f = \frac{\mu}{a}(v_1 - v_0).$$

Hence

$$v = f\left(\frac{1}{\sigma} + \frac{1}{\sigma'} + \frac{a}{\mu}\right).$$

If we make $\frac{\mu}{\sigma} = \beta$, and $\frac{\mu}{\sigma'} = \beta'$, then

$$v = f\left(\frac{a + \beta + \beta'}{\mu}\right),$$

---

* *Sitzungsberichte der k. k. Akad.* April 1860.

or the friction is equal to what it would have been if there had been no slipping, and if the interval between the planes had been increased by $\beta + \beta'$. By changing the interval between the planes, a may be made to vary while $\beta + \beta'$ remains constant, and thus the value of $\beta + \beta'$ may be determined. In the case of air, the amount of slipping is so small that it produces no appreciable effect on the results of experiments. In the case of glass surfaces rubbing on air, the probable value of β, deduced from the experiments, was $\beta = \cdot 0027$ inch. The distance between the moving surfaces cannot be measured so accurately as to give this value of β the character of an ascertained quantity. The probability is rather in favour of the theory that there is no slipping between air and glass, and that the value of β given above results from accidental discrepancy in the observations. I have therefore preferred to calculate the value of. μ on the supposition that there is no slipping between the air and the glass in contact with it.

The value of μ depends on the nature of the gas and on its physical condition. By making experiments in gas of different densities, it is shewn that μ remains constant, so that its value is the same for air at 0·5 inch and at 30 inches pressure, provided the temperature remains the same. This will be seen by examining Table IV., where the value of L, the logarithm of the decrement of arc in ten single vibrations, is the same for the same temperature, though the density is sixty times greater in some cases than in others. In fact the numbers in the column headed L' were calculated on the hypothesis that the viscosity is independent of the density, and they agree very well with the observed values.

It will be seen, however, that the value of L rises and falls with the temperature, as given in the second column of Table IV. These temperatures range from 51° to 74° Fahr., and were the natural temperatures of the room on different days in May 1865. The results agree with the hypothesis that the viscosity is proportional to $(461° + \theta)$, the temperature measured from absolute zero of the air-thermometer. In order to test this proportionality, the temperature was raised to 185° Fahr. by a current of steam sent round the space between the glass receiver and the tin vessel. The temperature was kept up for several hours, till the thermometer in the receiver became stationary, before the disks were set in motion. The ratio of the upper temperature (185° F.) to the lower (51°), measured from $-461°$ F., was

1·2605.

The ratio of the viscosity at the upper temperature to that at the lower was

1·2624,

which shews that the viscosity is proportional to the absolute temperature very nearly.   The simplicity of the other known laws relating to gases warrants us in concluding that the viscosity is really proportional to the temperature, measured from the absolute zero of the air-thermometer.

These relations between the viscosity of air and its pressure and temperature are the more to be depended on, since they agree with the results deduced by Mr. Graham from experiments on the transpiration of gases through tubes of small diameter.   The constancy of the viscosity for all changes of density when the temperature is constant is a result of the Dynamical Theory of Gases*, whatever hypothesis we adopt as to the mode of action between the molecules when they come near one another.   The relation between viscosity and temperature, however, requires us to make a particular assumption with respect to the force acting between the molecules.   If the molecules act on one another only at a determinate distance by a kind of impact, the viscosity will be as the square root of the absolute temperature.   This, however, is certainly not the actual law.   If, as the experiments of Graham and those of this paper shew, the viscosity is as the first power of the absolute temperature, then in the dynamical theory, which is framed to explain the facts, we must assume that the force between two molecules is proportional inversely to the fifth power of the distance between them.   The present paper, however, does not profess to give any explanation of the cause of the viscosity of air, but only to determine its value in different cases.

Experiments were made on a few other gases besides dry air.

Damp air, over water at 70° F. and 4 inches pressure, was found by the mean of three experiments to be about one-sixtieth part less viscous than dry air at the same temperature.

Dry hydrogen was found to be much less viscous than air, the ratio of its viscosity to that of air being ·5156.

A small proportion of air mixed with hydrogen was found to produce a large increase of viscosity, and a mixture of equal parts of air and hydrogen has a viscosity nearly equal to $\frac{15}{8}$ of that of air.

The ratio of the viscosity of dry carbonic acid to that of air was found to be ·859.

---

* " Illustrations of the Dynamical Theory of Gases," *Philosophical Magazine*, Jan. 1860.

It appears from the experiments of Mr. Graham that the ratio of the transpiration time of hydrogen to that of air is ·4855, and that of carbonic acid to air ·807. These numbers are both smaller than those of this paper. I think that the discrepancy arises from the gases being less pure in my experiments than in those of Graham, owing to the difficulty of preventing air from leaking into the receiver during the preparation, desiccation, and admission of the gas, which always occupied at least an hour and a half before the experiment on the moving disks could be begun.

It appears to me that for comparative estimates of viscosity, the method of transpiration is the best, although the method here described is better adapted to determine the absolute value of the viscosity, and is less liable to the objection that in fine capillary tubes the influence of molecular action between the gas and the surface of the tube may possibly have some effect.

The actual value of the coefficient of viscosity in inch-grain-second measure, as determined by these experiments, is

$$·00001492(461° + \theta).$$

At 62° F. $\mu = ·007802.$

Professor Stokes has deduced from the experiments of Baily on pendulums

$$\sqrt{\frac{\mu}{\rho}} = ·116,$$

which at ordinary pressures and temperatures gives

$$\mu = ·00417,$$

or not much more than half the value as here determined. I have not found any means of explaining this difference.

In metrical units and Centigrade degrees

$$\mu = ·01878(1 + ·00365\theta).$$

M. O. E. Meyer gives as the value of μ in centimetres, grammes, and seconds, at 18° C.,

$$·000360.$$

This, when reduced to metre-gramme-second measure, is

$$\mu = ·0360.$$

I make μ, at 18° C., $= ·0200.$

Hence the value given by Meyer is 1·8 times greater than that adopted in this paper.

M. Meyer, however, has a different method of taking account of the disturbance of the air near the edge of the disk from that given in this paper. He supposes that when the disk is very thin the effect due to the edge is proportional to the thickness, and he has given in Crelle's *Journal* a vindication of this supposition.   I have not been able to obtain a mathematical solution of the case of a disk oscillating in a large extent of fluid, but it can easily be shewn that there will be a finite increase of friction near the edge of the disk due to the want of continuity, even if the disk were infinitely thin.   I think therefore that the difference between M. Meyer's result and mine is to be accounted for, at least in part, by his having under-estimated the effect of the edge of the disk.   The effect of the edge will be much less in water than in air, so that any deficiency in the correction will have less influence on the results for liquids which are given in M. Meyer's very valuable paper.

Mathematical Theory of the Experiment.

A disk oscillates in its own plane about a vertical axis between two fixed horizontal disks, the amplitude of oscillation diminishing in geometrical progression, to find what part of the retardation is due to the viscosity of the air between it and the fixed disks.

That part of the surface of the disk which is not near the edge may be treated as part of an infinite disk, and we may assume that each horizontal stratum of the fluid oscillates as a whole.   In fact, if the motion of every part of each stratum can be accounted for by the actions of the strata above and below it, there will be no mutual action between the parts of the stratum, and therefore no relative motion between its parts.

Let θ be the angle which defines the angular position of the stratum which is at the distance y from the fixed disk, and let r be the distance of a point of that stratum from the axis, then its velocity will be $r\dfrac{d\theta}{dt}$, and the tangential force on its lower surface arising from viscosity will be on unit of surface

$$-\mu r\,\frac{d^2\theta}{dy\,dt}=f\dots\dots\dots\dots\dots\dots\dots\dots(1).$$

14   ON THE VISCOSITY OR INTERNAL FRICTION

The tangential force on the upper surface will be

$$\mu r \left(\frac{d^2\theta}{dydt} + \frac{d^3\theta}{dy^2dt} dy \right) ;$$

and the mass of the stratum per unit of surface is ρdy, so that the equation of motion of each stratum is

$$\rho \frac{d^2\theta}{dt^2} = \mu \frac{d^3\theta}{dy^2dt} \quad \dots\dots\dots\dots\dots\dots(2),$$

which is independent of r, shewing that the stratum moves as a whole.

The conditions to be satisfied are, that when $y = 0$, $\theta = 0$; and that when $y = b$,

$$\theta = Ce^{-lt} \cos(nt + a) \quad \dots\dots\dots\dots\dots\dots(3).$$

The disk is suspended by a wire whose elasticity of torsion is such that the moment of torsion due to a torsion θ is $I\omega^2\theta$, where I is the moment of inertia of the disks. The viscosity of the wire is such that an angular velocity $\dfrac{d\theta}{dt}$ is resisted by a moment $2Ik \dfrac{d\theta}{dt}$. The equation of motion of the disks is then

$$I \left(\frac{d^2\theta}{dt^2} + 2k \frac{d\theta}{dt} + \omega^2\theta \right) + NA\mu \frac{d^2\theta}{dydt} = 0 \quad \dots\dots\dots\dots (4),$$

where $A = \int 2\pi r^3 dr = \frac{1}{2}\pi r^4$, the moment of inertia of each surface, and N is the number of surfaces exposed to friction of air.

The equation for the motion of the air may be satisfied by the solution

$$\theta = e^{-lt} \{ e^{py} \cos(nt + qy) - e^{-py} \cos(nt - qy) \} \quad \dots\dots\dots\dots(5),$$

provided

$$2pq = \frac{\rho n}{\mu} \quad \dots\dots\dots\dots\dots\dots (6),$$

and

$$q^2 - p^2 = \frac{\rho l}{\mu} \quad \dots\dots\dots\dots\dots\dots (7);$$

and in order to fulfil the conditions (3) and (4),

$$2In(l-k)(e^{2pb} + e^{-2pb} - 2\cos 2qb) = NA\mu\{(pn - lq)(e^{2pb} - e^{-2pb}) + (qn + lp)2\sin 2qb\}\dots(8).$$

Expanding the exponential and circular functions, we find

$$2Ib(l-k) = NA\mu\{1 - \tfrac{1}{3}cl + \tfrac{1}{6}c^2(n^2 - 3l^2) + \tfrac{1}{7}\tfrac{2}{3}c^3(n^2l - l^3) + \tfrac{1}{10}c^4(\tfrac{7}{16}n^4 + \tfrac{215}{16}n^2l^2 - \tfrac{145}{4}l^4)\} \ (9),$$

where $c = \dfrac{4b^2\rho}{\mu}$,

$l =$ observed Napierian logarithmic decrement of the amplitude in unit of time,

$k =$ the part of the decrement due to the viscosity of the wire.

When the oscillations are slow as in these experiments, when the disks are near one another, and when the density is small and the viscosity large, the series on the right-hand side of the equation is rapidly convergent.

When the time from rest to rest was thirty-six seconds, and the interval between the disks 1 inch, then for air of pressure 29·9 inches, the successive terms of the series were

$$1 \cdot 0 - 0 \cdot 00508 \quad + 0 \cdot 24866 \quad + 0 \cdot 00072 \quad + 0 \cdot 00386 = 1 \cdot 24816;$$

but when the pressure was reduced to 1·44 inch, the series became

$$1 \cdot 0 - 0 \cdot 0002448 \quad + \cdot 0005768 + \cdot 00000008 + \cdot 00000002 = 1 \cdot 0003321.$$

The series is also made convergent by diminishing the distance between the disks. When the distance was ·1847 inch, the first two terms only were sensible. When the pressure was 29·29, the series was

$$1 - \cdot 000858 \quad + \cdot 000278 = 1 - \cdot 00058.$$

At smaller pressures the series became sensibly $= 1$.

The motion of the air between the two disks is represented in fig. 8, where the upper disk is supposed fixed and the lower one oscillates. A row of particles of air which when at rest form a straight line perpendicular to the disks, will when in motion assume in succession the forms of the curves 1, 2, 3, 4, 5, 6. If the ratio of the density to the viscosity of the air is very small, or if the time of oscillation is very great, or if the interval between the disks is very small, these curves approach more and more nearly to the form of straight lines.

The chief mathematical difficulty in treating the case of the moving disks arises from the necessity of determining the motion of the air in the neighbourhood of the edge of the disk. If the disk were accompanied in its motion by an indefinite plane ring surrounding it and forming a continuation of its surface, the motion of the air would be the same as if the disk were of indefinite extent; but if the ring were removed, the motion of the air in the neighbourhood of the edge would be diminished, and therefore the effect of its viscosity on the parts of the disk near the edge would be increased. The actual effect of the air on the disk may be considered equal to that on a disk of greater radius forming part of an infinite plane.

Since the correction we have to consider is confined to the space immediately surrounding the edge of the disk, we may treat the edge as if it were

16   ON THE VISCOSITY OR INTERNAL FRICTION

the straight edge of an infinite plane parallel to xz, oscillating in the direction of z between two planes infinite in every direction at distance b. Let w be the velocity of the fluid in the direction of z, then the equation of motion is[*]

$$\rho \frac{dw}{dt} = \mu \left(\frac{d^2w}{dx^2} + \frac{d^2w}{dy^2} \right) \dots\dots\dots\dots\dots\dots(10),$$

with the conditions

$$w = 0 \text{ when } y = \pm b \dots\dots\dots\dots\dots\dots(11),$$

and

$$w = C \cos nt \text{ when } y = 0, \text{ and } x \text{ is positive} \dots (12).$$

I have not succeeded in finding the solution of the equation as it stands, but in the actual experiments the time of oscillation is so long, and the space between the disks is so small, that we may neglect $\dfrac{b^2 \rho n}{\mu}$, and the equation is reduced to

$$\frac{d^2w}{dx^2} + \frac{d^2w}{dy^2} = 0 \dots\dots\dots\dots\dots\dots(13)$$

with the same conditions. For the method of treating these conditions I am indebted to Professor W. Thomson, who has shewn me how to transform these conditions into another set with which we are more familiar, namely, $w = 0$ when $x = 0$, and $w = 1$ when $y = 0$, and x is greater than $+1$, and $w = -1$ when x is less than -1. In this case we know that the lines of equal values of w are hyperbolas, having their foci at the points $y = 0$, $x = \pm 1$, and that the solution of the equation is

$$w = \frac{2}{\pi} \sin^{-1} \frac{r_1 - r_2}{2} \dots\dots\dots\dots\dots\dots (14)$$

where r_1, r_2 are the distances from the foci.

If we put

$$\phi = \frac{2}{\pi} \log \{ \sqrt{(r_1 + r_2)^2 - 4} + r_1 + r_2 \} \dots\dots\dots\dots(15),$$

then the lines for which ϕ is constant will be ellipses orthogonal to the hyperbolas, and

$$\frac{d^2\phi}{dx^2} + \frac{d^2\phi}{dy^2} = 0; \dots\dots\dots\dots\dots\dots(16);$$

and the resultant of the friction on any arc of a curve will be proportional

---

[*] Professor Stokes "On the Theories of the Internal Friction of Fluids in Motion, &c.," *Cambridge Phil. Trans.* Vol. VIII.

to $\phi_1 - \phi_0$, where ϕ_0 is the value of ϕ at the beginning, and ϕ_1 at the end of the given arc.

In the plane $y = 0$, when x is very great, $\phi = \dfrac{2}{\pi} \log 4x$, and when $x = 1$, $\phi = \dfrac{2}{\pi} \log 2$, so that the whole friction between $x = 1$ and a very distant point is $\dfrac{2}{\pi} \log 2x$.

Now let w and ϕ be expressed in terms of r and θ, the polar co-ordinates with respect to the origin as the pole; then the conditions may be stated thus:

When $\theta = \pm \dfrac{\pi}{2}$, $w = 0$. When $\theta = 0$ and r greater than 1, $w = 1$. When $\theta = \pi$ and r greater than 1, $w = -1$.

Now let x', y' be rectangular co-ordinates, and let

$$y' = \frac{2}{\pi} b\theta \quad \text{and} \quad x' = \frac{2}{\pi} b \log r \quad \text{...................... (17).}$$

and let w and ϕ be expressed in terms of x' and y'; the differential equations (13) and (16) will still be true; and when $y' = \pm b$, $w = 0$, and when $y' = 0$ and x' positive, $w = 1$.

When x' is great, $\phi = \dfrac{x'}{b} + \dfrac{2}{\pi} \log 4$, and when $x' = 0$, $\phi = \dfrac{2}{\pi} \log 2$, so that the whole friction on the surface is

$$\frac{x'}{b} + \frac{2}{\pi} \log 2 \quad \text{................................(18),}$$

which is the same as if a portion whose breadth is $\dfrac{2b}{\pi} \log_e 2$ had been added to the surface at its edge.

The curves of equal velocity are represented in fig. 9 at u, v, w, x, y. They pass round the edge of the moving disk AB, and have a set of asymptotes U, V, W, X, Y, arranged at equal distances parallel to the disks.

The curves of equal friction are represented at o, p, q, r, s, t. The form of these curves approximates to that of straight lines as we pass to the left of the edge of the disk.

18   ON THE VISCOSITY OR INTERNAL FRICTION

The dotted vertical straight lines O, P, Q, R, S, T represent the position of the corresponding lines of equal friction if the disk AB had been accompanied by an extension of its surface in the direction of B. The total friction on AB, or on any of the curves $u, v, w,$ &c., is equal to that on a surface extending to the point C, on the supposition that the moving surface has an accompanying surface which completes the infinite plane.

In the actual case the moving disk is not a mere surface, but a plate of a certain thickness terminated by a slightly rounded edge. Its section may therefore be compared to the curve uu' rather than to the axis AB.

The total friction on the curve is still equal to that on a straight line extending to C, but the velocity corresponding to the curve u is less than that corresponding to the line AB.

If the thickness of the disk is 2β, and the distance between the fixed disks $= 2b$, so that the distance of the surfaces is $b - \beta$, the breadth of the strip which must be supposed to be added to the surface at the edge will be

$$a = \frac{2b}{\pi} \log_e 10 \left\{ \log_{10} 2 + \log_{10} \sin \frac{\pi (b - \beta)}{2b} \right\}^* \dots \dots \dots \dots (19).$$

In calculating the moment of friction on this strip, we must suppose it to be at the same distance from the axis as the actual edge of the disk. Instead of $A = \frac{\pi}{2} r^4$ in equation (9), we must therefore put $A = \frac{\pi}{2} r^4 + 2\pi r^3 a$, and instead of b we must put $b - \beta$.

The actual value of $\frac{\pi}{2} r^4$ for each surface in inches $= 1112 \cdot 8$.

The value of I in inches and grains was 175337.

It was determined by comparing the times of oscillation of the axis and disks without the little magnet, with the times of the brass ring (fig. 4) and of the tube and weights (fig. 7). Four different suspension wires were used in these experiments.

The following Table gives the numbers required for the calculation of each of the five arrangements of the disks.

---

* This result is applicable to the calculation of the electrical capacity of a condenser in the form of a disk between two larger disks at equal distance from it.

Arrangement	Case 1	Case 2	Case 3	Case 4	Case 5
N = number of surfaces	2	2	6	6	6
$b - \beta$ = distance of surfaces	1·0	0·5	0·683	0·425	0·1847
$2\pi r^3 a$ = effect of edge	446·09	235·0	292·95	186·7	86·1
A = whole moment of each surface ...	1558·9	1347·8	1405·75	1299·5	1198·9
$\dfrac{N}{2}\dfrac{A}{Ib \log_e 10} = Q =$	·003815	·007398	·015110	·022448	·047640

If l is the Napierian logarithmic decrement per second, and L the observed decrement of the common logarithm (to base 10) of the arc in time T, then

$$L = lT \log_{10} e \quad.................................(20).$$

If n is the coefficient of t in the periodic terms, and T the time of five complete vibrations,

$$nT = 10\pi \quad.....................................(21).$$

Let
$$K = kT \log_{10} e \quad...............................(22),$$

then K is the part of the observed logarithmic decrement due to the viscosity of the wire, the yielding of the instrument, and the friction of the air on the axis, and is the same for all experiments as long as the wire is unaltered.

Let μ_0 be the value of μ at temperature zero, μ that at any other temperature θ, then if μ is proportional to the temperature from absolute zero,

$$\mu = (1 + a\theta)\,\mu_0 \quad.............................(23),$$

where a is the coefficient of expansion of air per degree.

Equation (9) may now be written in the form

$$\mu_0 Q (1 + x)(1 + a\theta)\,T + K = L........................(24),$$

where $1 + x$ is the series in equation (9), x being in most cases small, and may be calculated from an approximate value of μ_0.

The values of Q are to be taken from the Table according to the arrangement of disks in the experiment.

In this way I have combined the results of forty experiments on dry air in order to determine the values of μ_0 and K. Seven of these had the first arrangement, six had the second, six the third, nine the fourth, and twelve the fifth.

The values of Q for the five cases are roughly in the proportions of 1, 2, 4, 6, 12, so that it is easy to eliminate K and find μ_0. I had reason, however, to believe that the value of K was altered at a certain stage of the experiments when steam was first used to heat the air in the receiver. I therefore introduced two values of K, K_1 and K_2, into the experiments before and after this change respectively. The values of K_1 and K_2 deduced from these experiments were

In ten single vibrations.

$K_1 = \cdot 01568$

$K_2 = \cdot 01901.$

The value of μ in inch-grain-second measure at temperature θ° Fahrenheit is for air

$$\mu = \cdot 00001492 \, (461^\circ + \theta^\circ).$$

The value of L was then calculated for each experiment and compared with the observed value. In this way the error of mean square of a single experiment was found. The probable error of μ, as determined from the equations, was calculated from this and found to be $0\cdot 36$ per cent. of its value.

In order to estimate the value of the evidence in favour of there being a finite amount of slipping between the disks and the air in contact with them, the value of L for each of the forty experiments was found on the supposition that

$$\beta = \cdot 0027 \text{ inch and } \mu = (\cdot 000015419) \, (461^\circ + \theta^\circ).$$

The error of mean square for each observation was found to be slightly greater than in the former case; the probable error of β was 40 per cent., and that of $\mu = 1\cdot 6$ per cent.

I have no doubt that the true value of β is zero, that is, there is no slipping, and that the original value of μ is the best.

As the actual observations were very numerous, and the reduction of them would occupy a considerable space in this paper, I have given a specimen of the actual working of one experiment.

Table I. shews the readings of the scale as taken down at the time of observation, with the times of transit of the middle point of the scale after the fifth and sixth readings, with the sum of ten successive amplitudes deduced therefrom.

Table II. shews the results of this operation as extended to the rest of experiment 62, and gives the logarithmic decrement for each successive period of ten semivibrations, with the mean time and corresponding mean logarithmic decrement.

Table III. shews the method of combining forty experiments of different kinds. The observed decrement depends on two unknown quantities, the viscosity of air and that of the wire.

The experiments are grouped together according to the coefficients of μ and K that enter into them, and when the final results have been obtained, the decrements are calculated and compared with the results of observation. The calculated sums of the decrements are given in the last column.

Table IV. shews the results of the twelve experiments with the fifth arrangement. They are arranged in groups according to the pressure of the air, and it will be seen that the observed values of L are as independent of the pressure as the calculated values, in which the pressure is taken into account only in calculating the value of x in the fifth column. By arranging the values of $L - L'$ in order of temperature, it was found that within the range of atmospheric temperature during the course of the experiments the relation between the viscosity of air and its temperature does not perceptibly differ from that assumed in the calculation. Finally, the experiments were arranged in order of time, to determine whether the viscosity of the wire increased during the experiments, as it did when steam was first used to heat the apparatus. There did not appear any decided indication of any alteration in the wire.

Table V. gives the resultant value of μ in terms of the different units which are employed in scientific measurements.

Note, added February 6, 1866.—In the calculation of the results of the experiments, I made use of an erroneous value of the moment of inertia of the disks and axis = 1·012 of the true value, as determined by six series of experiments with four suspension wires and two kinds of auxiliary weights. The numbers in the coefficients of m in Table IV. are therefore all too large, and the value of μ is also too large in the same proportion, and should be

$$\mu = \cdot00001492 \left(461° + \theta\right).$$

The same error ran through all the absolute values in other parts of the paper as sent in to the Royal Society, but to save trouble to the reader I have corrected them where they occur.

TABLE I.—Experiment 62.   Arrangement 5.   Dry air at pressure 0·55 inch.
Temperature 68° F.   May 9, 1865.

	Greater scale reading	Time 3ʰ +	Less scale reading	Time
		m s		m s
	8309	....	1740	
	8071	....	1968	
	7852	27 42·4	2180	28 18·8
	7650	....	2377	
	7460	....	2561	
Sum of greater readings	39342	....	10826	
Sum of less do.	10826			
Difference 28516 = sum of 10 amplitudes.				

The observations were continued in the same way till five sets of readings of this kind were obtained. The following were the results.

TABLE II.

Times	Sum of ten amplitudes	Logarithm	Log. decrement
h m s			
3 28 0·6	28516	4·4550886	0·1587745
34 3·2	19784	4·2963141	0·1585171
40 5·8	13734	4·1377970	0·1587041
46 8·6	9530	3·9790929	0·1596806
52 11·2	6598	3·8194123	

Results of experiment 62.   Mean time of ten vibrations $= 362\cdot66$
Mean log. decrement . . . $=$ $0\cdot1588574.$

TABLE III.

Equations from which μ for air was determined; $m = \dfrac{\mu}{461^\circ + \theta}$.

Number of experiments	Arrangement	Result of observation	Result of calculation
3.	1	$6\cdot3647\,m + \ 3K_1 = \cdot00023167$	$\cdot00022779$
3.	2	$11\cdot2893\,m + \ 3K_1 = \cdot00028280$	$\cdot00030214$
6.	4	$71\cdot2412\,m + \ 6K_1 = \cdot00135467$	$\cdot00133897$
4.	1	$8\cdot7221\,m + \ 4K_2 = \cdot00034562$	$\cdot00034127$
3.	2	$11\cdot6680\,m + \ 3K_2 = \cdot00031505$	$\cdot00033335$
12.	5	$297\cdot7880\,m + 12K_2 = \cdot00511708$	$\cdot00512666$
3.	4	$36\cdot0551\,m + \ 3K_2 = \cdot00069607$	$\cdot00070159$
6.	3	$48\cdot8911\,m + \ 6K_2 = \cdot00108215$	$\cdot00105333$

Final result $\mu^* = \cdot00001510\ (461^\circ + \theta)$ with probable error $0\cdot36$ per cent.

$$K_1 = \cdot0000439, \qquad\qquad K_2 = \cdot0000524.$$

TABLE IV.—Experiments with Arrangement 5.

No. of experiment	Absolute temperature $461^\circ + \theta$	Pressure, in inches of mercury	Time of five double swings, in seconds	Correction for inertia of air $(1+x)$ in equation (24)	L = decrement of logarithm of arc in ten single vibrations		Diff. $L - L'$
					L' calculated	L observed	
62.	529	0·55	362·66		·15719	·15886	+ 167
77.	516	0·50	362·80	$1 - \cdot0000157$	·15378	·15260	− 118
80.	527	0·56	364·04		·15648	·15946	+ 279
63.	527·5	5·57	362·72		·15680	·15628	− 52
64.	535	5·97	362·94	$1 - \cdot000157$	·15875	·15838	− 37
81.	516	5·52	363·80		·15379	·15389	+ 10
65.	524	25·58	362·64		·15555	·15422	− 133
71.	513	19·87	362·50	$1 - \cdot000486$	·15299	·15144	− 155
72.	514·5	20·31	362·86		·15338	·15269	− 69
75.	517	29·90	363·8		·15398	·15377	− 21
76.	512·5	29·76	362·89	$1 - \cdot00058$	·15280	·15146	− 134
79.	521	28·22	363·9		·15502	·15510	+ 8

* This is the result derived from these equations, which is $1\cdot2$ per cent. too large.

TABLE V.—Results.

Coefficient of viscosity in dry air. Units—the inch, grain, and second, and Fahrenheit temperature,

$$\mu = {\cdot}00001492 \, (461 + \theta) = {\cdot}006876 + {\cdot}0000149\theta.$$

At 60° F. the mean temperature of the experiments, $\mu = {\cdot}007763$. Taking the foot as unit instead of the inch, $\mu = {\cdot}000179 \, (461 + \theta)$. In metrical units (metre, gramme, second, and Centigrade temperature),

$$\mu = {\cdot}01878(1 + {\cdot}003650\theta).$$

The coefficient of viscosity of other gases is to be found from that of air by multiplying μ by the ratio of the transpiration time of the gas to that of air as determined by Graham*.

POSTSCRIPT.—Received December 7, 1865.

Since the above paper was communicated to the Royal Society, Professor Stokes has directed my attention to a more recent memoir of M. O. E. Meyer, "Ueber die innere Reibung der Gase," in Poggendorff's *Annalen*, cxxv. (1865). M. Meyer has compared the values of the coefficient of viscosity deduced from the experiments of Baily by Stokes, with those deduced from the experiments of Bessel, and of Girault. These values are ·000104, ·000275, and ·000384 respectively, the units being the centimetre, the gramme, and the second. M. Meyer's own experiments were made by swinging three disks on a vertical axis in an air-tight vessel. The disks were sometimes placed in contact, and sometimes separate, so as to expose either two or six surfaces to the action of the air. The difference of the logarithmic decrement of oscillation in these two arrangements was employed to determine the viscosity of the air.

The effects of the resistance of the air on the axis, mirror, &c., and of the viscosity of the suspending wires are thus eliminated.

The calculations are made on the supposition that the moving disks are so far from each other and from the surface of the receiver which contains them, that the effect of the air upon each is the same as if it were in an infinite space.

At the distance of 30 millims., and with a period of oscillation of fourteen seconds, the mutual effect of the disks would be very small in air at the

---

* *Philosophical Transactions*, 1849.

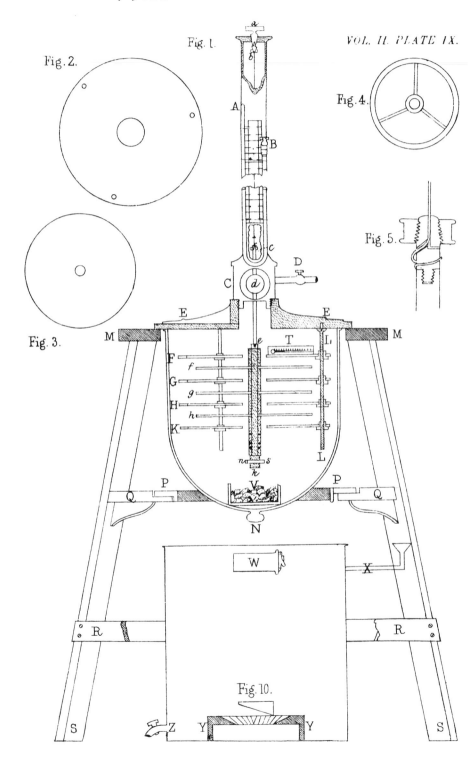

Fig. 1.

Fig. 2.

Fig. 4.

Fig. 5.

Fig. 3.

Fig. 10.

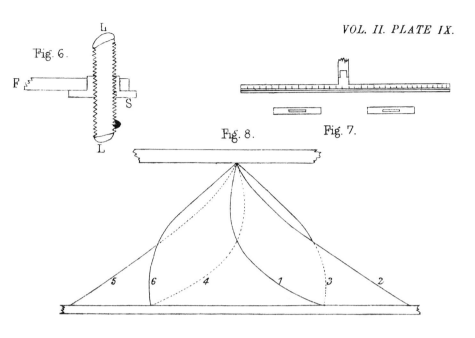

Fig. 6.

F

L

S

L

Fig. 8.

Fig. 7.

5   6   4   1   3   2

Fig. 9.

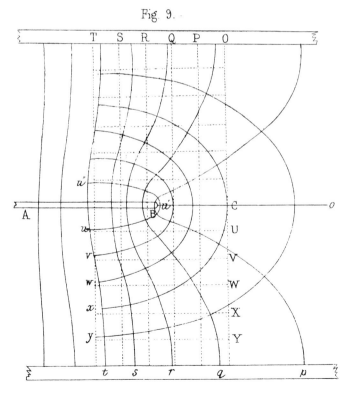

T   S   R   Q   P   O

u''

A   B u' C   O

u U

v V

w W

x X

y Y

t s r q p

ordinary pressure. In November 1863 I made a series of experiments with an arrangement of three brass disks placed on a vertical axis exactly as in M. Meyer's experiments, except that I had then no air-tight apparatus, and the disks were protected from currents of air by a wooden box only.

I attempted to determine the viscosity of air by means of the observed mutual action between the disks at various distances. I obtained the values of this mutual action for distances under 2 inches, but I found that the results were so much involved with the unknown motion of the air near the edge of the disks, that I could place no dependence on the results unless I had a complete mathematical theory of the motion near the edge.

In M. Meyer's experiments the time of vibration is shorter than in most of mine. This will diminish the effect of the edge in comparison with the total effect, but in rarefied air both the mutual action and the effect of the edge are much increased. In his calculations, however, the effect of the three edges of the disks is supposed to be the same, whether they are in contact or separated. This, I think, will account for the large value which he has obtained for the viscosity, and for the fact that with the brass disks which vibrate in 14 seconds, he finds the apparent viscosity diminish as the pressure diminishes, while with the glass disks which vibrate in 8 seconds it first increases and then diminishes.

M. Meyer concludes that the viscosity varies much less than the pressure, and that it increases slightly with increase of temperature. He finds the value of μ in metrical units (centimetre-gramme-second) at various temperatures,

Temperature.	Viscosity.
8°·3 C.	·000333
21°·5 C.	·000323
34°·4 C.	·000366

In my experiments, in which fixed disks are interposed between the moving ones, the calculation is not involved in so great difficulties; and the value of μ is deduced directly from the observations, whereas the experiments of M. Meyer give only the value of $\sqrt{\mu\rho}$, from which μ must be determined. For these reasons I prefer the results deduced from experiments with fixed disks interposed between the moving ones.

M. Meyer has also given a mathematical theory of the internal friction of gases, founded on the dynamical theory of gases. I shall not say anything of this part of his paper, as I wish to confine myself to the results of experiment.

19. [Notes for the Section "On the Mutual Action of Two Molecules" in "On the Dynamical Theory of Gases" (1867 Memoir)]

Cambridge University Library, Maxwell manuscripts, box 4.

Now let the components of the velocity of M_1 be ξ_1, η_1, ζ_1, ⟨and⟩ those of M_2, ξ_2, η_2, ζ_2 and those of the centre of gravity of M_1 & M_2 ⟨will-be⟩ $\bar{\xi}\,\bar{\eta}\,\bar{\zeta}$. Then we know by Dynamics that the velocity of the centre of gravity will be unchanged by the mutual action, and that the motion of either molecule relatively to the centre of gravity will not be affected by the circumstance that the whole system is in motion.

The components of V_1 the velocity of M_1 relative to the centre of gravity are $\xi_1 - \bar{\xi}, \eta_1 - \bar{\eta}, \zeta_1 - \bar{\zeta}$. After M_1 and M_2 have met and deflected each other from their courses, the value of V_1 will be the same but its direction will be turned through an angle 2θ in the plane ⟨through⟩ containing the directions of V_1 and b. Let φ be the angle which this plane makes with the plane containing the direction of V and parallel to the axis of x and let ⟨ξ_1 be⟩ $\xi_1 + \delta\xi_1$ be the component velocity of M_1 in the direction of x after the mutual action of the molecules

$$\xi_1 + \delta\xi_1 = \frac{M_1\xi_1 + M_2\xi_2}{M_1 + M_2} + \frac{M_2}{M_1 + M_2}\{(\xi_1 - \xi_2)\cos 2\theta$$

$$+ \sqrt{(\eta_1 - \eta_2)^2 + (\zeta_1 - \zeta_2)^2}\,\sin 2\theta \cos\varphi\}$$

or $\quad \xi_1 + \delta\xi_1 = \xi_1 + \dfrac{M_2}{M_1 + M_2}\{(\xi_2 - \xi_2)^2 \sin^2\theta$

$$+ \sqrt{(\eta_2 - \eta_1)^2 + (\zeta_2 - \zeta_1)^2}\,\sin 2\theta \cos\varphi$$

There will be similar expressions for the components of the mean velocity of M_1 in the other coordinate directions.

[The rest of this sheet is blank. The following material starts a new sheet.]

A molecule M_1 whose velocity components are ξ_1, η_1, ζ_1 encounters another, M_2, whose velocity components are ξ_2, η_2, ζ_2. Let V be the velocity of M_1 relative to M_2, and let a straight line be drawn through M_1 parallel to V before the encounter and let b be the perpendicular from M_2 on this line.

After the encounter, if no kinetic energy of translation is lost, the relative velocity will still be $V' = V$ in magnitude but turned through an angle 2θ in the plane of V and b.

Hence if this plane makes an angle φ with the plane of V and x, the value of ξ_1 after the encounter will be

$$\xi_1 = \xi_1 + \frac{M_2}{M_1 + M_2}\{(\xi_2 - \xi_1)^2 \sin^2\theta + ((\eta_2 - \eta_1)^2$$

$$+ (\zeta_2 - \zeta_1)^2)^{1/2} \sin 2\theta \cos \varphi\}$$

The angle θ will be a function of b and V, vanishing except for ⟨very small⟩ values of b less than the ⟨distance⟩ limiting distance of molecular action.

The number in unit volume of molecules M_2 whose component velocities lie within the limits ξ_2 and $\xi_2 + d\xi_2$, η_2 and $\eta_2 + d\eta_2$ and ζ_2 and $\zeta_2 + d\zeta_2$ is

$$N_2 f \, d\xi_2 \, d\eta_2 \, d\zeta_2$$

where N_2 is the whole number of molecules of the second kind ⟨and⟩ in unit of volume and f is a function of $\xi_2 \eta_2 \zeta_2$ the probable form of which is

$$\frac{h_1 h_2 h_3}{\pi^{3/2}} e^{-(h_1^2 \xi^2 + h_2^2 \eta^2 + h_3^2 \zeta^2 + 2g_1 \eta\zeta + 2g_2 \zeta\xi + 2g_3 \xi\eta)}$$

The number of encounters of N_1 molecules of the first kind with those in which b is between b and $b + db$ and φ between φ and $\varphi + d\varphi$ is

$$V b \, db \, d\varphi \, \delta t \, N_1 N_2 f \, d\xi_2 \, d\eta_2 \, d\zeta_2$$

If Q is any property of M_1 which after the encounter becomes Q' and if $\dfrac{\delta}{\delta t} Q$ denotes the rate of change of the mean

$$\left\langle \frac{\delta}{\delta t} \sum Q = Q' - Q \right\rangle$$

value of Q for all the N_1 molecules

$$\frac{\delta}{\delta t} Q = N_2 \iiiint (Q' - Q) V b \, db \, d\varphi f \, d\xi_2 \, d\eta_2 \, d\zeta_2$$

We can integrate with respect to φ at once in the following cases in which, for simplicity, we suppose $M_1 = M_2$

(α) Let $Q = \xi_1$

$$\frac{\delta}{\delta t} \xi_1 = N_2 \iiiint \frac{M_2}{M_1 + M_2} 4\pi \sin^2\theta \, V b \, db f \, d\xi_2 \, d\eta_2 \, d\zeta_2$$

$$\frac{\delta}{\delta t} \xi_1^2 = N_2 \iiiint \frac{M_2}{M_1 + M_2}$$

$$\frac{\delta}{\delta t} \xi_1 = N_2 2\pi \iiiint (\xi_2 - \xi_1) \sin^2\theta \, V b \, db f \, d\xi_2 \, d\eta_2 \, d\zeta_2$$

For a single gas $\dfrac{\delta}{\delta t} \xi_1 = 0$

(β) Let $Q = \xi_1^2$

$$\frac{\delta}{\delta t} \xi_1^2 = N_2 2\pi \iiiint \left[(\xi_2^2 - \xi_1^2) \sin^2 \theta - \frac{1}{2} (2(\xi_2 - \xi_1)^2 - (\eta_2 - \eta_1)^2 \right.$$
$$\left. - (\zeta_2 - \zeta_1)^2) \times \sin^2 \theta \cos^2 \theta \right] Vb \, db f \, d\xi_2 \, d\eta_2 \, d\zeta_2$$

For a single gas

$$\frac{\delta}{\delta t} \xi_1^2 = N_2 2\pi \iiiint [\eta^2 + \zeta^2 - 2\xi^2] \sin^2 \theta \cos^2 \theta \, Vb \, db f \, d\xi \, d\eta \, d\zeta$$

(γ) Let $Q = \xi_1^3$

$$\frac{\delta}{\delta t} \xi_1^3 =$$

[The rest of this sheet is blank. The following material is a fragment of another draft of the same paper.]

The number of such encounters in unit of time in which b lies between b and $b + db$ and φ between φ and $\varphi + d\varphi$ is

$$Vb \, db \, d\varphi \, dN_1 \, dN_2$$

Let us determine the time-variation of $\sum (M\xi_1)$, $\sum (M\xi_1^2)$ and $\sum (M\xi_1^3)$ arising from these encounters

⟨We may⟩

$$\frac{\delta \sum (M\xi_1)}{\delta t} = \iiiint M(\xi_1' - \xi_1) V \, db \, d\varphi \, dN_1 \, dN_2$$

We can always integrate with respect to φ. In this case we find

$$\frac{\delta \sum M\xi}{\delta t} = 2\pi M \iiint (\xi_2 - \xi_1) \sin^2 \theta \, V \, db \, dN_1 \, dN_2$$

In the case of a simple gas ⟨there is⟩ for every value of ξ_2 there is another negative value of the same magnitude, so that this term disappears altogether. In the case of two or more gases diffusing through each other the term is important.

Similarly we find

$$\frac{\delta \sum (M\xi^2)}{\delta t} = 2\pi M \iiint (\eta^2 + \zeta^2 - 2\xi^2) \sin^2 \theta \cos^2 \theta \, V \, db \, dN_1 \, dN_2$$

and

$$\frac{\delta \sum M\xi^2}{\delta t} = 2\pi M \iiint 3\xi(\eta^2 + \zeta^2 - 2\xi^2)\sin^2\theta \cos^2\theta \, V db \, dN_1 \, dN_2$$

Since θ is a function of b and V

Let $2\pi \int_0^\infty b \, db \sin^2\theta \cos^2\theta = B$

then B will be a function of V, and

$$\left\langle \frac{\delta \sum (M\xi^2)}{\delta t} = MN^2 \iiiiii (\eta^2 + \zeta^2 - 2\xi^2)VBf_1 f_2 \, d\xi_1 \, d\eta_1 \, d\zeta_1 \, d\xi_2 \, d\eta_2 \, d\zeta_2 \right\rangle$$

$$\frac{\delta \sum (M\xi^2)}{\delta t} = \rho N \iiint (\eta^2 + \zeta^2 - 2\xi^2)VBf_2 \, d\xi_2 \, d\eta_2 \, d\zeta_2$$

and $\dfrac{\delta M}{\delta t} = 3\rho N \iiint \xi(\eta^2 + \zeta^2 - 2\xi^2)VBf_2 \, d\xi_2 \, d\eta_2 \, d\zeta_2$

As we do not know the form of the function B we cannot perform these integrations. We may however for rough provisional purposes assume that

$$\frac{\delta \sum (M\xi^2)}{\delta t} = \rho^2 C(\eta^2 + \zeta^2 - 2\xi^2)$$

$$\frac{\delta \sum (M\xi^3)}{\delta t} = 3\rho^2 C\xi(\eta^2 + \zeta^2 - 2\xi^2)$$

where C is a function of the temperature.

If X, Y, Z are the components of the external force and the variation of the quantities arising from this force be indicated by the symbol δ'

$$\frac{\delta' \sum (M\xi)}{\delta' t} = \rho X$$

$$\frac{\delta' \sum (M\xi^2)}{\delta' t} = 2\rho u X$$

$$\frac{\delta' \sum M\xi^3}{\delta' t} = 3\rho u^2 X + 3\sum (M\xi^2)X$$

[The rest of this sheet is blank. The following material is a fragment of another draft of the same paper.]

We have next to find the part of $\dfrac{\delta}{\delta t}\sum (M\xi^2)$ and $\dfrac{\delta}{\delta t}\sum (M\xi^3)$ ⟨arising from⟩ ⟨the⟩ which depends on the encounters of the molecules.

If a quantity Q belonging to a molecule becomes Q' after an encounter

$$\frac{\delta \sum Q}{\delta t} = S(Q' - Q)$$

where S denotes a summation of all the cases which occur in unit of volume in unit of time.

Let us suppose that there are in every unit of volume dN_1 molecules having velocity-components between ξ_1 and $\xi_1 + d\xi_1$, η_1 and $\eta_1 + d\eta_1$, ζ_1 and $\zeta_1 + d\zeta_1$. If N is the whole number of molecules in unit of volume

$$dN_1 = Nf d\xi_1 d\eta_1 d\zeta_1$$

where f is a function of ξ, η, and ζ which ⟨when there is no disturbing cause⟩ in the ⟨state of⟩ ultimate state of the system is

$$\frac{1}{\alpha^3 \pi^{3/2}} e^{-(\xi^2 + \eta^2 + \zeta^2)/\alpha^2}$$

Let us also suppose that there are dN_2 molecules in unit of volume having velocities nearly ξ_2, η_2, ζ_2.

Then the velocity of M_1, one of the first set, relatively to M_2, one of the second set will be V and the components of V will be $\xi_1 - \xi_2, \eta_1 - \eta_2, \zeta_1 - \zeta_2$. Draw a straight line through M_1 in the direction of V and from M_2 draw a perpendicular, b, to this line.

⟨Af⟩ The value of b will determine the nature of the encounter and the angle, 2θ, through which the relative velocity is deflected. The relative velocity after the encounter is of the same magnitude as before but its direction is turned through an angle 2θ in the plane of V and b. If ξ_1' is the x-component of the velocity of M_1 after the encounter (M_1 and M_2 being equal)

$$\xi_1' = \xi_1 + (\xi_2 - \xi_1)\sin^2\theta + [(\eta_2 - \eta_1)^2 + (\zeta_2 - \zeta_1)^2]^{1/2} \sin\theta \cos\theta \cos\varphi$$

where φ is the angle between the plane of V and b and the plane of V and x.

20. "To Find the Equations of Motion..." [Notes for 1867 Memoir][a]

Cambridge University Library, Maxwell manuscripts, box 4.

To find the Equations of Motion of a Medium composed of Molecules in motion acting on one another with forces which are insensible at ⟨sensible⟩ distances which are small compared with the average distance of the Molecules.

Let us begin by considering two particles whose masses are M_1 M_2 at a considerable distance from each other so that the force between them is insensible. Let their velocities resolved in the coordinate directions be $\xi_1 \eta_1 \zeta_1$ and $\xi_2 \eta_2 \zeta_2$. Let V be the velocity of M_1 relative to M_2 and let G be the velocity of the centre of gravity of M_1 & M_2. Also let b be minimum distance to which the particles would approach if there were no action between them.

In consequence of the mutual action between M_1 & M_2 when they come near each other, each will describe a curve about G their centre of gravity in the plane of V and b and when they have passed out of each others influence each will have ⟨the same⟩ a velocity relative to the centre of gravity equal in magnitude to its former value, but in a direction inclined 2θ to its former direction in the plane of V & b, 2θ being the angle between the asymptotes of the orbit. Let us assume that the moving force between the particles is as the nth power of the distance inversely and that its value at distance unity is K and repulsive, then the well known equation

$$\langle\text{equation}\rangle \quad \frac{d^2u}{d\theta^2} + u + \frac{p}{n^2u^2} = 0$$

becomes by integration and putting $u = x/b$ and $b = \alpha\left(\dfrac{K}{r^2}\,\dfrac{(M_1 + M_2)}{M_2}\right)^{1/(n-1)}$

$$\theta\left\langle d\theta - \frac{d\theta}{d\alpha}\right\rangle = \int \frac{dx}{\sqrt{\left(1 - x^2 - \dfrac{2}{n-1}\dfrac{x}{\alpha}\right)^{n-1}}}$$

Let θ be the value of this integral taken between the limits $x = 0$ and x a root of the equation

$$\left(1 - x^2 - \frac{2}{n-1}\frac{x}{\alpha}\right)^{n-1} = 0$$

a. Cf, document III-30; *Scientific Papers*, vol. 2, 40.

21. "Law of Volumes" [Notes for Section "Equilibrium of Temperature Between Two Gases" (1867 Memoir)]

Cambridge University Library, Maxwell manuscripts, box 4.

Law of Volumes

To determine the variation in the quantity of ⟨heat⟩ energy, put

$$Q_1 = \tfrac{1}{2}M_1\langle(1 + \beta)\rangle(\xi_1^2 + \eta_1^2 + \zeta_1^2)$$

and let us first consider the value of $\dfrac{\delta Q}{\delta t}$ due to the ⟨mutual⟩ action of the molecules of the ⟨first⟩ second kind on those of the first in a mixed medium in equilibrium when u, v, w are $= 0$.

Then $\dfrac{\delta Q_1}{\delta t} = \dfrac{k\rho_2}{M_2} \Theta_1 \{\tfrac{1}{2}M_2(\xi_2^2 + \eta_2^2 + \zeta_2^2) - \tfrac{1}{2}M_1(\xi_1^2 + \eta_1^2 + \zeta_1^2)$

$$+ \tfrac{1}{2}(M_1 - M_2)(\xi_1\xi_2 + \eta_1\eta_2 + \zeta_1\zeta_2)\}$$

Since ξ_1 is independent of ξ_2 and since the mean values of both are zero the product $\xi_1\xi_2$ has zero for its mean value so that the terms $\xi_1\xi_2$, $\eta_1\eta_2$, $\zeta_1\zeta_2$ disappear and we may write the result

$$\frac{\delta Q_1}{\delta t} = \frac{k\rho_2}{M_2} \Theta_1 \{Q_2 - Q_1\}$$

Similarly

$$\frac{\delta Q_2}{\delta t} = \frac{k\rho_1}{M_1} \Theta_1 \{Q_1 - Q_2\}$$

whence

$$\frac{\delta T}{\delta t}(Q_1 - Q_2) = -k\Theta_1 \left(\frac{\rho_1}{M_1} + \frac{\rho_2}{M_2}\right)(Q_1 - Q_2)$$

or

$$Q_1 - Q_2 = Ce^{-t/T} \quad \text{where} \quad \frac{1}{T} = k\Theta_1 \left(\frac{\rho_1}{M_1} + \frac{\rho_2}{M_2}\right)$$

Hence if Q_1 is originally different from Q_2 the values of Q_1 and Q_2 will rapidly approach to equality and will be sensibly equal in a few multiples of the very short time T, and in all movements of the media except the most violent Q_1 & Q_2 will remain equal. Now Q_1 is the actual energy of a single molecule of the first system due to the motion of agitation of its centre of gravity and Q_2 is the same for a single molecule of the second system. The energies of single molecules of mixed system therefore tend to equality whatever the mass of each molecule may be.

Now when two gases are such that neither communicates energy to the other, their temperatures are the same and we have seen that $Q_1 = Q_2$ in this case. But for either gas $NQ = \tfrac{3}{2}p$ [Table 1] therefore if both the temperature and the pressure be the same in two different gases, then N the number of molecules in unit of volume is also the same in the two gases. This is the law of Volumes of gases, first discovered by Gay-Lussac from chemical considerations. It is a necessary result of the Dynamical Theory of Gases.

In the case of a single gas in motion, let Q be the total energy of a single molecule then

$$Q = \tfrac{1}{2}M((u + \xi)^2 + (v + \eta)^2 + (w + \zeta)^2 + \beta(\xi^2 + \eta^2 + \zeta^2))$$

[Table 1. This is a calculation in the margin of the manuscript.]

$$
\begin{array}{ll}
A_2 & A_1 \\
\left. 1.3682 \right) & 2.6595(2 - .056 \\
& 2.7366
\end{array}
$$

$$
\begin{array}{l}
\hline
.076900 \\
.6841 \\
\hline
.859 \\
.820 \\
\hline
39
\end{array}
$$

and

$$\frac{\delta Q}{\delta t} = M(uX + vY + wZ)$$

The general equation (   ) becomes

$$
\frac{1}{2}\rho\frac{\partial}{\partial t}(u^2 + v^2 + w^2 + (1 + \beta)(\xi^2 + \eta^2 + \zeta^2) + \frac{d}{dx}(u\rho\xi^2 + v\rho\xi\eta + w\rho\xi\zeta)
$$

$$
+ \frac{d}{dy}(u\rho\xi\eta + v\rho\eta^2 + w\rho\eta\xi) + \frac{d}{dz}(u\rho\xi\zeta + v\rho\eta\zeta + w\rho\zeta^2)
$$

$$
+ \frac{1}{2}\frac{d}{dx}(1 + \beta)\rho(\xi^3 + \xi\eta^2 + \xi\zeta^2) + \frac{1}{2}\frac{d}{dy}(1 + \beta)\rho(\xi^2\eta + \eta^3 + \eta\zeta^2)
$$

$$
+ \frac{1}{2}\frac{d}{dz}(1 + \beta\rho)(\xi^2\zeta + \eta^2\zeta + \zeta^3) = \rho(uX + vY + wZ)
$$

Substituting the values of $\rho X \; \rho Y \; \rho Z$

$$
\frac{1}{2}\rho\frac{\partial}{\partial t}(1 + \beta)(\xi^2 + \eta^2 + \zeta^2) + \rho\xi^2\frac{du}{dx} + \rho\eta^2\frac{dv}{dy} + \rho\zeta^2\frac{dw}{dz} + \rho\eta\zeta\left(\frac{dv}{dz} + \frac{dw}{dy}\right)
$$

$$
+ \rho\zeta\xi\left(\frac{dw}{dx} + \frac{du}{dz}\right) + \rho\xi\eta\left(\frac{du}{dy} + \frac{dv}{dx}\right) + \frac{1}{2}(1 + \beta)\left\{\frac{d}{dx}\rho(\xi^3 + \xi\eta^2 + \xi\zeta^2)\right.
$$

$$
+ \frac{d}{dy}\rho(\xi^2\eta + \eta^3 + \eta\zeta^2) + \frac{d}{dz}\rho(\xi^2\zeta + \eta^2\zeta + \zeta^3)\Big\} = 0
$$

**22. [Draft of Sections "Specific Heat of Unit of Mass at Constant Volume"
and "Specific Heat of Unit of Mass at Constant Pressure" (1867 Memoir)]**

Cambridge University Library, Maxwell manuscripts, box 4.

The total energy of agitation of unit of volume of the medium is $\frac{3}{2}(1 + \beta)\rho$
hence the total energy of agitation of unit of mass is

$$E = \frac{3}{2}(1 + \beta)\frac{p}{\rho}$$

If now additional energy be communicated to it in the form of heat without
altering its density

$$\partial E = \frac{3}{2}(1 + \beta)\frac{\partial p}{\rho} = \frac{3}{2}(1 + \beta)\frac{p}{\rho}\frac{d\theta}{\theta}$$

Hence the specific heat of unit of mass at constant volume is in dynamical
measure

$$\frac{\partial E}{\partial \theta} = \frac{3}{2}(1 + \beta)\frac{p}{\rho\theta}$$

If the gas be now allowed to expand without receiving more heat from without
till the pressure sinks to p the temperature will sink by a quantity $\partial\theta'$ such that

$$\frac{\partial\theta'}{\theta} = \frac{2}{5 + 3\beta}\frac{\partial p}{p} = \frac{2}{5 + 3\beta}\frac{\partial\theta}{\theta}$$

The total change of temperature is therefore $\partial\theta - \partial\theta' = \dfrac{3 + 3\beta}{5 + 3\beta}\partial\theta$ and the
specific heat of unit of mass at constant pressure is

$$\frac{\partial E}{\partial \theta'} = \frac{5 + 3\beta}{2}\frac{p}{\rho\theta}$$

The ratio of the specific heat at constant pressure to that at constant volume is
$\dfrac{5 + 3\beta}{3 + 3\beta}$ a quantity which is generally denoted by the symbol γ. We have then

$$\beta = \frac{5 - 3\gamma}{3\gamma - 3}$$

and $\dfrac{dE}{d\theta} = \dfrac{1}{\gamma - 1}\dfrac{p}{\rho\theta}$ the specific heat at constant volume

$$\frac{dE}{d\theta'} = \frac{\gamma}{\gamma - 1}\frac{p}{\rho\theta} \quad \text{the specific heat at constant pressure}$$

expressions from which the specific heat of air has been calculated by Professor Rankine and found to agree with the values determined experimentally by M. Regnault.

23. [Notes for Section "Determination of the Inequality of Pressure..." and Part of Following Section on Viscosity for 1867 Memoir]

Cambridge University Library, Maxwell manuscripts, box 4.

Let us next determine the variation of the pressure in the direction of x in a simple medium and make $Q = M(u + \xi)^2$ then by equation (   ) we find

$$\frac{\partial Q}{\partial t} = 2k\rho\Theta_2(\eta^2 + \zeta^2 - 2\xi^2) + 2M\xi X$$

whence

$$\frac{\partial \xi^2}{\partial t} + 2\left(\xi^2\rho\frac{du}{dx} + \xi\eta\rho\frac{du}{dy} + \xi\zeta\rho\frac{du}{dz} + \frac{d}{dx}\xi^3\rho + \frac{d}{dy}\xi^2\eta\rho + \frac{d}{dz}\xi^2\zeta\rho\right)$$

$$= \frac{6k}{M}\rho\Theta_2(p - \xi^2\rho)$$

Omitting for the present the terms involving three dimensions in ξ η ρ which refer to conduction of heat and taking the value of $\dfrac{\partial p}{\partial t}$ from equation (   )

$$\frac{\partial}{\partial t}(\rho\xi^2 - p) + 2p\frac{du}{dx} - \frac{2}{3}p\left(\frac{du}{dx} + \frac{dv}{dy} + \frac{dw}{dz}\right) = \frac{6k\rho}{M}\Theta_2(p - \xi^2\rho)$$

⟨the terms involving⟩ putting p for $\rho\xi^2$ and omitting $\xi\eta\rho$ and $\xi\zeta\rho$ in terms not involving the large coefficient $6k\rho\Theta_2$. If the motion is not very violent we may also neglect $\dfrac{\partial}{\partial t}(\rho\xi^2 - p)$ and thus we have

$$\xi^2\rho = p - \frac{M}{9k\rho\Theta_2}p\left(2\frac{du}{dx} - \frac{dv}{dy} - \frac{dw}{dz}\right)$$

with similar expressions for $\eta^2\rho$ and $\zeta^2\rho$. By transformation of coordinates we can easily obtain the expressions for $\xi\eta\rho$, $\eta\zeta\rho$ and $\zeta\xi\rho$. They are of the form

$$\eta\zeta\rho = -\frac{M}{6k\rho\Theta_2}p\left(\frac{dv}{dz} + \frac{dw}{dy}\right)$$

Having thus obtained the values of the pressures in different directions we may substitute in the equation of motion

$$\frac{\partial u}{\partial t} + \frac{d}{dx}(\rho\xi^2) + \frac{d}{dy}(\rho\xi\eta) + \frac{d}{dz}(\rho\xi\zeta) = X$$

which becomes

$$\frac{\partial u}{\partial t} + \frac{dp}{dx} - \frac{pM}{6k\rho\Theta_2}\left\{\frac{d^2u}{dx^2} + \frac{d^2u}{dy^2} + \frac{d^2u}{dz^2} + \frac{1}{3}\frac{d}{dx}\left(\frac{du}{dx} + \frac{dv}{dy} + \frac{dw}{dz}\right)\right\}$$

$$= X\rho$$

This is the equation of motion in the direction of x. The other equations may be written down by symmetry. The form of the equations is identical with that deduced by Poisson[a] from the theory of elasticity by supposing the strain to be constantly relaxed at a given rate and the ratio of the coefficients of $\nabla^2 u$ and $\frac{d}{dx}\frac{1}{\rho}$ [?] agrees with that given by Professor Stokes.[b]

The quantity $\frac{pM}{6k}$ is the coefficient of viscosity or of internal friction and is denoted by μ in the writings of Professor Stokes and in my paper on the Viscosity of Air and Other Gases.

In this expression Θ_2 is a numerical quantity.[c]

k is a quantity depending on the intensity of the action between two molecules at unit of distance. The ratio of p to ρ is proportional to the temperature from absolute zero and is independent of the density. Hence in a given gas, μ is independent of the pressure and proportional to the temperature, as is found by experiment.[d]

Putting $k = (K/8M)^{1/2}$ $p = MN\xi$ $\rho = MN$ σ = specific gravity compared with air

$$\mu = \frac{2^{3/2}}{6\Theta_2}\frac{M^{1/2}}{K^{1/2}}M\xi^2$$

Now $M\xi^2$ is the same for all gases at the same temperature, therefore μ varies as $(M/K)^{1/2}$ for different gases at the same temperature.

a. Maxwell's footnote: Journal de l'Ecole Polytechnique 1829 tom XIII cah. 20 p. 139. The title of the paper was "Memoire sur les Équations Generales de l'Équilibre et du Mouvement des Corps Solids, Élastiques et des Fluides"; it occupies the entire Cahier 20 (174 pp.). The work of Siméon Denis Poisson (1781–1840), one of the leading mathematical physicists of the early nineteenth century, was recently discussed at length in a ten part series of articles by D. H. Arnold, "The Mécanique Physique of Siméon Denis Poisson: The Evolution and Isolation in France of his Approach to Physical Theory (1800–1840)," *Arch. Hist. Exact Sci.* 28 (1983)–29 (1984).

b. Maxwell's footnote: Cambridge Phil. Trans. vol. VIII (1845). See: "On the Theories of the Internal Friction of Fluids in Motion, and of the Equilibrium and Motion of Elastic Solids," *Trans. Cambridge Phil. Soc.* 8 (1845), 287–319; *Mathematical and Physical Papers by the Late Sir George Gabriel Stokes*, second edition with a new preface by C. Truesdell (New York: Johnsone Reprint Corp., 1966), vol. 1, 75–129.

c. This seems to be the quantity denoted by A_2 in the published version of this paper (document III-30; *Scientific Papers*, vol. 2, 41).

d. There is a double dagger here but the footnote is left blank. Presumably Maxwell intended to cite his own paper on the viscosity of air (document III-18).

24. "Encounter of Two Molecules"

Cambridge University Library, Maxwell Collection.

If two molecules act on each other only when at a very small distance apart, and if the kinetic energy of the system is not altered by the encounter [Diagram 1] then if OA represents the velocity of the first and OB that of the second, BA will represent the velocity of A with respect to B. ⟨If we divide AB in G⟩ If G is the centre of gravity of the two molecules when placed at A and at B respectively, OG will represent the velocity of the centre of ⟨gravity matter⟩ mass of the two molecules which is not altered by their mutual action.

During the whole motion, therefore, in whatever manner the velocities of the two molecules ⟨alter the point G in the diagram⟩ represented by the lines OA and OB may alter the ⟨point G⟩ the centre of ⟨inertia⟩ mass, G, ⟨of the points A and B⟩ will remain fixed. ⟨If⟩ We shall also assume that the action between the molecules is such that after the encounter the kinetic energy of the system ⟨remains⟩ is the same as before. ⟨We obtain further We may⟩ The kinetic energy of the system of two molecules may be divided into two parts the first being the kinetic energy of a mass equal to that of the system and having the velocity of its centre of ⟨gravity⟩ mass, and the second being the kinetic energy arising from the motion of the parts of the system relatively to the centre of ⟨inertia⟩ mass.

The first part is necessarily unaffected by any material action of the parts of

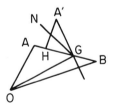

[Diagram 1]

the system. Hence if the whole kinetic energy is the same before and after the encounter the second part must be so.

Now the kinetic energy of the motion relative to the centre of inertia is $\frac{1}{2}(A \cdot \overline{GA}^2 + B \cdot \overline{GB}^2)$ before the encounter or since $A \cdot \overline{GA} = B \cdot \overline{BG}$ the energy is $\frac{1}{2} \frac{A \cdot B}{A + B} \overline{AB}^2$ and if OA', OB' represent the velocities after the encounter the kinetic energy is $\frac{1}{2}(A\overline{GA'}^2 + B\overline{GB'}^2) = \frac{1}{2} \frac{AB}{A + B} \overline{A'B'}^2$. Now the ratio of GA to GB is the same before and after the encounter so that if this part of the kinetic energy remains the same after the encounter we must have

$$AB = A'B' \quad GA = GA' \ \& \ GB = GB'$$

⟨or the line.⟩ The velocity of the molecules relative to the centre of inertia is therefore unaltered and the result of the encounter is therefore completely defined if the *direction* of this relative velocity be given.

⟨This direction is determined by the angle $AGA' = 2\theta$, and the angle, φ, which the plane of AGA' makes with a plane through AG ⟨parallel and a line through⟩ parallel to the axis of x. This plane which may be called the plane of the encounter may be determined by drawing through ⟨the centre of inertia of the two⟩ the line AB a ⟨plane⟩ line parallel to the line joining the molecules at any instant. The plane through these two lines is the plane of the encounter. It is manifest that all ⟨the possible positions⟩ values of the angle φ which determines the angular position of this plane round the line AB are equally probable.⟩

⟨The angle AGA' or 2θ, between the directions of the line GA before and after the encounter depends on the mode in which the force is exerted between the bodies and on their angular momentum about G.⟩

It is easy to see that if encounters take place among a greater number of molecules, their velocities even if originally equal will become unequal for except under conditions which can be only rarely satisfied two molecules having equal velocities before their encounter will have unequal velocities after the encounter.

⟨But though the velocities of different molecules at any instant are unequal and though the same molecule ⟨is continually changing⟩ changes its velocity at every encounter in a manner in which we can perceive no regularity⟩

Every molecule changes both its velocity and its direction of motion at every encounter, so that unless we are supposed to be able to calculate ⟨the exact circumstances of all the molecules the motion⟩ the elements of the motion of every other molecule which it encounters these changes of motion must appear to us very irregular if we follow the course of a single molecule.

⟨If however we adopt the statistical principle of describing the state of motion of the system not by estimating the velocity of particular molecules but by counting the number of molecules whose velocities are within given limits, we meet with a new kind of regularity—the regularity of the averages of large numbers of events.⟩

⟨In order⟩

As long as we have to deal with only two molecules and have all the data of the encounter given us, we can calculate the result of their mutual action, but when we have to deal with millions of molecules each of which has millions of encounters in a second, the complexity of the problem seems to shut out all hope of a legitimate solution.

We are therefore obliged to abandon the strictly kinetic method and to adopt the statistical method.

According to the strict kinetic or historical method as applied to the case before us ⟨the primary subject matter⟩ we follow the whole course of every individual molecule. We ⟨give⟩ arrange our symbols so as to be able to identify every molecule throughout its whole motion.

In passing from the considertion of the motions of individual molecules to that of the medium which consists of multitudes of moving molecules we are forced on account of ⟨the⟩ our limited powers of observation and even of imagination, to abandon the strict dynamical method of tracing the course of every molecule and to adopt the statistical method of dividing the molecules into groups according to some system, and then confining our attention to the number of molecules in each group. ⟨The consequences of our taking this step will⟩ This is a step the philosophical importance of which cannot be over-estimated. It is equivalent to the change from absolute certainty to high probability. ⟨If we admit that⟩ The kinetic theorems by which the motion of a single molecule ⟨depends⟩ is expressed no doubt are founded on axioms absolutely certain but as soon as we lose sight of the individual molecule and assert anything of groups of molecules which are continually exchanging molecules one with another our assertions can lay claim to nothing more than a high probability.

In the first place let us form a group of molecules by confining our attention to those which at a given instant are within a given region bounded by a closed surface of any form but large enough to contain a very great number of molecules.The mass of this group of molecules is the sum of the masses of the individual molecules which at the given instant are within the given region. The only way in which the mass of the group can change is by molecules entering or leaving the given region. The numerical value of the density of the medium within this region is obtained by dividing the number representing the mass by the number representing the volume of the region. ⟨The density so found is not⟩ The density of matter at any mathematical point within the region is [sentence not completed] There is no point within the region at which the actual density of matter has this value for ⟨within a⟩ if the point is within a molecule the density is much greater and if it ⟨lies out⟩ is not within a molecule the density is zero. The density we have obtained is therefore an average density and is the first example of a statistical quantity forming a mixture of several media. If there are several kinds of molecules within the region the mass and the density of each ⟨kind⟩ medium may be estimated separately.

We have thus made our selection of a group of molecules, namely those which happen at a given instant to be within a given region, marked out by an imaginary closed surface which forms the boundary between the given region and external space. We have defined what is meant by the mass of this group of molecules and by the density of tne medium within the given region. We have also defined the momentum of the group, the velocity of the medium and the motion of agitation of the individual molecules. Lastly we have ⟨defined⟩ distinguished the energy of the motion of the medium, and the energy of agitation of the molecules.

⟨The whole of the kinetic theory of gases⟩

On the study of these sets of quantities depends the whole kinetic theory of gases.

We have next to consider the action which goes on between the group of molecules and that which lies outsdie the group. This action takes place through the bounding surface. Let us therefore consider that part of the action which takes place through a small portion of the bounding surface, which, for convenience, we may suppose to be plane.

On account of the motion of the molecules there will be a continual interchange of molecules between the ⟨enclosed⟩ medium within the region bounded by the given surface and the medium outside. If as many molecules pass ⟨through the surface⟩ out of the enclosed space as pass into it the exchange ⟨may be said to be at par⟩ of molecules constitutes what has been called a moveable equilibrium. The state of the medium within the region as to density and other sensible properties remains the same though the individual molecules of which ⟨compose⟩ the medium consists are continually being exchanged for others exactly similar to them.

If we consider the case of a quantity of gas enclosed in a vessel and left to itself till there is no apparent motion or alteration of temperature in any part of the medium, then the exchange of molecules through any surface fixed with respect to the vessel will be equal in all respects in both directions and will constitute a moveable equilibrium as regards the matter, the momentum, and the energy of the moving molecules.

Let us now consider the alteration in the path of a molecule in consequence of the action of another molecule which comes near it in its course.

Let us suppose the two molecules moving with equal momentum in opposite directions so that their centre of gravity is at rest. Let their masses be M_1, M_2, their initial velocities V_1, V_2, and their distances from the centre of gravity r_1, r_2. We shall suppose them initially so distant that ⟨their⟩ force between them vanishes and they are moving sensibly in parallel straight lines at distances b_1, b_2 from the centre of gravity. In consequence of the mutual action between the molecules each will describe a plane curve, the two curves being symmetrical with respect to the centre of gravity, and when the ⟨particles⟩ molecules are again out of reach of their mutual action they will be found moving from one another with velocities V_1, V_2 on straight lines distant b_1, b_2 from the centre of gravity but inclined to the directions of the original lines of motion by an angle 2θ where θ is the angle between the asymptotes and the apse of the orbit.

If the molecules are not mere centres of force but bodies capable of internal motions their paths after the encounter will be different, ⟨but⟩ and their velocities may not be exactly equal before and after but as we do not know the constitution of molecules we shall treat them as if they were mere centres of force.

The angle θ may be calculated when we know V, b, and the law of force where $V = V_1 + V_2 \ \& \ b = b_1 + b_2$.

The elements therefore by which the circumstances of an encounter are completely defined are as follows.

1. The motion of the centre of mass of the two molecules. This motion is not affected by the encounter and the other elements of the encounter are not affected by it. We need not therefore ⟨consider⟩ take it into consideration till we have to do with the ⟨motion⟩ relation of the ⟨two⟩ encountering molecules to other bodies.

⟨⟨The velocity of the molecules relative to their centre of mass of the two molecules.⟩ The direction of this velocity is called the axis of the encounter and its magnitude is determined by the relative velocity of the two molecules. The velocity of the molecule A with respect to B is called the relative velocity of the encounter and its direction is called the axis of the encounter. The distance between⟩

In what follows we shall consider the centre of mass of the two molecules as the origin of the diagram of configurations [Diagram 2] and of that of

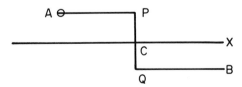

[Diagram 2]

velocities. The lines of motion of the two molecules will then be parallel and in opposite directions.

⟨The velocity of the molecule *A*⟩ with respect to *B* before the encounter is called the relative velocity of the encounter. A line parallel to the direction of this velocity is called the axis of the encounter direction of approach⟩

The velocity of the molecule *A* with respect to *B* before the encounter is called the velocity of approach and the direction is this velocity is called the direction of approach. The velocity of *A* with respect to *B* after the encounter is called the velocity of separation and its direction the direction of separation.

⟨The line *PQ* drawn through *G* the distance⟩

The distance, *PQ* between the lines of motion of the molecules before the encounter is called the arm of approach. The corresponding distance after the encounter is called the arm of separation.

The plane containing the direction of approach and the arm of approach is called the plane of the encounter. ⟨It also contains the⟩ During the whole ⟨motion⟩ encounter the line joining the molecules remains parallel to this plane.

The angle between the direction of approach and the direction of separation is called the deviation.

The direction and velocity of approach is determined when we know the velocities of the two molecules.

To determine the arm of approach we require also to know the relative position of the two molecules ⟨before⟩ at any instant before the encounter.

The plane of the encounter is thus determined. It is manifest that ⟨all positions of this plane passing⟩ of all the planes passing through the direction of approach any one is equally likely to be the plane of the encounter. The angle of deviation depends on the velocity of approach and on the arm of approach. ⟨If the molecules are not symmetrical abo⟩ If the force with which the molecules act on each other does not pass through their centres of mass the angular position of the molecules at the instant of encounter may affect the deviation, but we do not attempt to take account of this kind of irregularity, otherwise than by classing those encounters as of the same kind in which this irregularity as well as the ⟨more⟩ other elements has the same value.

Neglecting the irregularity the velocity of separation is equal to that of approach but its direction is turned through the angle of deviation in the plane of the encounter.

Whatever be the circumstances of an encounter it is always possible to arrange the positions of the molecules for a second encounter such that the velocities of approach ⟨in the one⟩ and separation in the second encounter shall be those of separation and approach in the first.

For if we suppose the whole figure of the encounter turned round through two right angles in the plane of the encounter, and the directions of motion

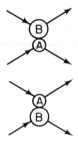

[Diagram 3]

⟨their⟩ reversed, the velocity of approach in the ⟨one⟩ original figure will be equal and parallel to the velocity of separation in the inverted figure and *vice versa* [Diagram 3].

Hence the arm of approach in the one figure is equal and parallel to that in the other figure but it is drawn in the opposite direction.

25. [Heat Conduction for the Case of Varying Temperature and Pressure]

Cambridge University Library, Maxwell Collection.

We have next to determine the rate of variation of the quantity $\xi^3 + \xi\eta^2 + \xi\zeta^2$ in a simple medium.

Putting $Q = M(u + \xi)(u^2 + v^2 + w^2 + 2u\xi + 2v\eta + 2w\zeta + (1 + \beta)(\xi^2 + \eta^2 + \zeta^2)$ and making u, v & w zero after the differentiations and neglecting terms of the form $\xi\eta$ in comparison with those of the form ξ^2 and remembering that terms of the forms ξ^3 and $\xi\eta^2$ are also very small we get

$$N\frac{\delta Q}{\delta t} = -\frac{\sqrt{2}}{M}\sqrt{\frac{K}{M}}(1 + \beta)\Theta_2\rho^2(\xi^3 + \xi\eta^2 + \xi\zeta^2) + 2\rho\xi^2\bar{X}$$

$$+ 2\rho\xi\eta\bar{Y} + 2\rho\xi\zeta Z + (1 + \beta)\rho(\xi^2 + \eta^2 + \zeta^2)\bar{X}$$

Hence equation [space] becomes

$$\rho(1 + \beta)(\xi^2 + \eta^2 + \zeta^2)\frac{\partial u}{\partial t} + 2\rho\xi^2\frac{\partial u}{\partial t} + 2\rho\xi\eta\frac{\partial v}{\partial t} + 2\rho\xi\zeta\frac{\partial w}{\partial t}$$

$$+ \rho\frac{\partial}{\partial t}(1 + \beta)(\xi^3 + \xi\eta^2 + \xi\zeta^2) + (1 + \beta)\rho(\xi^2 + \eta^2 + \zeta^2)$$

$$+ (1 + \beta)\frac{d}{dx}\rho\xi^2(\xi^2 + \eta^2 + \zeta^2)$$

$$= -\frac{\sqrt{2}}{M}\sqrt{\frac{K}{M}}(1 + \beta)\Theta_2\rho^2(\xi^3 + \xi\eta^2 + \xi\zeta^2) + 2\rho\xi^2\bar{X} + 2\rho\xi\eta\bar{Y}$$

$$+ 2\rho\xi\zeta Z + (1 + \beta)\rho(\xi^2 + \eta^2 + \zeta^2)\bar{X}$$

Putting $\rho\dfrac{\partial u}{\partial t} + \dfrac{dp}{dx}$ for ρX and omitting terms in ξ, η, &c.

$$(1 + \beta)\rho\frac{\partial}{\partial t}(\xi^2 + \xi\eta^2 + \xi\zeta^2) + (1 + \beta)\frac{d}{dx}\rho(\xi^4 + \xi^2\eta^2 + \xi^2\zeta^2)$$

$$- (1 + \beta)(\xi^2 + \eta^2 + \zeta^2)\frac{dp}{dx} - 2\xi^2\frac{dp}{dx}$$

$$= -\frac{\sqrt{2}}{M}\sqrt{\frac{K}{M}}\,\Theta_2(1 + \beta)\rho^2(\xi^3 + \xi\eta^2 + \xi\zeta^2)$$

When the motion is steady or when it is not very violent the first term will disappear and the equation may be written

$$3(1 + \beta)\frac{d}{dx}\cdot\frac{p^2}{\rho} - (5 + 3\beta)\frac{p}{\rho}\frac{dp}{dx}$$

$$= -\sqrt{2}\sqrt{\frac{K}{M^3}}\,\Theta_2(1 + \beta)\rho^2(\xi^3 + \xi\eta^2 + \xi\zeta^2)$$

or

$$(1 + \beta)\rho(\xi^3 + \xi\eta^2 + \xi\zeta^2)$$

$$= \sqrt{\frac{M^3}{2K}}\frac{1}{\Theta_2}\frac{p^2}{\rho^3}\left(2\frac{1}{p}\frac{dp}{dx} - 3(1 + \beta)\frac{d\theta}{\theta\,dx}\right)\frac{5\beta}{2}\mu\frac{p}{\rho\theta}$$

$$= \frac{3}{2}\mu\frac{p}{\rho}\left(2\frac{1}{p}\frac{dp}{dx} - 3(1 + \beta)\frac{1}{\theta}\frac{d\theta}{dx}\right)$$

where μ is the coefficient of viscosity. This is the quantity of heat, measured as mechanical energy, which is carried over unit of area in unit of time when the pressure and temperature vary from point to point.

26. [Miscellaneous Notes and Calculations]

Cambridge University Library, Maxwell Collection.

Let us suppose that a material system A, in the state a' (defined as to configuration and motion) comes into mutual action with another system b', and that the result is that after a time t, A is in the state a and B in the state b.

We may suppose that this process may be investigated by purely dynamical processes when all the circumstances are given. If however they are not completely given there are still seven conditions which the states a and b must satisfy namely the equation of energy, the three equations of linear momentum and the three equations of angular momentum.

Let us now suppose that in unit of volume there are $f(a')$ systems in the state a', and $f(b')$ in the state b'. The number of these which in unit of time encounter each other in such a way as to pass into the states a and b may be expressed by

$$C_{(a',b')}f(a')f(b')$$

where C is a quantity which depends on the particular states a' and b'. On account of the reversibility of the path of any motion the number which pass from the states a and b to the states a' and b' will be

$$C_{(a,b)}f(a)f(b)$$

where C is the same quantity as before, that is

$$C_{(a,b)} = C_{(a',b')}$$

It is manifest that there will be an equilibrium $\langle \text{bet} \rangle$ of exchange between the states a, b and $a'b'$ provided

$$f(a')f(b') = f(a)f(b)$$

the relations between a', b' and a, b being such as to satisfy the seven conditions above mentioned.

$$N\frac{\partial \beta \langle ? \rangle (V^2)}{\partial t} + \frac{d}{dx}(\beta\xi^2 V^2 N) + \frac{d}{dy}(\beta\xi\eta V^3 N) + \frac{d}{dr}(\beta\xi\zeta V^2 N)$$

$$= (N\langle ? \rangle (\beta\xi V^2)$$

$$0 + \beta\frac{d}{dx}(\overline{\xi^4} + \overline{\xi^2}(\overline{\eta} + \overline{\zeta^2})\rho = \rho\beta\frac{\delta}{\delta t}\xi V^2$$

$$= -3\rho^2\beta k_1 A_2 \xi V^2 + x(3\xi^2 + \eta^2 + \zeta^2)\rho\beta$$

$$\beta\frac{d}{dx}\frac{5p^2}{\rho\theta}\theta = -3\rho^2\beta k_1\alpha_2\xi V^2 + (\beta X\,5p)/5\beta\frac{dp}{dx}\frac{p}{\rho}$$

$$5\frac{d}{dx}\frac{p}{\rho\theta}p\theta - 5\frac{p}{\rho u}\theta\frac{dp}{dx} = -3\rho^2 k_1 A_2 \xi V^2$$

$$Q = \xi^{\langle ? \rangle} \qquad \frac{d}{dx}\int \xi^2\,dN \qquad NX$$

$$Q = \xi^3 \qquad \frac{d}{dx}\int \xi^4\, dN \qquad NX\, 3\xi^2 \qquad 3\bar{\xi}\frac{d}{dx}\int \xi^2\, dN$$

$$Q = \xi^5 \qquad \frac{d}{dx}\int \xi^6\, dN \qquad NX\, 5\xi^4 \qquad 5\bar{\xi}^4\frac{d}{dx}\int \xi^2\, dN$$

$$\overline{\xi^4} = 3\overline{\xi^2} \qquad \overline{\xi^6} = 15\overline{\xi^2} \qquad \overline{\xi^8} = 7\cdot 5\cdot 3\overline{\xi^2}$$

$$\frac{d}{dx}\xi^n N - \xi^{n-2}\frac{d}{dx}\xi^2\, dN$$

$$= \frac{(n-2)}{2}\frac{d}{dx}\xi^{n-4}\xi$$

$$Q = V^2 M \qquad \xi^2 + \eta^2 + \zeta^2$$

$$N\frac{\partial\langle ?\rangle}{\partial t}\overline{V^2}M + \frac{d}{dx}\overline{\xi V^2}\rho = N_2\xi\frac{d\xi}{dt} + kA_2\rho_1(\eta^2 + \zeta^2 - 2\xi^2)$$

$$N\frac{\partial}{\partial t}\overline{M} + \frac{d}{dx}\overline{\xi}M$$

$$N\frac{\partial}{\partial t}\overline{\xi}M + \frac{d}{dx}\overline{\xi^2}M = NX$$

$$\frac{\partial}{\partial t}\int Q\, dN + \frac{d}{dx}\int \xi Q\, dN = \int \frac{\delta}{\delta t}Q\, dN$$

$$Q = M \qquad \frac{\partial}{\partial t}MN + \frac{d}{dx}\int \xi M\, dN = 0$$

$$Q = \xi M \qquad \frac{\partial}{\partial t}\int \xi M\, dN + \frac{d}{dx}\int \xi^2 M\, dN = \int XM\, dN$$

$$Q = \xi^2 M \qquad \frac{\partial}{\partial t}\int \xi^2 M\, dN + \frac{d}{dx}\int \xi^3 M\, dN = \int X\xi M\, dN + k\xi^2$$

$$Q = \xi^3 M \qquad \frac{\partial}{\partial t}\int \xi^3 M\, dN + \frac{d}{dx}\int \xi^4 M\, dN = \int X\zeta^2 N + k\xi^3$$

[End of sheet]

$$\delta\xi_1 = 0$$

$$\delta\xi_1^2 = -2\sin^2\theta\xi^2 + \tfrac{1}{4}\{4\sin^2\theta\, 2\xi^2 + \sin^2 2\theta(\eta^2 + \zeta^2)\}$$

$$= \tfrac{1}{4}\sin^2 2\theta(\eta^2 + \zeta^2 - 2\xi^2)$$

$$\delta\xi_1^3 = -3\sin^2\theta(2u\xi^2 + \xi^3) + \tfrac{3}{4}\{4\sin^4\theta(2u\xi_1^2 + \xi^3)$$

$$+ \tfrac{1}{2}\sin^2 2\theta(2u(\eta^2 + \zeta^2 + \xi\zeta^2)\}$$

$$= -\tfrac{3}{4}\sin^2 2\theta(2u\xi^2 + \xi^3) + \tfrac{3}{8}\sin^2\theta(2u(\eta^2 + \zeta^2) + \xi\eta^2 + \xi\zeta^2)$$

$$= \tfrac{3}{8}\sin^2 2\theta\{2u(\eta^2 + \zeta^3 - 2\xi^2) + \xi\eta^2 + \xi\zeta^2 - 2\xi^3\}$$

$$\delta\xi^2\eta = \tfrac{1}{8}\sin^2 2\theta\{(2v + \eta)(\eta^2 + \zeta^2 - 2\xi^2) - 6(2u + \xi)\xi\eta\}$$

$$\delta(\xi\eta) = -\tfrac{3}{4}\sin^2 2\theta(\xi\eta)$$

$$\delta(\xi(\xi^2 + \eta^2 + \zeta^2))$$

$$= \tfrac{1}{8}\sin^2 2\theta\left\{2u + \xi)\begin{pmatrix}3(\eta + \zeta^2 - 2\xi^2 \\ + \zeta^2 + \xi^2 - 2\eta^2 \\ + \xi^2 + \eta^2 - 2\xi^2\end{pmatrix}\begin{matrix}- 6(2v + \eta)\xi\eta \\ - 6(2w + \zeta)\xi\zeta\end{matrix}\right\}$$

$$2(\eta^2 + \zeta^2 - 2\xi^2) - 6\xi(2(v\eta + w\zeta) + \eta^2 + \zeta^2)$$

$$= \tfrac{1}{8}\sin^2 2\theta\{4u(\eta^2 + \zeta^2 - 2\xi^2) - 12\xi(v\eta + w\zeta) - 4\xi(\xi^2 + \eta^2 + \zeta^2)\}$$

$$\xi\frac{d}{dx}\,\xi\theta\left\langle\xi\frac{\delta\theta}{\delta t} + \theta\frac{\delta\xi}{\delta t}\right\rangle\xi^2\frac{d\theta}{dx} + \zeta\eta\frac{d\theta}{dy} + \xi\zeta\frac{d\theta}{dt} + \xi\frac{d\theta}{dz} + \theta\frac{(\delta\xi)}{(\delta t)}$$

$$= -\frac{1}{2}\sin^2 2\theta(\xi\theta)$$

$$\xi\theta = -\frac{p}{\frac{1}{2}\sin^2 2\theta}\cdot\frac{d\theta}{dx} \qquad 4\sin^4\theta = 4\sin^2\theta - \sin^2 2\theta$$

$$\langle\sin^6\theta = \sin^4\theta - s\rangle \quad \sin 3\theta = \sin 2\theta\cos\theta + \cos 2\theta\sin\theta$$

$$= 3\sin\theta\cos^2\theta - \sin^3\theta$$

$$\langle\sin^6\theta = 9\sin^2\theta\cos^4\rangle \quad 16\sin^6\theta = 16\sin^4\theta - 4\sin^2\theta\sin^2 2\theta$$

$$= 16\sin^2\theta - 4\sin^2 2\theta - 4\sin^2\theta\sin^2 2\theta$$

$$- (\cos\theta - \cos 3\theta)^2$$

$$- \cos^2\theta - \cos^2 3\theta + \cos 2\theta + \cos 4\theta$$

$$\sin^2\theta + \sin^2 3\theta - 2\sin^2\theta - 2\sin^2 2\theta$$

$$16\sin^6\theta = 15\sin^2\theta - 6\sin^2 2\theta + \sin^2 3\theta$$

$$4\sin^2\theta\sin^2 2\theta = \sin^2\theta + 2\sin^2 2\theta - \sin^2 3\theta$$

$$\delta\xi_1^3 = +3m\,2\sin^2\theta(\xi_2 - \xi_1)\xi_1^2 + 3m^2\{(4\sin^2\theta - \sin^2 2\theta)(\xi_2 - \xi_1)^2\xi_1$$

$$+ \tfrac{1}{2}\sin^2 2\theta\xi_1((\eta_1 - \eta_2)^2 + (\zeta_1 - \zeta_2)^2)\}$$

$$+ \tfrac{1}{2}m^3\{(15\sin^2\theta - 6\sin^2 2\theta + \sin^2 3\theta)(\xi_2 - \xi_1)^3$$

$$+ \tfrac{3}{2}(\sin^2\theta + 2\sin^2 2\theta - \sin^2 3\theta)(\xi_2 - \xi_1)((\eta_1 - \eta_2)^2 + (\zeta_1 - \zeta_2)^2)\}$$

$$\partial_{\xi_1} V_1^2 = 3m\, 2\sin^2\theta\{\tfrac{1}{3}(\xi_2 - \xi_1)V_1^2 + \tfrac{2}{3}\xi_1(V_{12} - V_1^2)\}$$

$$+ 3m^2(4\sin^2\theta - \sin^2 2\theta)(\tfrac{1}{3}\xi_1(V_1^2 + v_2^2 - 2V_{12})$$

$$+ \tfrac{2}{3}(\xi_2 - \xi_1)(V_{12} - V_1^2)\} + \tfrac{3}{2}m^2\sin^2 2\theta\,\tfrac{2}{3}\xi_1(V_1^2 + V_2^2 - 2V_1 V_2)$$

$$+ \tfrac{1}{2}m^3(15\sin^2\theta - 6\sin^2 2\theta + \sin^2 3\theta)(\xi_2 - \xi_1)(V_1^2 + V_2^2 - 2V_{12})$$

$$+ \tfrac{1}{2}m^3(\sin^2\theta + 2\sin^2 2\theta - \sin^2 3\theta)(\xi_2 - \xi_1)(V_1^2 + V_2^2 - 2V_1 V_2)$$

$$= \sin^2\theta\{2m\{(\xi_2 - 3\xi_1)V_1^2 + 2\xi_1 V_{12}\} + 4m^2\{(3\xi_1 - 2\xi_2)V_1^2 + \xi_1 V_2^2$$

$$+ (4\xi_1 - 1\xi_2)V_{12}\} + 8m^3(\xi_2 - \xi_1)(V_1^2 + V_2^2 - 2V_{12})$$

$$+ \sin^2 2\theta\{-m^2((3\xi_1 - 2\xi_2)V_1^2 + \xi_1 V_2^2 - (4\xi_1 - 2\xi_2)V_{12})$$

$$- 2m^3(\xi_2 - \xi_1)(V_1^2 + V_2^2 - 2V_{12} + \xi_1 V_1^2 + \xi_1 V_2^2 - 2\xi_1 V_{12}$$

$$m^2\{2(\xi_2 - \xi_1)(V_1^2 - V_{12})\sin^2\theta\{V_1^2\{(-6m + 12m^2 - 8m^3)\xi_1$$

$$+ (2m - 8m^2 + 8m^3)\xi_2\} + V_2^2\{(4m^2 - 8m^3)\xi_1 + 8m^3\xi_2\}$$

$$+ V_{12}\langle 56m\rangle 4m - 16m^2 + 16m^3)\xi_1 + (8m^2 - 16m^3)\xi_2)$$

φ	$\sqrt{\log \sin 2\varphi}$ u	$\sqrt{\log \cos 2\varphi}$		3.536274	518			
30	9.9687653	9.849485	0.226793[a] 3.612552	0.076278 8682	1.1920	68° 18'	40978	68.18
32	9.976835	9.820921	0.231173 3.588368	0.052094 1043	1.12744	64° 36	387587	64.36
34	9.983588	0.786788	0.235879 3.558941	0.022667 638 52	1.05358	60° 22	362194	60.22
36	9.989103	9.744991	0.240923	9.985914 3.522188	96809		332801	55.28
38	9.993452	9.691838	0.246315	9.937153 3.473327	86528		297390	49.34
40	9.996676	9.61835	0.252068	9.871903 3.408177	74457		255963	42.40
42	9.998807	9.509617	0.258197	9.767814 3.304088	58589		201413	33.34

θ	2θ	φ	log sin θ	1/sin 2φ	log sin 2φ			√A₁	√A₂
21.42	43.24	30	9.567904	0.062469	9.837012	9.630373	9.899481	.42695	.79338
25.24	50.48	32	9.632392	0.046340	9.889271	9.678732	9.935611	.47723	.86221
29.38	59.16	34	9.694120	0.032834	9.934274	9.726954	9.967108	.53328	.92706
34.32	69.04	36	9.753495	0.021793	9.970345	9.775288	9.992138	.59606	.98206
40.26	80.52	38	9.911952	0.013096	9.994459	9.825038	0.007555	.66840	1.01755
47.20	94.40	40	9.866470	0.006649	9.998558	9.873119	0.005207	.74665	1.01206
56.26	112.52	42	9.920771	0.002386	9.964454	9.923157	9.966840	.83783	.92649

30	9.260746	9.79862	.182283	.629451	1	$\dfrac{3h}{10}(u_1 + u_3 + u_5 + u_7 + 5(u_2 + u_6) + 6u_6)$
32	9.357464	9.871222	.227750	.743400	2	
34	9.453908	9.934216	.284386	.859440	3	
36	9.550576	9.984276	.355285	.96440	4	
38	9.650076	0.015110	.446762	⟨.858380⟩		
40	9.746238	0.010414	.557491	1.035405	5	
42	9.986314	9.933680	.701963	1.024270	6	
				.858380	7	

	A_1	A_2			A_1	A_2
u_1	.182283	.629451	u_1	.002696	000000	000000	000000
u_3	.284386	.859440	$5u_2$	.073200	013480	013480	053925
u_5	.446762	1.035405	u_3	.027814	$0\langle?\rangle81$	014640	.036892
u_7	.701963	.858380	$6u_4$	.280045		.166884	.662904
$5u_2$	1.138750	3.717000	u_5	.103254		.056009	.218852
$5u_6$	2.787455	5.121350	$5u_6$			.516265	1.939940
$6u_4$	2.131710	5.786649	u_7			.182283	.629451
30° to 42°	7.673309	18.007666	$h=2$		0° to 30°	.949561	3.541964	$h=5$

	A_1	A_2			A_1	A_2
u_1	0.70196	0.85838				
$5u_2$	3.72375	3.81155	0 to 30		4.757805	17.709820
u_3	.78872	.67868	30 to 42		15.346618	36.015332
$6u_4$	5.01858	3.32964			8.414980	5.090790
u_5	.88745	.40338	$A_1 = \dfrac{4\pi^2\sqrt{2}}{180}\cdot$	28.50940	58.815942	$\dfrac{\pi^2\sqrt{\langle?\rangle}2}{180} = A_2$
u_6	4.70950	1.09995	log	1.4549881	1.7694048
u_7	1.00000	0.00000		1.144815	1.144814
2° to 45°	16.82996	10.18158	$\pi^2\sqrt{2}$	1.599803	2.914310
			90	39.7929	820.938
			$h=1/2$	4.43252	9.12153
			$A_1 =$	8.86504	$A_2 =$	4.56076

a. See the next two documents for the explanation of these numbers.

27. Letter from Maxwell to Peter Guthrie Tait, April 4, 1866

Cambridge University Library, Maxwell Collection, Maxwell-Tait
Correspondence.

Glenlair

Dalbeattie

1866 April 4

Dear Tait

I have not access to Legendre here. Could you get for me or ask one of your
students to get for me the following values of $\log F_c$ or

$$\int_0^{\pi/2} \frac{d\psi}{\sqrt{1 - \sin^2 \varphi \sin^2 \psi}} \text{ for these values of } \varphi$$

$$
\begin{array}{ll}
 & \log F_c \\
\varphi = 30° & 0.226793 \\
\quad 32° & \\
\quad 34° & \\
\quad 36° & \\
\quad 38° & \\
\quad 40° & 0.252068 \\
\quad 42° & \\
\end{array}
$$

I have put down the values for 30° & 40° that there may be no mistake as to
the table of F_c. It is the 1st or 2nd table of Complete Functions. I have got all
the values required except these which occur in a place I want more accuracy.

I hope to hear soon of you and Thomson coming out.[a] If you do not
come out soon I shall not be able to tickle the Questionists next Jan[ry] with
the Scotch School in a lawful manner.[b] I suppose you know that Laplace's
Coeffts[c] and Fig $\oplus$[d] are lawful.[e] If you were out I could set things that I can
only set now with ten lines of explanation or problem upon problem, that is
if Thomson quotes correct in his papers.

The dynamical theory of Viscosity of Gases, Conduction of Heat in do[f]
and interdiffusion of do[f] with absolute measures of most things will soon be
out. That is what F_c is for.[g]

Stewart is buzzing away with the Chimaera at Kew.[h] He should have a
sulphuric acid vacuum gauge such as you and Andrews used to observe
changes of pressure in the rarified air (indicating changes of temp. due to
friction of air) which will be proportionally greater as the air is rarer.

Yours truly,

J. Clerk Maxwell

a. This refers to the publication of Thomson and Tait's *Treatise of Natural Philosophy*. Maxwell's hopes were doomed to disappointment: over a year was to pass before the appearance of the work.

b. "Tickle the Questionists next January," i.e., set questions for the Mathematical Tripos examination of January 1867, for which Maxwell and Joseph Wolstenholme (1829–1891) were the examiners.

c. "Laplace's Coefficients" was the common name then for spherical harmonics. The term spherical harmonic was introduced in fact by Thomson and Tait in their treatise.

d. Figure of the earth.

e. This refers to a recent change in the rules for the Mathematical Tripos which allowed the introduction of questions on physical subjects. Maxwell's co-option as an examiner was in part a consequence and in part a cause of the change of rules.

f. ditto, i.e., gases.

g. The above integral, the complete elliptic function of the first kind, is needed to compute the angle of deflection in an encounter of particles interacting with forces inversely as the fifth power of the distance. See document III-30; *Scientific Papers*, vol. 2, 41–42.

h. Stewart's Chimaera: this refers to the experiment started by Balfour Stewart (1828–1887) and continued for several years by Stewart and Tait to look for friction in the ether by rotating a disk at high speed in a vacuum. See the papers by Stewart and Tait in *Proc. R. Soc. London*, 1865, 1867, and 1873. Another card from Maxwell to Tait calls it Stewart's "Bombylaris Chimaera in vacuo." Balfour Stewart was the Director of the Observatory at Kew. Tait's reply will appear in our next volume.

28. Letter from Peter Guthrie Tait to Maxwell, April 6, 1866

Tait Letters, Cambridge University.

Coll. Library, Edin[r]
6/4/66

ϕ	$\log F'$	Diff I	For 0°.1 II	III
30°	0.226 793 259 758	211 349 731	796 387	1060
32	0.231 172 806 867	227 486 050	818 789	1188
34	0.235 879 485 458	244 095 463	843 883	1332
36	0.240 923 287 876	261 235 001	872 015	1492
38	0.246 315 415 669	278 969 140	903 590	1677
40	0.252 068 441 749	297 371 255	939 099	1886
42	0.258 196 504 876	316 525 426	979 129	2133

Dear Maxwell,

There they are, as large as life.[a] Legendre gives them for every tenth of a degree so you may get any amount more if you want them.

Thomson & I will be out certainly in May—& you will find ample materials to justify you in giving P_2 & Fig. $\oplus$.[b] But perhaps you would like a few sheets now—if so, say so.

I'll take Stewart to task about the Chimaera next week—He is here, but too busy at present.

Yours truly,

P. G. Tait[c]

$$A_1 = 4\pi \int \sqrt{2 \frac{\sin^2 \theta}{\sin^2 2\varphi}} \, d\varphi$$

π	0.497 150
10860	4 033 424
	$\overline{4\,463\,726}$
	3.536 274

$$a_1 = \int \langle \sin^2 \theta \rangle \sin 2\phi \frac{180}{\pi} \, d\phi$$

multiply by $\dfrac{\pi\sqrt{2}}{180}$

a. Values of the complete elliptic function of the first kind, requested by Maxwell in his letter of April 4, 1866 (preceding document). See p. 410 for their use in calculation. Tait has written sideways to the right of column III: "Splendid exercise in interpolation. When are we (if ever) to see you in Edinburgh?"

b. Laplace coefficients (P_2) and figure of the Earth (symbol $\oplus$) mentioned in letter of April 4 as possible examination questions.

c. The following calculation is written on the back of the letter, presumably by Maxwell. 0.497150 is the logarithm of π. For the explanation of A_1, see document III-30; *Scientific Papers*, vol. 2, 41.

29. "On the Dynamical Theory of Gases" [Abstract]

Proceedings of the Royal Society of London 15 (1867), 167–171; *Phil. Magazine* [4] 32 (1866), 390–393.

Gases in this theory are supposed to consist of molecules in motion, acting on one another with forces which are insensible, except at distances which are small in comparison with the average distance of the molecules. The path of each molecule is therefore sensibly rectilinear, except when two molecules come within a certain distance of each other, in which case the direction of motion is rapidly changed, and the path becomes again sensibly rectilinear as soon as the molecules have separated beyond the distance of mutual action.

Each molecule is supposed to be a small body consisting in general of parts capable of being set into various kinds of motion relative to each other, such as rotation, oscillation, or vibration, the amount of energy existing in this form bearing a certain relation to that which exists in the form of the agitation of the molecules among each other.

The mass of a molecule is different in different gases, but in the same gas all the molecules are equal.

The pressure of the gas is on this theory due to the impact of the molecules on the sides of the vessel, and the temperature of the gas depends on the velocity of the molecules.

The theory as thus stated is that which has been conceived, with various degrees of clearness, by D. Bernoulli, Le Sage and Prevost, Herapath, Joule, and Krönig, and which owes its principal developments to Professor Clausius. The action of the molecules on each other has been generally assimilated to that of hard elastic bodies, and I have given some application of this form of the theory to the phenomena of viscosity, diffusion, and conduction of heat in the Philosophical Magazine for 1860. M. Clausius has since pointed out several errors in the part relating to conduction of heat, and the part relating to diffusion also contains errors. The dynamical theory of viscosity in this form has been investigated by M. O. E. Meyer, whose experimental researches on the viscosity of fluids have been very extensive.

In the present paper the action between the molecules is supposed to be that of bodies repelling each other at a distance, rather than of hard elastic bodies acting by impact; and the law of force is deduced from experiments on the viscosity of gases to be that of the inverse fifth power of the distance, any other law of force being at variance with the observed fact that the viscosity is proportional to the absolute temperature. In the mathematical application of the theory, it appears that the assumption of this law of force leads to a great simplification of the results, so that the whole subject can be treated in a more general way than has hitherto been done.

I have therefore begun by considering, first, the mutual action of two molecules; next that of two systems of molecules, the motion of all the molecules in each system being originally the same. In this way I have determined the rate of variation of the mean values of the following functions of the velocity of molecules of the first system:—

α, the resolved part of the velocity in a given direction.

β, the square of this resolved velocity.

γ, the resolved velocity multiplied by the square of the whole velocity.

It is afterwards shown that the velocity of translation of the gas depends on α, the pressure on β, and the conduction of heat on γ.

The final distribution of velocities among the molecules is then considered, and it is shown that they are distributed according to the same law as the errors are distributed among the observations in the theory of "Least Squares;" and that if several systems of molecules act on one another, the average *vis viva* of each molecule is the same, whatever be the mass of the molecule. The demonstration is of a more strict kind than that which I formerly gave, and this is the more necessary, as the "Law of Equivalent Volumes," so important in the chemistry of gases, is deduced from it.

The rate of variation of the quantities α, β, γ in an element of the gas is then considered, and the following conclusions are arrived at.

(α) 1st. In a mixture of gases left to itself for a sufficient time under the action of gravity, the density of each gas at any point will be the same as if the other gases had not been present.

2nd. When this condition is not fulfilled, the gases will pass through each other by diffusion. When the composition of the mixed gases varies slowly from one point to another, the velocity of each gas will be so small that the effects due to inertia may be neglected. In the quiet diffusion of two gases, the volume of either gas diffused through unit of area in unit of time is equal to the rate of diminution of pressure of that gas as we pass in the direction of the normal to the plane, multiplied by a certain coefficient, called the coefficient of interdiffusion of these two gases. This coefficient must be determined experimentally for each pair of gases. It varies directly as the square of the absolute temperature, and inversely as the total pressure of the mixture. Its value for carbonic acid and air, as deduced from experiments given by Mr. Graham in his paper on the Mobility of Gases,[*a] is

$$D = 0.0235,$$

the inch, the grain, and the second being units. Since, however, air is itself a mixture, this result cannot be considered as final, and we have no experiments from which the coefficient of interdiffusion of two pure gases can be found.

3rd. When two gases are separated by a thin plate containing a small hole, the rate at which the composition of the mixture varies in and near the hole will depend on the thickness of the plate and the size of the hole. As the thickness of the plate and the diameter of the hole are diminished, the rate of variation will increase, and the effect of the mutual action of the molecules of the gases in impeding each other's motions will diminish relatively to the moving force due to the variation of pressure. In the limit when the dimensions of the hole are indefinitely small, the velocity of either gas will be the same as if the other gas were absent. Hence the volumes diffused under equal pressures will be inversely as the square roots of the specific gravities of the gases, as was first established by Graham[tb]; and the quantity of a gas which passes through a thin plug into another gas will be nearly the same as that which passes into a vacuum in the same time.

(β) By considering the variation of the total energy of motion of the molecules, it is shown that,

1st. In a mixture of two gases the mean energy of translation will become the same for a molecule of either gas. From this follows the law of Equivalent Volumes, discovered by Gay-Lussac from chemical considerations; namely, that equal volumes of two gases at equal pressures and temperatures contain the same numbers of molecules.

2nd. The law of cooling by expansion is determined.

3rd. The specific heats at constant volume and at constant pressure are determined and compared. This is done merely to determine the value of a constant in the dynamical theory for the agreement between theory and experiment with respect to the values of the two specific heats, and their ratio is a consequence of the general theory of thermodynamics, and does not depend on the mechanical theory which we adopt.

4th. In quiet diffusion the heat produced by the interpenetration of the gases is exactly neutralized by the cooling of each gas as it passes from a dense to a rare state in its progress through the mixture.

5th. By considering the variation of the difference of pressures in different directions, the coefficient of viscosity or internal friction is determined and the equations of motion of the gas are formed. These are of the same form as those obtained by Poisson by conceiving an elastic solid the strain on which is continually relaxed at a rate proportional to the strain itself.

As an illustration of this view of the theory, it is shown that any strain existing in air at rest would diminish according to the values of an exponential term the modulus of which is $\frac{1}{5,100,000,000}$ second, an excessively small time, so that the equations are applicable, even to the case of the most acute audible sounds, without any modification on account of the rapid change of motion.

This relaxation is due to the mutual deflection of the molecules from their paths. It is then shown that if the displacements are instantaneous, so that no time is allowed for relaxation, the gas would have an elasticity of form, or "rigidity," whose coefficient is equal to the pressure.

It is also shown that if the molecules were mere points, not having any mutual action, there would be no such relaxation, and that the equations of motion would be those of an elastic solid, in which the coefficient of cubic and linear elasticity have the same ratio as that deduced by Poisson from the theory of molecules at rest acting by central forces on one another. This coincidence of the results of two theories so opposite in their assumptions is remarkable.

6th. The coefficient of viscosity of a mixture of two gases is then deduced from the viscosity of the pure gases, and the coefficient of interdiffusion of the two gases. The latter quantity has not as yet been ascertained for any pair of pure gases, but it is shown that sufficiently probable values may be assumed, which being inserted in the formula agree very well with some of the most remarkable of Mr. Graham's experiments on the Transpiration of Mixed Gases*.c The remarkable experimental result that the viscosity is independent of the pressure and proportional to the absolute temperature is a necessary consequence of the theory.

(γ) The rate of conduction of heat is next determined, and it is shown,

1st. That the final state of a quantity of gas in a vessel will be such that the temperature will increase according to a certain law from the bottom to the top.[d] The atmosphere, as we know, is colder above. This state would be produced by winds alone, and is no doubt greatly increased by the effects of radiation. A perfectly calm and sunless atmosphere would be coldest below.

2nd. The conductivity of a gas for heat is then deduced from its viscosity, and found to be

$$\frac{5}{3} \frac{1}{\gamma - 1} \frac{p_0}{\rho_0 \theta_0} \frac{\mu}{S},$$

where γ is the ratio of the two specific heats, p_0 the pressure, and ρ_0 the density of the standard gas at absolute temperature θ_0, S the specific gravity of the gas in question, and μ its viscosity. The conductivity is, like the viscosity, independent of the pressure and proportional to the absolute temperature. Its value for air is about 3500 times less than that of wrought iron, as determined by Principal Forbes. Specific gravity is .0069.

For oxygen, nitrogen, and carbonic oxide, the theory gives the conductivity equal to that of air. Hydrogen according to the theory should have a conductivity seven times that of air, and carbonic acid about 7/9 of air.

a. The asterisk refers to a footnote at the bottom of the page: *Philosophical Transactions*, 1863. See document III-14, note c.

b. This refers to a footnote at the bottom of the page: "On the Law of the Diffusion of Gases," *Transactions of the Royal Society of Edinburgh*, vol. xii. (1831).

c. Footnote at the bottom of the page: *Philosophical Transactions*, 1846. (See document III-14, note b.)

d. This result is incorrect. See the "Addition made December 17, 1866" near the end of document III-30; *Scientific Papers*, vol. 2, 75–76.

30. "On the Dynamical Theory of Gases"

Phil. Trans. R. Soc. London 157 (1867), 49–88; *Phil. Mag.* [4] 35 (1868), 129–145, 185–217; *Scientific Papers*, vol. 2, 26–78.

[From the *Philosophical Transactions*, Vol. CLVII.]

XXVIII.   *On the Dynamical Theory of Gases.*

(Received May 16,—Read May 31, 1866.)

THEORIES of the constitution of bodies suppose them either to be continuous and homogeneous, or to be composed of a finite number of distinct particles or molecules.

In certain applications of mathematics to physical questions, it is convenient to suppose bodies homogeneous in order to make the quantity of matter in each differential element a function of the co-ordinates, but I am not aware that any theory of this kind has been proposed to account for the different properties of bodies. Indeed the properties of a body supposed to be a uniform *plenum* may be affirmed dogmatically, but cannot be explained mathematically.

Molecular theories suppose that all bodies, even when they appear to our senses homogeneous, consist of a multitude of particles, or small parts the mechanical relations of which constitute the properties of the bodies. Those theories which suppose that the molecules are at rest relative to the body may be called statical theories, and those which suppose the molecules to be in motion, even while the body is apparently at rest, may be called dynamical theories.

If we adopt a statical theory, and suppose the molecules of a body kept at rest in their positions of equilibrium by the action of forces in the directions of the lines joining their centres, we may determine the mechanical properties of a body so constructed, if distorted so that the displacement of each molecule is a function of its co-ordinates when in equilibrium. It appears from the mathematical theory of bodies of this kind, that the forces called into play by a small change of form must always bear a fixed proportion to those excited by a small change of volume.

Now we know that in fluids the elasticity of form is evanescent, while that of volume is considerable. Hence such theories will not apply to fluids. In solid bodies the elasticity of form appears in many cases to be smaller in proportion to that of volume than the theory gives*, so that we are forced to give up the theory of molecules whose displacements are functions of their co-ordinates when at rest, even in the case of solid bodies.

The theory of moving molecules, on the other hand, is not open to these objections. The mathematical difficulties in applying the theory are considerable, and till they are surmounted we cannot fully decide on the applicability of the theory. We are able, however, to explain a great variety of phenomena by the dynamical theory which have not been hitherto explained otherwise.

The dynamical theory supposes that the molecules of solid bodies oscillate about their positions of equilibrium, but do not travel from one position to another in the body. In fluids the molecules are supposed to be constantly moving into new relative positions, so that the same molecule may travel from one part of the fluid to any other part. In liquids the molecules are supposed to be always under the action of the forces due to neighbouring molecules throughout their course, but in gases the greater part of the path of each molecule is supposed to be sensibly rectilinear and beyond the sphere of sensible action of the neighbouring molecules.

I propose in this paper to apply this theory to the explanation of various properties of gases, and to shew that, besides accounting for the relations of pressure, density, and temperature in a single gas, it affords a mechanical explanation of the known chemical relation between the density of a gas and its equivalent weight, commonly called the Law of Equivalent Volumes. It also explains the diffusion of one gas through another, the internal friction of a gas, and the conduction of heat through gases.

The opinion that the observed properties of visible bodies apparently at rest are due to the action of invisible molecules in rapid motion is to be found in Lucretius. In the exposition which he gives of the theories of Democritus as modified by Epicurus, he describes the invisible atoms as all moving downwards with equal velocities, which, at quite uncertain times and places, suffer an imperceptible change, just enough to allow of occasional collisions taking place

---

* In glass, according to Dr Everett's second series of experiments (1866), the ratio of the elasticity of form to that of volume is greater than that given by the theory. In brass and steel it is less.— March 7, 1867.

28   THE DYNAMICAL THEORY OF GASES.

between the atoms. These atoms he supposes to set small bodies in motion by an action of which we may form some conception by looking at the motes in a sunbeam. The language of Lucretius must of course be interpreted according to the physical ideas of his age, but we need not wonder that it suggested to Le Sage the fundamental conception of his theory of gases, as well as his doctrine of ultramundane corpuscles.

Professor Clausius, to whom we owe the most extensive developments of the dynamical theory of gases, has given* a list of authors who have adopted or given countenance to any theory of invisible particles in motion. Of these, Daniel Bernoulli, in the tenth section of his *Hydrodynamics*, distinctly explains the pressure of air by the impact of its particles on the sides of the vessel containing it.

Clausius also mentions a book entitled *Deux Traités de Physique Mécanique*, publiés par Pierre Prevost, comme simple Éditeur du premier et comme Auteur du second, Genève et Paris, 1818. The first memoir is by G. Le Sage, who explains gravity by the impact of "ultramundane corpuscles" on bodies. These corpuscles also set in motion the particles of light and various ethereal media, which in their turn act on the molecules of gases and keep up their motions. His theory of impact is faulty, but his explanation of the expansive force of gases is essentially the same as in the dynamical theory as it now stands. The second memoir, by Prevost, contains new applications of the principles of Le Sage to gases and to light. A more extensive application of the theory of moving molecules was made by Herapath†. His theory of the collisions of perfectly hard bodies, such as he supposes the molecules to be, is faulty, inasmuch as it makes the result of impact depend on the absolute motion of the bodies, so that by experiments on such hard bodies (if we could get them) we might determine the absolute direction and velocity of the motion of the earth‡. This author, however, has applied his theory to the numerical results of experiment in many cases, and his speculations are always ingenious, and often throw much real light on the questions treated. In particular, the theory of temperature and pressure in gases and the theory of diffusion are clearly pointed out.

---

* Poggendorff's *Annalen*, Jan. 1862. Translated by G. C. Foster, B.A., *Phil. Mag.* June, 1862.

† *Mathematical Physics*, &c., by John Herapath, Esq. 2 vols. London: Whittaker and Co., and Herapath's *Railway Journal* Office, 1847.

‡ *Mathematical Physics*, &c., p. 134.

Dr Joule* has also explained the pressure of gases by the impact of their molecules, and has calculated the velocity which they must have in order to produce the pressure observed in particular gases.

It is to Professor Clausius, of Zurich, that we owe the most complete dynamical theory of gases. His other researches on the general dynamical theory of heat are well known, and his memoirs *On the kind of Motion which we call Heat*, are a complete exposition of the molecular theory adopted in this paper. After reading his investigation† of the distance described by each molecule between successive collisions, I published some propositions‡ on the motions and collisions of perfectly elastic spheres, and deduced several properties of gases, especially the law of equivalent volumes, and the nature of gaseous friction. I also gave a theory of diffusion of gases, which I now know to be erroneous, and there were several errors in my theory of the conduction of heat in gases which M. Clausius has pointed out in an elaborate memoir on that subject §.

M. O. E. Meyer‖ has also investigated the theory of internal friction on the hypothesis of hard elastic molecules.

In the present paper I propose to consider the molecules of a gas, not as elastic spheres of definite radius, but as small bodies or groups of smaller molecules repelling one another with a force whose direction always passes very nearly through the centres of gravity of the molecules, and whose magnitude is represented very nearly by some function of the distance of the centres of gravity. I have made this modification of the theory in consequence of the results of my experiments on the viscosity of air at different temperatures, and I have deduced from these experiments that the repulsion is inversely as the *fifth* power of the distance.

If we suppose an imaginary plane drawn through a vessel containing a great number of such molecules in motion, then a great many molecules will cross the plane in either direction. The excess of the mass of those which traverse the plane in the positive direction over that of those which traverse it in the negative direction, gives a measure of the flow of gas through the plane in the positive direction.

---

* *Some Remarks on Heat and the Constitution of Elastic Fluids*, Oct. 3, 1848.
† *Phil. Mag.* Feb. 1859.
‡ "Illustrations of the Dynamical Theory of Gases," *Phil. Mag.* 1860, January and July.
§ Poggendorff, Jan. 1862; *Phil. Mag.* June, 1862.
‖ "Ueber die innere Reibung der Gase" (Poggendorff, Vol. cxxv. 1865).

If the plane be made to move with such a velocity that there is no excess of flow of molecules in one direction through it, then the velocity of the plane is the mean velocity of the gas resolved normal to the plane.

There will still be molecules moving in both directions through the plane, and carrying with them a certain amount of momentum into the portion of gas which lies on the other side of the plane.

The quantity of momentum thus communicated to the gas on the other side of the plane during a unit of time is a measure of the force exerted on this gas by the rest. This force is called the pressure of the gas.

If the velocities of the molecules moving in different directions were independent of one another, then the pressure at any point of the gas need not be the same in all directions, and the pressure between two portions of gas separated by a plane need. not be perpendicular to that plane. Hence, to account for the observed equality of pressure in all directions, we must suppose some cause equalizing the motion in all directions. This we find in the deflection of the path of one particle by another when they come near one another. Since, however, this equalization of motion is not instantaneous, the pressures in all directions are perfectly equalized only in the case of a gas at rest, but when the gas is in a state of motion, the want of perfect equality in the pressures gives rise to the phenomena of viscosity or internal friction. The phenomena of viscosity in all bodies may be described, independently of hypothesis, as follows :—

A distortion or strain of some kind, which we may call S, is produced in the body by displacement. A state of stress or elastic force which we may call F is thus excited. The relation between the stress and the strain may be written $F = ES$, where E is the coefficient of elasticity for that particular kind of strain. In a solid body free from viscosity, F will remain $= ES$, and

$$\frac{dF}{dt} = E \frac{dS}{dt} .$$

If, however, the body is viscous, F will not remain constant, but will tend to disappear at a rate depending on the value of F, and on the nature of the body. If we suppose this rate proportional to F, the equation may be written

$$\frac{dF}{dt} = E \frac{dS}{dt} - \frac{F}{T} ,$$

THE DYNAMICAL THEORY OF GASES.   31

which will indicate the actual phenomena in an empirical manner.   For if S be constant,

$$F = ESe^{-\frac{t}{T}},$$

shewing that F gradually disappears, so that if the body is left to itself it gradually loses any internal stress, and the pressures are finally distributed as in a fluid at rest.

If $\dfrac{dS}{dt}$ is constant, that is, if there is a steady motion of the body which continually increases the displacement,

$$F = ET\frac{dS}{dt} + Ce^{-\frac{t}{T}},$$

shewing that F tends to a constant value depending on the rate of displacement.   The quantity ET, by which the rate of displacement must be multiplied to get the force, may be called the coefficient of viscosity.   It is the product of a coefficient of elasticity, E, and a time T, which may be called the "time of relaxation" of the elastic force.   In mobile fluids T is a very small fraction of a second, and E is not easily determined experimentally.   In viscous solids T may be several hours or days, and then E is easily measured.   It is possible that in some bodies T may be a function of F, and this would account for the gradual untwisting of wires after being twisted beyond the limit of perfect elasticity.   For if T diminishes as F increases, the parts of the wire furthest from the axis will yield more rapidly than the parts near the axis during the twisting process, and when the twisting force is removed, the wire will at first untwist till there is equilibrium between the stresses in the inner and outer portions.   These stresses will then undergo a gradual relaxation; but since the actual value of the stress is greater in the outer layers, it will have a more rapid rate of relaxation, so that the wire will go on gradually untwisting for some hours or days, owing to the stress on the interior portions maintaining itself longer than that of the outer parts.   This phenomenon was observed by Weber in silk fibres, by Kohlrausch in glass fibres, and by myself in steel wires.

In the case of a collection of moving molecules such as we suppose a gas to be, there is also a resistance to change of form, constituting what may be called the linear elasticity, or "rigidity" of the gas, but this resistance gives way and diminishes at a rate depending on the amount of the force and on the nature of the gas.

Suppose the molecules to be confined in a rectangular vessel with perfectly elastic sides, and that they have no action on one another, so that they never strike one another, or cause each other to deviate from their rectilinear paths. Then it can easily be shewn that the pressures on the sides of the vessel due to the impacts of the molecules are perfectly independent of each other, so that the mass of moving molecules will behave, not like a fluid, but like an elastic solid. Now suppose the pressures at first equal in the three directions perpendicular to the sides, and let the dimensions a, b, c of the vessel be altered by small quantities, δa, δb, δc.

Then if the original pressure in the direction of a was p, it will become

$$p \left(1 - 3 \frac{\delta a}{a} - \frac{\delta b}{b} - \frac{\delta c}{c} \right) ;$$

or if there is no change of volume,

$$\frac{\delta p}{p} = - 2 \frac{\delta a}{a} ,$$

shewing that in this case there is a "longitudinal" elasticity of form of which the coefficient is $2p$. The coefficient of "Rigidity" is therefore $= p$.

This rigidity, however, cannot be directly observed, because the molecules continually deflect each other from their rectilinear courses, and so equalize the pressure in all directions. The rate at which this equalization takes place is great, but not infinite; and therefore there remains a certain inequality of pressure which constitutes the phenomenon of viscosity.

I have found by experiment that the coefficient of viscosity in a given gas is independent of the density, and proportional to the absolute temperature, so that if ET be the viscosity, $ET \propto \dfrac{p}{\rho}$.

But $E = p$, therefore T, the time of relaxation, varies inversely as the density and is independent of the temperature. Hence the number of collisions producing a given deflection which take place in unit of time is independent of the temperature, that is, of the velocity of the molecules, and is proportional to the number of molecules in unit of volume. If we suppose the molecules hard elastic bodies, the number of collisions of a given kind will be proportional to the velocity, but if we suppose them centres of force, the angle of deflection will be smaller when the velocity is greater; and if the force is inversely as the fifth power of the distance, the number of deflections of a given kind will be independent of the velocity. Hence I have adopted this law in making my calculations.

The effect of the mutual action of the molecules is not only to equalize the pressure in all directions, but, when molecules of different kinds are present, to communicate motion from the one kind to the other. I formerly shewed that the final result in the case of hard elastic bodies is to cause the average *vis viva* of a molecule to be the same for all the different kinds of molecules. Now the pressure due to each molecule is proportional to its *vis viva*, hence the whole pressure due to a given number of molecules in a given volume will be the same whatever the mass of the molecules, provided the molecules of different kinds are permitted freely to communicate motion to each other.

When the flow of *vis viva* from the one kind of molecules to the other is zero, the temperature is said to be the same. Hence equal volumes of different gases at equal pressures and temperatures contain equal numbers of molecules.

This result of the dynamical theory affords the explanation of the "law of equivalent volumes" in gases.

We shall see that this result is true in the case of molecules acting as centres of force. A law of the same general character is probably to be found connecting the temperatures of liquid and solid bodies with the energy possessed by their molecules, although our ignorance of the nature of the connexions between the molecules renders it difficult to enunciate the precise form of the law.

The molecules of a gas in this theory are those portions of it which move about as a single body. These molecules may be mere points, or pure centres of force endowed with inertia, or the capacity of performing work while losing velocity. They may be systems of several such centres of force, bound together by their mutual actions, and in this case the different centres may either be separated, so as to form a group of points, or they may be actually coincident, so as to form one point.

Finally, if necessary, we may suppose them to be small solid bodies of a determinate form; but in this case we must assume a new set of forces binding the parts of these small bodies together, and so introduce a molecular theory of the second order. The doctrines that all matter is extended, and that no two portions of matter can coincide in the same place, being deductions from our experiments with bodies sensible to us, have no application to the theory of molecules.

34   THE DYNAMICAL THEORY OF GASES.

The actual energy of a moving body consists of two parts, one due to the motion of its centre of gravity, and the other due to the motions of its parts relative to the centre of gravity. If the body is of invariable form, the motions of its parts relative to the centre of gravity consist entirely of rotation, but if the parts of the body are not rigidly connected, their motions may consist of oscillations of various kinds, as well as rotation of the whole body.

The mutual interference of the molecules in their courses will cause their energy of motion to be distributed in a certain ratio between that due to the motion of the centre of gravity and that due to the rotation, or other internal motion. If the molecules are pure centres of force, there can be no energy of rotation, and the whole energy is reduced to that of translation; but in all other cases the whole energy of the molecule may be represented by $\frac{1}{2}Mv^2\beta$, where β is the ratio of the total energy to the energy of translation. The ratio β will be different for every molecule, and will be different for the same molecule after every encounter with another molecule, but it will have an average value depending on the nature of the molecules, as has been shown by Clausius. The value of β can be determined if we know either of the specific heats of the gas, or the ratio between them.

The method of investigation which I shall adopt in the following paper, is to determine the mean values of the following functions of the velocity of all the molecules of a given kind within an element of volume :—

(a) the mean velocity resolved parallel to each of the coordinate axes;

(β) the mean values of functions of two dimensions of these component velocities ;

(γ) the mean values of functions of three dimensions of these velocities.

The rate of translation of the gas, whether by itself, or by diffusion through another gas, is given by (a), the pressure of the gas on any plane, whether normal or tangential to the plane, is given by (β), and the rate of conduction of heat through the gas is given by (γ).

I propose to determine the variations of these quantities, due, 1st, to the encounters of the molecules with others of the same system or of a different system; 2nd, to the action of external forces such as gravity; and 3rd, to the passage of molecules through the boundary of the element of volume.

I shall then apply these calculations to the determination of the statical cases of the final distribution of two gases under the action of gravity, the equilibrium of temperature between two gases, and the distribution of temperature in a vertical column. These results are independent of the law of force between the molecules. I shall also consider the dynamical cases of diffusion, viscosity, and conduction of heat, which involve the law of force between the molecules.

On the Mutual Action of Two Molecules.

Let the masses of these molecules be M_1, M_2, and let their velocities resolved in three directions at right angles to each other be ξ_1, η_1, ζ_1 and ξ_2, η_2, ζ_2. The components of the velocity of the centre of gravity of the two molecules will be

$$\frac{\xi_1 M_1 + \xi_2 M_2}{M_1 + M_2}, \qquad \frac{\eta_1 M_1 + \eta_2 M_2}{M_1 + M_2}, \qquad \frac{\zeta_1 M_1 + \zeta_2 M_2}{M_1 + M_2}.$$

The motion of the centre of gravity will not be altered by the mutual action of the molecules, of whatever nature that action may be. We may therefore take the centre of gravity as the origin of a system of coordinates moving parallel to itself with uniform velocity, and consider the alteration of the motion of each particle with reference to this point as origin.

If we regard the molecules as simple centres of force, then each molecule will describe a plane curve about this centre of gravity, and the two curves will be similar to each other and symmetrical with respect to the line of apses. If the molecules move with sufficient velocity to carry them out of the sphere of their mutual action, their orbits will each have a pair of asymptotes inclined at an angle $\frac{\pi}{2} - \theta$ to the line of apses. The asymptotes of the orbit of M_1 will be at a distance b_1 from the centre of gravity, and those of M_2 at a distance b_2, where

$$M_1 b_1 = M_2 b_2.$$

The distance between two parallel asymptotes, one in each orbit, will be

$$b = b_1 + b_2.$$

If, while the two molecules are still beyond each other's action, we draw a straight line through M_1 in the direction of the relative velocity of M_1 to M_2, and draw from M_2 a perpendicular to this line, the length of this perpendicular will be b, and the plane including b and the direction of relative motion will be the plane of the orbits about the centre of gravity.

When, after their mutual action and deflection, the molecules have again reached a distance such that there is no sensible action between them, each will be moving with the same velocity relative to the centre of gravity that it had before the mutual action, but the direction of this relative velocity will be turned through an angle 2θ in the plane of the orbit.

The angle θ is a function of the relative velocity of the molecules and of b, the form of the function depending on the nature of the action between the molecules.

If we suppose the molecules to be bodies, or systems of bodies, capable of rotation, internal vibration, or any form of energy other than simple motion of translation, these results will be modified. The value of θ and the final velocities of the molecules will depend on the amount of internal energy in each molecule before the encounter, and on the particular form of that energy at every instant during the mutual action. We have no means of determining such intricate actions in the present state of our knowledge of molecules, so that we must content ourselves with the assumption that the value of θ is, on an average, the same as for pure centres of force, and that the final velocities differ from the initial velocities only by quantities which may in each collision be neglected, although in a great many encounters the energy of translation and the internal energy of the molecules arrive, by repeated small exchanges, at a final ratio, which we shall suppose to be that of 1 to $\beta - 1$.

We may now determine the final velocity of M_1 after it has passed beyond the sphere of mutual action between itself and M_2.

Let V be the velocity of M_1 relative to M_2, then the components of V are

$$\xi_1 - \xi_2, \quad \eta_1 - \eta_2, \quad \zeta_1 - \zeta_2.$$

The plane of the orbit is that containing V and b. Let this plane be inclined ϕ to a plane containing V and parallel to the axis of x; then, since the direction of V is turned round an angle 2θ in the plane of the orbit, while its magnitude remains the same, we may find the value of ξ_1 after the encounter. Calling it ξ'_1,

$$\xi'_1 = \xi_1 + \frac{M_2}{M_1 + M_2} \{ (\xi_2 - \xi_1)\, 2 \sin^2\theta + \sqrt{(\eta_2 - \eta_1)^2 + (\zeta_2 - \zeta_1)^2} \sin 2\theta \cos \phi \} \quad \ldots\ldots(1).$$

There will be similar expressions for the components of the final velocity of M_1 in the other coordinate directions.

If we know the initial positions and velocities of M_1 and M_2 we can determine V, the velocity of M_1 relative to M_2; b the shortest distance between M_1 and M_2 if they had continued to move with uniform velocity in straight lines; and ϕ the angle which determines the plane in which V and b lie. From V and b we can determine θ, if we know the law of force, so that the problem is solved in the case of two molecules.

When we pass from this case to that of two systems of moving molecules, we shall suppose that the time during which a molecule is beyond the action of other molecules is so great compared with the time during which it is deflected by that action, that we may neglect both the time and the distance described by the molecules during the encounter, as compared with the time and the distance described while the molecules are free from disturbing force. We may also neglect those cases in which three or more molecules are within each other's spheres of action at the same instant.

On the Mutual Action of Two Systems of Moving Molecules.

Let the number of molecules of the first kind in unit of volume be N_1, the mass of each being M_1. The velocities of these molecules will in general be different both in magnitude and direction. Let us select those molecules the components of whose velocities lie between

$$\xi_1 \text{ and } \xi_1 + d\xi_1, \quad \eta_1 \text{ and } \eta_1 + d\eta_1, \quad \zeta_1 \text{ and } \zeta_1 + d\zeta_1,$$

and let the number of these molecules be dN_1. The velocities of these molecules will be very nearly equal and parallel.

On account of the mutual actions of the molecules, the number of molecules which at a given instant have velocities within given limits will be definite, so that

$$dN_1 = f_1\left(\xi_1\eta_1\zeta_1\right) d\xi_1 d\eta_1 d\zeta_1 \quad \dots\dots\dots\dots\dots\dots (2).$$

We shall consider the form of this function afterwards.

Let the number of molecules of the second kind in unit of volume be N_2, and let dN_2 of these have velocities between ξ_2 and $\xi_2 + d\xi_2$, η_2 and $\eta_2 + d\eta_2$, ζ_2 and $\zeta_2 + d\zeta_2$, where

$$dN_2 = f_2\left(\xi_2\eta_2\zeta_2\right) d\xi_2 d\eta_2 d\zeta_2.$$

The velocity of any of the dN_1 molecules of the first system relative to the dN_2 molecules of the second system is V, and each molecule M_1 will in the time δt describe a relative path $V\delta t$ among the molecules of the second system. Conceive a space bounded by the following surfaces. Let two cylindrical surfaces have the common axis $V\delta t$ and radii b and $b+db$. Let two planes be drawn through the extremities of the line $V\delta t$ perpendicular to it. Finally, let two planes be drawn through $V\delta t$ making angles ϕ and $\phi+d\phi$ with a plane through V parallel to the axis of x. Then the volume included between the four planes and the two cylindric surfaces will be $Vb\,db\,d\phi\,\delta t$.

If this volume includes one of the molecules M_2, then during the time δt there will be an encounter between M_1 and M_2, in which b is between b and $b+db$, and ϕ between ϕ and $\phi+d\phi$.

Since there are dN_1 molecules similar to M_1 and dN_2 similar to M_2 in unit of volume, the whole number of encounters of the given kind between the two systems will be

$$Vb\,db\,d\phi\,\delta t\,dN_1\,dN_2.$$

Now let Q be any property of the molecule M_1, such as its velocity in a given direction, the square or cube of that velocity or any other property of the molecule which is altered in a known manner by an encounter of the given kind, so that Q becomes Q' after the encounter, then during the time δt a certain number of the molecules of the first kind have Q changed to Q', while the remainder retain the original value of Q, so that

$$\delta Q\,dN_1 = (Q'-Q)\,Vb\,db\,d\phi\,\delta t\,dN_1\,dN_2,$$

or

$$\frac{\delta Q\,dN_1}{\delta t} = (Q'-Q)\,Vb\,db\,d\phi\,dN_1\,dN_2 \quad\ldots\ldots\ldots\ldots\ldots\ldots(3).$$

Here $\dfrac{\delta Q\,dN_1}{\delta t}$ refers to the alteration in the sum of the values of Q for the dN_1 molecules, due to their encounters of the given kind with the dN_2 molecules of the second sort. In order to determine the value of $\dfrac{\delta Q N_1}{\delta t}$, the rate of alteration of Q among all the molecules of the first kind, we must perform the following integrations :—

1st, with respect to ϕ from $\phi=0$ to $\phi=2\pi$.

2nd, with respect to b from $b=0$ to $b=\infty$. These operations will give the results of the encounters of every kind between the dN_1 and dN_2 molecules.

3rd, with respect to dN_2, or $f_2(\xi_2\eta_2\zeta_2)\,d\xi_2 d\eta_2 d\zeta_2$.

4th, with respect to dN_1, or $f_1(\xi_1\eta_1\zeta_1)\,d\xi_1 d\eta_1 d\zeta_1$.

These operations require in general a knowledge of the forms of f_1 and f_2.

1st. *Integration with respect to ϕ.*

Since the action between the molecules is the same in whatever plane it takes place, we shall first determine the value of $\int_0^{2\pi}(Q'-Q)\,d\phi$ in several cases, making Q some function of ξ, η, and ζ.

(a) Let $Q=\xi_1$ and $Q'=\xi'_1$, then

$$\int_0^{2\pi}(\xi'_1-\xi_1)\,d\phi=\frac{M_2}{M_1+M_2}(\xi_2-\xi_1)\,4\pi\sin^2\theta \dots\dots\dots\dots (4).$$

(β) Let $Q=\xi_1^2$ and $Q'=\xi'_1{}^2$,

$$\int_0^{2\pi}(\xi'_1{}^2-\xi_1^2)\,d\phi=\frac{M_2}{(M_1+M_2)^2}[(\xi_2-\xi_1)(M_1\xi_1+M_2\xi_2)\,8\pi\sin^2\theta+M_2\{(\eta_2-\eta_1)^2$$
$$+(\zeta_2-\zeta_1)^2-2(\xi_2-\xi_1)^2\}\pi\sin^2 2\theta]\dots\dots(5).$$

By transformation of coordinates we may derive from this

$$\int_0^{2\pi}(\xi'_1\eta'_1-\xi_1\eta_1)\,d\phi=\frac{M_2}{(M_1+M_2)^2}[\{M_2\xi_2\eta_2-M_1\xi_1\eta_1+\tfrac{1}{2}(M_1-M_2)(\xi_1\eta_2+\xi_2\eta_1)\}\,8\pi\sin^2\theta$$
$$-3M_2(\xi_2-\xi_1)(\eta_2-\eta_1)]\dots\dots(6),$$

with similar expressions for the other quadratic functions of ξ, η, ζ.

(γ) Let $Q=\xi_1(\xi_1^2+\eta_1^2+\zeta_1^2)$, and $Q'=\xi'_1(\xi'_1{}^2+\eta'_1{}^2+\zeta'_1{}^2)$; then putting

$$\xi_1^2+\eta_1^2+\zeta_1^2=V_1^2,\quad \xi_1\xi_2+\eta_1\eta_2+\zeta_1\zeta_2=U,\quad \xi_2^2+\eta_2^2+\zeta_2^2=V_2^2,$$

and $(\xi_2-\xi_1)^2+(\eta_2-\eta_1)^2+(\zeta_2-\zeta_1)^2=V^2$, we find

$$\left.\begin{array}{l}\displaystyle\int_0^{2\pi}(\xi'_1 V'_1{}^2-\xi_1 V_1^2)\,d\phi=\frac{M_2}{M_1+M_2}4\pi\sin^2\theta\,\{(\xi_2-\xi_1)V_1^2+2\xi_1(U-V_1^2)\}\\[2ex]
\qquad+\left(\frac{M_2}{M_1+M_2}\right)^2(8\pi\sin^2\theta-3\pi\sin^2 2\theta)\,2(\xi_2-\xi_1)(U-V_1^2)\\[2ex]
\qquad+\left(\frac{M_2}{M_1+M_2}\right)^2(8\pi\sin^2\theta+2\pi\sin^2 2\theta)\,\xi_1 V^2\\[2ex]
\qquad+\left(\frac{M_2}{M_1+M_2}\right)^3(8\pi\sin^2\theta-2\pi\sin^2 2\theta)\,2(\xi_2-\xi_1)V^2\end{array}\right\}\dots (7).$$

These are the principal functions of ξ, η, ζ whose changes we shall have to consider; we shall indicate them by the symbols α, β, or γ, according as the function of the velocity is of one, two, or three dimensions.

2nd. *Integration with respect to* b.

We have next to multiply these expressions by $b\,db$, and to integrate with respect to b from $b = 0$ to $b = \infty$. We must bear in mind that θ is a function of b and V, and can only be determined when the law of force is known. In the expressions which we have to deal with, θ occurs under two forms only, namely, $\sin^2\theta$ and $\sin^2 2\theta$. If, therefore, we can find the two values of

$$B_1 = \int_0^\infty 4\pi b\,db \, \sin^2\theta, \quad \text{and} \quad B_2 = \int_0^\infty \pi b\,db \, \sin^2 2\theta \quad \dots\dots\dots\dots (8),$$

we can integrate all the expressions with respect to b.

B_1 and B_2 will be functions of V only, the form of which we can determine only in particular cases, after we have found θ as a function of b and V.

Determination of θ *for certain laws of Force.*

Let us assume that the force between the molecules M_1 and M_2 is repulsive and varies inversely as the nth power of the distance between them, the value of the moving force at distance unity being K, then we find by the equation of central orbits,

$$\frac{\pi}{2} - \theta = \int_0^{x'} \frac{dx}{\sqrt{1 - x^2 - \dfrac{2}{n-1}\left(\dfrac{x}{a}\right)^{n-1}}} \quad \dots\dots\dots\dots\dots (9),$$

where $x = \dfrac{b}{r}$, or the ratio of b to the distance of the molecules at a given time: x is therefore a numerical quantity; a is also a numerical quantity and is given by the equation

$$a = b\left\{\frac{V^2 M_1 M_2}{K(M_1 + M_2)}\right\}^{\frac{1}{n-1}} \quad \dots\dots\dots\dots\dots\dots (10).$$

The limits of integration are $x = 0$ and $x = x'$, where x' is the least positive root of the equation

$$1 - x^2 - \frac{2}{n-1}\left(\frac{x}{a}\right)^{n-1} = 0 \quad \dots\dots\dots\dots\dots (11).$$

It is evident that θ is a function of a and n, and when n is known θ may be expressed as a function of a only.

Also
$$b\,db = \left\{ \frac{K\,(M_1 + M_2)}{V^2 M_1 M_2} \right\}^{\frac{2}{n-1}} a\,da \dots\dots\dots\dots (12);$$

so that if we put
$$A_1 = \int_0^\infty 4\pi a\,da \, \sin^2 \theta, \quad A_2 = \int_0^\infty \pi a\,da \, \sin^2 2\theta \dots\dots\dots (13),$$

A_1 and A_2 will be definite numerical quantities which may be ascertained when n is given, and B_1 and B_2 may be found by multiplying A_1 and A_2 by

$$\left\{ \frac{K\,(M_1 + M_2)}{M_1 M_2} \right\}^{\frac{2}{n-1}} V^{\frac{-4}{n-1}}.$$

Before integrating further we have to multiply by V, so that the form in which V will enter into the expressions which have to be integrated with respect to dN_1 and dN_2 will be

$$V^{\frac{n-5}{n-1}}.$$

It will be shewn that we have reason from experiments on the viscosity of gases to believe that $n = 5$. In this case V will disappear from the expressions of the form (3), and they will be capable of immediate integration with respect to dN_1 and dN_2.

If we assume $n = 5$ and put $a^4 = 2 \cot^2 2\phi$ and $x = \sqrt{1 - \tan^2 \phi} \cos \psi$,

$$\left. \begin{aligned} \frac{\pi}{2} - \theta &= \sqrt{\cos 2\phi} \int_0^{\frac{\pi}{2}} \frac{d\psi}{\sqrt{1 - \sin^2 \phi \, \sin^2 \psi}} \\ &= \sqrt{\cos 2\phi} \, F_{\sin \phi}, \end{aligned} \right\} \dots\dots\dots\dots\dots (14),$$

where $F_{\sin \phi}$ is the complete elliptic function of the first kind and is given in Legendre's Tables. I have computed the following Table of the distance of the asymptotes, the distance of the apse, the value of θ, and of the quantities whose summation leads to A_1 and A_2.

ϕ	b	Distance of apse	θ	$\dfrac{\sin^2\theta}{\sin^2 2\phi}$	$\dfrac{\sin^2 2\theta}{\sin^2 2\phi}$
0 0	infinite	infinite	0 0	0	0
5 0	2381	2391	0 31	·00270	·01079
10 0	1658	1684	1 53	·01464	·03689
15 0	1316	1366	4 47	·02781	·11048
20 0	1092	1172	8 45	·05601	·21885
25 0	916	1036	14 15	·10325	·38799
30 0	760	931	21 42	·18228	·62942
35 0	603	845	31 59	·31772	·71433
40 0	420	772	47 20	·55749	1·02427
41 0	374	758	51 32	·62515	·96763
42 0	324	745	56 26	·70197	·85838
43 0	264	732	62 22	·78872	·67868
44 0	187	719	70 18	·88745	·40338
44 30	132	713	76 1	·94190	·21999
45 0	0	707	90 0	1·00000	·00000

$$A_1 = \int 4\pi a\, da\, \sin^2\theta = 2\cdot6595 \dots\dots\dots\dots\dots\dots(15),$$

$$A_2 = \int \pi a\, da\, \sin^2 2\theta = 1\cdot3682 \dots\dots\dots\dots\dots\dots(16).$$

The paths described by molecules about a centre of force S, repelling inversely as the fifth power of the distance, are given in the figure.

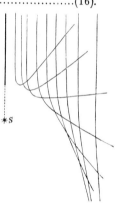

The molecules are supposed to be originally moving with equal velocities in parallel paths, and the way in which their deflections depend on the distance of the path from S is shewn by the different curves in the figure.

3rd.   *Integration with respect to* dN_2.

We have now to integrate expressions involving various functions of ξ, η, ζ, and V with respect to all the molecules of the second sort. We may write the expression to be integrated

$$\iiint Q V^{\frac{n-5}{n-1}} f_2 \left(\xi_2 \eta_2 \zeta_2 \right) d\xi_2\, d\eta_2\, d\zeta_2,$$

where Q is some function of ξ, η, ζ, &c., already determined, and f_2 is the function which indicates the distribution of velocity among the molecules of the second kind.

In the case in which $n = 5$, V disappears, and we may write the result of integration $\overline{Q}N_2$,

where $\overline{Q}$ is the mean value of Q for all the molecules of the second kind, and N_2 is the number of those molecules.

If, however, n is not equal to 5, so that V does not disappear, we should require to know the form of the function f_2 before we could proceed further with the integration.

The only case in which I have determined the form of this function is that of one or more kinds of molecules which have by their continual encounters brought about a distribution of velocity such that the number of molecules whose velocity lies within given limits remains constant. In the *Philosophical Magazine* for January 1860, I have given an investigation of this case, founded on the assumption that the probability of a molecule having a velocity resolved parallel to x lying between given limits is not in any way affected by the knowledge that the molecule has a given velocity resolved parallel to y. As this assumption may appear precarious, I shall now determine the form of the function in a different manner.

On the Final Distribution of Velocity among the Molecules of Two Systems acting on one another according to any Law of Force.

From a given point O let lines be drawn representing in direction and magnitude the velocities of every molecule of either kind in unit of volume. The extremities of these lines will be distributed over space in such a way that if an element of volume dV be taken anywhere, the number of such lines which will terminate within dV will be $f(r)\,dV$, where r is the distance of dV from O.

Let $OA = a$ be the velocity of a molecule of the first kind, and $OB = b$ that of a molecule of the second kind before they encounter one another, then

BA will be the velocity of A relative to B; and if we divide AB in G inversely as the masses of the molecules, and join OG, OG will be the velocity of the centre of gravity of the two molecules.

Now let $OA' = a'$ and $OB' = b'$ be the velocities of the two molecules after the encounter, $GA = GA'$ and $GB = GB'$, and $A'GB'$ is a straight line not necessarily in the plane of OAB. Also $AGA' = 2\theta$ is the angle through which the relative velocity is turned in the encounter in question. The relative motion of the molecules is completely defined if we know BA the relative velocity before the encounter, 2θ the angle through which BA is turned during the encounter, and ϕ the angle which defines the direction of the plane in which BA and $B'A'$ lie. All encounters in which the magnitude and direction of BA, and also θ and ϕ, lie within certain almost contiguous limits, we shall class as encounters of the given kind. The number of such encounters in unit of time will be

$$n_1 n_2 F de \dots\dots\dots\dots\dots\dots\dots\dots\dots\dots\dots\dots\dots(17),$$

where n_1 and n_2 are the numbers of molecules of each kind under consideration, and F is a function of the relative velocity and of the angle θ, and de depends on the limits of variation within which we class encounters as of the same kind.

Now let A describe the boundary of an element of volume dV while AB and $A'B'$ move parallel to themselves, then B, A', and B' will also describe equal and similar elements of volume.

The number of molecules of the first kind, the lines representing the velocities of which terminate in the element dV at A, will be

$$n_1 = f_1(a)\, dV \dots\dots\dots\dots\dots\dots\dots\dots\dots\dots\dots(18).$$

The number of molecules of the second kind which have velocities corresponding to OB will be

$$n_2 = f_2(b)\, dV \dots\dots\dots\dots\dots\dots\dots\dots (19);$$

and the number of encounters of the given kind between these two sets of molecules will be

$$f_1(a) f_2(b)\, (dV)^2 F de \dots\dots\dots\dots\dots\dots\dots(20).$$

The lines representing the velocities of these molecules after encounters of the given kind will terminate within elements of volume at A' and B', each equal to dV.

In like manner we should find for the number of encounters between molecules whose original velocities corresponded to elements equal to dV described about A' and B', and whose subsequent velocities correspond to elements equal to dV described about A and B,

$$f_1(a')f_2(b')\,(dV)^2F'de\dots\dots\dots\dots\dots\dots\dots\dots (21),$$

where F' is the same function of $B'A'$ and $A'GA$ that F is of BA and AGA'. F is therefore equal to F'.

When the number of pairs of molecules which change their velocities from OA, OB to OA', OB' is equal to the number which change from OA', OB' to OA, OB, then the final distribution of velocity will be obtained, which will not be altered by subsequent exchanges. This will be the case when

$$f_1(a)f_2(b)=f_1(a')f_2(b')\dots\dots\dots\dots\dots\dots(22).$$

Now the only relation between a, b and a', b' is

$$M_1a^2+M_2b^2=M_1a'^2+M_2b'^2,\dots\dots\dots\dots\dots\dots(23),$$

whence we obtain $$f_1(a)=C_1e^{-\frac{a^2}{a^2}},\; f_2(b)=C_2e^{-\frac{b^2}{\beta^2}}\dots\dots\dots\dots\dots(24),$$

where $$M_1a^2=M_2\beta^2\dots\dots\dots\dots\dots\dots\dots\dots(25).$$

By integrating $\iiint C_1 e^{-\frac{\xi^2+\eta^2+\zeta^2}{a^2}}d\xi\,d\eta\,d\zeta$, and equating the result to N_1, we obtain the value of C_1. If, therefore, the distribution of velocities among N_1 molecules is such that the number of molecules whose component velocities are between ξ and $\xi+d\zeta$, η and $\eta+d\eta$, and ζ and $\zeta+d\zeta$ is

$$dN_1=\frac{N_1}{a^3\pi^{\frac{3}{2}}}e^{-\frac{\xi^2+\eta^2+\zeta^2}{a^2}}d\xi\,d\eta\,d\zeta\dots\dots\dots\dots\dots(26),$$

then this distribution of velocities will not be altered by the exchange of velocities among the molecules by their mutual action.

This is therefore a possible form of the final distribution of velocities. It is also the only form; for if there were any other, the exchange between velocities represented by OA and OA' would not be equal. Suppose that the number of molecules having velocity OA' increases at the expense of OA. Then since the total number of molecules corresponding to OA' remains constant, OA' must communicate as many to OA'', and so on till they return to OA.

Hence if OA, OA', OA'', &c. be a series of velocities, there will be a tendency of each molecule to assume the velocities OA, OA', OA'', &c. in order, returning to OA. Now it is impossible to assign a reason why the successive velocities of a molecule should be arranged in this cycle, rather than in the reverse order. If, therefore, the direct exchange between OA and OA' is not equal, the equality cannot be preserved by exchange in a cycle. Hence the direct exchange between OA and OA' is equal, and the distribution we have determined is the only one possible.

This final distribution of velocity is attained only when the molecules have had a great number of encounters, but the great rapidity with which the encounters succeed each other is such that in all motions and changes of the gaseous system except the most violent, the form of the distribution of velocity is only slightly changed.

When the gas moves in mass, the velocities now determined are compounded with the motion of translation of the gas.

When the differential elements of the gas are changing their figure, being compressed or extended along certain axes, the values of the mean square of the velocity will be different in different directions. It is probable that the form of the function will then be

$$f_1(\xi\eta\zeta) = \frac{N_1}{\alpha\beta\gamma\pi^{\frac{3}{2}}} e^{-\left(\frac{\xi^2}{\alpha^2} + \frac{\eta^2}{\beta^2} + \frac{\zeta^2}{\gamma^2}\right)} \quad\ldots\ldots\ldots\ldots\ldots\ldots (27),$$

where α, β, γ are slightly different. I have not, however, attempted to investigate the exact distribution of velocities in this case, as the theory of motion of gases does not require it.

When one gas is diffusing through another, or when heat is being conducted through a gas, the distribution of velocities will be different in the positive and negative directions, instead of being symmetrical, as in the case we have considered. The want of symmetry, however, may be treated as very small in most actual cases.

The principal conclusions which we may draw from this investigation are as follows. Calling α the modulus of velocity,

1st. The mean velocity is $\bar{v} = \dfrac{2}{\sqrt{\pi}}\alpha$ $\ldots\ldots\ldots\ldots\ldots$ (28).

2nd. The mean square of the velocity is $\overline{v^2} = \dfrac{3}{2}\alpha^2$ $\ldots\ldots\ldots\ldots\ldots$ (29).

3rd. The mean value of ξ^2 is $\overline{\xi^2} = \dfrac{1}{2}\alpha^2$ $\ldots\ldots\ldots\ldots\ldots$ (30).

4th. The mean value of ξ^4 is $\overline{\xi^4} = \tfrac{3}{4}a^4$ ....(31).

5th. The mean value of $\xi^2\eta^2$ is $\overline{\xi^2\eta^2} = \tfrac{1}{4}a^4$ ....(32).

6th. When there are two systems of molecules

$$M_1a^2 = M_2\beta^2 \quad\text{................................ (33),}$$

whence
$$M_1v_1^2 = M_2v_2^2 \quad\text{................................ (34),}$$

or the mean *vis viva* of a molecule will be the same in each system. This
is a very important result in the theory of gases, and it is independent of the
nature of the action between the molecules, as are all the other results relating
to the final distribution of velocities. We shall find that it leads to the law
of gases known as that of Equivalent Volumes.

Variation of Functions of the Velocity due to encounters between the Molecules.

We may now proceed to write down the values of $\dfrac{\delta\overline{Q}}{\delta t}$ in the different
cases. We shall indicate the mean value of any quantity for all the molecules
of one kind by placing a bar over the symbol which represents that quantity
for any particular molecule, but in expressions where all such quantities are to
be taken at their mean values, we shall, for convenience, omit the bar. We
shall use the symbols δ_1 and δ_2 to indicate the effect produced by molecules of
the first kind and second kind respectively, and δ_3 to indicate the effect of
external forces. We shall also confine ourselves to the case in which $n = 5$,
since it is not only free from mathematical difficulty, but is the only case which
is consistent with the laws of viscosity of gases.

In this case V disappears, and we have for the effect of the second system
on the first,

$$\frac{\delta Q}{\delta t} = N_2 \left\{ \frac{K(M_1 + M_2)}{M_1 M_2} \right\}^{\frac{1}{2}} A \int_0^\pi (Q' - Q)\, d\phi \quad\text{.................(35),}$$

where the functions of ξ, η, ζ in $\int (Q' - Q)\, d\phi$ must be put equal to their mean
values for all the molecules, and A_1 or A_2 must be put for A according as
$\sin^2\theta$ or $\sin^2 2\theta$ occurs in the expressions in equations (4), (5), (6), (7). We
thus obtain

$$(a) \quad \frac{\delta_2\xi_1}{\delta t} = \left\{ \frac{K}{M_1 M_2 (M_1 + M_2)} \right\}^{\frac{1}{2}} N_2 M_2 A_1 (\xi_2 - \xi_1) \quad\text{................................(36);}$$

(β) $\dfrac{\delta_2 \xi_1^2}{\delta t} = \left\{ \dfrac{K}{M_1 M_2 (M_1 + M_2)} \right\}^{\frac{1}{4}} \dfrac{N_2 M_2}{M_1 + M_2} \left\{ 2A_1 (\xi_2 - \xi_1)(M_1 \xi_1 + M_2 \xi_2) \right.$

$\qquad\qquad\qquad\qquad\qquad \left. + A_2 M_2 \left(\overline{\eta_2 - \eta_1}^2 + \overline{\zeta_2 - \zeta_1}^2 - 2\overline{\xi_2 - \xi_1}^2 \right) \right\}$ ...... (37);

$\dfrac{\delta_2 \xi_1 \eta_1}{\delta t} = \left\{ \dfrac{K}{M_1 M_2 (M_1 + M_2)} \right\}^{\frac{1}{4}} \dfrac{N_2 M_2}{M_1 + M_2} \left[A_1 \{ 2M_2 \xi_2 \eta_2 - 2M_1 \xi_1 \eta_1 \right.$

$\qquad\qquad\qquad\qquad \left. + (M_1 - M_2)(\xi_1 \eta_2 + \xi_2 \eta_1) \} - 3A_2 M_2 (\xi_2 - \xi_1)(\eta_2 - \eta_1) \right]$ ...(38);

(γ) $\dfrac{\delta_2 \xi_1 V_1^2}{\delta t} = \left\{ \dfrac{K}{M_1 M_2 (M_1 + M_2)} \right\}^{\frac{1}{4}} N_2 M_2 \left[A_1 \{ \overline{\xi_2 - \xi_1} V_1^2 + 2\xi_1 (U - V_1^2) \} \right.$

$\qquad\qquad\qquad\qquad + \dfrac{M_2}{M_1 + M_2} (2A_1 - 3A_2) 2 (\xi_2 - \xi_1)(U - V_1^2)$

$\qquad\qquad\qquad\qquad + \dfrac{M_2}{M_1 + M_2} (2A_1 + 2A_2) \xi_1 V^2$

$\qquad\qquad\qquad\qquad \left. + \left(\dfrac{M_2}{M_1 + M_2} \right)^2 (2A_1 - 2A_2) 2 (\xi_2 - \xi_1) V^2 \right]$ .... (39),

using the symbol δ_2 to indicate variations arising from the action of molecules of the second system.

These are the values of the rate of variation of the mean values of ξ_1, ξ_1^2, $\xi_1 \eta_1$, and $\xi_1 V_1^2$, for the molecules of the first kind due to their encounters with molecules of the second kind. In all of them we ·must multiply up all functions of ξ, η, ζ, and take the mean values of the products so found. As this has to be done for all such functions, I have omitted the bar over each function in these expressions.

To find the rate of variation due to the encounters among the particles of the same system, we have only to alter the suffix $_{(2)}$ into $_{(1)}$ throughout, and to change K, the coefficient of the force between M_1 and M_2 into K_1, that of the force between two molecules of the first system. We thus find

(a) $\dfrac{\delta_1 \overline{\xi_1}}{\delta t} = 0$ ....(40);

(β) $\dfrac{\delta_1 \overline{\xi_1^2}}{dt} = \left(\dfrac{K_1}{2M_1^3} \right)^{\frac{1}{4}} M_1 N_1 A_2 \{ \overline{\eta_1^2} + \overline{\zeta_1^2} - 2\overline{\xi_1^2} - (\overline{\eta_1 . \eta_1} + \overline{\zeta_1 . \zeta_1} - 2\overline{\xi_1 . \xi_1}) \}$ .... (41);

$\dfrac{\delta_1 \overline{\xi_1 \eta_1}}{\delta t} = \left(\dfrac{K_1}{2M_1^3} \right)^{\frac{1}{4}} M_1 N_1 A_2 3 \{ \overline{\xi_1 . \eta_1} - \overline{\xi_1 \eta_1} \}$ ....(42);

(γ) $\dfrac{\delta_1 \overline{\xi_1 V_1^2}}{\delta t} = \left(\dfrac{K_1}{2M_1^3}\right) M_1 N_1 A_2 3\left(\bar{\xi}_1 . \overline{V_1^2} - \overline{\xi_1 V_1^2}\right)$ .... (43).

These quantities must be added to those in equations (36) to (39) in order to get the rate of variation in the molecules of the first kind due to their encounters with molecules of both systems. When there is only one kind of molecules, the latter equations give the rates of variation at once.

On the Action of External Forces on a System of Moving Molecules.

We shall suppose the external force to be like the force of gravity, producing equal acceleration on all the molecules. Let the components of the force in the three coordinate directions be X, Y, Z. Then we have by dynamics for the variations of ξ, ξ^2, and ξV^2 due to this cause,

(a) $\qquad \dfrac{\delta_3 \xi}{\delta t} = X$ ....(44);

(β) $\qquad \dfrac{\delta_3 . \xi^2}{\delta t} = 2\xi X$ ....(45);

$\qquad \dfrac{\delta_3 . \xi \eta}{\delta t} = \eta X + \xi Y$ .... (46);

(γ) $\dfrac{\delta_3 . \xi V^2}{\delta t} = 2\xi(\xi X + \eta Y + \zeta Z) + X V^2$ .... (47);

where δ_3 refers to variations due to the action of external forces.

On the Total rate of change of the different functions of the velocity of the molecules of the first system arising from their encounters with molecules of both systems and from the action of external forces.

To find the total rate of change arising from these causes, we must add

$$\frac{\delta_1 Q}{\delta t}, \quad \frac{\delta_2 Q}{\delta t}, \quad \text{and} \quad \frac{\delta_3 Q}{\delta t},$$

the quantities already found. We shall find it, however, most convenient in the remainder of this investigation to introduce a change in the notation, and to substitute for

$$\xi, \ \eta, \text{ and } \zeta, \quad u+\xi, \quad v+\eta, \text{ and } w+\zeta \dots \dots \dots \dots (48),$$

where u, v, and w are so chosen that they are the mean values of the components of the velocity of all molecules of the same system in the immediate neighbourhood of a given point. We shall also write

$$M_1 N_1 = \rho_1, \quad M_2 N_2 = \rho_2 \dots\dots\dots\dots\dots\dots\dots(49),$$

where ρ_1 and ρ_2 are the densities of the two systems of molecules, that is, the mass in unit of volume. We shall also write

$$\left(\frac{K_1}{2M_1^2}\right)^{\frac{1}{2}} = k_1, \quad \left(\frac{K}{M_1 M_2 (M_1 + M_2)}\right)^{\frac{1}{2}} = k, \quad \text{and} \quad \left(\frac{K_2}{2M_2^2}\right)^{\frac{1}{2}} = k_2 \dots\dots(50);$$

ρ_1, ρ_2, k_1, k_2, and k are quantities the absolute values of which can be deduced from experiment. We have not as yet experimental data for determining M, N, or K.

We thus find for the rate of change of the various functions of the velocity,

(α) $\quad \dfrac{\delta u_1}{\delta t} = k A_1 \rho_2 (u_2 - u_1) + X \dots\dots\dots\dots\dots\dots\dots\dots\dots\dots (51);$

(β) $\quad \dfrac{\delta . \xi_1^2}{\delta t} = k_1 A_2 \rho_1 \{\eta_1^2 + \zeta_1^2 - 2\xi_1^2\} + k\rho_2 \dfrac{M_2}{M_1 + M_2} \{2A_1 (u_2 - u_1)^2$

$$+ A_2 (\overline{v_2 - v_1}^2 + \overline{w_2 - w_1}^2 - 2\overline{u_2 - u_1}^2)\} + \frac{k\rho_2}{M_1 + M_2} \{2A_1 (M_2 \xi_2^2 - M_1 \xi_1^2)$$

$$+ A_2 M_2 (\eta_1^2 + \zeta_1^2 - 2\xi_1^2 + \eta_2^2 + \zeta_2^2 - 2\xi_2^2)\} \quad \Bigg\} \dots (52);$$

also $\dfrac{\delta . \xi \eta}{\delta t} = - 3k_1 A_2 \rho_1 \xi_1 \eta_1 + k\rho_2 \dfrac{M_2}{M_1 + M_2} (2A_1 - 3A_2)(u_2 - u_1)(v_2 - v_1)$

$$+ \frac{k\rho_2}{M_1 + M_2} \{2A_1 (M_2 \xi_2 \eta_2 - M_1 \xi_1 \eta_1) - 3A_2 M_2 (\xi_1 \eta_1 + \xi_2 \eta_2)\} \quad \Bigg\} \dots\dots\dots(53).$$

(γ) As the expressions for the variation of functions of three dimensions in mixed media are complicated, and as we shall not have occasion to use them, I shall give the case of a single medium,

$$\frac{\delta}{\delta t}(\xi_1^3 + \xi_1 \eta_1^2 + \xi_1 \zeta_1^2) = - 3k_1 \rho_1 A_2 (\xi_1^3 + \xi_1 \eta_1^2 + \xi_1 \zeta_1^2) + X (3\xi_1^2 + \eta_1^2 + \zeta_1^2)$$

$$+ 2Y \xi_1 \eta_1 + 2Z \xi_1 \zeta_1 \dots\dots\dots\dots(54).$$

Theory of a Medium composed of Moving Molecules.

We shall suppose the position of every moving molecule referred to three rectangular axes, and that the component velocities of any one of them, resolved in the directions of x, y, z, are

$$u + \xi, \quad v + \eta, \quad w + \zeta,$$

where u, v, w are the components of the mean velocity of all the molecules which are at a given instant in a given element of volume, and ξ, η, ζ are the components of the relative velocity of one of these molecules with respect to the mean velocity.

The quantities u, v, w may be treated as functions of x, y, z, and t, in which case differentiation will be expressed by the symbol d. The quantities ξ, η, ζ, being different for every molecule, must be regarded as functions of t for each molecule. Their variation with respect to t will be indicated by the symbol δ.

The mean values of ξ^2 and other functions of ξ, η, ζ for all the molecules in the element of volume may, however, be treated as functions of x, y, z, and t.

If we consider an element of volume which always moves with the velocities u, v, w, we shall find that it does not always consist of the same molecules, because molecules are continually passing through its boundary. We cannot therefore treat it as a mass moving with the velocity u, v, w, as is done in hydrodynamics, but we must consider separately the motion of each molecule. When we have occasion to consider the variation of the properties of this element during its motion as a function of the time we shall use the symbol ∂.

We shall call the velocities u, v, w the velocities of translation of the medium, and ξ, η, ζ the velocities of agitation of the molecules.

Let the number of molecules in the element $dx\,dy\,dz$ be $N\,dx\,dy\,dz$, then we may call N the number of molecules in unit of volume. If M is the mass of each molecule, and ρ the density of the element, then

$$MN = \rho \quad\quad\quad\quad\quad\quad\quad\quad\quad\quad\quad (55).$$

Transference of Quantities across a Plane Area.

We must next consider the molecules which pass through a given plane of unit area in unit of time, and determine the quantity of matter, of momentum, of heat, &c. which is transferred from the negative to the positive side of this plane in unit of time.

We shall first divide the N molecules in unit of volume into classes according to the value of ξ, η, and ζ for each, and we shall suppose that the number of molecules in unit of volume whose velocity in the direction of x lies between ξ and $\xi+d\xi$, η and $\eta+d\eta$, ζ and $\zeta+d\zeta$ is dN, dN will then be a function of the component velocities, the sum of which being taken for all the molecules will give N the total number of molecules. The most probable form of this function for a medium in its state of equilibrium is

$$dN = \frac{N}{a^3\pi^{\frac{3}{2}}} e^{-\frac{\xi^2+\eta^2+\zeta^2}{a^2}} d\xi\, d\eta\, d\zeta \dots\dots\dots\dots\dots(56).$$

In the present investigation we do not require to know the form of this function.

Now let us consider a plane of unit area perpendicular to x moving with a velocity of which the part resolved parallel to x is u'. The velocity of the plane relative to the molecules we have been considering is $u'-(u+\xi)$, and since there are dN of these molecules in unit of volume it will overtake

$$\{u'-(u+\xi)\}\, dN$$

such molecules in unit of time, and the number of such molecules passing from the negative to the positive side of the plane, will be

$$(u+\xi-u')\, dN.$$

Now let Q be any property belonging to the molecule, such as its mass, momentum, *vis viva*, &c., which it carries with it across the plane, Q being supposed a function of ξ or of ξ, η, and ζ, or to vary in any way from one molecule to another, provided it be the same for the selected molecules whose number is dN, then the quantity of Q transferred across the plane in the positive direction in unit of time is

$$\int (u-u'+\xi)\, QdN,$$

or

$$(u-u')\int QdN + \int \xi QdN \dots\dots\dots\dots\dots(57).$$

If we put $\overline{Q}N$ for $\int QdN$, and $\overline{\xi Q}N$ for $\int \xi QdN$, then we may call $\overline{Q}$ the mean value of Q, and $\overline{\xi Q}$ the mean value of ξQ, for all the particles in the element of volume, and we may write the expression for the quantity of Q which crosses the plane in unit of time

$$(u-u')\,\overline{Q}N + \overline{\xi Q}N \dots\dots\dots\dots\dots(58).$$

THE DYNAMICAL THEORY OF GASES.   53

(*a*)   *Transference of Matter across a Plane—Velocity of the Fluid.*

To determine the quantity of matter which crosses the plane, make Q equal to M the mass of each molecule; then, since M is the same for all molecules of the same kind, $\overline{M} = M$; and since the mean value of ξ is zero, the expression is reduced to

$$(u - u')\,MN = (u - u')\,\rho \ \dots\dots\dots\dots\dots (59).$$

If $u = u'$, or if the plane moves with velocity u, the whole excess of matter transferred across the plane is zero; the velocity of the fluid may therefore be defined as the velocity whose components are u, v, w.

(*β*)   *Transference of Momentum across a Plane—System of Pressures at any point of the Fluid.*

The momentum of any one molecule in the direction of x is $M(u + \xi)$. Substituting this for Q, we get for the quantity of momentum transferred across the plane in the positive direction

$$(u - u')\,u\rho + \overline{\xi^2}\rho \ \dots\dots\dots\dots\dots (60).$$

If the plane moves with the velocity u, this expression is reduced to $\overline{\xi^2}\rho$. where $\overline{\xi^2}$ represents the mean value of ξ^2.

This is the whole momentum in the direction of x of the molecules projected from the negative to the positive side of the plane in unit of time. The mechanical action between the parts of the medium on opposite sides of the plane consists partly of the momentum thus transferred, and partly of the direct attractions or repulsions between molecules on opposite sides of the plane. The latter part of the action must be very small in gases, so that we may consider the pressure between the parts of the medium on opposite sides of the plane as entirely due to the constant bombardment kept up between them. There will also be a transference of momentum in the directions of y and z across the same plane,

$$(u - u')\,v\rho + \overline{\xi\eta}\rho \dots\dots\dots\dots\dots(61),$$

and

$$(u - u')\,w\rho + \overline{\xi\zeta}\rho \dots\dots\dots\dots\dots(62),$$

where $\overline{\xi\eta}$ and $\overline{\xi\zeta}$ represent the mean values of these products.

If the plane moves with the mean velocity u of the fluid, the total force exerted on the medium on the positive side by the projection of molecules into it from the negative side will be

a normal pressure $\overline{\xi^2}\rho$ in the direction of x,

a tangential pressure $\overline{\xi\eta}\rho$ in the direction of y,

and a tangential pressure $\overline{\xi\zeta}\rho$ in the direction of z.

If X, Y, Z are the components of the pressure on unit of area of a plane whose direction cosines are l, m, n,

$$\left.\begin{aligned}X &= l\overline{\xi^2}\rho + m\overline{\xi\eta}\rho + n\overline{\xi\zeta}\rho\\ Y &= l\overline{\xi\eta}\rho + m\overline{\eta^2}\rho + n\overline{\eta\zeta}\rho\\ Z &= l\overline{\xi\zeta}\rho + m\overline{\eta\zeta}\rho + n\overline{\zeta^2}\rho\end{aligned}\right\}\quad\ldots\ldots\ldots\ldots\ldots\ldots(63).$$

When a gas is not in a state of violent motion the pressures in all directions are nearly equal, in which case, if we put

$$\overline{\xi^2}\rho + \overline{\eta^2}\rho + \overline{\zeta^2}\rho = 3p \ldots\ldots\ldots\ldots\ldots\ldots\ldots(64),$$

the quantity p will represent the mean pressure at a given point, and $\overline{\xi^2}\rho$, $\overline{\eta^2}\rho$, and $\overline{\zeta^2}\rho$ will differ from p only by small quantities; $\overline{\eta\zeta}\rho$, $\overline{\zeta\xi}\rho$, and $\overline{\xi\eta}\rho$ will then be also small quantities with respect to p.

Energy in the Medium—Actual Heat.

The actual energy of any molecule depends partly on the velocity of its centre of gravity, and partly on its rotation or other internal motion with respect to the centre of gravity. It may be written

$$\tfrac{1}{2}M\{(u+\xi)^2 + (v+\eta)^2 + (w+\zeta)^2\} + \tfrac{1}{2}EM\ldots\ldots\ldots\ldots\ldots(65),$$

where $\tfrac{1}{2}EM$ is the internal part of the energy of the molecule, the form of which is at present unknown. Summing for all the molecules in unit of volume, the energy is

$$\tfrac{1}{2}\left(u^2 + v^2 + w^2\right)\rho + \tfrac{1}{2}\left(\overline{\xi^2} + \overline{\eta^2} + \overline{\zeta^2}\right)\rho + \tfrac{1}{2}\overline{E}\rho\ldots\ldots\ldots\ldots\ldots(66).$$

The first term gives the energy due to the motion of translation of the medium in mass, the second that due to the agitation of the centres of gravity of the molecules, and the third that due to the internal motion of the parts of each molecule.

If we assume with Clausius that the ratio of the mean energy of internal motion to that of agitation tends continually towards a definite value $(\beta - 1)$, we may conclude that, except in very violent disturbances, this ratio is always preserved, so that

$$\overline{E} = (\beta - 1)\left(\xi^2 + \eta^2 + \zeta^2\right) \quad\text{......................} (67).$$

The total energy of the invisible agitation in unit of volume will then be

$$\tfrac{1}{2}\beta\left(\xi^2 + \eta^2 + \zeta^2\right)\text{................................}(68),$$

or

$$\tfrac{3}{2}\beta p \text{...}(69).$$

This energy being in the form of invisible agitation, may be called the total heat in the unit of volume of the medium.

(γ) *Transference of Energy across a Plane—Conduction of Heat.*

Putting

$$Q = \tfrac{1}{2}\beta\left(\xi^2 + \eta^2 + \zeta^2\right) M, \quad\text{and}\quad u = u' \text{......................} (70),$$

we find for the quantity of heat carried over the unit of area by conduction in unit of time

$$\tfrac{1}{2}\beta\left(\overline{\xi^3} + \overline{\xi\eta^2} + \overline{\xi\zeta^2}\right)\rho \text{..........................} (71),$$

where $\overline{\xi^3}$, &c. indicate the mean values of ξ^3, &c. They are always small quantities.

On the Rate of Variation of Q *in an Element of Volume,* Q *being any property of the Molecules in that Element.*

Let Q be the value of the quantity for any particular molecule, and $\overline{Q}$ the mean value of Q for all the molecules of the same kind within the element.

The quantity $\overline{Q}$ may vary from two causes. The molecules within the element may by their mutual action or by the action of external forces produce an alteration of $\overline{Q}$, or molecules may pass into the element and out of it, and so cause an increase or diminution of the value of $\overline{Q}$ within it. If we employ the symbol δ to denote the variation of Q due to actions of the first kind on the individual molecules, and the symbol ∂ to denote the actual variation of Q in an element moving with the mean velocity of the system of molecules under consideration, then by the ordinary investigation of the increase or diminution of matter in an element of volume as contained in treatises on Hydrodynamics,

$$\left.\begin{aligned}\frac{\partial \overline{Q}N}{\partial t} = \frac{\delta \overline{Q}}{\delta t}\, N - \frac{d}{dx}\left\{(u-u')\,\overline{Q}N + \overline{\xi Q N}\right\}\\[2mm]-\frac{d}{dy}\left\{(v-v')\,\overline{Q}N + \eta\overline{Q}N\right\} - \frac{d}{dz}\left\{(w-w')\,\overline{Q}N + \overline{\zeta Q N}\right\}\end{aligned}\right\} \dots\dots\dots(72),$$

where the last three terms are derived from equation (59) and two similar equations, and denote the quantity of Q which flows out of an element of volume, that element moving with the velocities u', v', w'. If we perform the differentiations and then make $u'=u$, $v'=v$, and $w'=w$, then the variation will be that in an element which moves with the actual mean velocity of the system of molecules, and the equation becomes

$$\frac{\partial \overline{Q}N}{\partial t} + \overline{Q}N\left(\frac{du}{dx}+\frac{dv}{dy}+\frac{dw}{dz}\right) + \frac{d}{dx}(\overline{\xi Q}N) + \frac{d}{dy}(\overline{\eta Q}N) + \frac{d}{dz}(\overline{\zeta Q}N) = \frac{\delta Q}{\delta t}N \dots (73).$$

Equation of Continuity.

Put $Q = M$ the mass of a molecule; M is unalterable, and we have, putting $MN = \rho$,

$$\frac{\partial \rho}{\partial t} + \rho\left(\frac{du}{dx}+\frac{dv}{dy}+\frac{dw}{dz}\right) = 0 \dots\dots\dots\dots\dots (74),$$

which is the ordinary equation of continuity in hydrodynamics, the element being supposed to move with the velocity of the fluid. Combining this equation with that from which it was obtained, we find

$$N\frac{\partial \overline{Q}}{\partial t} + \frac{d}{dx}(\overline{\xi Q}N) + \frac{d}{dy}(\overline{\eta Q}N) + \frac{d}{dz}(\overline{\zeta Q}N) = N\frac{\delta Q}{\delta t} \dots\dots\dots\dots (75),$$

a more convenient form of the general equation.

Equations of Motion (a).

To obtain the Equation of Motion in the direction of x, put $Q = M_1(u_1 + \xi_1)$, the momentum of a molecule in the direction of x.

We obtain the value of $\dfrac{\delta Q}{\delta t}$ from equation (51), and the equation may be written

$$\rho_1\frac{\partial u_1}{\partial t} + \frac{d}{dz}(\rho_1\overline{\xi_1^2}) + \frac{d}{dy}(\rho_1\overline{\xi_1\eta_1}) + \frac{d}{dz}(\rho_1\overline{\xi_1\zeta_1}) = kA_1\rho_1\rho_2(u_2 - u_1) + X\rho_1 \dots (76).$$

In this equation the first term denotes the efficient force per unit of volume, the second the variation of normal pressure, the third and fourth the variations of tangential pressure, the fifth the resistance due to the molecules of a different system, and the sixth the external force acting on the system.

The investigation of the values of the second, third, and fourth terms must be deferred till we consider the variations of the second degree.

Condition of Equilibrium of a Mixture of Gases.

In a state of equilibrium u_1 and u_2 vanish, $\rho_1 \xi_1^2$ becomes p_1, and the tangential pressures vanish, so that the equation becomes

$$\frac{dp_1}{dx} = X \rho_1 \dots\dots\dots\dots\dots\dots\dots\dots\dots\dots\dots (77),$$

which is the equation of equilibrium in ordinary hydrostatics.

This equation, being true of the system of molecules forming the first medium independently of the presence of the molecules of the second system, shews that if several kinds of molecules are mixed together, placed in a vessel and acted on by gravity, the final distribution of the molecules of each kind will be the same as if none of the other kinds had been present. This is the same mode of distribution as that which Dalton considered to exist in a mixed atmosphere in equilibrium, the law of diminution of density of each constituent gas being the same as if no other gases were present.

This result, however, can only take place after the gases have been left for a considerable time perfectly undisturbed. If currents arise so as to mix the strata, the composition of the gas will be made more uniform throughout.

The result at which we have arrived as to the final distribution of gases, when left to themselves, is independent of the law of force between the molecules.

Diffusion of Gases.

If the motion of the gases is slow, we may still neglect the tangential pressures. The equation then becomes for the first system of molecules

$$\rho_1 \frac{\partial u_1}{\partial t} + \frac{dp_1}{dx} = kA_1\rho_1\rho_2(u_2 - u_1) + X\rho_1 \dots\dots\dots\dots\dots (78),$$

and for the second,

$$\rho_2 \frac{\partial u_2}{\partial t} + \frac{dp_2}{dx} = kA_1\rho_1\rho_2(u_1 - u_2) + X\rho_2 \dots\dots\dots\dots\dots (79).$$

In all cases of quiet diffusion we may neglect the first term of each equation. If we then put $p_1 + p_2 = p$, and $\rho_1 + \rho_2 = \rho$, we find by adding,

$$\frac{dp}{dx} = X\rho \dots\dots\dots\dots\dots\dots\dots\dots\dots (80).$$

If we also put $p_1u_1 + p_2u_2 = pu$, then the volumes transferred in opposite directions across a plane moving with velocity u will be equal, so that

$$p_1(u_1 - u) = p_2(u - u_2) = \frac{p_1p_2}{p\rho_1\rho_2kA_1} \cdot \left(X\rho_1 \frac{dp_1}{dx}\right) \dots\dots\dots\dots (81).$$

Here $p_1(u_1 - u)$ is the volume of the first gas transferred in unit of time across unit of area of the plane reduced to pressure unity, and at the actual temperature; and $p_2(u - u_2)$ is the equal volume of the second gas transferred across the same area in the opposite direction.

The external force X has very little effect on the quiet diffusion of gases in vessels of moderate size. We may therefore leave it out in our definition of the coefficient of diffusion of two gases.

When two gases not acted on by gravity are placed in different parts of a vessel at equal pressures and temperatures, there will be mechanical equilibrium from the first, and u will always be zero. This will also be approximately true of heavy gases, provided the denser gas is placed below the lighter. Mr Graham has described in his paper on the Mobility of Gases*, experiments which were made under these conditions. A vertical tube had its lower tenth part filled with a heavy gas, and the remaining nine-tenths with a lighter gas. After the lapse of a known time the upper tenth part of the tube was shut off, and the gas in it analyzed, so as to determine the quantity of the heavier gas which had ascended into the upper tenth of the tube during the given time.

In this case we have $\qquad\qquad u = 0 \dots\dots\dots\dots\dots\dots\dots (82),$

$$p_1u_1 = -\frac{p_1p_2}{\rho_1\rho_2kA_1} \frac{1}{p} \frac{dp_1}{dx} \dots\dots\dots\dots\dots (83),$$

* *Philosophical Transactions*, 1863.

and by the equation of continuity,

$$\frac{dp_1}{dt} + \frac{d}{dx}(p_1 u_1) = 0 \quad \dotfill (84),$$

whence

$$\frac{dp_1}{dt} = \frac{p_1 p_2}{\rho_1 \rho_2 k A_1} \frac{1}{p} \frac{d^2 p_1}{dx^2} \dotfill (85);$$

or if we put $D = \dfrac{p_1 p_2}{\rho_1 \rho_2 k A_1} \dfrac{1}{p}$,

$$\frac{dp_1}{dt} = D \frac{d^2 p_1}{dx^2} \dotfill (86).$$

The solution of this equation is

$$p_1 = C_1 + C_2 e^{-n^2 D t} \cos(nx + a) + \&c. \dotfill (87).$$

If the length of the tube is a, and if it is closed at both ends,

$$p_1 = C_1 + C_2 e^{-\frac{\pi^2 D}{a^2} t} \cos \frac{\pi x}{a} + C_3 e^{-4 \frac{\pi^2 D}{a^2} t} \cos 2 \frac{\pi x}{a} + \&c. \dotfill (88),$$

where C_1, C_2, C_3 are to be determined by the condition that when $t = 0$, $p_1 = p$, from $x = 0$ to $x = \frac{1}{10} a$, and $p_1 = 0$ from $x = \frac{1}{10} a$ to $x = a$. The general expression for the case in which the first gas originally extends from $x = 0$ to $x = b$, and in which after a time t the gas from $x = 0$ to $x = c$ is collected, is

$$\frac{p_1}{p} = \frac{b}{a} + \frac{2a}{\pi^2 c} \left\{ e^{-\frac{\pi^2 D}{a^2} t} \sin \frac{\pi b}{a} \cdot \sin \frac{\pi c}{a} + \frac{1}{2^2} e^{-4 \frac{\pi^2 D}{a^2} t} \sin \frac{2\pi b}{a} \sin \frac{2\pi c}{a} + \&c. \right\} \dotfill (89),$$

where $\dfrac{p_1}{p}$ is the proportion of the first gas to the whole in the portion from $x = 0$ to $x = c$.

In Mr Graham's experiments, in which one-tenth of the tube was filled with the first gas, and the proportion of the first gas in the tenth of the tube at the other end ascertained after a time t, this proportion will be

$$\frac{p_1}{p} = \frac{1}{10} - \frac{20}{\pi^2} \left\{ e^{-\frac{\pi^2 D}{a^2} t} \sin^2 \frac{\pi}{10} - e^{-2^2 \frac{\pi^2 D}{a^2} t} \sin^2 2 \frac{\pi}{10} + e^{-3^2 \frac{\pi^2 D}{a^2} t} \sin^2 3 \frac{\pi}{10} - \&c. \right\} \dotfill (90).$$

We find for a series of values of $\dfrac{p_1}{p}$ taken at equal intervals of time T,

where

$$T = \frac{\log_e 10}{10 \pi^2} \frac{a^2}{D}.$$

Time.	$\frac{p_1}{p}$.
0	0
T	·01193
$2T$	·02305
$3T$	·03376
$4T$	·04366
$5T$	·05267
$6T$	·06072
$8T$	·07321
$10T$	·08227
$12T$	·08845
∞	·10000

Mr Graham's experiments on carbonic acid and air, when compared with this Table, give $T = 500$ seconds nearly for a tube 0·57 metre long. Now

$$D = \frac{\log_e 10}{10\pi^2} \frac{a^2}{T} \quad\ldots\ldots\ldots\ldots\ldots\ldots\ldots\ldots\ldots\ldots(91),$$

whence $D = ·0235$

for carbonic acid and air, in inch-grain-second measure.

Definition of the Coefficient of Diffusion.

D is the volume of gas reduced to unit of pressure which passes in unit of time through unit of area when the total pressure is uniform and equal to p, and the pressure of either gas increases or diminishes by unity in unit of distance. D may be called the coefficient of diffusion. It varies directly as the square of the absolute temperature, and inversely as the total pressure p.

The dimensions of D are evidently $L^2 T^{-1}$, where L and T are the standards of length and time.

In considering this experiment of the interdiffusion of carbonic acid and air, we have assumed that air is a simple gas. Now it is well known that the constituents of air can be separated by mechanical means, such as passing them through a porous diaphragm, as in Mr Graham's experiments on Atmolysis.

The discussion of the interdiffusion of three or more gases leads to a much more complicated equation than that which we have found for two gases, and it is not easy to deduce the coefficients of interdiffusion of the separate gases. It is therefore to be desired that experiments should be made on the inter-diffusion of every pair of the more important pure gases which do not act chemically on each other, the temperature and pressure of the mixture being noted at the time of experiment.

Mr Graham has also published in Brande's *Journal* for 1829, pt. 2, p. 74, the results of experiments on the diffusion of various gases out of a vessel through a tube into air. The coefficients of diffusion deduced from these ex-periments are—

Air and Hydrogen......................	·026216
Air and Marsh-gas	·010240
Air and Ammonia......................	·00962
Air and Olefiant gas..................	·00771
Air and Carbonic acid	·00682
Air and Sulphurous acid	·00582
Air and Chlorine	·00486

The value for carbonic acid is only one third of that deduced from the experiment with the vertical column. The inequality of composition of the mixed gas in different parts of the vessel is, however, neglected; and the dia-meter of the tube at the middle part, where it was bent, was probably less than that given.

Those experiments on diffusion which lasted ten hours, all give smaller values of D than those which lasted four hours, and this would also result from the mixture of the gases in the vessel being imperfect.

Interdiffusion through a small hole.

When two vessels containing different gases are connected by a small hole, the mixture of gases in each vessel will be nearly uniform except near the hole; and the inequality of the pressure of each gas will extend to a distance from the hole depending on the diameter of the hole, and nearly proportional to that diameter.

Hence in the equation

$$\rho_1 \frac{\partial u_1}{dt} + \frac{dp_1}{dx} = kA\rho_1\rho_2\left(u_2 - u_1\right) + X\rho \dots\dots\dots\dots(92)$$

the term $\frac{dp_1}{dx}$ will vary inversely as the diameter of the hole, while u_1 and u_2 will not vary considerably with the diameter.

Hence when the hole is very small the right-hand side of the equation may be neglected, and the flow of either gas through the hole will be independent of the flow of the other gas, as the term $kA\rho_1\rho_2\left(u_2 - u_1\right)$ becomes comparatively insignificant.

One gas therefore will escape through a very fine hole into another nearly as fast as into a vacuum; and if the pressures are equal on both sides, the volumes diffused will be as the square roots of the specific gravities inversely, which is the law of diffusion of gases established by Graham*.

Variation of the invisible agitation (β).

By putting for Q in equation (75)

$$Q = \frac{M}{2}\left\{\left(u_1 + \xi_1\right)^2 + \left(v_1 + \eta_1\right)^2 + \left(w_1 + \zeta_1\right)^2 + \left(\beta - 1\right)\left(\xi_1^2 + \eta_1^2 + \zeta_1^2\right)\right\}\dots\dots(93),$$

and eliminating by means of equations (76) and (52), we find

$$\begin{aligned}
&\tfrac{1}{2}\rho_1\frac{\partial}{\partial t}\beta_1\left(\xi_1^2 + \eta_1^2 + \zeta_1^2\right) + \rho_1\xi_1^2\frac{du_1}{dx} + \rho_1\eta_1^2\frac{dv_1}{dy} + \rho_1\zeta_1^2\frac{dw_1}{dz} + \rho_1\eta_1\zeta_1\left(\frac{dv_1}{dz} + \frac{dw_1}{dy}\right) \\
&+ \rho_1\zeta_1\xi_1\left(\frac{dw_1}{dx} + \frac{du_1}{dz}\right) + \rho_1\xi_1\eta_1\left(\frac{du_1}{dy} + \frac{dv_1}{dx}\right) + \beta_1\left\{\frac{d}{dx}(\rho_1\xi_1^3 + \rho_1\xi_1\eta_1^2 + \rho_1\xi_1\zeta_1^2)\right. \\
&+ \frac{d}{dy}\left(\rho_1\eta_1\xi_1^2 + \rho_1\eta_1^3 + \rho_1\eta_1\zeta_1^2\right) + \frac{d}{dz}\left.\left(\rho_1\zeta_1\xi_1^2 + \rho_1\zeta_1\eta_1^2 + \rho_1\zeta_1^3\right)\right\} \\
&= \frac{k\rho_1\rho_2 A_1}{M_1 + M_2}\left[M_2\left\{\left(u_2 - u_1\right)^2 + \left(v_2 - v_1\right)^2 + \left(w_2 - w_1\right)^2\right\}\right. \\
&\qquad + M_2\left(\xi_2^2 + \eta_2^2 + \zeta_2^2\right) - M_1\left(\xi_1^2 + \eta_1^2 + \zeta_1^2\right)\big]
\end{aligned}\quad\dots(94).$$

In this equation the first term represents the variation of invisible agitation or heat; the second, third, and fourth represent the cooling by expansion; the

* *Trans. Royal Society of Edinburgh*, Vol. XII. p. 222.

THE DYNAMICAL THEORY OF GASES.   63

fifth, sixth, and seventh the heating effect of fluid friction or viscosity; and the last the loss of heat by conduction. The quantities on the other side of the equation represent the thermal effects of diffusion, and the communication of heat from one gas to the other.

The equation may be simplified in various cases, which we shall take in order.

1st. *Equilibrium of Temperature between two Gases.—Law of Equivalent Volumes.*

We shall suppose that there is no motion of translation, and no transfer of heat by conduction through either gas. The equation (94) is then reduced to the following form,

$$\tfrac{1}{2}\rho_1 \frac{\partial}{\partial t} \beta_1 (\xi_1^2+\eta_1^2+\zeta_1^2) = \frac{k\rho_1\rho_2 A_1}{M_1+M_2} \{M_2(\xi_2^2+\eta_2^2+\zeta_2^2) - M_1(\xi_1^2+\eta_1^2+\zeta_1^2)\} \dots (95).$$

If we put

$$\frac{M_1}{M_1+M_2}(\xi_1^2+\eta_1^2+\zeta_1^2) = Q_1, \text{ and } \frac{M_2}{M_1+M_2}(\xi_2^2+\eta_2^2+\zeta_2^2) = Q_2 \dots (96),$$

we find

$$\frac{\partial}{\partial t}(Q_2-Q_1) = -\frac{2kA_1}{M_1+M_2}(M_2\rho_2\beta_1+M_1\rho_1\beta_2)(Q_2-Q_1) \dots (97),$$

or

$$Q_2-Q_1 = Ce^{-nt}, \text{ where } n = \frac{2kA_1}{M_1+M_2}(M_2\rho_2\beta_2+M_1\rho_1\beta_1)\frac{1}{\beta_1\beta_2} \dots (98).$$

If, therefore, the gases are in contact and undisturbed, Q_1 and Q_2 will rapidly become equal. Now the state into which two bodies come by exchange of invisible agitation is called equilibrium of heat or equality of temperature. Hence when two gases are at the same temperature,

$$Q_1 = Q_2 \dots (99),$$

or

$$1 = \frac{Q_1}{Q_2} = \frac{M_1(\xi_1^2+\eta_1^2+\zeta_1^2)}{M_2(\xi_2^2+\eta_2^2+\zeta_2^2)}$$

$$= \frac{M_1\dfrac{p_1}{\rho_1}}{M_2\dfrac{p_2}{\rho_2}}.$$

Hence if the pressures as well as the temperatures be the same in two gases,

$$\frac{M_1}{\rho_1} = \frac{M_2}{\rho_2} \dots\dots\dots\dots\dots\dots\dots\dots\dots(100),$$

or the masses of the individual molecules are proportional to the density of the gas.

This result, by which the relative masses of the molecules can be deduced from the relative densities of the gases, was first arrived at by Gay-Lussac from chemical considerations. It is here shewn to be a necessary result of the Dynamical Theory of Gases; and it is so, whatever theory we adopt as to the nature of the action between the individual molecules, as may be seen by equation (34), which is deduced from perfectly general assumptions as to the nature of the law of force.

We may therefore henceforth put $\dfrac{s_1}{s_2}$ for $\dfrac{M_1}{M_2}$, where s_1, s_2 are the specific gravities of the gases referred to a standard gas.

If we use θ to denote the temperature reckoned from absolute zero of a gas thermometer, M_0 the mass of a molecule of hydrogen, V_0^2 its mean square of velocity at temperature unity, s the specific gravity of any other gas referred to hydrogen, then the mass of a molecule of the other gas is

$$M = M_0 s \dots\dots\dots\dots\dots\dots\dots\dots(101).$$

Its mean square of velocity, $$V^2 = \frac{1}{s} V_0^2 \theta \dots\dots\dots\dots\dots\dots\dots(102).$$

Pressure of the gas, $$p = \tfrac{1}{3} \frac{\rho}{s} \theta V_0^2 \dots\dots\dots\dots\dots\dots(103).$$

We may next determine the amount of cooling by expansion.

Cooling by Expansion.

Let the expansion be equal in all directions, then

$$\frac{du}{dx} = \frac{dv}{dy} = \frac{dw}{dz} = -\frac{1}{3\rho} \frac{\partial \rho}{\partial t} \dots\dots\dots\dots(104),$$

and $\dfrac{du}{dy}$ and all terms of unsymmetrical form will be zero.

If the mass of gas is of the same temperature throughout there will be no conduction of heat, and the equation (94) will become

$$\tfrac{1}{2}\rho\beta\,\frac{\partial \overline{V^2}}{\partial t} - \tfrac{1}{3}\,\overline{V^2}\,\frac{\partial \rho}{\partial t} = 0 \dots\dots\dots\dots\dots\dots\dots\dots (105),$$

or

$$2\,\frac{\partial \rho}{\rho} = 3\beta\,\frac{\partial \overline{V^2}}{\overline{V^2}} = 3\beta\,\frac{\partial \theta}{\theta} \dots\dots\dots\dots\dots\dots\dots(106),$$

or

$$\frac{\partial \theta}{\theta} = \frac{2}{3\beta}\,\frac{\partial \rho}{\rho} \dots\dots\dots\dots\dots\dots\dots\dots\dots (107),$$

which gives the relation between the density and the temperature in a gas expanding without exchange of heat with other bodies. We also find

$$\frac{\partial p}{p} = \frac{\partial \rho}{\rho} + \frac{\partial \theta}{\theta}$$

$$= \frac{2+3\beta}{3\beta}\,\frac{\partial \rho}{\rho} \dots\dots\dots\dots\dots\dots\dots\dots(108),$$

which gives the relation between the pressure and the density.

Specific Heat of Unit of Mass at Constant Volume.

The total energy of agitation of unit of mass is $\tfrac{1}{2}\beta V^3 = \tfrac{1}{2}E$, or

$$E = \frac{3\beta}{2}\,\frac{p}{\rho} \dots\dots\dots\dots\dots\dots\dots\dots (109).$$

If, now, additional energy in the form of heat be communicated to it without changing its density,

$$\partial E = \frac{3\beta}{2}\,\frac{\partial p}{\rho} = \frac{3\beta}{2}\,\frac{p}{\rho}\,\frac{\partial \theta}{\theta} \dots\dots\dots\dots\dots\dots\dots(110).$$

Hence the specific heat of unit of mass of constant volume is in dynamical measure

$$\frac{\partial E}{\partial \theta} = \frac{3\beta}{2}\,\frac{p}{\rho\theta} \dots\dots\dots\dots\dots\dots\dots\dots (111).$$

Specific Heat of Unit of Mass at Constant Pressure.

By the addition of the heat ∂E the temperature was raised $\partial\theta$ and the pressure ∂p. Now, let the gas expand without communication of heat till the pressure sinks to its former value, and let the final temperature be $\theta + \partial'\theta$. The temperature will thus sink by a quantity $\partial\theta - \partial'\theta$, such that

$$\frac{\partial\theta - \partial'\theta}{\theta} = \frac{2}{2+3\beta}\frac{\partial p}{p} = \frac{2}{2+3\beta}\frac{\partial\theta}{\theta},$$

whence

$$\frac{\partial'\theta}{\theta} = \frac{3\beta}{2+3\beta}\frac{\partial\theta}{\theta} \quad\dots\dots\dots\dots\dots\dots\dots(112);$$

and the specific heat of unit of mass at constant pressure is

$$\frac{\partial E}{\partial'\theta} = \frac{2+3\beta}{2}\frac{p}{\rho\theta} \quad\dots\dots\dots\dots\dots\dots\dots(113).$$

The ratio of the specific heat at constant pressure to that of constant volume is known in several cases from experiment. We shall denote this ratio by

$$\gamma = \frac{2+3\beta}{3\beta} \quad\dots\dots\dots\dots\dots\dots\dots\dots(114),$$

whence

$$\beta = \tfrac{2}{3}\frac{1}{\gamma - 1} \quad\dots\dots\dots\dots\dots\dots\dots(115).$$

The specific heat of unit of volume in ordinary measure is at constant volume

$$\frac{1}{\gamma - 1}\frac{p}{J\theta} \quad\dots\dots\dots\dots\dots\dots\dots(116),$$

and at constant pressure

$$\frac{\gamma}{\gamma - 1}\frac{p}{J\theta} \quad\dots\dots\dots\dots\dots\dots\dots(117),$$

where J is the mechanical equivalent of unit of heat.

From these expressions Dr Rankine[*] has calculated the specific heat of air, and has found the result to agree with the value afterwards determined experimentally by M. Regnault[†].

[*] *Transactions of the Royal Society of Edinburgh,* Vol. xx. (1850).
[†] *Comptes Rendus,* 1853.

Thermal Effects of Diffusion.

If two gases are diffusing into one another, then, omitting the terms relating to heat generated by friction and to conduction of heat, the equation (94) gives

$$\frac{1}{2}\rho_1 \frac{\partial}{\partial t}\,\beta_1\left(\xi_1^2+\eta_1^2+\zeta_1^2\right)+\frac{1}{2}\rho_2\frac{\partial}{\partial t}\,\beta_2\left(\xi_2^2+\eta_2^2+\zeta_2^2\right)+p_1\left(\frac{du_1}{dx}+\frac{dv_1}{dy}+\frac{dw_1}{dz}\right) \left.\right\}$$
$$+p_2\left(\frac{du_2}{dx}+\frac{dv_2}{dy}+\frac{dw_2}{dz}\right)=k\rho_1\rho_2 A_1\{(u_1-u_2)^2+(v_1-v_2)^2+(w_1-w_2)^2\} \quad \left.\right\}\ldots(118).$$

By comparison with equations (78), and (79), the right-hand side of this equation becomes

$$X\left(\rho_1 u_1+\rho_2 u_2\right)+Y\left(\rho_1 v_1+\rho_2 v_2\right)+Z\left(\rho_1 w_1+\rho_2 w_2\right)$$

$$-\left(\frac{dp_1}{dx}u_1+\frac{dp_1}{dy}v_1+\frac{dp_1}{dz}w_1\right)-\left(\frac{dp_2}{dx}u_2+\frac{dp_2}{dy}v_2+\frac{dp_2}{dz}w_2\right)$$

$$-\tfrac{1}{2}\rho_1\frac{\partial}{\partial t}\left(u_1^2+v_1^2+w_1^2\right)-\tfrac{1}{2}\rho_2\frac{\partial}{\partial t}\left(u_2^2+v_2^2+w_2^2\right).$$

The equation (118) may now be written

$$\frac{1}{2}\rho_1\frac{\partial}{\partial t}\{u_1^2+v_1^2+w_1^2+\beta_1(\xi_1^2+\eta_1^2+\zeta_1^2)\}+\frac{1}{2}\rho_2\frac{\partial}{\partial t}\{u_2^2+v_2^2+w_2^2+\beta_2(\xi_2^2+\eta_2^2+\zeta_2^2)\} \left.\right\}$$
$$=X(\rho_1 u_1+\rho_2 u_2)+Y(\rho_1 v_1+\rho_2 v_2)+Z(\rho_1 w_1+\rho_2 w_2)-\left(\frac{d.pu}{dx}+\frac{d.pv}{dy}+\frac{d.pw}{dz}\right) \quad \left.\right\}\ldots(119).$$

The whole increase of energy is therefore that due to the action of the external forces *minus* the cooling due to the expansion of the mixed gases. If the diffusion takes place without alteration of the volume of the mixture, the heat due to the mutual action of the gases in diffusion will be exactly neutralized by the cooling of each gas as it expands in passing from places where it is dense to places where it is rare.

Determination of the Inequality of Pressure in different directions due to the Motion of the Medium.

Let us put $\qquad \rho_1 \xi_1^2 = p_1 + q_1$ and $\rho_2 \xi_2^2 = p_2 + q_2$ .... (120).

Then by equation (52),

$$\frac{\delta q_1}{\delta t} = -3k_1 A_2 \rho_1 q_1 - \frac{k}{M_1 + M_2} (2M_1 A_1 + 3M_2 A_2)\rho_2 q_1 - k(3A_2 - 2A_1)\frac{M_1}{M_1 + M_2}\,\rho_1 q_2 \left.\right\}$$
$$\left. - k\rho_1\rho_2 \frac{M_2}{M_1 + M_2}(A_2 - \tfrac{2}{3}A_1)\left(\overline{2u_1 - u_2}^{\,2} - \overline{v_1 - v_2}^{\,2} - \overline{w_1 - w_2}^{\,2}\right)\right\} \quad ...(121),$$

the last term depending on diffusion; and if we omit in equation (75) terms of three dimensions in ξ, η, ζ, which relate to conduction of heat, and neglect quantities of the form $\xi\eta\rho$ and $\rho\xi^2 - p$, when not multiplied by the large coefficients k, k_1, and k_2, we get

$$\frac{\partial q}{\partial t} + 2p\frac{du}{dx} - \tfrac{2}{3}p\left(\frac{du}{dx} + \frac{dv}{dy} + \frac{dw}{dz}\right) = \frac{\delta q}{\delta t} \quad .... (122).$$

If the motion is not subject to any very rapid changes, as in all cases except that of the propagation of sound, we may neglect $\frac{\partial q}{\partial t}$. In a single system of molecules

$$\frac{\delta q}{\delta t} = -3kA_2\rho q \quad ....(123),$$

whence $\qquad q = -\frac{2p}{3kA_2\rho}\left\{\frac{du}{dx} - \tfrac{1}{3}\left(\frac{du}{dx} + \frac{dv}{dy} + \frac{dw}{dz}\right)\right\} ....(124).$

If we make $\qquad \tfrac{1}{3}\frac{1}{kA_2}\frac{p}{\rho} = \mu$ .... (125),

μ will be the coefficient of viscosity, and we shall have by equation (120),

$$\rho\xi^2 = p - 2\mu\left\{\frac{du}{dx} - \tfrac{1}{3}\left(\frac{du}{dx} + \frac{dv}{dy} + \frac{dw}{dz}\right)\right\} \left.\right]$$
$$\rho\eta^2 = p - 2\mu\left\{\frac{dv}{dy} - \tfrac{1}{3}\left(\frac{du}{dx} + \frac{dv}{dy} + \frac{dw}{dz}\right)\right\} \left.\right\} \quad ....(126);$$
$$\rho\zeta^2 = p - 2\mu\left\{\frac{dw}{dz} - \tfrac{1}{3}\left(\frac{du}{dx} + \frac{dv}{dy} + \frac{dw}{dz}\right)\right\} \left.\right]$$

THE DYNAMICAL THEORY OF GASES.   69

and by transformation of co-ordinates we obtain

$$\rho\eta\zeta = -\mu\left(\frac{dv}{dz} + \frac{dw}{dy}\right)$$

$$\rho\zeta\xi = -\mu\left(\frac{dw}{dx} + \frac{du}{dz}\right) \Bigg\} \quad \dots\dots\dots\dots\dots\dots (127).$$

$$\rho\xi\eta = -\mu\left(\frac{du}{dy} + \frac{dv}{dx}\right)$$

These are the values of the normal and tangential stresses in a simple gas when the variation of motion is not very rapid, and when μ, the coefficient of viscosity, is so small that its square may be neglected.

Equations of Motion corrected for Viscosity.

Substituting these values in the equation of motion (76), we find

$$\rho\frac{\partial u}{\partial t} + \frac{dp}{dx} - \mu\left\{\frac{d^2u}{dx^2} + \frac{d^2u}{dy^2} + \frac{d^2u}{dz^2}\right\} - \tfrac{1}{3}\mu\frac{d}{dx}\left(\frac{du}{dx} + \frac{dv}{dy} + \frac{dw}{dz}\right) = X\rho\dots(128),$$

with two other equations which may be written down with symmetry. The form of these equations is identical with that of those deduced by Poisson[*] from the theory of elasticity, by supposing the strain to be continually relaxed at a rate proportional to its amount. The ratio of the third and fourth terms agrees with that given by Professor Stokes[†].

If we suppose the inequality of pressure which we have denoted by q to exist in the medium at any instant, and not to be maintained by the motion of the medium, we find, from equation (123),

$$q_1 = Ce^{-3kA_2\rho t} \quad \dots\dots\dots\dots\dots\dots\dots\dots\dots (129)$$

$$= Ce^{-\frac{t}{T}} \text{ if } T = \frac{1}{3kA_2\rho} = \frac{\mu}{p}\dots\dots\dots\dots(130);$$

the stress q is therefore relaxed at a rate proportional to itself, so that

$$\frac{\delta q}{q} = \frac{\delta t}{T} \quad \dots\dots\dots\dots\dots\dots\dots\dots\dots(131).$$

We may call T the modulus of the time of relaxation.

[*] *Journal de l'École Polytechnique*, 1829, Tom. XIII. Cah. XX. p. 139.

[†] "On the Friction of Fluids in Motion and the Equilibrium and Motion of Elastic Solids," *Cambridge Phil. Trans.* Vol. VIII. (1845), p. 297, equation (2).

If we next make $k=3$, so that the stress q does not become relaxed, the medium will be an elastic solid, and the equation

$$\frac{\partial \left(\rho \xi^{2}-p\right)}{\partial t}+2p\frac{du}{dx}-\tfrac{2}{3}p\left(\frac{du}{dx}+\frac{dv}{dy}+\frac{dw}{dz}\right)=0 \ldots\ldots\ldots\ldots(132)$$

may be written $\dfrac{\partial}{\partial t}\left\{\left(p_{xx}-p\right)+2p\dfrac{da}{dx}-\tfrac{2}{3}p\left(\dfrac{da}{dx}+\dfrac{d\beta}{dy}+\dfrac{d\gamma}{dz}\right)\right\}=0 \ldots\ldots\ldots\ldots(133),$

where a, β, γ are the displacements of an element of the medium, and p_{xx} is the normal pressure in the direction of x. If we suppose the initial value of this quantity zero, and p_{xx} originally equal to p, then, after a small displacement,

$$p_{xx}=p-p\left(\frac{da}{dx}+\frac{d\beta}{dy}+\frac{d\gamma}{dz}\right)-2p\frac{da}{dx}\ldots\ldots\ldots\ldots\ldots(134);$$

and by transformation of co-ordinates the tangential pressure

$$p_{xy}=-p\left(\frac{da}{dy}+\frac{d\beta}{dx}\right)\ldots\ldots\ldots\ldots\ldots\ldots(135).$$

The medium has now the mechanical properties of an elastic solid, the rigidity of which is p, while the cubical elasticity is $\tfrac{5}{3}p$*.

The same result and the same ratio of the elasticities would be obtained if we supposed the molecules to be at rest, and to act on one another with forces depending on the distance, as in the statical molecular theory of elasticity. The coincidence of the properties of a medium in which the molecules are held in equilibrium by attractions and repulsions, and those of a medium in which the molecules move in straight lines without acting on each other at all, deserve notice from those who speculate on theories of physics.

The fluidity of our medium is therefore due to the mutual action of the molecules, causing them to be deflected from their paths.

The coefficient of instantaneous rigidity of a gas is therefore p

The modulus of the time of relaxation is T $\qquad\Bigg\}\ldots(136).$

The coefficient of viscosity is $\mu=pT$

Now p varies as the density and temperature conjointly, while T varies inversely as the density.

Hence μ varies as the absolute temperature, and is independent of the density.

* *Camb. Phil. Trans.* Vol. VIII. (1845), p. 311, equation (29).

THE DYNAMICAL THEORY OF GASES.   7 1

This result is confirmed by the experiments of Mr Graham on the Transpiration of Gases*, and by my own experiments on the Viscosity or Internal Friction of Air and other Gases†.

The result that the viscosity is independent of the density, follows from the Dynamical Theory of Gases, whatever be the law of force between the molecules. It was deduced by myself ‡ from the hypothesis of hard elastic molecules, and M. O. E. Meyer § has given a more complete investigation on the same hypothesis.

The experimental result, that the viscosity is proportional to the absolute temperature, requires us to abandon this hypothesis, which would make it vary as the square root of the absolute temperature, and to adopt the hypothesis of a repulsive force inversely as the fifth power of the distance between the molecules, which is the only law of force which gives the observed result.

Using the foot, the grain, and the second as units, my experiments give for the temperature of 62° Fahrenheit, and in dry air,

$$\mu = 0\cdot0936.$$

If the pressure is 30 inches of mercury, we find, using the same units,

$$p = 477360000.$$

Since $pT = \mu$, we find that the modulus of the time of relaxation of rigidity in air of this pressure and temperature is

$$\frac{1}{5099100000} \text{ of a second.}$$

This time is exceedingly small, even when compared with the period of vibration of the most acute audible sounds ; so that even in the theory of sound we may consider the motion as steady during this very short time, and use the equations we have already found, as has been done by Professor Stokes ‖.

---

* *Philosophical Transactions*, 1846 and 1849.

† *Proceedings of the Royal Society*, February 8, 1866 ; *Philosophical Transactions*, 1866, p. 249.

‡ *Philosophical Magazine*, January 1860. [Vol. I. xx.]

§ Poggendorff's *Annalen*, 1865.

‖ "On the effect of the Internal Friction of Fluids on the motion of Pendulums," *Cambridge Transactions*, Vol. IX. (1850), art. 79.

Viscosity of a Mixture of Gases.

In a complete mixture of gases, in which there is no diffusion going on, the velocity at any point is the same for all the gases.

Putting

$$\tfrac{2}{3}\left(2\frac{du}{dx}-\frac{dv}{dy}-\frac{dw}{dz}\right)=U \dots\dots\dots\dots(137),$$

equation (122) becomes

$$p_1 U = -3k_1 A_2\rho_1 q_1 - \frac{k}{M_1+M_2}\left(2M_1 A_1 + 3M_2 A_2\right)\rho_2 q_1 - k\left(3A_2 - 2A_1\right)\frac{M_2}{M_1+M_2}\rho_1 q_2\dots(138).$$

Similarly,

$$p_2 U = -3k_2 A_2\rho_2 q_2 - \frac{k}{M_1+M_2}\left(2M_2 A_1 + 3M_1 A_2\right)\rho_1 q_2 - k\left(3A_2 - 2A_1\right)\frac{M_1}{M_1+M_2}\rho_2 q_1\dots(139).$$

Since $p = p_1 + p_2$ and $q = q_1 + q_2$, where p and q refer to the mixture, we shall have

$$\mu U = -q = -(q_1 + q_2),$$

where μ is the coefficient of viscosity of the mixture.

If we put s_1 and s_2 for the specific gravities of the two gases, referred to a standard gas, in which the values of p and ρ at temperature θ_0 and p_0 and ρ_0,

$$\mu = \frac{p_0\theta}{\rho_0\theta_0}\cdot\frac{Ep_1^2 + Fp_1p_2 + Gp_2^2}{3A_1k_1s_1Ep_1^2 + Hp_1p_2 + 3A_2k_2s_2Gp_2^2}\dots\dots\dots\dots(140),$$

where μ is the coefficient of viscosity of the mixture, and

$$\left.\begin{array}{l}
E = \dfrac{ks_1}{s_1+s_2}\left(2s_2 A_1 + 3s_1 A_2\right) \\[2ex]
F = 3A_2\left(k_1s_1 + k_2s_2\right) - \left(3A_2 - 2A_1\right)k\,\dfrac{2s_1s_2}{s_1+s_2} \\[2ex]
G = \dfrac{ks_2}{s_1+s_2}\left(2s_1 A_1 + 3s_2 A_2\right) \\[2ex]
H = 3A_2 s_1 s_2\left(3k_1 k_2 A_2 + 2k^2 A_1\right)
\end{array}\right\}\dots\dots\dots(141).$$

This expression is reduced to μ_1 when $p_2 = 0$, and to μ_2 when $p_1 = 0$. For other values of p_1 and p_2 we require to know the value of k, the coefficient

THE DYNAMICAL THEORY OF GASES.

of mutual interference of the molecules of the two gases. This might be deduced from the observed values of μ for mixtures, but a better method is by making experiments on the interdiffusion of the two gases. The experiments of Graham on the transpiration of gases, combined with my experiments on the viscosity of air, give as values of k_1 for air, hydrogen, and carbonic acid,

$$\text{Air} \dots\dots\dots\dots k_1 = \quad 4\cdot81 \times 10^{10},$$
$$\text{Hydrogen} \dots\dots k_1 = 142\cdot8 \ \times 10^{10},$$
$$\text{Carbonic acid} \dots k_1 = \quad 3\cdot9 \ \times 10^{10}.$$

The experiments of Graham in 1863, referred to at page 58, on the interdiffusion of air and carbonic acid, give the coefficient of mutual interference of these gases,

$$\text{Air and carbonic acid} \dots\dots k = 5\cdot2 \times 10^{10};$$

and by taking this as the absolute value of k, and assuming that the ratios of the coefficients of interdiffusion given at page 76 are correct, we find

$$\text{Air and hydrogen} \dots\dots k = 29\cdot8 \times 10^{10}.$$

These numbers are to be regarded as doubtful, as we have supposed air to be a simple gas in our calculations, and we do not know the value of k between oxygen and nitrogen. It is also doubtful whether our method of calculation applies to experiments such as the earlier observations of Mr Graham.

I have also examined the transpiration-times determined by Graham for mixtures of hydrogen and carbonic acid, and hydrogen and air, assuming a value of k roughly, to satisfy the experimental results about the middle of the scale. It will be seen that the calculated numbers for hydrogen and carbonic acid exhibit the peculiarity observed in the experiments, that a small addition of hydrogen *increases* the transpiration-time of carbonic acid, and that in both series the times of mixtures depend more on the slower than on the quicker gas.

The assumed values of k in these calculations were—

$$\text{For hydrogen and carbonic acid} \quad k = 12\cdot5 \times 10^{10},$$
$$\text{For hydrogen and air} \dots\dots\dots \quad k = 18\cdot8 \times 10^{10};$$

and the results of observation and calculation are, for the times of transpiration of mixtures of—

Hydrogen and Carbonic acid		Observed	Calculated	Hydrogen and Air		Observed	Calculated
100	0	·4321	·4375	100	0	·4434	·4375
97·5	2·5	·4714	·4750	95	5	·5282	·5300
95	5	·5157	·5089	90	10	·5880	·6028
90	10	·5722	·5678	75	25	·7488	·7438
75	25	·6786	·6822	50	50	·8179	·8488
50	50	·7339	·7652	25	75	·8790	·8946
25	75	·7535	·7468	10	90	·8880	·8983
10	90	·7521	·7361	5	95	·8960	·8996
0	100	·7470	·7272	0	100	·9000	·9010

The numbers given are the ratios of the transpiration-times of mixtures to that of oxygen as determined by Mr Graham, compared with those given by the equation (140) deduced from our theory.

Conduction of Heat in a Single Medium (γ).

The rate of conduction depends on the value of the quantity

$$\tfrac{1}{2}\beta\rho\left(\xi^3 + \xi\eta^2 + \xi\zeta^2\right),$$

where ξ^3, $\xi\eta^2$, and $\xi\zeta^2$ denote the mean values of those functions of ξ, η, ζ for all the molecules in a given element of volume.

As the expressions for the variations of this quantity are somewhat complicated in a mixture of media, and as the experimental investigation of the conduction of heat in gases is attended with great difficulty, I shall confine myself here to the discussion of a single medium.

Putting

$$Q = M\left(u + \xi\right)\left\{u^2 + v^2 + w^2 + 2u\xi + 2v\eta + 2w\zeta + \beta\left(\xi^2 + \eta^2 + \zeta^2\right)\right\}\ldots\ldots\ldots(142),$$

and neglecting terms of the forms $\xi\eta$ and ξ^3 and $\xi\eta^2$ when not multiplied by the large coefficient k_1, we find by equations (75), (77), and (54),

$$\left.\begin{aligned}\rho\,\frac{\partial}{\partial}\,\beta\left(\xi^3 + \xi\eta^2 + \xi\zeta^2\right) + \beta\,\frac{d}{dx}\cdot\rho\left(\xi^4 + \xi^2\eta^2 + \xi^2\zeta^2\right) - \beta\left(\xi^2 + \eta^2 + \zeta^2\right)\frac{dp}{dx}\\[4pt]- 2\beta\xi^2\,\frac{dp}{dx} = -3k_1\rho^2 A_2\beta\left\{\xi^3 + \xi\eta^2 + \xi\zeta^2\right\}\end{aligned}\right\}\ldots\ldots\ldots(143).$$

The first term of this equation may be neglected, as the rate of conduction will rapidly establish itself. The second term contains quantities of four dimensions in ξ, η, ζ, whose values will depend on the distribution of velocity among the molecules. If the distribution of velocity is that which we have proved to exist when the system has no external force acting on it and has arrived at its final state, we shall have by equations (29), (31), (32),

$$\overline{\xi^4} = \overline{3\xi^2} \cdot \overline{\xi^2} = 3\frac{p^2}{\rho^2} \dots\dots\dots\dots\dots (144),$$

$$\overline{\xi^2\eta^2} = \overline{\xi^2} \cdot \overline{\eta^2} = \frac{p^2}{\rho^2} \dots\dots\dots\dots\dots (145),$$

$$\overline{\xi^2\zeta^2} = \overline{\xi^2} \cdot \overline{\zeta^2} = \frac{p^2}{\rho^2} \dots\dots\dots\dots\dots (146);$$

and the equation of conduction may be written

$$5\beta\,\frac{p^2}{\rho\theta}\frac{d\theta}{dx} = -3k_1\rho^2A_2\beta\left\{\xi^3 + \xi\eta^2 + \xi\zeta^2\right\} \dots\dots\dots\dots (147).$$

[Addition made December 17, 1866.]

[*Final Equilibrium of Temperature.*]

[The left-hand side of equation (147), as sent to the Royal Society, contained a term $2\,(\beta-1)\dfrac{p}{\rho}\dfrac{dp}{dx}$, the result of which was to indicate that a column of air, when left to itself, would assume a temperature varying with the height, and greater above than below. The mistake arose from an error* in equation (143). Equation (147), as now corrected, shews that the flow of heat depends on the variation of temperature only, and not on the direction of the variation of pressure. A vertical column would therefore, when in thermal equilibrium, have the same temperature throughout.

When I first attempted this investigation I overlooked the fact that $\overline{\xi^4}$ is not the same as $\overline{\xi^2}\cdot\overline{\xi^2}$, and so obtained as a result that the temperature diminishes as the height increases at a greater rate than it does by expansion when air is carried up in mass. This leads at once to a condition of instability,

---

* The last term on the left-hand side was not multiplied by β.

which is inconsistent with the second law of thermodynamics. I wrote to Professor Sir W. Thomson about this result, and the difficulty I had met with, but presently discovered *one* of my mistakes, and arrived at the conclusion that the temperature would increase with the height. This does not lead to mechanical instability, or to any self-acting currents of air, and I was in some degree satisfied with it. But it is equally inconsistent with the second law of thermodynamics. In fact, if the temperature of any substance, when in thermic equilibrium, is a function of the height, that of any other substance must be the same function of the height. For if not, let equal columns of the two substances be enclosed in cylinders impermeable to heat, and put in thermal communication at the bottom. If, when in thermal equilibrium, the tops of the two columns are at different temperatures, an engine might be worked by taking heat from the hotter and giving it up to the cooler, and the refuse heat would circulate round the system till it was all converted into mechanical energy, which is in contradiction to the second law of thermo-dynamics.

The result as now given is, that temperature in gases, when in thermal equilibrium, is independent of height, and it follows from what has been said that temperature is independent of height in all other substances.

If we accept this law of temperature as the actual one, and examine our assumptions, we shall find that unless $\bar{\xi}^4 = 3\bar{\xi}^2 . \bar{\xi}^2$, we should have obtained a different result. Now this equation is derived from the law of distribution of velocities to which we were led by independent considerations. We may there-fore regard this law of temperature, if true, as in some measure a confirmation of the law of distribution of velocities.]

Coefficient of Conductivity.

If C is the coefficient of conductivity of the gas for heat, then the quantity of heat which passes through unit of area in unit of time measured as me-chanical energy, is

$$C\frac{d\theta}{dx} = \tfrac{5}{6}\frac{\beta}{k_1 A_2}\frac{p^2}{\rho^2\theta}\frac{d\theta}{dx} \quad\dots\dots\dots\dots\dots (148).$$

by equation (147).

Substituting for β its value in terms of γ by equation (115), and for k_1 its value in terms of μ by equation (125), and calling p_0, ρ_0, and θ_0 the simultaneous pressure, density, and temperature of the standard gas, and s the specific gravity of the gas in question, we find

$$C = \frac{5}{3\,(\gamma-1)}\ \frac{p_0}{\rho_0 \theta_0}\ \frac{\mu}{s}\,.........................(149).$$

For air we have $\gamma = 1{\cdot}409$, and at the temperature of melting ice, or $274^{\circ}{\cdot}6\ \mathrm{C.}$ above absolute zero, $\sqrt{\dfrac{p}{\rho}} = 918{\cdot}6$ feet per second, and at $16^{\circ}{\cdot}6\ \mathrm{C.}$, $\mu = 0{\cdot}0936$ in foot-grain-second measure. Hence for air at $16^{\circ}{\cdot}6\ \mathrm{C.}$ the conductivity for heat is

$$C = 1172\ ....\,(150).$$

That is to say, a horizontal stratum of air one foot thick, of which the upper surface is kept at $17^{\circ}\,\mathrm{C.}$, and the lower at $16^{\circ}\,\mathrm{C.}$, would in one second transmit through every square foot of horizontal surface a quantity of heat the mechanical energy of which is equal to that of 2344 grains moving at the rate of one foot per second.

Principal Forbes[*] has deduced from his experiments on the conduction of heat in bars, that a plate of wrought iron one foot thick, with its opposite surfaces kept $1^{\circ}\,\mathrm{C.}$ different in temperature, would, when the mean temperature is $25^{\circ}\,\mathrm{C.}$, transmit in one minute through every square foot of surface as much heat as would raise one cubic foot of water $0^{\circ}{\cdot}0127\ \mathrm{C.}$

Now the dynamical equivalent in foot-grain-second measure of the heat required to raise a cubic foot of water $1^{\circ}\,\mathrm{C.}$ is $1{\cdot}9157 \times 10^{10}$.

It appears from this that iron at $25^{\circ}\,\mathrm{C.}$ conducts heat 3525 times better than air at $16^{\circ}{\cdot}6\ \mathrm{C.}$

M. Clausius, from a different form of the theory, and from a different value of μ, found that lead should conduct heat 1400 times better than air. Now iron is twice as good a conductor of heat as lead, so that this estimate is not far different from that of M. Clausius in actual value.

In reducing the value of the conductivity from one kind of measure to another, we must remember that its dimensions are MLT^{-3}, when expressed in absolute dynamical measure.

[*] " Experimental Inquiry into the Laws of the Conduction of Heat in Bars," *Edinburgh Transactions*, 1861—62.

Since all the quantities which enter into the expression for C are constant except μ, the conductivity is subject to the same laws as the viscosity, that is, it is independent of the pressure, and varies directly as the absolute temperature. The conductivity of iron diminishes as the temperature increases.

Also, since γ is nearly the same for air, oxygen, hydrogen, and carbonic oxide, the conductivity of these gases will vary as the ratio of the viscosity to the specific gravity. Oxygen, nitrogen, carbonic oxide, and air will have equal conductivity, while that of hydrogen will be about seven times as great.

The value of γ for carbonic acid is 1·27, its specific gravity is $\frac{11}{8}$ of oxygen, and its viscosity $\frac{8}{11}$ of that of oxygen. The conductivity of carbonic acid for heat is therefore about $\frac{7}{9}$ of that of oxygen or of air.

31. Letter from Maxwell to Peter Guthrie Tait, March 12, 1868

Cambridge University Library, Maxwell Collection, Maxwell-Tait Correspondence.

8 P[alace] G[ardens] T[errace]
12 March 1868

Dr Tait

Yours received. I dispatched the proofs to you yesterday. As regards conduction of heat I have not considered it enough to know whether a deductive method like yours would predict anything about it. I have come to a knowledge of my ignorance of the nature of electrical conduction in metals which is a phenomenon like that of heat, and both very easy to formulate but difficult to conceive.

As regards Clausius he pointed out *gross* mistakes in M. [Maxwell].[a] I have no doubt he has some of his own but I have not had patience to find them out except that he stuck to uniform velocity in the molecules though I proved it impossible and pointed out the only true distribution of velocity. Clausius's uniform velocity leads (by sound mathematics) to an expression for the mean relative velocity which is unsymmetrical with respect to the components so that you need to know which is the greater of the two velocities and to put it in the right place of the formula.

With respect to Riemann[b] for whom I have great respect and regret I only lately got either Pogg. or Phil. Mag.[c] from the binder and wrote you a rough note for yourself. I now have him more distinct. Weber says that electrical force depends on the distance and its 1st and 2nd derivatives with respect to t.[d]

Riemann says that this is due to the fact that the potential at a point is due to the distribution of electricity elsewhere not at that instant but at times before depending on the distance.

In other words potential is propagated through space at a certain rate and he actually expresses this by a partial diff. eq^n appropriate to propagation. Hence either (1) space contains a medium capable of dynamical actions which go on during transmission independently of the causes which excited them (and this is no more or less than my theory divested of particular assumptions), or (2) if we consider the hypothesis as a fact without any etherial substratum ⟨and if A and B are two bodies each of which can vary in electrical power, say each a pair of equal magnets [Diagram 1] one of which revolves about the middle of the other so that the combination is alternately $=2$ and $=0$.

[Diagram 1]

Now let things be so arranged that the time of propagation from A to $B = tT$ then if the magnetism of A be $A\cos(nt + \alpha)$ and that of B $B\cos(nt + \beta)$ the action of B on A will be $A\cos(nt + \alpha)B\cos(nt + \beta - nT)$ into a function of the distance and that of A on B $A\cos(nt + \alpha - nT)B\cos(nt + \beta)$ into same function. The difference of these is $F(r)AB\sin nT(\sin(nt + \beta) - \sin(nt + \alpha))$ that is, action and reaction are not equal and opposite. (I mean pushes and pulls not Hamiltonian action.)

Weber's action and reaction are equal but his energy is unreclaimable⟩.[e] Riemann's action and reaction between the gross bodies are unequal and his energy is nowhere unless he admits a medium which he does not do explicitly. My action and reaction are equal only between things in contact not between the gross bodies till they have been in position for a sensible time, and my energy is and remains in the medium including the gross bodies which are among it.

Instead of the part about A and B read as follows [Diagram 2]. Let X and Y be travelling to the right with velocity v at a distance a then the force of X on Y will be $\dfrac{XY}{a^2}\left(1 - \dfrac{v}{V}\right)^2$ and that of Y on X $\dfrac{XY}{a^2}\left(1 + \dfrac{v}{V}\right)^2$, where V is the velocity of transmission of force. If the force is an attraction and if X and Y are connected by a rigid rod ⟨which⟩ X will be pulled forward more than Y is pulled back and the system will be a locomotive engine fit to carry you through space with continually increasingly velocity. See Gulliver's Travels in Laputa.[f]

Yours Truly
J. Clerk Maxwell.

X
•
———→

Y
•
   ———→

[Diagram 2]

a. Maxwell is referring to the criticisms Clausius made of Maxwell's 1860 paper (document III-6) in 1862 ("Ueber die Wärmeleitung ..." cited in note 45 of chapter I). Clausius criticized Maxwell's use of a spherical distribution of velocity in situations where the velocity distribution could not be spherically symmetric in the gas.

b. Georg Friedrich Bernhard Riemann (1826–1866) is known primarily for his work in mathematics but also wrote a few papers on theoretical physics.

c. This is a reference to Riemann's "Ein Beitrag zur Electrodynamik," *Ann. Phys.* [2] 131 (1867), translated in *Phil. Mag.* [4] 34 (1867), 368–372. This paper, giving Riemann's propagated action theory of electromagnetic forces, was originally presented to the Göttingen Königliche Akademie in 1858 but was withdrawn from publication and remained unpublished until after Riemann's death.

Maxwell's criticism of propagated action theories, sketched in the remaining part of this letter, was published in his "Note on the Electromagnetic Theory of Light" (see *Scientific Papers*, vol. 2, 137–143). See also *Treatise on Electricity and Magnetism*, vol. 2, sec. 862–866. A number of people mistakenly supposed that Maxwell's criticism of the propagated action theories is invalidated by his reference (*Scientific Papers*, vol. 2, 138) to the nonrelativistic concept of a rigid rod (cf. the last paragraph of this letter). In fact, the criticism, being based on a static distribution of forces in the rod or train, remains sound and even allows for relativistic dynamics. Static distribution can only be evaded in the propagated action theories by assuming the peculiar combination of advanced and retarded potentials proposed by J. A. Wheeler and R. P. Feynman, "Interaction with the Absorber as the Mechanism of Radiation," *Reviews of Modern Physics* 17 (1945), 157–181.

d. Wilhelm Weber (1804–1851). His law was published in "Electrodynamische Maasbestimmungen inbesondere Widerstandsmessungen," *Abhandlungen der Mathematisch-Physischen Classe der K. Sächsischen Gesellschaft der Wissenschaften, Leipzig* 1 (1852), 197–382; translated in *Scientific Memoirs*, Richard Taylor, ed. (London, 1852; New York: Johnson Reprint Corp., 1966), vol. 5.

e. See Maxwell's *Treatise on Electricity and Magnetism*, vol. 2, sec. 851–854 for his mature criticism of Weber's electromagnetic theory.

f. Jonathan Swift (1667–1745), in his *Gulliver's Travels* (1726), "Voyage to Laputa," chapter 3, describes a fantastic floating island propelled through the air by a lodestone that attracts the earth at one end and repels it at the other. Cf. *Maxwell on Saturn's Rings*, 1, for another accomplishment of the Laputan scientists described in the same chapter, the discovery of two satellites of Mars.

32. Letter from Maxwell to Lewis Campbell, October 19, 1872

Lewis Campbell and William Garnett, *Life of James Clerk Maxwell*, 383–384.

Glenlair, Dalbeattie, 19th October, 1872.

[…] Lectures begin 24th. Laboratory rising,[a] I hear, but I have no place to erect my chair, but move about like the cuckoo, depositing my notions in the chemical lecture-room 1st term; in the Botanical in Lent, and in Comparative Anatomy in Easter.

I am continually engaged in stirring up the Clarendon Press,[b] but they have been tolerably regular for two months. I find nine sheets in thirteen weeks is their average. Tait gives me great help in detecting absurdities. I am getting converted to Quarternions, and have put some in my book, in a heretical form, however, for as the Greek alphabet was used up, I have used German capitals from 𝔄 to 𝔉 to stand for vectors, and, of course ∇ occurs continually. This letter is called "Nabla," and the investigation a Nablody.[c] You will be glad to hear that the theory of gases is being experimented on by Profs. Loschmidt and Stefan of Vienna, and that the conductivity of air and hydrogen are within 2 per cent of the value calculated from my experiments on friction of gases, though the diffusion of one gas into another is "*in erglanzender ubereinstimmung mit* $\frac{dp}{dt}$ *schen Theorie.*"[d]

a. The Cavendish Laboratory was not completed until 1874, but Maxwell's appointment as Professor was in 1871, and he began lecturing in the autumn of the same year.

b. Maxwell was seeing his treatise on electricity and magnetism through the press at this time. It was not published until 1873.

c. ∇ is the operator $\frac{d}{dx} + \frac{d}{dy} + \frac{d}{dz}$. The term nabla for ∇, which is still used by some British mathematicians in preference to Gibbs's del, was coined by Tait's student W. Robertson Smith, the Assyrian scholar, from the resemblance of ∇ to the shape of the Assyrian harp of that name. See C. G. Knott, *Life and Scientific Work of P. G. Tait* (1911), 143–147, 172–173, which gives a poem from Maxwell to Tait with a facsimile of the Hebrew inscription to Tait as "The Chief Musician on Nabla" written out by Smith.

d. As noted in chapter I, the apparent agreement of experimental data with Maxwell's theory was destroyed when Boltzmann found an error in the latter.

33. [Notes on Boltzmann's *H* Theorem][a]

Cambridge University Library, Maxwell Collection, Scientific Papers 5.

Let N_A be the number of molecules which are in the state A and let the ⟨number of⟩ *proportion* of these which in unit of time pass into the state B be denoted by C_{AB}, then the number which pass from the state A to the state B in time dt is $N_A C_{AB} dt$.

Similarly the number of those which pass from the state B to the state A in the time dt is $N_B C_{BA} dt$.

It may be shown by Hamilton's principle of variable action that in whatever way the states A and B are defined

$$C_{AB} = C_{BA}$$

The whole number which passes from the state A into any other state is

$$dt \, N_A \sum_B C_{AB}$$

The whole number which enter the state A from any other is

$$dt \sum_B N_B C_{BA}$$

The rate of increment of N_A is

$$\frac{dN_A}{dt} = \sum_B (N_B - N_A) C_{AB}$$

Now consider the quantity $E = \sum_A N_A \log N_A$

$$\frac{dE}{dt} = \sum_A \frac{dN_A}{dt}(\log N_A + 1)$$

$$= \sum_A (\log N_A + 1)\sum_B (N_B - N_A)C_{AB}$$

$$= \sum_B (\log N_B + 1)\sum_A (N_A - N_B)C_{BA}$$

$$= \sum_A \sum_B (\log N_A - \log N_B)(N_B - N_A)C_{AB}$$

a. Maxwell uses Boltzmann's original symbol E for the quantity now called H; see Brush, *Kind of Motion we Call Heat*, 619, 626n.

34. [Draft of "On Loschmidt's Experiments on Diffusion in Relation to the Kinetic Theory of Gases"][a]

Cambridge University Library.

[...] by prof Loschmidt[b] of Vienna, ⟨and it was in comparing his results with the kinetic theory of gases⟩ and in making a revision of the kinetic theory of gases ⟨I was⟩ and comparing these results of Prof Loschmidt with that theory I obtained evidence of the consistency of these results with each other and with the other experimental data of the theory which I consider encouraging both to experimenters and speculators. Besides the diffusion of matter ⟨of which⟩ with which we are now dealing, there are two other kinds of diffusion on which experiments have been made—that of momentum or the lateral communication of ⟨visible⟩ sensible motion from one portion of a gas to another, and that of kinetic energy. The diffusion of momentum gives rise to internal friction or viscosity, that of energy to the conduction of heat. The investigations on viscosity have been made by Graham,[c] O. E. Meyer[d] and myself.[e] They involve the construction of the ⟨effects of visco⟩ mutual action between the moving gas and the ⟨solid⟩ surfaces of the solids over which it moves. Hence in all these ⟨experiments⟩ investigations it is only by carefully arranged method of comparison that trustworthy result can be obtained. The conduction of heat in air has recently been investigated experimentally by Prof Stephan[f] of Vienna who finds his results in striking agreement with the kinetic theory but the ⟨difficul⟩ practical difficulties of the investigation are even greater than in the case of viscosity.

[End of sheet]

In Prof Loschmidt's experiments on diffusion, on the other hand, everything appears favourable to the accuracy of the results. He appears to have got rid of all disturbing currents, and his methods of measurement, founded on those of Bunsen[g] are most precise. The interdiffusing gases are left to themselves and are not disturbed by the presence of any solid body. The results of different experiments with the same pair of gases are very consistent with each other. They prove conclusively that the coefficient of diffusion varies inversely as the pressure, a result in accordance with the kinetic theory ⟨in⟩ whatever hypothesis we assume for the mode of action between the molecules.

They also show that the coefficient of diffusion increases as the temperature rises, but the range of temperature in the experiments was ⟨somewhat⟩ too small to enable us to decide whether it varies as T^2 which it does according to the theory of a force inversely as the ⟨square⟩ fifth power of the distance or as $T^{3/2}$ as it does according to the theory of the elastic spherical moleclues.

In comparing ⟨experiments⟩ the coefficients of diffusion of different pairs of gases Prof Loschmidt has adopted a formula which is simple enough but which does not appear to me to agree ⟨either⟩ with the kinetic theory ⟨or with the results of his⟩ and ⟨which assign values⟩ from which he deduces values for a quantity k',[h] which according to him should be constant for all gases but these values do not agree together ⟨in a way at all comparable with the agreement of⟩ in the manner which we should expect from the accuracy of the experiments.

[End of sheet]

According to the kinetic theory deduced from the collisions of elastic spheres the coefficient of diffusion between two gases at standard pressure and temperature

$$D_{12} = \frac{p_1 p_2}{\rho_1 \rho_2} \frac{M_1 + M_2}{\pi s^2 V}$$

$$D_{12} = \sqrt{\frac{1}{w_1} + \frac{1}{w_2}} \frac{M}{N} \frac{1}{2\sqrt{6\pi}} \frac{1}{s_{12}^2}$$

where w_1 and w_2 are the molecular weights of the gases, that of hydrogen being unity, M is the mass of a molecule of hydrogen, V the velocity of mean square of hydrogen $= (3p/\rho)$ at standard pressure and temperature, N the number of molecules in unit of volume (the same for all gases) and s_{12} the distance between the centres at collision. Hence if we make $A^2 = (MV/N2\sqrt{6\pi})$ and

$$\Sigma_{12}^2 = \frac{1}{D_{12}} \sqrt{\frac{1}{w_1} + \frac{1}{w_2}} \text{ we find } s_{12} = A\Sigma_{12}$$

The quantity A which is constant for all gases contains two hitherto unknown quantities, the mass of a molecule of hydrogen and the number of molecules in unit of volume. The product of these quantities is the density of hydrogen, a known quantity but their ratio which appears in A is not fully ascertained though prof Loschmidt himself was the first to estimate it roughly which has been since repeated independently by Mr. Stoney[i] and Sir William Thomson.[j]

[End of sheet]

The quantity Σ_{12} is proportional to the distance between the centres of the molecules at the distance of collision or to $r_1 + r_2$ if r_1 and r_2 are the radii of the molecules. Now from Prof. Loschmidt's data we can deduce the values of Σ for the six pairs of gases which may be made up from the four gases H, O, CO and CO_2. These, according to our theory, are not independent, being the six sums of pairs of the four independent quantities r_1, r_2, r_3, r_4. Accordingly, assuming

$2r\,(H) = \langle 1.043 \rangle$	$2r(O) = 2.283$
1.734	$\langle 1.370 \rangle$
$2r(CO) = \langle 1.477 \rangle$	$2r\,(CO_2) = \langle 1.665 \rangle$
2.461	2.775

we find,

	By calculation	By Loschmidt
	$\Sigma_{12} = r_1 + r_2$	$\Sigma_{12} = \sqrt{\frac{1}{w_1} + \frac{1}{w_2}} \frac{1}{D_{12}}$
$\Sigma\,(H, O)$	2011	1992
$\Sigma\,(H, CO)$	2100	2116
$\Sigma\,(H, CO_2)$	2257	2260
$\Sigma\,(O, CO)$	2372	2375
$\Sigma\,(O, CO_2)$	2529	2545
$\Sigma\,(CO, CO_2)$	2618	2599

The agreement of these numbers furnishes I think $\langle a \rangle$ very strong evidence in favour of the kinetic theory of gases. But we may derive further evidence of a higher order by a comparison between experiments of two entirely different kinds, those on diffusion, already spoken of, and those in viscosity.

If μ is the viscosity of a gas and ρ its density, the theory gives,

$$\frac{\mu}{\rho} = A \sqrt{\frac{2}{w}} \frac{1}{(2r)^2}$$

so that we have the following relation between the viscosities of the two gases and their coefficient of diffusion

$$2D_{12} = \frac{\mu_1}{\rho_1} + \frac{\mu_2}{\rho_2}$$

Calculating on this system the viscosities of the gases experimented on by Loschmidt in centimetre gramme second measure and comparing them with those deduced by E. Meyer[k] from his own experiments and those of Graham and with my own we find,

	the experiments μ	Meyer	Maxwell
H	0.000116	0.000134	0.0000971
O	270	306	
CO	217	266	
CO_2	214	231	0.000161

The numbers do not appear to agree very well with each other. The numbers given by Meyer are all greater than those deduced from diffusion. Mine, on the other hand, are smaller. I have no doubt, however, that the best method of determining all these quantities is by the comparison of the diffusion coefficients of a great many pairs of gases.

I have shown that by this method we may obtain a set of numbers which are proportional to the diameters of the molecules of different gases. ⟨We already⟩ From these we can determine their relative volumes and since we already known their relative masses we can determine the⟨ir⟩ relative densities of the molecules. These densities have been compared by Loschmidt and Lorenz Meyer[l] with the densities of the same substances in the liquid condition and with the "molecular volumes" of the ⟨various⟩ substances in ⟨their⟩ its compounds as estimated by Kopp.[m] It appears from these comparisons that the relative molecular volumes of the gases whose viscosities have been determined are on the whole
[End of sheet]
roughly proportional to those of the same substance in the liquid state or in their liquid compounds.

It is manifest, however, that the density of the molecules must be greater than that of the liquefied substance, for not only for even if we could determine the density of the substance at $-273°$ C and at an infinite pressure there would still be interstices between the spherical molecules. Indeed the method of estimating ⟨specific volume⟩ molecular volume by observations on liquids at the boiling point under a pressure of 76 cm of mercury seems a very arbitary one, for there is no reason why the average pressure of our atmosphere at the level of the sea should be ⟨taken as an absolute standard⟩ placed on any very high rank as a physical constant. It is probable that at all observed temperatures the molecules of bodies are kept ⟨at a certain distance⟩ further apart by their motions of agitation than they would be if in actual contact. It is therefore more likely that we should obtain consistent results if we measured the molecular volume of substances when that volume is the smallest attainable. The ⟨phenomena⟩ volume relations of potassium its oxide and its hydrated oxide as described by Faraday[n] seem to indicate that the whole theory of molecular volume is not quite understood.

If, however, we assume as the molecular volume of oxygen that already deduced by Kopp from that ⟨volume⟩ of oxide of tin, namely, 2.7 when O = 16 which is the smallest ⟨that I am⟩ of those quoted by L. Meyer we find for the number of molecules of any gas in a cubic centimetre at 760 mm and 0° C $N = 19 \times 10^{18}$ Hence the side of a cube which would on an average contain one molecule is $N^{-1/3} = 37.67$ [sic] tenth metres.[o] The mass of a molecule of hydrogen is,

$$M = 4.607 \times 10^{-24} \text{ grammes}$$

and its diameter is ⟨12.67 twentieth metres⟩ 5.8 tenth metres

O	7.6
CO	8.3
CO_2	9.3

[End of sheet]

These estimates are much smaller than those of Prof. Loschmidt, Mr. Stoney, and Sir W. Thomson.[p] This arises from the molecular volume of oxygen being assumed much smaller than is usually done. There is another quantity, however, which plays a considerable part in the kinetic theory as developed by Clausius, namely the mean length of the uninterrupted path of the molecule. The determination of this quantity does not involve any estimate of such doubtful matters as the density of a molecule. We find the length of the mean path

for hydrogen	$l = 965$	tenth metres
Oxygen	= 560	
Carbonic Oxide	= 482	
Carbonic acid	= 430	

The length of a wave of green light is about ten times the mean length of path of a molecule of oxygen at 760 mm and 0° C

[End of sheet]

a. The draft is incomplete but the text is obviously continuous and not just rough notes. The order of the sheets was dictated by the order in which the subject matter is dealt with in the published paper (see document III-42). The sheets do not form a continuous draft but are housed in the same box in the collection of Maxwell manuscripts.

b. Johann Joseph Loschmidt (1821–1895) made the first good estimate of the size of a molecule (1865) using Maxwell's measured values for the mean free path and Hermann Kopp's condensation coefficients for gases (see chapter I, note 139). He published an extensive series of experiments on diffusion in 1870 (chapter I, note 128). Ironically he is now best known as a critic of kinetic theory because of his "reversibility paradox" against Boltzmann's *H* theorem; see Brush, *Kind of Motion We Call Heat*, sec. 14.5

c. See Graham's 1846 and 1849 papers, cited in chapter I, note 57.

d. Oskar Emil Meyer (1834–1909) published many papers on the theory of viscosity as well as reporting his own experiments; see Brush, *Kind of Motion*, 435–442, 749. Meyer's textbook *Die Kinetische Theorie der Gase* (Breslau: Maruschke & Berendt, 1877; second edition, 1889) helped to popularize the kinetic theory among physicists, though he retained the mean free path long after Maxwell's second transport theory should have superseded it.

e. Maxwell, "On the Viscosity or Internal Friction of Air and Other Gases," document III-18.

f. The reference should be to Josef Stefan (1835–1893); see his paper cited in chapter I, note 111.

g. Robert Wilhelm Eberhard Bunsen (1811–1899), *Gasometrische Methoden* (Brunswick, 1857), translated by Henry E. Roscoe as *Gasometry: Comprising the Leading Physical and Chemical Properties of Gases* (London, 1857); reviewed in *Phil. Mag.* [4] 14 (1857), 146–148.

h. Loschmidt examined the temperature and pressure behavior of several pairs of gases separately and wrote the diffusion coefficient k as $k = k_0 T^2/P$, where T is the absolute temperature, P the pressure, and k_0 the value for the diffusion coefficient at standard temperature and pressure. He also examined the diffusion of various pairs of gases and concluded that a coefficient k' that was constant for all gases could be constructed, $k' = k_0/\sqrt{m_1 m_2}$, where m_1 and m_2 are the masses of the molecules of the two gases. Even Loschmidt admitted in his account that k' could not be easily estimated. See his second 1870 paper cited in chapter I, note 128.

i. George Johnstone Stoney (1826–1911). The paper to which Maxwell refers (cited in chapter I, note 140) does not contain any estimate for the mass or size of a molecule. What Stoney was concerned with, as an astronomer and a theoretician interested in the nature of light, was comparing the velocities of molecules with that of light and the mean free path of the molecule and the number of collisions it suffers in a second with the period of vibration of light. By assuming that light waves arise from the vibrations of molecules, Stoney compared the two motions, translational and vibratory, of molecules. The times between collisions were extremely small for physicists to imagine, and Stoney was the first to publish any values for them. These time periods, however, were large compared with the period of the vibrations associated with light. It was to compare these numbers and introduce his colleagues to such minute, fast phenomena that Stoney published this paper, not to estimate the size of a molecule.

j. William Thomson (1824–1907), 1870 papers cited in chapter I, note 140. In the first of these papers Thomson estimates the size of a molecule (not an atom) based on the mean free path values Maxwell deduced from his experiments on viscosity, assuming that at condensation molecules are arranged in a cubic array in a fluid. He used capillarity arguments and considerations of surface tension to give him independent estimates for the size of molecules. They all gave him the same order of magnitude, so he was confident of his results. In the second paper he obtained another estimate from electrostatic effects in a Daniell's battery, confirming those based on "Clausius' and Maxwell's magnificent working out of the Kinetic Theory of Gases."

k. Oskar Emil Meyer, paper cited in chapter I, note 70. The last part of the paper (pp.353–425) contains the actual tabulations of viscosity for the gases listed.

l. Maxwell must have meant not "Lorenz Meyer" (there was no scientist with this name at the time) but (Julius) Lothar Meyer (1830–1895), brother of Oskar Emil Meyer. Lothar Meyer was one of the first chemists to accept Avogadro's hypothesis and Cannizzaro's method of determining atomic weights, presented at the Karlsruhe conference in 1860. Lothar Meyer attempted to systematize the elements on the basis of atomic weight, though not quite as early or as successfully as Mendeleev. His paper "Ueber die Molecularvolumina chemischer Verbindungen," *Annalen der Chemie und Pharmacie* (suppl. bd.) 5 (1867), 129–147, is cited by Maxwell in the published version of this paper (document III-42; *Scientific Papers*, vol. 2, 349), but the name is still given as Lorenz Meyer.

m. Hermann Kopp (1817–1892) was another physical chemist who tried to relate the physical properties of substances to their chemical nature. He developed the idea of "atomic volume" and the experimental methods to measure it. These consisted of measuring the change in volume of a substance at condensation, assuming that in a fluid the molecules are tightly packed. Similar considerations could be used to estimate the atomic volume during crystallization. He concluded that similar elements and isomorphous compounds had similar atomic volumes and tried to relate this to their crystal structure. See "Ueber Atomvolum, Isomorphismus and specifisches Gewicht," *Annalen der Chemie und Pharmacie* 75 (1839), 406–435. See also "On the Atomic Volume and Crystalline Condition of Bodies," *Phil. Mag.* [3] 18 (1841), 255–264.

n. Faraday's work on potassium was published as "On the Fluidity of Sulphur and Potassium at Common Temperatures," *Quart. J. Sci.* 2 (1827), 469–472.

o. A tenth-metre is 10^{-10} meters.

p. See the papers by Loschmidt, Stoney, and Thomson cited in chapter I, notes 139 and 140.

35. "*N* August 1, 1873" [Calculations][a]

Cambridge University Library, Maxwell Collection.

N August 1 1873

$$\varepsilon = \frac{\pi s^3 N}{6} = \frac{\rho_{gas}}{\rho_{max}} \qquad A = \frac{2\sqrt{6\pi N}}{v} \qquad \varepsilon = \frac{2\rho_n v}{N}$$

$$As^2 = \Sigma^2 \qquad \varepsilon^2 = \frac{\pi^2 N^2}{6^2}\frac{\Sigma^6}{A^3} = \frac{\Sigma^6 \pi^2 N^2 v^3}{6^2 8.6\pi\sqrt{6\pi}\cdot N^3}$$

$$N = \frac{\Sigma^6 \pi^{1/2} v^3}{6^3 8\sqrt{6\varepsilon^2}} = \frac{\sqrt{\pi}\,9\sqrt{6}v^3}{10^6 8\rho_n^2 v^2} \qquad v = \text{molecular } v$$

$\sqrt{\pi}$	0.248575		
9	0.954243	10^6	6.
$\sqrt{6}$	0.389075	8	0.103090
v^3	15.808128		5.952201 − 10
	17.400021 + 10		5.952201 − 10
Σ^6	8.807492		8.807492 − 10
$V^2 N$	18.592529		$\tau = \rho/N$
	9.739317 − 10		
$N = 2.147 \times 10^{18}$	18.331846		
	5.952201 − 10		
ρ	22.379645		

2.397 × 10²² molecules of hydrogen in a gramme
2.147 × 10¹⁸ molecules in a cubic centimetre at 0° C and 760 B
Size of cube   6.110615   1,209000 of a centimetre

$$A = \frac{2\sqrt{6\pi N}}{v}$$

$\sqrt{\pi}$	0.248575	
$\sqrt{6}$	0.389075	$s = \Sigma 10^{-7}$
2	0.301030	
N	18.331846	
	19.270526	
v	5.269376	
A	14.001150	
$\sqrt{A}$	7.000575	

a. N means Avogadro's number. This estimate of the number of hydrogen molecules in a cubic centimeter at standard temperature and pressure was based on kinetic theory, i.e., defining the mean free path in terms of the molecular diameter and using the condensation coefficient of Hermann Kopp. Loschmidt was the first to combine these two methods to get an estimate for the size of a molecule (1865), thereby making possible an explicit estimate of what is now called "Avogadro's number" (chapter I; note 139). The condensation coefficient is measured by the increase or decrease in volume at evaporation or condensation, respectively. If l is the mean free path and s the diameter of the molecule, then $(1/l) = (4/3)\pi s^2 N$, where N is the number of molecules in a unit of volume. This is taken from Clausius's kinetic theory. If the condensation coefficient is defined as the unit of volume take up by a molecule at rest, the condensation coefficient and the mean free path can be connected by considering the volume swept out by a molecule in moving through one mean free path length. Under such circumstances, $s = 8\varepsilon l$, where ε is the condensation coefficient. In these notes Maxwell uses the same notation as above with the addition that ρ_{gas} is the density of the substance as a gas and ρ_{max} is its density when compressed at absolute zero with no spaces left between the molecules. In addition, v denotes "molecular volume" and v the volume of the gas containing N molecules. In defining A, Maxwell has assumed that the mass of the hydrogen molecule is 1.

36. [Calculations for the Number of Molecules of Hydrogen in Unit Volume]

Cambridge University Library.

Air by Maxwell 0.0001878
0001967

Hydrogen

H_{CO_3}
2

ρ_{27} = 5.958161
1.431364
6.526797
0.480600

Σ^2 0.480600

μ = 0.0001112   6.046197
2.855   9.686189
                     6.360008

2291

Oxygen
0.602060
7.128857
0.717012
0.0002581   6.411845

CH
6.526797
0.451545
6.978342
0.831616
1402   6.146726

Oxygen 16   6.526797
             1.204120
             7.730917
             0.717012
0.001032   7.013905

CO 14   6.526797
0.573064
5.953733
7.099861
0.782224
.00276   .771509
NO_{22} 6.317637
             6.526797
             0.671211
             7.198008
             0.972296
1681   6.225714

CO_2 14   6.526797
              1.146128
              7.672925
              0.886526
.0006116   9.238413
              6.786399

CO_2
22   6.526797
      0.671211
      7.198008
      0.886526
.807 6.311482
.002049   9.906876   air   002539
SO_2 604608
32   6.526797
      0.752595
      7.279392
      0.985242
1968 6.294130

τ $\quad$ V $\qquad$ Almost condensation $= \frac{1}{8} s^3 N = \varepsilon$

SO_2 $\quad$ 64 $\quad$ 43.9 $\qquad$ $\sum s^2 = A\Sigma^2$ $\qquad$ $\frac{1}{6}\pi s \Sigma^2 \dfrac{\overline{v^2}\tau N}{9\pi\overline{v}} \dfrac{\theta^{3/2}}{p_0} = \varepsilon$

$\overline{v^2} = \dfrac{3p}{\rho}$

$\varepsilon = \Sigma^2 s \dfrac{\dfrac{1}{\pi}}{\sqrt{6.9}} \dfrac{9p\sqrt{\dfrac{3p}{\rho}}}{\dfrac{2}{\sqrt{\pi}}\sqrt{\dfrac{2}{3}}}$ $\qquad$ $\dfrac{1}{p} = s\Sigma^2 \sqrt{\pi}\sqrt{\dfrac{p}{\rho}}$

$A = \dfrac{\tau\left(\dfrac{p}{\rho}\right)^{3/2}}{2\sqrt{2\pi}} \dfrac{1}{p_0}$ $\qquad$ $\varepsilon^2 = \dfrac{1}{36}\pi^2 A^3 \Sigma^6 \dfrac{\rho_2^2}{\tau^2} = \dfrac{\pi^2}{6^2}\left(\dfrac{p}{\rho}\right)^{9/2} \dfrac{\tau}{16\sqrt{2\pi}\sqrt{\pi}} \dfrac{\rho^2}{p^3}$

$\varepsilon^2 = \dfrac{\sqrt{\pi}\tau}{6^2 4^2 \sqrt{2}}\left(\dfrac{p}{\rho}\right)^{1/2} \dfrac{p}{\rho^2}\Sigma^6$ $\qquad$ $\tau = \dfrac{6^2 4^2 \sqrt{2}}{\sqrt{\pi}}\left(\dfrac{\rho}{p}\right)^{1/2} \dfrac{\rho^2}{p}\dfrac{\varepsilon^2}{\Sigma^6}$

Let ρ_2 be density of hydrogen $w\rho$ of any gas $= \rho_2$(molec[ular] volume)

$\tau = \dfrac{6^2 4^2 \sqrt{2}}{\sqrt{\pi}}\left(\dfrac{\rho}{p}\right)^{3/2} \dfrac{\rho^4 (\text{molecular } V)^2}{\Sigma^6}$ $\qquad$ $\left(\dfrac{6}{10}\right)^6$

$6^2 4^2$ $\quad$ 2.760422

$\sqrt{2}$ $\quad$ 0.150515

37. [Comparison of Data on Diffusion]

Cambridge University Library.

	Hydrogen	Oxygen	Carbonic Oxide	Carbonic Acid
Density at 0° C and 760 mm				
Relative molecular mass (H = 1)	1	16	14	22
Velocity of mean square metres per second	1859	464.9	497.0	396.6
Mean velocity	1713	428.3	457.9	365.2
Mean path tenth-metres[a]	965.5	560.2	482.1	379.2
Collisions per second (millions)	17750	7646	9489	9720
Coefficient of Viscosity	0.000116	0.000210	0.000217	0.000214
Diameter of molecule in tenth-metres	5.8	7.6	8.3	9.3
Mass of molecule in twenty-fifth grammes[b]	46	736	644	1012
μ/ρ	1.299	0.1884	0.1748	0.1087
O. E. Meyer	1.49	0.213	0.212	0.117

Diffusion Coefficients

	metres^2/hour			cents^2/second[d]		
	Calculated	Observed[c]	difference	Calculated	Observed	difference
H and O	.2551	.2598	+.0047	.7086	.7214	+.0128
H and CO	.2347	.2312	−.0035	.6519	.6422	−.0097
H and CO$_2$	.2007	.2001	−.0006	.5585	.5558	−.0017
O and CO	.06505	.06488	−.00017	.1807	.1802	−.0005
O and CO$_2$	.05137	.05074	−.00063	.1427	.1409	−.0018
O and CO$_2$	.04988	.05060	.00072	.1386	.1406	+.0020

Table of Diffusion (centimetres)^2/second

	Observed	Calculated
H and O	0.7214	0.7086
H and CO	0.6422	0.6519
H and CO$_2$	0.5558	0.5575
O and CO	0.1802	0.1807
O and CO$_2$	0.1409	0.1427
CO and CO$_2$	0.1406	0.1386

H and H	Diffusion	1.299 ⎱
O and O	of momentum	0.1884
CO and CO	⟨into⟩	0.1748 ⎰ Diffusion of momentum
CO$_2$ and CO$_2$	⟨itself⟩	0.1087

Cane sugar in water 0.00000365 or in a day 0.3144 v oil
Salt and water   0.00000116
Air   0.256 ⎱
Copper   1.077 ⎰ Diffusion of Temperature Stefan
Iron   0.183

Regnault density of hydrogen grammes/cc

$$\bar{v} = \frac{2}{\sqrt{\pi}} \sqrt{\frac{2}{3} \overline{v^2}} = \sqrt{\frac{8}{3\pi}} \sqrt{\overline{v^2}}$$

log 8 = 0.903090	
3 = 0.477121	
π 0.497151	
2 9.928819	
9.964409	

	5.952201
p_0	6.013836
$\overline{v^2}$	10.061635
log 3	0.477121
$\overline{v^2}$	10.538756
$\sqrt{\overline{v^2}}$	5.269378
	9.964409
$\bar{v}$	5.233787

TABLE 1

	$\sqrt{\text{sp} \cdot \text{g}}$ [e]	Velocity mean sq	Mean velocity	metres per second	
				Vel mean square	Mean vel
Hydrogen	0	5.269376	5.233787	1859	1713
Oxygen	$\sqrt{16}$ 0.602060	4.667316	4.631727	464.8	428.3
CO	$\sqrt{14} = 0.573064$	4.696312	4.660723	497.0	457.9
CO$_2$	$\sqrt{22} = 0.671211$	4.598165	4.562576	396.4	365.2
CH	$\sqrt{8}$ 0.451545	4.817831	4.782242	657.4	605.7

TABLE I

		Hydrogen 1	Oxygen 16	Carbonic Oxide 14	Carbonic Acid 22
I	Mass of molecules compared with Hydrogen	1	16	14	22
	Velocity (of mean square) (metres per second) at 0° C	1859	465	497	396
II	Mean Path (tenth-metres)	965	560	482	379
	Collisions in a second (millions)	17750	7646	9489	9720
III	Diameter in centimetres ⟨tenth-metres⟩	5.8×10^{-8}	7.6	8.3	9.3
	Mass in grammes ⟨twenty fifth grammes⟩	46×10^{-25}	736	644	1012

a. A tenth-metre is 10^{-10} meter.

b. A twenty-fifth gramme is 10^{-25} gram.

c. The data in the "Observed" column are from Loschmidt's 1870 paper (cited in chapter I, note 128), 466–467. Maxwell included a comparison of data on the diffusion of various gases, taken from Loschmidt, at the end of the later editions of his *Theory of Heat*, e.g., the seventh edition (London: Longmans, Green, and Co., 1883), 332.

d. centimeters^2/second.

e. $\sqrt{\text{specific gravity}}$

38. [Notes on Diffusion]

Cambridge University Library.

Diffusion[a]
  Diffusion

$$\rho_1 \frac{\partial u_1}{\partial t} + \frac{dp_1}{dx} = \frac{\rho_1 \rho_2 \pi s^2 V}{M_1 + M_2}(u_2 - u_1) + X\rho_1 \tag{1}$$

Let $p_1 + p_2 = p$, $\rho_1 + \rho_2 = \rho$ $p_1 u_1 + p_2 u_2 = pu$

$$\rho_2 \frac{\partial u_2}{\partial t} + \frac{dp_2}{dx} = \frac{\rho_1 \rho_2 \pi s^2 V}{M_1 + M_2}(u_1 - u_2) + X\rho_2 \tag{2}$$

adding (1) + (2) $\dfrac{dp}{dx} = X\rho$

$$\frac{dp_1}{dx} = \frac{\rho_1 \rho_2 \pi s^2 V}{M_1 + M_2}\left\{ \frac{pu - p_1 u_1}{p_2} - u_1 \right\} + X\rho_1$$

$$\frac{dp_1}{dx} - X\rho_1 = \frac{\rho_1 \rho_2}{p_1 p_2} \frac{\pi s^2 V}{M_1 + M_2}\{pp_1(u - u_1)\} \quad p = \frac{1}{3}v\tau N$$

$$D = \frac{p_1 p_2}{\rho_1 \rho_2} \frac{M_1 + M_2}{\pi s^2 V} \frac{1}{P} \quad \text{where } V \text{ is the mean velocity of } M_1 \text{ relative to } M_2$$

$$= \frac{1}{q}v^4 \frac{1}{w_1 w_2} \frac{\tau(w_1 + w_2)}{\pi s^2 \sqrt{\dfrac{8}{3\pi}} \sqrt{\dfrac{1}{w_1} + \dfrac{1}{w_2}}} \frac{1}{vP}$$

$$= \frac{\tau\sqrt{\dfrac{1}{w_1} + \dfrac{1}{w_2}}}{6\sqrt{6\pi s^2}} \frac{v^3}{P} = \frac{\sqrt{\dfrac{1}{w_1} + \dfrac{1}{w_2}}}{2\sqrt{6\pi s^2}} \frac{v}{N}$$

when v = vel[ocity] of mean sq[uare] of hydrogen
Diffusion by 1860 proof XIV[b]

$$u_1 \rho = -\frac{1}{3}\frac{d}{dx}(\rho vl)$$

$$p_1 u_1 = -\frac{1}{3}\frac{d}{dx}(p_1 vl)$$

$$\frac{d}{dx}(pu) + \frac{dp}{dt} = 0 \qquad \frac{dp}{dt} = \frac{1}{3}\frac{d^2}{dx^2}(pvl) \qquad D = \frac{1}{3}vl$$

[End of sheet]

a. Maxwell's title.

b. Proposition XIV of Maxwell's paper "Illustrations of the Dynamical Theory of Gases," document III-6; *Scientific Papers*, vol. 1, 393–394. The proposition was to find the quantity of matter transferred across a plane by the motion of agitation of the molecules. If q was the matter transferred across a unit area in a unit of time, $q = -(1/2)(d/dx)(pvl)$, where v is the mean velocity of agitation and l the mean free path of the molecules.

39. [Notes on Viscosity]

Cambridge University Library.

Viscosity[a]

$$\cdot \; \frac{\delta q}{\delta t} = \frac{\pi s^2}{6}(\eta^2 + \zeta^2 - 2\xi^2)NV = \frac{\pi}{2}s^2NV(p - q)$$

$$\cdot \; \frac{1}{T} = \frac{\pi s^2 NV}{2} = \frac{p}{\mu} \qquad \mu = pT \qquad V = \text{relative velocity of 2 molecules}$$

$$= \frac{2p}{\pi s^2 NV}$$

Number of collisions per second

$$\cdot \; \pi s^2 NV = 2p/\mu$$

Mean path

$$\cdot \; \frac{1}{\sqrt{2\pi s^2 N}} = \frac{\sqrt{2p}}{\mu V} = \frac{p}{\mu\bar{v}} \qquad \bar{v} = \text{mean velocity}$$

$$\cdot \; \tau w = M \qquad \mu = \frac{2}{3}\frac{\overline{v^2}\tau w}{\pi s^2 \bar{v}} = \frac{2\tau\sqrt{w}}{3\pi s^2}\frac{\overline{v^2}}{\sqrt{2v}}$$

Let $v =$ velocity of mean square for hydrogen, mean velocity $= \sqrt{\dfrac{8}{3\pi}}v$

$$\cdot \; \mu = \frac{2\tau\sqrt{w}}{3\pi s^2\sqrt{2}}\frac{v}{\sqrt{\dfrac{8}{3\pi}}} = \frac{\tau v\sqrt{w}}{2\sqrt{3\pi}s^2}$$

$$\cdot \frac{\mu}{\rho} = \frac{1}{2\sqrt{3\pi}\sqrt{ws^2}} \frac{v}{N} = \frac{v}{2\sqrt{3\pi N}} \frac{\sqrt{\frac{1}{w}}}{s^2}$$

$$D = \frac{v}{2\sqrt{3\pi N}} \frac{\sqrt{\frac{1}{2}\left(\frac{1}{w_1} + \frac{1}{w_2}\right)}}{s^2}$$

$$\text{Let } A = \frac{2\sqrt{6\pi N}}{v} \qquad \Sigma^2_{12} = As^2_{12} = \frac{\sqrt{\frac{1}{w_1} + \frac{1}{w_2}}}{D_{12}}$$

$$\cdot \frac{\mu}{\rho} = \frac{\sqrt{\frac{1}{w}}}{\Sigma^2_{11}} \qquad \mu = \frac{\rho_2\sqrt{2w}}{\Sigma^2_1} \frac{100}{36} \qquad \log \rho_2 = \overline{5}.952201$$

$$100.\sqrt{2} \qquad 2.150515$$
$$\overline{\phantom{100.\sqrt{2}}}$$
$$2.102716$$
$$36 \qquad 1.556303$$
$$\overline{}$$
$$\overline{4}.546413 = \log\frac{100}{36} \rho_2\sqrt{2}$$

[End of sheet]

a. Maxwell's title.

40. Postcard from Maxwell to Peter Guthrie Tait, Summer 1873[a]

Cambridge University Library, Maxwell Collection, Maxwell-Tait Correspondence.

O.T. Can you supply me with the no [number] of grammes of Hydrogen in a litre at a named temperature and pressure. I require the value of p/ρ in absolute measure and I have not the data here.

Also do you know if L. Boltzmann has done anything new in electricity? Wiedemann in a letter seems to imply it.[b] I only know Boltzmann as a student of the ultimate distribution of vis viva in a swarm of molecules.

Viscosity in centimetre-gramme-second measure. Deduced from Loschmidt on Diffusion of gases by the elastic-sphere theory compared with dp/dt direct.

Loschmidt	dp/dt direct
Hyd[c] 0.000112	0.0000967
Ox[d] 0.0002581	
CO 0.0002076	
CO_2 0.0002049	0.0001612

Note on Ångström received today.[e] No explanation of α, α', etc.

Write $q_n = \sqrt{\dfrac{\pi n}{kT}}(1 - e)$

How long were the bars?

If you go 17 miles per minute and take a totally new course 1700,000,000 times in a second where will you be in an hour?[f]

a. Undated, presumably written in the summer of 1873 when Maxwell was composing his paper on Loschmidt's experiments.

b. We have not found the letter from Gustav Heinrich Wiedemann (1826–1899), professor of physical chemistry at Leipzig and the successor of Poggendorff as editor of *Annalen der Physik*. He is known for his work with Rudolph Franz in establishing the relation between electrical and thermal conductivity of metals.

Ludwig Boltzmann (1844–1906) had published a few papers on electricity and magnetism by this time: "Ueber die Bewegung der Electrizität in krummen Flächen," *Sitz. Math.-Naturwiss. Cl. Akad. Wiss., Wien* 52 (1865), 214–221; "Ueber die elektrodynamische Wechselwirkung der Teile eines elektrischen Stromes von veränderlicher Gestalt," *Sitz. Math.-Naturwiss. Cl. Akad. Wiss., Wien* 60 (1869), 69–87; "Resultate einer Experimentaluntersuchung über das Verhalten nicht leitender Körper unter dem Einflusse elektrischer Kräfte," *Sitz. Math.-Naturwiss. Cl. Akad. Wiss., Wien* 66 (1872), 256–263; "Experimentelle Bestimmung der Dielektrizitätskonstante von Isolatoren," *Sitz. Math.-Naturwiss. Cl. Akad. Wiss., Wien* 67 (1873), 17–80; "Experimentaluntersuchung über die elektrostatische Fernwirkung dielektrischer Körper," ibid. 68 (1873), 81–155. These are reprinted in Boltzmann's *Wissenschaftliche Abhandlungen*, Bd. I.

One notable experiment by Boltzmann about this time was his verification for a number of gases of Maxwell's relation between the refractive index and the dielectric constant. This was considered then as an important piece of evidence in support of the electromagnetic theory of light. See "Experimentelle Bestimmung der Dielektricitätsconstante einiger Gase," ibid. 69 (1874), 794–813.

c. Hydrogen.

d. Oxygen.

e. Tait was working on the thermal conductivity of iron and copper bars. He used a cooling method, and since he could find no simple form for the equation of condition he used Forbes's methods. His results were published as "Thermal and Electrical Conductivity," *Trans. R. Soc. Edinburgh* 28 (1878), 717–740, reprinted in Tait's *Scientific Papers* (Cambridge, 1900), vol. 1, 363–392. Tait did not think that the equation of condition Ångström had used to reduce his data on the thermal cooling of iron bars was correct. Anders Jonas Ångström (1814–1874) published several papers on the conductivity of metals, e.g., "Notiz über die latente und specifische Wärme des Eises," *Ann. Phys.* [2] 90 (1853), 509–512; "Ny method att bestämma kroppars lednings förmåga för värme," *Ofversigt af Kongliga Vetenskaps Academiens Förhandlingar, Stockholm* 18 (1861), 365–370, translated in *Phil. Mag.* [4] 25 (1863), 130–142, and in *Ann. Phys.* [2] 114 (1861), 512–530. This is the paper to which Tait refers in much of his correspondence with Maxwell and in his own paper of 1878. See also Ångström, "Recherches sur la conductibilité des corps pour la chaleur," *Nova Acta Regiae Societatis Scientiarum Upsaliensis* 3 (1861), 51–72; "Om Koppars och jernets lednings förmåga för värme vid olika temperatur," *Ofversigt ... Stockholm* 19 (1862), 21–28, translated in *Ann. Phys.* [2] 118 (1863), 423–431.

f. This innocent-sounding question predates by seven years Rayleigh's random-phase theorem and by 32 years Karl Pearson's formal publication of the "random walk problem." Pearson's statement (*Nature* 72 (1905), 294) was as follows:

A man starts from a point O and walks l yards in a straight line; he then turns through any angle whatever and walks another l yards in a second straight line. He repeats this process n times. I require the probability that after these n stretches he is at a distance between r and $r + \delta r$ from his starting point O. The problem is one of considerable interest, but I have only succeeded in obtaining an integrated solution for two stretches. I think, however, that a solution ought to be found, if only in the form of a series of power of $1/n$, when n is large.

For Rayleigh's response, see *Nature* 72 (1905), 318, and his *Scientific Papers*, vol. 5, 256. A comprehensive review of this and related problems was given by S. Chandrasekhar, "Stochastic Problems in Physics and Astronomy," *Reviews of Modern Physics* 15 (1943), 1–89.

Note that Maxwell's problem, given in the context of kinetic theory, refers to three dimensions, not to two as do Pearson's and Rayleigh's. We do not know if he had solved it; the tone of the question suggests that he had.

41. Postcard from Maxwell to Peter Guthrie Tait, Summer 1873 [?]

Cambridge University Library, Maxwell Collection, Maxwell-Tait Correspondence.

O.T'. $\theta\alpha\gamma\xi$[a] for the density of H. What do you expect me to do with Ewing and McGregor on Salts?[b] Figures and curves not sent to me but I see that the ordinates are densities. Now the best ordinates are Volumes of as much of the stuff as contains 1 of the original water. As for Å,[c] if he neglects H, he does so at his peril. How can I save him? Let him sink! Have you seen Clausius ueber einen neuen mechanischen Satz in Bezug auf stationare Bewegungen.[d]

For the absolute values of molecular constants to be used for viscosity, diffusion and conduction I think diffusion expts [experiments] as done by Loschmidt for the best and least interfered with by sides of vessel etc. Thus for diameters of molecules of H, O, CO, CO_2.

		Calculated	By diffusion	diff [difference][e]
H	1739	$\frac{1}{2}(H + O) = 2011$	1992	-19
O	2283	$\frac{1}{2}(H + CO) = 2100$	2116	$+16$
CO	2461	$\frac{1}{2}(H + CO_2) = 2257$	2260	$+3$
CO_2	2775	$\frac{1}{2}(O + CO) = 2372$	2375	$+3$
CH	2605	$\frac{1}{2}(O + CO_2) = 2529$	2545	$+16$
NO	3063	$\frac{1}{2}(CO + CO_2) = 2618$	2599	-19
SO_2	3109			

Mass of Hydrogen molecules not less than 10^{-27} grammes.

a. Thanks. (In classical Greek γ assumes the sound *ng* when placed in combination with κ, ξ or another γ.)

b. Tait had sent Maxwell a paper by James Alfred Ewing (1855–1935) and James Gordon McGregor (1852–1913) on the volumes of salt solutions. Tait asked Maxwell to review the paper for inclusion in the *Transactions* of the Royal Society of Edinburgh. The paper was ultimately published as an abstract, "Note on Volumes of Solutions," *British Association Report*, 1877, 40–41.

c. Ångström.

d. R. J. E. Clausius, "Ueber einer neuen mechanischen Satz in Bezug auf stationäre Bewegungen," *Sitzungsberichte der Niederrheinischen Gesellschaft für Natur-und Heilkunde zu Bonn* 30 (1873), 136–154, translated in *Phil. Mag.* [4] 46 (1873), 236–244, 266–273.

e. In an undated postcard to Maxwell, possibly in reply to this one, Tait wrote: "The diffusion numbers certainly show a splendid agreement—Why are the differences symmetrical? Is it a new Semal-bis-ter [?] Law???" (Cambridge University Library)

42. "On Loschmidt's Experiments on Diffusion in Relation to the Kinetic Theory of Gases"

Nature 8 (1873), 298–300; *Scientific Papers*, vol. 2, 343–350.

[From *Nature*, Vol. VIII.]

LIX. *On Loschmidt's Experiments on Diffusion in relation to the Kinetic Theory of Gases.*

THE kinetic theory asserts that a gas consists of separate molecules, each moving with a velocity amounting, in the case of hydrogen, to 1,800 metres per second. This velocity, however, by no means determines the rate at which a group of molecules set at liberty in one part of a vessel full of the gas will make their way into other parts. In spite of the great velocity of the molecules, the direction of their course is so often altered and reversed by collision with other molecules, that the process of diffusion is comparatively a slow one.

The first experiments from which a rough estimate of the rate of diffusion of one gas through another can be deduced are those of Graham*. Professor Loschmidt, of Vienna, has recently† made a series of most valuable and accurate experiments on the interdiffusion of gases in a vertical tube, from which he has deduced the coefficient of diffusion of ten pairs of gases. These results I consider to be the most valuable hitherto obtained as data for the construction of a molecular theory of gases.

There are two other kinds of diffusion capable of experimental investigation, and from which the same data may be derived, but in both cases the experimental methods are exposed to much greater risk of error than in the case of diffusion. The first of these is the diffusion of momentum, or the lateral communication of sensible motion from one stratum of a gas to another. This is the explanation, on the kinetic theory, of the viscosity or internal friction of gases. The investigation of the viscosity of gases requires experiments of great delicacy, and involving very considerable corrections before the true

---

* *Brande's Journal* for 1829, pt. ii. p. 74, "On the Mobility of Gases," *Phil. Trans.* 1863.

† *Sitzb. d. k. Akad. d. Wissench.* 10 März. 1870.

coefficient of viscosity is obtained. Thus the numbers obtained by myself in 1865 are nearly double of those calculated by Prof. Stokes from the experiments of Baily on pendulums, but not much more than half those deduced by O. E. Meyer from his own experiments. The other kind of diffusion is that of the energy of agitation of the molecules. This is called the conduction of heat. The experimental investigation of this subject is confessedly so difficult, that it is only recently that Prof. Stefan of Vienna*, by means of a very ingenious method, has obtained the first experimental determination of the conductivity of air. This result is, as he says, in striking agreement with the kinetic theory of gases.

The experiments on the interdiffusion of gases, as conducted by Prof. Loschmidt and his pupils, appear to be far more independent of disturbing causes than any experiments on viscosity or conductivity. The interdiffusing gases are left to themselves in a vertical cylindrical vessel, the heavier gas being underneath. No disturbing effect due to currents seems to exist, and the results of different experiments with the same pair of gases appear to be very consistent with each other.

They prove conclusively that the coefficient of diffusion varies inversely as the pressure, a result in accordance with the kinetic theory, whatever hypothesis we adopt as to the nature of the mutual action of the molecules during their encounters.

They also shew that the coefficient of diffusion increases as the temperature rises, but the range of temperature in the experiments appears to be too small to enable us to decide whether it varies as T^2, as it should be according to the theory of a force inversely as the fifth power of the distance adopted in my paper in the *Phil. Trans.* 1866, or as $T^{\frac{3}{2}}$ as it should do according to the theory of elastic spherical molecules, which was the hypothesis originally developed by Clausius, by myself in the *Phil. Mag.* 1860, and by O. E. Meyer.

In comparing the coefficients of diffusion of different pairs of gases, Prof. Loschmidt has made use of a formula according to which the coefficient of diffusion should vary inversely as the geometric mean of the atomic weights of the two gases. I am unable to see any ground for this hypothesis in the kinetic theory, which in fact leads to a different result, involving the diameters of the molecules, as well as their masses. The numerical results obtained by

---

* *Sitzb. d. k. Akad.* Feb. 22, 1872.

Prof. Loschmidt do not agree with his formula in a manner corresponding to the accuracy of his experiments. They agree in a very remarkable manner with the formula derived from the kinetic theory.

I have recently been revising the theory of gases founded on that of the collisions of elastic spheres, using, however, the methods of my paper on the dynamical theory of gases (*Phil. Trans.* 1866) rather than those of my first paper in the *Phil. Mag.*, 1860, which are more difficult of application, and which led me into great confusion, especially in treating of the diffusion of gases.

The coefficient of interdiffusion of two gases, according to this theory, is

$$D_{12} = \frac{1}{2\sqrt{6\pi}} \frac{V}{N} \sqrt{\frac{1}{w_1} + \frac{1}{w_2}} \frac{1}{s_{12}^2} \quad \dots\dots\dots\dots\dots (1),$$

where w_1 and w_2 are the molecular weights of the two gases, that of hydrogen being unity.

s_{12} is the distance between the centres of the molecules at collision in centimetres.

V is the "velocity of mean square" of a molecule of hydrogen at $0°$ C.

$$V = \sqrt{\frac{3p}{\rho}} = 185,900 \text{ centimetres per second.}$$

N is the number of molecules in a cubic centimetre at $0°$ C. and 76 cm. B. (the same for all gases).

D_{12} is the coefficient of interdiffusion of the two gases in $\dfrac{\text{(centimetre)}^2}{\text{second}}$ measure.

We may simplify this expression by writing

$$a^2 = \frac{1}{2\sqrt{6\pi}} \frac{V}{N}, \quad \sigma_{12}^2 = \frac{1}{D_{12}} \sqrt{\frac{1}{w_1} + \frac{1}{w_2}} \quad \dots\dots\dots\dots\dots(2).$$

Here a is a quantity the same for all gases, but involving the unknown number N.

σ is a quantity which may be deduced from the corresponding experiment of M. Loschmidt. We have thus

$$s_{12} = a\sigma_{12} \dots\dots\dots\dots\dots\dots\dots\dots(3),$$

or the distance between the centres of the molecules at collision is proportional to the quantity σ, which may be deduced from experiment.

If d_1 and d_2 are the diameters of the two molecules,

$$s_{12} = \tfrac{1}{2}(d_1 + d_2).$$

Hence if $d = a\delta \ldots \ldots \sigma_{12} = \tfrac{1}{2}(\delta_1 + \delta_2)$ .... (4).

Now M. Loschmidt has determined D for the six pairs of gases which can be formed from Hydrogen, Oxygen, Carbonic Oxide, and Carbonic Acid. The six values of σ deduced from these experiments ought not to be independent, since they may be deduced from the four values of δ belonging to the two gases. Accordingly we find, by assuming

TABLE I.

$\delta\,(\mathrm{H}) = 1\cdot739$
$\delta\,(\mathrm{O}) = 2\cdot283$
$\delta\,(\mathrm{CO}) = 2\cdot461$
$\delta\,(\mathrm{CO}_2) = 2\cdot775$

σ_{12}	Calculated $\tfrac{1}{2}(\delta_1 + \delta_2)$	Observed $\sqrt{\dfrac{1}{D}}\sqrt{\dfrac{1}{w_1} + \dfrac{1}{w_2}}$
For H and O	2·011	1·992
For H and CO	2·100	2·116
For H and CO$_2$	2·257	2·260
For O and CO	3·372	2·375
For O and CO$_2$	2·529	2·545
For CO and CO$_2$	2·618	2·599

Note.—These numbers must be multiplied by 0·6 to reduce them to (centimetre-second) measure from the (metre-hour) measure employed by Loschmidt.

The agreement of these numbers furnishes, I think, evidence of considerable strength in favour of this form of the kinetic theory, and if it should be confirmed by the comparison of results obtained from a greater number of pairs of gases it will be greatly strengthened.

Evidence, however, of a higher order may be furnished by a comparison between the results of experiments of entirely different kinds, as for instance, the coefficients of diffusion and those of viscosity. If μ denotes the coefficient of viscosity, and ρ the density of a gas at $0°$ C. and 760 mm. B., the theory gives

$$\frac{\mu}{\rho} = a^2 \sqrt{\frac{2}{w}} \frac{1}{d^2} \quad\dots\dots\dots\dots\dots\dots\dots\dots\dots\dots(5),$$

so that the following relation exists between the viscosities of two gases and their coefficient of interdiffusion—

$$D_{12} = \tfrac{1}{2}\left(\frac{\mu_1}{\rho_1} + \frac{\mu_2}{\rho_2}\right) \quad\dots\dots\dots\dots\dots\dots\dots\dots(6).$$

Calculating from the data of Table I., the viscosities of the gases, and comparing them with those found by O. E. Meyer and by myself, and reducing all to centimetre, gramme, second measure, and $0°$ C.—

Table II.

Coefficient of Viscosity.

Gas.	Loschmidt.	O. E. Meyer.	Maxwell.
H	0·000116	0·000134	0·000097
O	0·000270	0·000306	
CO	0·000217	0·000266	
CO₂	0·000214	0·000231	0·000161

The numbers given by Meyer are greater than those derived from Loschmidt. Mine, on the other hand, are much smaller. I think, however, that of the three, Loschmidt's are to be preferred as an estimate of the absolute value of the quantities, while those of Meyer, derived from Graham's experiments, may possibly give the ratios of the viscosities of different gases more correctly. Loschmidt has also given the coefficients of interdiffusion of four other pairs of gases, but as each of these contains a gas not contained in any other pair, I have made no use of them.

In the form of the theory as developed by Clausius, an important part is played by a quantity called the *mean length of the uninterrupted path of a*

44—2

molecule, or, more concisely, the *mean path.* Its value, according to my calculations, is

$$l = \frac{1}{\sqrt{2}\pi s^2 N} = \frac{\sqrt{12}}{\sqrt{\pi} V}\frac{1}{\delta^2}* \quad \dots\dots\dots\dots\dots\dots(7).$$

Its value in tenth-metres $(1 \text{ metre} \times 10^{-10})$ is

TABLE III.

For Hydrogen . . .	965	Tenth-metres at 0° C. and 760 B.
For Oxygen . . .	560	
For Carbonic Oxide . .	482	
For Carbonic Acid . .	430	

(The wave-length of the hydrogen ray F is 4,861 tenth-metres, or about ten times the mean path of a molecule of carbonic oxide.)

We may now proceed for a few steps on more hazardous ground, and inquire into the actual size of the molecules. Prof. Loschmidt himself in his paper "Zur Grösse der Luftmolecüle" (*Acad. Vienna,* Oct. 12, 1865), was the first to make this attempt. Independently of him and of each other, Mr G. J. Stoney (*Phil. Mag.,* Aug. 1868), and Sir W. Thomson (*Nature,* March 31, 1870), have made similar calculations. We shall follow the track of Prof. Loschmidt.

The volume of a spherical molecule is $\frac{\pi}{6} s^3$, where s is its diameter. Hence if N is the number of molecules in unit of volume, the space actually filled by the molecules is $\frac{\pi}{6} N s^3$.

This, then, would be the volume to which a cubic centimetre of the gas would be reduced if it could be so compressed as to leave no room whatever between the molecules. This, of course, is impossible; but we may, for the sake of clearness, call the quantity

$$\epsilon = \frac{\pi}{6} N s^3 \quad \dots\dots\dots\dots\dots\dots\dots\dots (8)$$

---

* The difference between this value and that given by M. Clausius in his paper of 1858, arises from his assuming that all the molecules have equal velocities, while I suppose the velocities to be distributed according to the "law of errors."

the ideal coefficient of condensation. The actual coefficient of condensation, when the gas is reduced to the liquid or even the solid form, and exposed to the greatest degree of cold and pressure, is of course greater than ϵ.

Multiplying equations (7) and (8), we find

$$s = 6\sqrt{2}\epsilon l \dots\dots\dots\dots\dots\dots\dots\dots\dots\dots (9),$$

where s is the diameter of a molecule, ϵ the coefficient of condensation, and l the mean path of a molecule.

Of these quantities, we know l approximately already, but with respect to ϵ we only know its superior limit. It is only by ascertaining whether calculations of this kind, made with respect to different substances, lead to consistent results, that we can obtain any confidence in our estimates of ϵ.

M. Lorenz Meyer[*] has compared the "molecular volumes" of different substances, as estimated by Kopp from measurements of the density of these substances and their compounds, with the values of s^3 as deduced from experiments on the viscosity of gases, and has shewn that there is a considerable degree of correspondence between the two sets of numbers.

The "molecular volume" of a substance here spoken of is the volume in cubic centimetres of as much of the substance in the liquid state as contains as many molecules as one gramme of hydrogen. Hence if ρ_0 denote the density of hydrogen, and $\mathfrak{b}$ the molecular volume of a substance, the actual coefficient of condensation is

$$\epsilon' = \rho_0 \mathfrak{b} \dots\dots\dots\dots\dots\dots\dots\dots\dots\dots(10).$$

These "molecular volumes" of liquids are estimated at the boiling-points of the liquids, a very arbitrary condition, for this depends on the pressure, and there is no reason in the nature of things for fixing on 760 mm. B. as a standard pressure merely because it roughly represents the ordinary pressure of our atmosphere. What would be better, if it were not impossible to obtain it, would be the volume at $-273°$ C. and ∞ B.

But the volume relations of potassium with its oxide and its hydrated oxide as described by Faraday seem to indicate that we have a good deal yet to learn about the volumes of atoms.

[*] *Annalen d. Chemie u. Pharmacie* v. Supp. bd. 2, Heft (1867).

350 EXPERIMENTS ON DIFFUSION IN RELATION TO THE KINETIC THEORY OF GASES.

If, however, for our immediate purpose, we assume the smallest molecular volume of oxygen given by Kopp as derived from a comparison of the volume of tin with that of its oxide and put

$$\mathfrak{b}\,(O = 16) = 2\text{·}7,$$

we find for the diameters of the molecules—

TABLE IV.

Hydrogen .	.	.	.	5·8 tenth-metres.
Oxygen	.	.	.	7·6
Carbonic Oxide		.	.	8·3
Carbonic Acid .		.	.	9·3

The mass of a molecule of hydrogen on this assumption is

$$4\text{·}6 \times 10^{-24} \text{ gramme.}$$

The number of molecules in a cubic centimetre of any gas at 0° C. and 760 mm. B. is

$$N = 19 \times 10^{18}.$$

Hence the side of a cube which, on an average, would contain one molecule would be

$$N^{-\frac{1}{3}} = 37 \text{ tenth-metres.}$$

43. Letter from Maxwell to John William Strutt, Third Baron Rayleigh, August 28, 1873[a]

Lord Rayleigh (John William Strutt), *Scientific Papers*, (Cambridge, 1903; reprint, New York: Dover Pubs., 1964), vol. 4, 397–398.

I have left your papers on the light of the sky, &c. at Cambridge, and it would take me, even if I had them, some time to get them assimilated sufficiently to answer the following question, which I think will involve less expense to the energy of the race if you stick the data into your formula and send me the result [...].

Suppose that there are N spheres of density ρ and diameter s in unit of volume of the medium. Find the index of refraction of the compound medium and the coefficient of extinction of light passing through it.

The object of the enquiry is, of course, to obtain data about the size of the molecules of air. Perhaps it may lead also to data involving the density of the aether. The following quantities are known, being combinations of the three unknowns,

M = mass of molecule of hydrogen;

N = number of molecules of any gas in a cubic centimetre at $0°$ C. and 760 B.

s = diameter of molecule in any gas:—

Known Combinations.

MN = density.

Ms^2 from diffusion or viscosity.

Conjectural Combination.

$\dfrac{6M}{\pi s^3}$ = density of molecule.

If you can give us (i) the quantity of light scattered in a given direction by a stratum of a certain density and thickness; (ii) the quantity cut out of the direct ray; and (iii) the effect of the molecules on the index of refraction, which I think ought to come out easily, we might get a little more information about these little bodies.

You will see by *Nature*, Aug. 14, 1873, that I make the diameter of molecules about 1/1000 of a wave-length.

The enquiry into scattering must begin by accounting for the great observed transparency of air. I suppose we have no numerical data about its absorption.

But the index of refraction can be numerically determined, though the observation is of a delicate kind, and a comparison of the result with the dynamical theory may lead to some new information.

a. Rayleigh published this letter (apparently omitting some material after the first paragraph) in his 1899 paper on the blue color of the sky, with the remark "My attention was specially directed to this question a long while ago by Maxwell in a letter which I may be pardoned for reproducing here." Rayleigh says that Maxwell wrote in a subsequent letter: "Your letter of Nov. 17 quite accounts for the observed transparency of any gas." But, according to Rayleigh, "So far as I remember, my argument was of a general character only."

Rayleigh's original work on the blue color of the sky was stimulated by the experiments of John Tyndall. Maxwell was impressed by it, writing to C. J. Monro on March 15, 1871: "I think Strutt on sky-blue is very good. It settles Clausius's earlier vesicular theory to explain the blue sky,

for, putting all his words together
'tis 3 blue beans in one blue bladder.—Mat. Prior

(Campbell and Garnett, *Life*, 380) For the vesicular theory, see Elizabeth Garber, "Rudolph Clausius' Work in Meteorological Optics," *Rete* 2 (1975), 323–337; Ivo Schneider, "Clausius' erste Anwendung der Wahrscheinlichkeitsrechnung im Rahmen der atmosphärischen Lichtstreuung," *Arch. Hist. Exact Sci.* 14 (1974), 143–158.

b. See document III-42, last page.

44. Letter from Alexander Crum Brown[a] to Maxwell, September 4, 1873

Cambridge University Library, Maxwell Collection, Miscellaneous Correspondence.

Lindau Bavaria
Sep. 4. 1873
Dear Sir,

Your letter of the 18th ult. was forwarded to me, but owing to some uncertainty as to our route it was long of coming.[b]

1st as to molecular (and atomic) volume of *solids*. The *atomic* volume of similar *elements* is often nearly the same, i.e., the sp. gr. [specific gravity] varies as the *atomic* weight, e.g., metals of the iron group. Manganese, Iron, Cobalt and also Copper, have atomic volumes nearly = 7. Platinum, Iridium, Palladium, Rhodium, about 9, Gold and Silver about 10, Sulphur and Selenium about 16. The molecular volume (or rather the mean atomic volume, ※ for we do not know the molecular weights of solids or liquids) is *nearly* the same for isomorphous ⟨substances⟩ compounds of similar composition and in the case of those substances which are only approximately isomorphous—such as rhombohedral carbonates—the molecular volumes are in the same order as the angles. I happen to have in a notebook here some of them—

	Mol. vol.	Angle of rhombohedron
Carbonate of Zinc	28.4	107° 40′
″ Magnesium	28.6	107° 25′
Iron	30.8	107° 0
Manganese	30.8	106° 51′
Calcium (Calc-spar)	36.8	105° 5′

By adding 5.2 for each atom of oxygen, to the atomic volume of a metal the molecular volume of the oxide is obtained in a *good many cases*. Thus—

Metal	Oxide	Atomic vol. metal	Mol vol oxide
Pb	PbO	18.2	23.8
Cd	CdO	13	18.2
Cu	CuO	7.2	12.4
Zn	ZnO	9.2	14.4
2Fe	Fe_2O_3	2(7.2)	30.6

But this breaks down in many cases besides that which you mention of the alkaline metals. In these as you mention the mol. vol. is diminished by the addition of oxygen.

2nd mol. vol. of liquids. Owing to the large coeff[icient] of expansion of liquids—no comparison can be made that is of any use unless at corresponding temperatures. Kopp compares them at temperatures at which they have the same vapour tension that is at the boiling points under the same pressure, and has made out a number of regularities or indications of regularities, but nothing like a law. You ask if the mol. volumists have a dodge to explain the anomalies. They have and I have indicated it above. It is that we do not know the molecular weight of solids or liquids, but only of gases. I don't see how that helps a pure mol. volumist, then again they say that we do not know at what temperatures diff^t [different] solids and liquids should be compared so that anomalies are to be expected.

I think Graham made out the sp. gr. of Hydrogenium as occluded in Paladium to be about 2. I understand the homoeopaths^c to mean by a stuff in the nth dilution—a substance containing $1/(100)^n$ of the active stuff. You take 1 part of your stuff and 99 mmgrams [?] of milk if the stuff be solid, 99 of water or spirit if it be liquid, mix *well*   the result is the first delusion or dilution, take 1 part of that and do it again etc. etc. I think they sometimes go to the 10th dilution! I have not any books with me except guide books, and a small notebook in which I keep my accounts and which oddly enough happened to contain some numbers connected with molecular volumes, so that I do not speak with authority. I am afraid this will be too late and too vague to be of use to you. I have some notions of my own about mol. volumes and molecular heat, but have not worked them fully out, and many of the experimental numbers are too untrustworthy to make it worth while to work at it without making new determinations. In particular we have not many determinations of sp. heat of vapours near the point of saturation. I hope your lecture will give us some light on the matter

Yours Truly,

Alexander Crum Brown

P.S. ※ By mean atomic volume I mean molecular volume ÷ number of atoms in molecule.

a. Alexander Crum Brown (1838–1922) was professor of Chemistry at Edinburgh from 1869 to 1908. He was one of the pioneers of the graphical system of describing chemical structures used today. See the *Dictionary of Scientific Biography* for his earlier "pragmatic attitude to the atomic theory." Brown was married to a sister of Mrs. Tait and was a member of Tait's golfing circle.

b. The character of the letter indicates that Maxwell had written to Brown asking for help in preparing his lecture "Molecules" for the British Association meeting of 1873 (document II-16).

c. The theory of homeopathic medicine, originated by Christian Friedrich Samuel Hahnemann (1755–1843), was then at the height of its popularity in Britain, being patronized even by Queen Victoria. The theory was to treat a disease by minute doses of drugs that in healthy persons would produce symptoms like those of the disease. Cf. the last part of document II-15, which is a draft of the "Molecules" lecture. Maxwell omitted the reference to homeopathy in the published version of the lecture (document II-16).

45. Letter from Maxwell to Peter Guthrie Tait, February 14, 1876

Cambridge University Library, Maxwell Collection, Maxwell-Tait Correspondence.

14/2/76

O.T'!

1. Clausius "Ueber die mittlere Lange" 1858 assumes uniform velocity and gets the number 8.[a]

2. Maxwell, *Phil Mag* 1860 ascertained the law of distribution of velocities and gives a result corresponding to $\sqrt{72}$ remarking that Clausius makes it different.[b]

3. Clausius supposing Maxwell's knowledge of the integral calculus is imperfect writes to *Phil Mag.* showing how to do the integration on the assumption $v = $ constant.[c]

4. Maxwell in 1866, going in for forces at a distance became hazy as to what constitutes a collision and ignored collisions in favour of encounters of various degrees of closeness.

5. Maxwell in 1873 attributed to Loschmidt by mistake the number 8, using, however $\sqrt{72}$ in his own calculations from Loschmidt's diffusion experiments.

6. Clausius on receipt of a "Molecule" returns thanks, but claims "8" as his own.

7. Maxwell ever after gives Clausius, as he deserves, the credit of the number 8.

8. T' discovers that $\sqrt{72}$ is not far from $8\frac{1}{2}$ a number better adapted for popular exposition.

dp/dt was actuated by the same motive when at Bradford[d] he mentioned 8, without the fraction, holding that the secret should be withheld from the knowledge of the people according to the principle of "reserve."

See Willard Gibbs on the "Equilibrium of Heterogeneous Substances" before you republish Thermodynamics.[e]

Ohms law has now been tested with currents that make the wire sway and swelter with heat, and it is now at least 10^5 to 1 that if Schuster observed anything it was not an error of Ohm's law.[f] Have you seen Lorentz (of Leiden) Over de Terugkaatsing &c. van het Licht.[g] He goes in for Electromagnetic vibrations and has his doubts of Ohm founded on the inertia of electricity.

$$\frac{dp}{dt}$$

a. See chapter I, note 19. The number 8 does not actually appear in this paper, but emerges only in Maxwell's transcription of Clausius's result into his own notation (see note b).

b. See document III-6; *Scientific Papers*, vol. 1, 387. In Maxwell's notation, Clausius found the value 8 for the ratio $6\alpha/\pi s^2 N$, whereas Maxwell found the value $\sqrt{72}$. In the formula for the mean free path, Clausius has a factor $4/3$, whereas Maxwell has a factor $\sqrt{2}$. See also note f to document III-12.

c. Clausius, "On the Dynamical Theory of Gases," *Phil. Mag.* [4] 19 (1860), 434–436.

d. This refers to Maxwell's lecture to the British Association meeting at Bradford in 1873; "8" does not appear in the published version of the lecture. See document II-16; *Scientific Papers*, vol. 2, 369.

e. Josiah Willard Gibbs, "On the Equilibrium of Heterogeneous Substances," *Trans. Connecticut Acad. Sci.* 3 (1876), 108–248. Maxwell was the first major scientist to recognize the value of Gibbs's work; he made a model of Gibbs's thermodynamic surface for water and presented it to him. See Lynde Phelps Wheeler, *Josiah Williard Gibbs* (New Haven: Yale University Press, 1962), 74.

f. This refers to the experiments by George Chrystal (1851–1911) at the Cavendish Laboratory on Ohm's law. See Maxwell's *Scientific Papers*, vol. 2, 537, and Campbell and Garnett, *Life*, 365–366, 392.

g. Hendrik Antoon Lorentz (1853–1928). Maxwell is referring to Lorentz's Ph.D. thesis, *Over de theorie der terugkaatsing en breking van het licht* (Leiden, 1875), later published as "Ueber die Theorie der Reflexion und Refraction des Lichtes," *Zeitschrift für Mathematik und Physik* 22 (1877), 1–30, 205–219; 23 (1878), 197–210. For a more extended assessment of Lorentz's thesis by Maxwell, see *Memoir and Scientific Correspondence of Stokes*, J. Larmor, ed. (Cambridge University Press, 1907; New York: Johnson Reprint Corp., 1971), vol. 2, 41. Lorentz had evidently sent Maxwell one of the original copies printed in 1875. Maxwell may have learned to read Dutch in order to study the work of J. D. van der Waals; see *Scientific Papers*, vol. 2, 407–415, 426.

46. "Diffusion of Gases through Absorbing Substances"

Review of *Ueber die Diffusion der Gase durch absorbirende Substanzen*, Habilitationsschrift der Mathematischen und Naturwissenschaftlichen Facultät der Universität Strassburg, vorgelegt von Dr. Sigmund v. Wroblewski, ersten Assistanten am physikalischen Institute (Strassburg: G. Fischbach, 1876). Published in *Nature* 14 (1876), 24–25, and in *Scientific Papers*, vol. 2, 501–504 (without the citation of the book under review).

[From *Nature*, Vol. xiv.]

LXXVII. *Diffusion of Gases through Absorbing Substances.*

THE importance of the exact study of the motions of gases, not only as a method of distinguishing one gas from another, but as likely to increase our knowledge of the dynamical theory of gases, was pointed out by Thomas Graham. Graham himself studied the most important phenomena, and distinguished from each other those in which the principal effect is due to different properties of gases.

The motion of large masses of the gas approximates to that of a perfect fluid having the same density and pressure as the gas. This is the case with the motion of a single gas when it flows through a large hole in a thin plate from one vessel into another in which the pressure is less. The result in this case is found to be in accordance with the principles of the dynamics of fluids. This was approximately established by Graham, and the more accurate formula, in which the thermodynamic properties of the gas are taken into account, has been verified by the experiments of Joule and Thomson. (*Proc. R. S.*, May, 1856.)

When the orifice is exceedingly small, it appears from the molecular theory of gases that the total discharge may be calculated by supposing that there are two currents in opposite directions, the quantity flowing in each current being the same as if it had been discharged into a vacuum.

For different gases the volume discharged in a given time, reduced to standard pressure and temperature, is proportional to—

$$\frac{p}{\sqrt{s\theta}}$$

where p is the actual pressure, s is the specific gravity, and θ the temperature reckoned from $-274°$ C.

When the gases in the two vessels are different, each gas is discharged according to this law independently of the other.

These phenomena, however, can be observed only when the thickness of the plate and the diameter of the aperture are very small.

When this is the case, the distance is very small between a point in the first vessel where the mixed gas has a certain composition, and a point in the second vessel where the mixed gas has a quite different composition, so that the velocity of diffusion through the hole between these two points is large compared with the velocity of flow of the mixed gas arising from the difference of the total pressures in the two vessels.

When the hole is of sensible magnitude this distance is larger, because the region of mixed gases extends further from the hole, and the effects of diffusion become completely masked by the effect of the current of the gas in mass, arising from the difference of the total pressures in the two vessels. In this latter case the discharge depends only on the nature of the gas in the vessel of greater pressure, and on the resultant pressures in the two vessels. It consists entirely of the gas of the first vessel, and there is no appreciable counter current of the gas of the other vessel.

Hence the experiments on the double current must be made either through a single very small aperture, as in Graham's first experiment with a glass vessel accidentally cracked, or through a great number of apertures, as in Graham's later experiments with porous septa of plaster of Paris or of plumbago.

With such septa the following phenomena are observed :—

When the gases on the two sides of the septum are different, but have the same pressure, the reduced volumes of the gases diffused in opposite directions through the septum are inversely as the square roots of their specific gravities.

If one or both of the vessels is of invariable volume, the interchange of gas will cause an inequality of pressure, the pressure becoming greater in the vessel which contains the heavier gas.

If a vessel contains a mixture of gases, the gas diffused from the vessel through a porous septum will contain a larger proportion of the lighter gas, and the proportion of the heavier gas remaining in the vessel will increase during the process.

The rate of flow of a gas through a long capillary tube depends upon the viscosity or internal friction of the gas, a property quite independent of its specific gravity.

The phenomena of diffusion studied by Dr v. Wroblewski are quite distinct from any of these. The septum through which the gas is observed to pass is apparently quite free from pores, and is indeed quite impervious to certain gases, while it allows others to pass.

It was the opinion of Graham that the substance of the septum is capable of entering into a more or less intimate combination with the substance of the gas; that on the side where the gas has greatest pressure the process of combination is always going on; that at the other side, where the pressure of the gas is smaller, the substance of the gas is always becoming dissociated from that of the septum; while in the interior of the septum those parts which are richer in the substance of the gas are communicating it to those which are poorer.

The rate at which this diffusion takes place depends therefore on the power of the gas to combine with the substance of the septum. Thus if the septum be a film of water or a soap-bubble, those gases will pass through it most rapidly which are most readily absorbed by water, but if the septum be of caoutchouc the order of the gases will be different. The fact discovered by St Claire-Deville and Troost that certain gases can pass through plates of red-hot metals, was explained by Graham in the same manner.

Franz Exner* has studied the diffusion of gases through soap-bubbles, and finds the rate of diffusion is directly as the absorption-coefficient of the gas, and inversely as the square root of the specific gravity.

Stefan† in his first paper on the diffusion of gases has shewn that a law of this form is to be expected, but he says that he will not go further into the problem of the motion of gases in absorbing medium, as it ought to form the subject of a separate investigation.

Dr v. Wroblewski has confined himself to the investigation of the relation between the rate of diffusion and the pressure of the diffusing gas on the two sides of the membrane. The membrane was of caoutchouc, 0·0034 cm. thick. It was almost completely impervious to air. The rate at which carbonic acid diffused through the membrane was proportional to the pressure of that gas, and was independent of the pressure of the air on the other side of the

---

* *Pogg. Ann.*, Bd. 155.

† *Ueber das Gleichgewicht u. d. Diffusion von Gasgemengen.* Sitzb. der k. Akad. (Wien), Jan. 5, 1871.

membrane, provided this air was from carbonic acid. The connexion between this result and Henry's law of absorption is pointed out.

The time of diffusion of hydrogen through caoutchouc is 3·6 times that of an equal volume of carbonic acid. The diffusion of a mixture of hydrogen and carbonic acid takes place as if each gas diffused independently of the other at a rate proportional to the part of the pressure which is due to that gas.

We hope that Dr v. Wroblewski will continue his researches, and make a complete investigation of the phenomena of diffusion through absorbing substances.

47. Letter from Maxwell to William Garnett,[a] June 30, 1877

Lewis Campbell and William Garnett, *Life of James Clerk Maxwell*, 570–571.

Dear Garnett—... I have been considering diffusion of gases, and the method of separating heavy gases from light ones and I find it hopeless to do it by gravity, but if a tube 10 cm. long with two bulbs, and the straight part stuffed with cotton-wool, were filled with equal volumes of H and CO_2, and spun 100 times round per second for about half an hour, then the ratio of CO_2 to H by volume would be greater in A than in B by about 1/150, which is measurable.[b] I have got a new light about equilibrium of temperature in two different gases. Let forces having potentials act on the molecules of two gases, but differently on each. Let the potential of forces acting on the gas a be zero in the region A and very large in B, diminishing continuously in the stratum C. Let the potential for gas b be zero in B and very great in A, diminishing continuously in C. Then the region A will contain the gas a nearly pure, and B gas b nearly pure, and in the stratum C there will be encounters between the two kinds of molecules. By Boltzmann and Watson the average kinetic energy of a single molecule is the same throughout the whole vessel.[c] Hence the condition of thermal equilibrium between two gases (not mixed, but kept pure though in contact) is that the mean kinetic energy is the same in each. And it is difficult to see where this method breaks down when applied to solids.

I find the electric conductivity of air supposed of conducting spheres to be $(1/18)\pi^2 s^2 N V$.

Where s = distance of centres at striking.

N Number in cubic centimetre.

V Mean velocity.

Now $\frac{\pi}{4}s^2 N = 17{,}700$ for air, and $V = 48{,}500$.

But this is in electrostatic measure. In electromagnetic measure the resistance is

$$\frac{4v^2}{\pi\, 48500000},$$

so that $r = 2 \cdot 10^{13}$ per cubic centimetre, or about 10^{10} greater than that of copper; but this is far smaller than that of guttapercha. Hence the insulating power of air is not consistent with its molecules being conducting spheres.[d]

But why should the molecules be conductors?—Yours very truly,

J. CLERK MAXWELL

a. William Garnett (1850–1932) was Demonstrator at the Cavendish Laboratory from 1874 to 1880. In 1884 he became the first principal of Durham College of

Science, Newcastle-upon-Tyne (subsequently King's College and now the University of Newcastle-upon-Tyne). From 1893 to 1915 Garnett served as Educational Adviser to the London County Council. He was coauthor with Lewis Campbell of *Life of James Clerk Maxwell* (cf. Robert Kargon's preface to the 1969 reprint) and of about a dozen other works, including *Heroes of Science* (London, 1844), which contains a separate short biography of Maxwell. His son, James Clerk Maxwell Garnett, born a few minutes after Maxwell's death, also became a physicist and worked in the Cavendish Laboratory under J. J. Thomson from 1902 to 1905. See B. M. Allen, *William Garnett: A Memoir* (Cambridge: Heffer and Sons, 1933).

b. This experiment is described in Maxwell's paper "On Boltzmann's Theorem on the Average Distribution of Energy in a System of Material Points," *Trans. Cambridge Phil. Soc.* 12 (1879), 547–570; see *Scientific Papers*, vol. 2, 739–741.

c. This is a generalization of the result that the temperature in a column of gas in thermal equilibrium is independent of height. See the note added December 17, 1866, to Maxwell's 1867 paper (document III-30, *Scientific Papers*, vol. 2, 75–76); L. Boltzmann, "Studien über das Gleichgewicht der lebendigen Kraft zwischen bewegten materiellen Punkten," *Sitz. Math.-Naturwiss. Cl. Akad. Wiss. Wien* 58 (1868), 517–560; Maxwell, "On the Equilibrium of Temperature of a Gaseous Column Subject to Gravity," *Nature* 8 (1873), 527–528; "On the Final State of a System of Molecules in Motion Subject to Forces of Any Kind," *Nature* 8 (1873), 537–538; Henry William Watson, *A Treatise on the Kinetic Theory of Gases* (Oxford: Clarendon Press, 1876), 13–17, 28–33. Maxwell's work on this subject is presented in our next volume.

d. Maxwell had stated in his *Treatise on Electricity and Magnetism* (1873), sec. 57, that the phenomena of electrical discharges in gases "are exceedingly important, and when they are better understood they will probably throw great light on the nature of electricity as well as on the nature of gases and of the medium pervading space. At present, however, they must be considered as outside the domain of the mathematical theory of electricity." His unpublished referee's report to the Royal Society on William Crookes's paper "On the Illumination of Lines of Molecular Pressure, and the Trajectory of Molecules" (*Phil. Trans. R. Soc. London* 170 (1879), 135–164) contains an intriguing attempt to calculate the masses of the particles studied by Crookes; the particles are now called electrons.

48. [Notes on Diffusion]

Cambridge University Library.

Let z be measured along the tube from 0 to T then p_0 is the value of p in the vessel A and p_a in B.

Also since the pressure is constant and temperature

$$k_1 Q_1 + k_2 Q_2 = 0$$

$$(k_1 \rho_1 + k_2 \rho_2)V - (\sqrt{k_1} - \sqrt{k_2})\frac{1}{2}\sqrt{\frac{\pi}{2}}\, l\frac{dp}{dz} = 0$$

Therefore $V = \dfrac{\sqrt{k_1} - \sqrt{k_2}}{P} \cdot \dfrac{1}{2}\sqrt{\dfrac{\pi}{2}} \, l \dfrac{dp}{dz}$

$$Q_1 = \left[\frac{\sqrt{k_1} - \sqrt{k_2}}{\sqrt{k_1}} \cdot \frac{p}{P} - 1\right] \frac{1}{2}\sqrt{\frac{\pi}{2k_1}} \, l \frac{dp}{dz}$$

$$Q_2 = \left[\frac{\sqrt{k_1} - \sqrt{k_2}}{\sqrt{k_2}}\left(1 - \frac{p}{P}\right) + 1\right] \frac{1}{2}\sqrt{\frac{\pi}{2k_2}} \, l \frac{dp}{dz}$$

Now since the vessels are large compared with the tube the motion of the gas may be considered "steady" at any instant and therefore Q_1 constant.

Let $\dfrac{1}{2}\sqrt{\dfrac{\pi}{2k_1}} \, l = R$ and $\left(1 - \sqrt{\dfrac{k_2}{k_1}}\right)\dfrac{1}{P} = S$

Then $Q_1 = (Sp - 1)R\dfrac{dp}{dz}$

$$\frac{dz}{dp} = (Sp - 1)\frac{R}{Q_1}$$

$$z = \left(\frac{1}{2}Sp^2 - p\right)\frac{R}{Q_1} + C$$

Putting $p = p_a$ when $z = 0$ and $p = p_b$ when $z = a$

$$0 = \left(\frac{1}{2}Sp_a^2 - p_a\right)\frac{R}{Q_1} + C$$

$$a = \left(\frac{1}{2}Sp_b^2 - p_b\right)\frac{R}{Q_1} + C$$

$$Q_1 = \left\{\frac{1}{2}S(p_b^2 - p_a^2) - (p_b - p_a)\right\}\frac{R}{a}$$

By this flow of gas from A to B p_a is decreased and p_b increased. If C be the section of the tube, then in unit of time a mass $+CQ$ leaves the vessel A, decreasing the density by CQ/A and pressure by CQk/A so that,

$$\frac{dp_a}{dt} = -\frac{CkQ}{A} \qquad \frac{dp_b}{dt} = \frac{CkQ}{B}$$

whence $Ap_a + Bp_b = Ap_0$ a constant

Therefore $p_b = \dfrac{A}{B}(p_0 - p_a)$

and $\dfrac{dp_a}{dt} = \dfrac{RCk}{2aAB^2}\{S[A^2(p_0 - p_a)^2 - B^2 p_a^2] + 2B(\overline{A + B}p_a - Ap_0)\}$

We may now leave out the suffix p_a and put p for the pressure in the vessel A at any time. We find that

$$\frac{RCk}{aAB}(A + B - SAp_0)t = \log\left\{\frac{2B + S(\overline{A - Bp} - Ap_0)}{(A + B)p - Ap_0} \times \frac{p_0}{2 - Sp_0}\right\}$$

Case I $A = B$ and $p_0 = P$

$$\frac{RCk}{Aa}\left(1 + \sqrt{\frac{k_2}{k_1}}\right)t = -\log_e\left(2\frac{p}{P} - 1\right)$$

Case II $B = \infty$ and $p_0 = P$

$$\frac{RCk}{Aa}t = \log\frac{2\frac{P}{p} + \sqrt{\frac{k_2}{k_1}} - 1}{\sqrt{\frac{k_2}{k_1}} + 1}$$

[End of sheet]

49. "Quiet Diffusion July 3, 1877"

Cambridge University Library.

$$p = k_1\frac{dm_1}{DX} + k_2\frac{dm_2}{dx} \qquad \frac{dp}{dx} + g\frac{dm_1}{dx} + g\frac{dm_2}{dx} = 0$$

$$\frac{dm_1}{dx} = \frac{gp - k_2(dp/dx)}{g(k_2 - k_1)} \qquad \frac{dm_2}{dx} = \frac{k_1(dp/dx) - gp}{g(k_1 - k_2)}$$

$$\frac{1}{D} = \left\{\frac{dm_1}{dt}\frac{dm_2}{dx} - \frac{dm_2}{dt}\frac{dm_1}{dx}\right\} = k_1\frac{d^2m_1}{dx^2} - g\frac{dm_1}{dx} = g\frac{dm_2}{dx} - k_2\frac{d^2m_2}{dx^2}$$

Let $p = \frac{dy}{dx}g(k_1 - k_2)$, $m_1 = gy - k_2\frac{dy}{dx}$, $m_2 = -gy + k_1\frac{dy}{dx}$

$$\frac{1}{D}g(k_1 - k_2)\left[\frac{dy}{dt}\frac{d^2y}{dx^2} - \frac{dy}{dx}\frac{d^2y}{dx\,dt}\right] = \frac{d}{dx}\left(g - k_1\frac{d}{dx}\right)\left(k_2\frac{d}{dx} - g\right)y$$

[End of sheet.]

50. "On Steady Diffusion"[a]

Cambridge University Library.

According to the kinetic theory of gases, the equations of motion of one gas of a mixture are of the form,

$$\rho_1 \frac{\partial_1 u_1}{\partial t} - x_1 \rho_1 + k_1 \frac{\partial \rho_1}{\partial x} + C_{12} \rho_1 \rho_2 (u_1 - u_2) + C_{13} \rho_1 \rho_3 (u_1 - u_3) + \text{etc.} = 0$$

(1)

In this equation the first term represents the rate of increase of momentum (resolved along x) of unit of volume of the first gas. The symbol of operation $\partial_1 / \partial t$ prefixed to any quantity denotes the time variation of that quantity at a point which moves with the velocity whose components are u_1 along with the ⟨medium whose⟩ gas which is distinguished by the suffix 1, or more explicitly,

$$\frac{\partial_1}{\partial t} = \frac{d}{dt} + u_1 \frac{d}{dx} + v_1 \frac{d}{dy} + w_1 \frac{d}{dz}.$$

The second term denotes the effect of a force like that of gravity acting on the gas. For the sake of generality we shall suppose that ⟨this inter⟩ external force is different for each gas of the mixture.

In the third term k_1 is the ratio of the pressure to the temperature and depends on the temperature. The third term may be written dp_1/dx and denotes the effect of the space-variation of pressure.

The remaining terms denote the resistance which the gas meets with in percolating through the other gases of the mixture arising from encounters between their molecules. Each term is proportional to the relative velocity of the two gases, to the product of their densities and to a coefficient depending on the nature of the two gases and also on the temperature.

If there are n gases there will be $n - 1$ terms of this kind in each of the n equations, and there will be $\frac{1}{2}n(n - 1)$ different coefficients.

If we add together the corresponding equations for all the different gases, the terms depending on their mutual action destroy each other and we find,

$$\rho_1 \frac{\partial_1 u_1}{\partial t} + \rho_2 \frac{\partial_2 u_2}{\partial t} + \rho_3 \frac{\partial_3 u_3}{\partial t} - (x_1 \rho_1 + x_2 \rho_2 + x_3 \rho_3) + k_1 \frac{d\rho_1}{dx} + k_2 \frac{d\rho_2}{dx}$$

$$+ k_3 \frac{d\rho_3}{dx} = 0.$$

If we write,

$$\rho_1 + \rho_2 + \rho_3 = \rho$$

$$\rho_1 u_1 + \rho_2 u_2 + \rho_3 u_3 = \rho u$$

$$k_1 \rho_1 + k_2 \rho_2 + k_3 \rho_3 = p$$

and if we use the symbol of operation $\partial/\partial t$ to denote the time variation at a point moving with the mixed medium, that is to say with the velocity whose components are u, v, w then,

$$\frac{\partial_1 u_1}{\partial t} = \frac{\partial u_1}{\partial t} + (u_1 - u)\frac{du_1}{dx} + (v_1 - v)\frac{du_1}{dy} + (w_1 - w)\frac{du_1}{dz}$$

Hence

$$\rho_1 \frac{\partial_1 u_1}{\partial t} + \rho_2 \frac{\partial_2 u_2}{\partial t} + \frac{\partial_3 u_3}{\partial t}$$

$$= \rho_1 \frac{\partial u_1}{\partial t} + \rho_2 \frac{\partial u_2}{\partial t} + \rho_3 \frac{\partial u_3}{\partial t} + (u_1 - u)\rho_1 \frac{du_1}{dx} + (u_2 - u)\frac{du_2}{dx}\rho_2 + \text{etc.}$$

and

$$\left\langle \rho_1 \frac{\partial u_1}{\partial t} + \rho_2 \frac{\partial u_2}{\partial t} + \rho_3 \frac{\partial u_3}{\partial t} = \rho \frac{\partial u}{\partial t} + u\frac{\partial \rho}{\partial t} - u_1 \frac{\partial \rho_1}{\partial t} - u_2 \frac{\partial \rho_2}{\partial t} - u_3 \frac{\partial \rho_3}{\partial t} \right.$$

$$= \rho \frac{\partial u}{\partial t} - (u_1 - u)\frac{\partial \rho_1}{\partial t} - (u_2 - u)\frac{\partial \rho_2}{\partial t} - (u_3 - u)\frac{\partial \rho_3}{\partial t}$$

Since in all actual cases of diffusion except those in which the $\rangle$

$$= \rho \frac{\partial u}{\partial t} - (u_1 - u)\frac{\partial \rho_1}{\partial t} - (u_2 - u)\frac{\partial \rho_2}{\partial t} - (u_3 - u)\frac{\partial \rho_3}{\partial t}$$

$$\left\langle + \rho_1(u_1 - u)\frac{d}{dx}(u_1 - u_2) + \rho_2(u_2 - u)d \right\rangle$$

$$+ \rho_1(u_1 - u)\frac{d}{dx}(u_1 - u) + \rho_2(u_2 - u)\frac{d}{dx}(u_2 - u) + \rho_3(u_3 - u)\frac{d}{dx}(u_3 - u)$$

$$+ \rho_1(v_1 - v)\frac{d}{dy}(u_1 - u) + \rho_2(v_2 - v)\frac{d}{dx}(u_2 - u) + \rho_3(v_3 - v)\frac{d}{dx}(u_3 - u)$$

$$+ \rho_1(w_1 - w)\frac{d}{dz}(u_1 - u) + \rho_2(w_2 - w)\frac{d}{dx}(u_2 - u) + \rho_3(w_3 - w)\frac{d}{dx}(u_3 - u)$$

In this expression the quantities $u_1 - u$ etc represent the velocities of diffusion and are except in certain cases very small. The rate of change of the density of each gas is also in most cases very small. Hence the above expression is, except in cases of extremely rapid diffusion, sensibly equal to its first term and equation becomes,

$$\rho \frac{\partial u}{\partial t} + \frac{dp}{dx} = X_1 \rho_1 + X_2 \rho_2 + X_3 \rho_3$$

This equation is of the same form as the ordinary equations of Hydrokinetics and indicates that the mixture as a whole moves in a manner sensibly the same as that of a continuous fluid whose density at any point is the same as that of the mixture and which is acted upon by forces ⟨whose resultant is⟩ equivalent to those acting on the mixture.

If we now write for shortness,

$$X\rho - k\frac{d\rho}{dx} - \rho\frac{\partial u}{\partial t} = p$$

then the equations for the first two gases are,

$$(C_{12}\rho_2 + C_{13}\rho_3)\rho_1 u_1 - C_{12}\rho_1\rho_2 u_2 - C_{13}\rho_1\rho_3 u_3 = P_1$$

$$-C_{12}\rho_2\rho_1 u_1 + (C_{23}\rho_3 + C_{21}\rho_1)\rho_2 u_2 - C_{23}\rho_2\rho_3 u_3 = P_2$$

Combining these with the equations,

$$\rho_1 u_1 + \rho_2 u_2 + \rho_3 u_3 = \rho u$$

we find,

$$\rho_1(u_1 - u) = \frac{C_{23}\rho P_1 - \rho_1(C_{23}P_1 + C_{31}P_2 + C_{13}P_3)}{C_{12}C_{31}\rho_1 + C_{23}C_{12}\rho_2 + C_{31}C_{23}\rho_3}$$

with similar expressions for the other two gases.

If we write D_{23}, D_{31}, D_{12} for the reciprocals of C_{23}, C_{31}, C_{12} respectively, these expressions become,

$$\rho_1(u_1 - u) = \frac{D_{12}D_{31}\rho P_1 - \rho_1(D_{12}D_{31}P_1 + D_{23}D_{12}P_2 + D_{31}D_{23}P_3)}{D_{23}\rho_1 + D_{31}\rho_2 + D_{12}\rho_3}$$

If there are only two gases, $P_3 = 0$ and $P_1 + P_2 = 0$ and $\Big\langle \rho_1(u_1 - u) =$

$\dfrac{1}{C_{12}}P_1 = -\dfrac{1}{C_{12}}P_2 = \rho_2(u - u_1)\Big\rangle$ and $\rho_1(u_1 - u) = D_{12}P_1 = -D_{12}P_2 =$

$-\rho_2(u_2 - u)$

Hence D_{12} may be called the coefficient of ⟨diffusion⟩ interdiffusion of the two gases.

[End of sheet]

In the case of two gases not acted on by gravity or in a vessel so small that we may neglect the differences of density ⟨arising from the indifferent⟩ arising from the differences of pressures in different parts of the vessel, we have,

$$\frac{dp_1}{dx} + C\rho_1\rho_2(u_1 - u_2) = 0$$

$$\frac{dp_2}{dx} + C\rho_1\rho_2(u_2 - u_1) = 0$$

$$p_1 = k_1\rho_1 \quad p_2 = k_2\rho_2 \quad p = p_1 + p_2$$

If the diffusion takes place ⟨along⟩ parallel to the axis of x, as it does in a vertical cylinder, the equation of continuity is,

$$\frac{d\rho_1}{dt} + \frac{d}{dx}(\rho_1 u_1) = 0 \quad \text{or multiplying by } k_1$$

$$\frac{dp_1}{dt} + \frac{d}{dx}(p_1 u_1) = 0$$

Since the diffusion takes place by interchange of equal volumes,

$$p_1 u_1 + p_2 u_2 = 0$$

Hence by equation (1)

$$p_1 u_1 = -\frac{p_1 p_2}{p \rho_1 \rho_2 C}\frac{dp_1}{dx}$$

$$= -\frac{k_1 k_2}{pC}\frac{dp_1}{dx}$$

and the differential equation of p_1 is

$$\frac{dp_1}{dt} = \frac{k_1 k_2}{pC}\frac{dp_1}{dx}$$

The quantity $(k_1 k_2/pC) = D$ is called the coefficient of diffusion of the two gases. ⟨If⟩ The dimensions of this coefficient are evidently $L^2 T^{-1}$ where L is the unit of length and T the unit of time.

The coefficient of diffusion varies inversely as the total pressure of the medium and is also ⟨a funct depends on⟩ increases with the temperature, but in what ratio is not yet ascertained.

a. The draft is dated July 1877. This is a draft for Maxwell's *Encyclopedia Britannica* article on "Diffusion," reprinted as document III-52.

51. [Notes on the Diffusion of Gases]

Cambridge University Library.

Thus in the case of two gases not acted on by gravity or in so small a vessel that we may neglect the effect of gravity in producing differences of density in different parts of the vessel the ordinary hydrodynamical equations are of the form,

$$\rho_1 \frac{\partial_1 u}{\partial t} - X_1 \rho_1 + k_1 \frac{d}{dx}p_1 + C\rho_1 \rho_2(u_1 - u_2) = 0$$

for the one gas, and,

$$\rho_2 \frac{\partial_2 u}{\partial t} - X_2 \rho_2 + k_2 \frac{d}{dx} \rho_2 + C \rho_2 \rho_1 (u_2 - u_1) = 0$$

for the other. If we add the two corresponding equations we find,

$$\rho_1 \frac{\partial_1 u_1}{\partial t} + \rho_2 \frac{\partial_2 u_2}{\partial t} - X_1 \rho_1 - X_2 \rho_2 + k_1 \frac{d}{dx} \rho_1 + k_2 \frac{d}{dX} \rho_2 = 0$$

Since the velocities u_1 and u_2 are never very different from each other or from u the velocity of the mixed medium, this equation shows that the more rapid movements of the mixed medium go on very nearly as if it were a continuous fluid, ⟨the density bei⟩

If we now write $\rho_1 \frac{\partial_1}{\partial t} u - X_1 \rho_1 + k_1 \frac{d}{dx} \rho_1 = P_1$

we find,

$$P_1 = C \rho_1 \rho_2 (u_2 - u_1) = -P_2$$

and if we write,

$$\langle \rho_1 u_1 + \rho_2 u_2 = \rho u \rangle$$

$$\rho_1 + \rho_2 = \rho$$

and,

$$\rho_1 u_1 + \rho_2 u_2 = \rho u$$

$$\rho_1 (u_1 - u) = -\frac{P_1}{C} = \frac{P_2}{C} = \rho_2 (u - u_2)$$

These equations determine the rate of flow of the two gases with respect to the ⟨medi⟩ mixed medium.

The equation of continuity is,

$$\frac{\partial \rho_1}{\partial t} + \frac{d}{dx} \rho_1 (u_1 - u) + \frac{d}{dy} \rho_1 (v_1 - v) + \frac{d}{dx} \rho_1 (w_1 - w) = 0$$

or,

$$\frac{\partial \rho_1}{\partial t} = \frac{1}{C} \left[\frac{dP_1}{dx} + \frac{dQ_1}{dy} + \frac{dR_1}{dz} \right]$$

[End of sheet]

52. "Diffusion"

Encyclopedia Britannica, ninth edition (1878), vol. 7, 214–221; *Scientific Papers*, vol. 2, 625–646.

[From the *Encyclopædia Britannica*.]

LXXXIX. *Diffusion.*

SOME liquids, such as mercury and water, when placed in contact with each other do not mix at all, but the surface of separation remains distinct, and exhibits the phenomena described under CAPILLARY ACTION. Other pairs of liquids, such as chloroform and water, mix, but only in certain proportions. The chloroform takes up a little water, and the water a little chloroform; but the two mixed liquids will not mix with each other, but remain in contact separated by a surface shewing capillary phenomena. The two liquids are then in a state of equilibrium with each other. The conditions of the equilibrium of heterogeneous substances have been investigated by Professor J. Willard Gibbs in a series of papers published in the *Transactions of the Connecticut Academy of Arts and Sciences*, Vol. III. part I. p. 108. Other pairs of liquids, and all gases, mix in all proportions.

When two fluids are capable of being mixed, they cannot remain in equilibrium with each other; if they are placed in contact with each other the process of mixture begins of itself, and goes on till the state of equilibrium is attained, which, in the case of fluids which mix in all proportions, is a state of uniform mixture.

This process of mixture is called diffusion. It may be easily observed by taking a glass jar half full of water and pouring a strong solution of a coloured salt, such as sulphate of copper, through a long-stemmed funnel, so as to occupy the lower part of the jar. If the jar is not disturbed we may trace the process of diffusion for weeks, months, or years, by the gradual rise of the colour into the upper part of the jar, and the weakening of the colour in the lower part.

This, however, is not a method capable of giving accurate measurements of the composition of the liquid at different depths in the vessel. For more exact determinations we may draw off a portion from a given stratum of the mixed liquid, and determine its composition either by chemical methods or by its specific gravity, or any other property from which its composition may be deduced.

But as the act of removing a portion of the fluid interferes with the process of diffusion, it is desirable to be able to ascertain the composition of any stratum of the mixture without removing it from the vessel. For this purpose Sir W. Thomson places in the jar a number of glass beads of different densities, which indicate the densities of the strata in which they are observed to float. The principal objection to this method is, that if the liquids contain air or any other gas, bubbles are apt to form on the glass beads, so as to make them float in a stratum of less density than that marked on them.

M. Voit has observed the diffusion of cane-sugar in water by passing a ray of plane-polarized light horizontally through the vessel, and determining the angle through which the plane of polarization is turned by the solution of sugar. This method is of course applicable only to those substances which cause rotation of the plane of polarized light.

Another method is to place the diffusing liquids in a hollow glass prism, with its refracting edge vertical, and to determine the deviation of a ray of light passing through the prism at different depths. The ray is bent downwards on account of the variable density of the mixture, as well as towards the thicker part of the prism; but by making it pass as near the edge of the prism as possible, the vertical component of the refraction may be made very small; and by placing the prism within a vessel of water having parallel sides of glass, we can get rid of the constant part of the deviation, and are able to use a prism of large angle, so as to increase the part due to the diffusing substance. At the same time we can more easily control and register the temperature.

The laws of diffusion were first investigated by Graham. The diffusion of gases has recently been observed with great accuracy by Loschmidt, and that of liquids by Fick and by Voit.

Diffusion as a Molecular Motion.

If we observe the process of diffusion with our most powerful microscopes, we cannot follow the motion of any individual portions of the fluids. We

DIFFUSION.   627

cannot point out one place in which the lower fluid is ascending, and another in which the upper fluid is descending. There are no currents visible to us, and the motion of the material substances goes on as imperceptibly as the conduction of heat or of electricity. Hence the motion which constitutes diffusion must be distinguished from those motions of fluids which we can trace by means of floating motes. It may be described as a motion of the fluids, not *in mass* but by *molecules*.

When we reason upon the hypothesis that a fluid is a continuous homogeneous substance, it is comparatively easy to define its density and velocity; but when we admit that it may consist of molecules of different kinds, we must revise our definitions. We therefore define these quantities by considering that part of the medium which at a given instant is within a certain small region surrounding a given point. This region must be so small that the properties of the medium as a whole are sensibly the same throughout the region, and yet it must be so large as to include a large number of molecules. We then define the density of the medium at the given point as the mass of the medium within this region divided by its volume, and the velocity of the medium as the momentum of this portion of the medium divided by its mass.

If we consider the motion of the medium relative to an imaginary surface supposed to exist within the region occupied by the medium, and if we define the flow of the medium through the surface as the mass of the medium which in unit of time passes through unit of area of the surface, then it follows from the above definitions that the velocity of the medium resolved in the direction of the normal to the surface is equal to the flow divided by the density. If we suppose the surface itself to move with the same velocity as the fluid, and in the same direction, there will be no flow through it.

Having thus defined the density, velocity, and flow of the medium as a whole, or, as it is sometimes expressed, "in mass," we may now consider one of the fluids which constitute the medium, and define its density, velocity, and flow in the same way. The velocity of this fluid may be different from that of the medium in mass, and its velocity relative to that of the medium is the velocity of diffusion which we have to study.

Diffusion of Gases according to the Kinetic Theory.

So many of the phenomena of gases are found to be explained in a consistent manner by the kinetic theory of gases, that we may describe with considerable probability of correctness the kind of motion which constitutes diffusion in gases. We shall therefore consider gaseous diffusion in the light of the kinetic theory before we consider diffusion in liquids.

A gas, according to the kinetic theory, is a collection of particles or molecules which are in rapid motion, and which, when they encounter each other, behave pretty much as elastic bodies, such as billiard balls, would do if no energy were lost in their collisions. Each molecule travels but a very small distance between one encounter and another, so that it is every now and then altering its velocity both in direction and magnitude, and that in an exceedingly irregular manner.

The result is that the velocity of any molecule may be considered as compounded of two velocities, one of which, called the velocity of the medium, is the same for all the molecules, while the other, called the velocity of agitation, is irregular both in magnitude and in direction, though the average magnitude of the velocity may be calculated, and any one direction is just as likely as any other.

The result of this motion is, that if in any part of the medium the molecules are more numerous than in a neighbouring region, more molecules will pass from the first region to the second than in the reverse direction, and for this reason the density of the gas will tend to become equal in all parts of the vessel containing it, except in so far as the molecules may be crowded towards one direction by the action of an external force such as gravity. Since the motion of the molecules is very swift, the process of equalization of density in a gas is a very rapid one, its velocity of propagation through the gas being that of sound.

Let us now consider two gases in the same vessel, the proportion of the gases being different in different parts of the vessel, but the pressure being everywhere the same. The agitation of the molecules will still cause more molecules of the first gas to pass from places where that gas is dense to places where it is rare than in the opposite direction, but since the second gas is dense where the first one is rare, its molecules will be for the most part travelling in the opposite direction. Hence the molecules of the two gases will encounter each other, and every encounter will act as a check to the process of equalization of the density of each gas throughout the mixture.

The interdiffusion of two gases in a vessel is therefore a much slower process than that by which the density of a single gas becomes equalized, though it appears from the theory that the final result is the same, and that each gas is distributed through the vessel in precisely the same way as if no other gas had been present, and this even when we take into account the effect of gravity.

If we apply the ordinary language about fluids to a single gas of the mixture, we may distinguish the forces which act on an element of volume as follows :—

1st. Any external force, such as gravity or electricity.

2nd. The difference of the pressure *of the particular gas* on opposite sides of the element of volume. [The pressure due to other gases is to be considered of no account.]

3rd. The resistance arising from the percolation of the gas through the other gases which are moving with different velocity.

The resistance due to encounters with the molecules of any other gas is proportional to the velocity of the first gas relative to the second, to the product of their densities, and to a coefficient which depends on the nature of the gases and on the temperature. The equations of motion of one gas of a mixture are therefore of the form

$$\rho_1 \frac{\delta_1 u_1}{\delta t} + \frac{dp_1}{dx} - X_1\rho_1 + C_{12}\rho_1\rho_2(u_1 - u_2) + C_{13}\rho_1\rho_3(u_1 - u_3) + \&c. = 0,$$

where the symbol of operation $\frac{\delta_1}{\delta t}$ prefixed to any quantity denotes the time-variation of that quantity at a point which moves along with that medium which is distinguished by the suffix $(_1)$, or more explicitly

$$\frac{\delta_1}{\delta t} = \frac{d}{dt} + u_1\frac{d}{dx} + v_1\frac{d}{dy} + w_1\frac{d}{dz}.$$

In the state of ultimate equilibrium $u_1 = u_2 = \&c. = 0$, and the equation is reduced to

$$\frac{dp_1}{dx} - X\rho_1 = 0,$$

which is the ordinary form of the equations of equilibrium of a single fluid.

Hence, when the process of diffusion is complete, the density of each gas at any point of the vessel is the same as if no other gas were present.

If V_1 is the potential of the force which acts on the gas, and if in the equation $p_1 = k_1\rho_1$, k_1 is constant, as it is when the temperature is uniform, then the equation of equilibrium becomes

$$k_1 \frac{d\rho_1}{dx} + \frac{dV_1}{dx}\rho_1 = 0,$$

the solution of which is

$$\rho_1 = A_1 e^{-\frac{V_1}{k_1}}.$$

Hence if, as in the case of gravity, V is the same for all gases, but k is different for different gases, the composition of the mixture will be different in different parts of the vessel, the proportion of the heavier gases, for which k is smaller, being greater at the bottom of the vessel than at the top. It would be difficult, however, to obtain experimental evidence of this difference of composition except in a vessel more than 100 metres high, and it would be necessary to keep the vessel free from inequalities of temperature for more than a year, in order to allow the process of diffusion to advance to a state even half-way towards that of ultimate equilibrium. The experiment might, however, be made in a few minutes by placing a tube, say 10 centimetres long, on a whirling apparatus, so that one end shall be close to the axis, while the other is moving at the rate, say, of 50 metres per second. Thus if equal volumes of hydrogen and carbonic acid were used, the proportion of hydrogen to carbonic acid would be about $\frac{1}{134}$ greater at the end of the tube nearest the axis. The experimental verification of the result is important, as it establishes a method of effecting the partial separation of gases without the selective action of chemical agents.

Let us next consider the case of diffusion in a vertical cylinder. Let m_1 be the mass of the first gas in a column of unit area extending from the bottom of the vessel to the height x, and let v_1 be the volume which this mass would occupy at unit pressure, then

$$k_1 m_1 = v_1,$$

$$\rho_1 = \frac{dm_1}{dx}, \qquad \rho_1 u_1 = -\frac{dm_1}{dt},$$

$$p_1 = \frac{dv_1}{dx}, \qquad p_1 u_1 = -\frac{dv_1}{dt};$$

and the equation of motion becomes

$$\frac{1}{k_1\left|\frac{dv_1}{dx}\right|^2}\left\{\left|\frac{d^2v_1}{dx^2}\frac{dv_1}{dt}\right|^2 - \frac{d^2v_1}{dt^2}\left|\frac{dv_1}{dx}\right|^2\right\} + \frac{d^2v_1}{dx^2} - \frac{X}{k_1}\frac{dv_1}{dx}$$

$$+ \frac{C_{12}}{k_1k_2}\left\{\frac{dv_2}{dt}\frac{dv_1}{dx} - \frac{dv_1}{dt}\frac{dv_2}{dx}\right\} + \&c. = 0.$$

If we add the corresponding equations together for all the gases, we find that the terms in C_{12} destroy each other, and that if the medium is not affected with sensible currents the first term of each equation may be neglected. In ordinary experiments we may also neglect the effect of gravity, so that we get

$$\frac{d^2}{dx^2}(v_1 + v_2) = 0,$$

or
$$v_1 + v_2 = px,$$

where p is the uniform pressure of the mixed medium. Hence

$$\frac{dv_2}{dt} = -\frac{dv_1}{dt} \quad \text{and} \quad \frac{dv_2}{dx} = p - \frac{dv_1}{dx},$$

and the equation becomes

$$\frac{d^2v_1}{dx^2} = \frac{C_{12}}{k_1k_2}p\frac{dv_1}{dt},$$

an equation, the form of which is identical with the well-known equation for the conduction of heat. We may write it

$$\frac{dv_1}{dt} = D\frac{d^2v_1}{dx^2}.$$

D is called the coefficient of diffusion. It is equal to

$$\frac{k_1k_2}{C_{12}p}.$$

It therefore varies inversely as the total pressure of the medium, and if the coefficient of resistance, C_{12}, is independent of the temperature, it varies directly as the product k_1k_2, i.e., as the square of the absolute temperature. It is probable, however, that the effect of temperature is not so great as this would make it.

In liquids D probably depends on the proportion of the ingredients of the mixed medium as well as on the temperature. The dimensions of D are L^2T^{-1}, where L is the unit of length and T the unit of time.

The values of the coefficients of diffusion of several pairs of gases have been determined by Loschmidt[*]. They are referred in the following table to the centimetre and the second as units, for the temperature 0°C. and the pressure of 76 centimetres of mercury.

	D
Carbonic acid and air	0·1423
Carbonic acid and hydrogen . . .	0·5558
Oxygen and hydrogen	0·7214
Carbonic acid and oxygen	0·1409
Carbonic acid and carbonic oxide . .	0·1406
Carbonic acid and marsh gas . . .	0·1586
Carbonic acid and nitrous oxide . . .	0·0983
Sulphurous acid and hydrogen . . .	0·4800
Oxygen and carbonic oxide . . .	0·1802
Carbonic oxide and hydrogen . . .	0·6422

Diffusion in Liquids.

The nature of the motion of the molecules in liquids is less understood than in gases, but it is easy to see that if there is any irregular displacement among the molecules in a mixed liquid, it must, on the whole, tend to cause each component to pass from places where it forms a large proportion of the mixture to places where it is less abundant. It is also manifest that any relative motion of two constituents of the mixture will be opposed by a resistance arising from the encounters between the molecules of these components. The value of this resistance, however, depends, in liquids, on more complicated conditions than in gases, and for the present we must regard it as a function of all the physical properties of the mixture at the given place, that is to say, its temperature and pressure, and the proportions of the different components of the mixture.

[*] Imperial Academy of Vienna, 10th March, 1870.

The coefficient of interdiffusion of two liquids must therefore be considered as depending on all the physical properties of the mxiture according to laws which can be ascertained only by experiment.

Thus Fick has determined the coefficient of diffusion for common salt in water to be 0·00000116, and Voit has found that of cane-sugar to be 0·00000365.

It appears from these numbers that in a vessel of the same size the process of diffusion of liquids requires a greater number of days to reach a given stage than the process of diffusion of gases in the same vessel requires seconds.

When we wish to mix two liquids, it is not sufficient to place them in the same vessel, for if the vessel is, say, a metre in depth, the lighter liquid will lie above the denser, and it will be many years before the mixture becomes even sensibly uniform. We therefore stir the two liquids together, that is to say, we move a solid body through the vessel, first one way, then another, so as to make the liquid contents eddy about in as complicated a manner as possible. The effect of this is that the two liquids, which originally formed two thick horizontal layers, one above the other, are now disposed in thin and excessively convoluted strata, which, if they could be spread out, would cover an immense area. The effect of the stirring is thus to increase the area over which the process of diffusion can go on, and to diminish the distance between the diffusing liquids; and since the time required for diffusion varies as the square of the thickness of the layers, it is evident that by a moderate amount of stirring the process of mixture which would otherwise require years may be completed in a few seconds. That the process is not instantaneous is easily ascertained by observing that for some time after the stirring the mixture appears full of streaks, which cause it to lose its transparency. This arises from the different indices of refraction of different portions of the mixture which have been brought near each other by stirring. The surfaces of separation are so drawn out and convoluted, that the whole mass has a woolly appearance, for no ray of light can pass through it without being turned many times out of its path.

Graham observed that the diffusion both of liquids and gases takes place through porous solid bodies, such as plugs of plaster of Paris or plates of pressed plumbago, at a rate not very much less than when no such body is interposed, and this even when the solid partition is amply sufficient to check all ordinary currents, and even to sustain a considerable difference of pressure on its opposite sides.

But there is another class of cases in which a liquid or a gas can pass through a diaphragm, which is not, in the ordinary sense, porous. For instance, when carbonic acid gas is confined in a soap bubble it rapidly escapes. The gas is absorbed at the inner surface of the bubble, and forms a solution of carbonic acid in water. This solution diffuses from the inner surface of the bubble, where it is strongest, to the outer surface, where it is in contact with air, and the carbonic acid evaporates and diffuses out into the atmosphere. It is also found that hydrogen and other gases can pass through a layer of caoutchouc. Graham shewed that it is not through pores, in the ordinary sense, that the motion takes place, for the ratios are determined by the chemical relations between the gases and the caoutchouc, or the liquid film.

According to Graham's theory, the caoutchouc is a colloïd substance,—that is, one which is capable of combining, in a temporary and very loose manner, with indeterminate proportions of certain other substances, just as glue will form a jelly with various proportions of water. Another class of substances, which Graham called crystalloïd, are distinguished from these by being always of definite composition, and not admitting of these temporary associations. When a colloïd body has in different parts of its mass different proportions of water, alcohol, or solutions of crystalloïd bodies, diffusion takes place through the colloïd body, though no part of it can be shewn to be in the liquid state.

On the other hand, a solution of a colloïd substance is almost incapable of diffusion through a porous solid, or another colloïd body. Thus, if a solution of gum and salt in water is placed in contact with a solid jelly of gelatine and alcohol, alcohol will be diffused into the gum, and salt and water will be diffused into the gelatine, but the gum and the gelatine will not diffuse into each other.

There are certain metals whose relations to certain gases Graham explained by this theory. For instance, hydrogen can be made to pass through iron and palladium at a high temperature, and carbonic oxide can be made to pass through iron. The gases form colloïdal unions with the metals, and are diffused through them as water is diffused through a jelly. Root has lately found that hydrogen can pass through platinum, even at ordinary temperatures.

By taking advantage of the different velocities with which different liquids and gases pass through parchment-paper and other solid bodies, Graham was enabled to effect many remarkable analyses. He called this method the method of Dialysis.

Diffusion and Evaporation, Condensation, Solution, and Absorption.

The rate of evaporation of liquids is determined principally by the rate of diffusion of the vapour through the air or other gas which lies above the liquid. Indeed, the coefficient of diffusion of the vapour of a liquid through air can be determined in a rough but easy manner by placing a little of the liquid in a test tube, and observing the rate at which its weight diminishes by evaporation day by day. For at the surface of the liquid the density of the vapour is that corresponding to the temperature, whereas at the mouth of the test tube the air is nearly pure. Hence, if p be the pressure of the vapour corresponding to the temperature, and $p = k\rho$, and if m be the mass evaporated in time t, and diffused into the air through a distance h *, then

$$D = \frac{khm}{pt} .$$

This method is not, of course, applicable to vapours which are rarer than the superincumbent gas.

The solution of a salt in a liquid goes on in the same way, and so does the absorption of a gas by a liquid.

These processes are all accelerated by currents, for the reason already explained.

The processes of evaporation and condensation go on much more rapidly when no air or other non-condensible gas is present. Hence the importance of the air-pump in the steam engine.

Relation between Diffusion of Matter and Diffusion of Heat.

The same motion of agitation of the molecules of gases which causes two gases to diffuse through each other also causes two portions of the same gas to diffuse through each other, although we cannot observe this kind of diffusion, because we cannot distinguish the molecules of one portion from those of the

---

* h should be taken equal to the height of the tube above the surface of the liquid, together with about $\frac{2}{8}$ of the diameter of the tube.—See Clerk Maxwell's *Electricity*, Art. 309.

other when they are once mixed. If, however, the molecules of one portion have any property whereby they can be distinguished from those of the other, then that property will be communicated from one part of the medium to an adjoining part, and that either by convection—that is by the molecules themselves passing out of one part into the other, carrying the property with them—or by transmission—that is by the property being communicated from one molecule to another during their encounters. The chemical properties by which different substances are recognized are inseparable from their molecules, so that the diffusion of such properties can take place only by the transference of the molecules themselves, but the momentum of a molecule in any given direction and its energy are also properties which may be different in different molecules, but which may be communicated from one molecule to another. Hence the diffusion of momentum and that of energy through the medium can take place in two different ways, whereas the diffusion of matter can take place only in one of these ways.

In gases the great majority of the particles, at any instant, are describing free paths, and it is therefore possible to shew that there is a simple numerical relation between the coefficients of the three kinds of diffusion,—the diffusion of matter, the lateral diffusion of velocity (which is the phenomenon known as the internal friction or viscosity of fluids), and the diffusion of energy (which is called the conduction of heat). But in liquids the majority of the molecules are engaged at close quarters with one or more other molecules, so that the transmission of momentum and of energy takes place in a far greater degree by communication from one molecule to another, than by convection by the molecules themselves. Hence the ratios of the coefficient of diffusion to those of viscosity and thermal conductivity are much smaller in liquids than in gases.

Theory of the Wet Bulb Thermometer.

The temperature indicated by the wet bulb thermometer is determined in great part by the relation between the coefficients of diffusion and thermal conductivity. As the water evaporates from the wet bulb heat must be supplied to it by convection, conduction, or radiation. This supply of heat will not be sufficient to maintain the temperature constant till the temperature of the wet bulb has sunk so far below that of the surrounding air and other

DIFFUSION.

bodies that the flow of heat due to the difference of temperature is equal to the latent heat of the vapour which leaves the bulb.

The use of the wet bulb thermometer as a means of estimating the humidity of the atmosphere was employed by Hutton[*] and Leslie[†], but the formula by which the dew-point is commonly deduced from the readings of the wet and dry thermometers was first given by Dr Apjohn[‡].

Dr Apjohn assumes that, when the temperature of the wet bulb is stationary, the heat required to convert the water into vapour is given out by portions of the surrounding air in cooling from the temperature of the atmosphere to that of the wet bulb, and that the air thus cooled becomes saturated with the vapour which it receives from the bulb.

Let m be the mass of a portion of air at a distance from the wet bulb, θ_0 its temperature, p_0 the pressure due to the aqueous vapour in it, and P the whole pressure.

If σ is the specific gravity of aqueous vapour (referred to air), then the mass of water in this portion of air is $\frac{p_0}{P}\sigma m$.

Let this portion of air communicate with the wet bulb till its temperature sinks to θ_1, that of the wet bulb, and the pressure of the aqueous vapour in it rises to p_1, that corresponding to the temperature θ_1.

The quantity of vapour which has been communicated to the air is

$$(p_1 - p_0)\frac{\sigma m}{P},$$

and if L is the latent heat of vapour at the temperature θ_1, the quantity of heat required to produce this vapour is

$$(p_1 - p_0)\frac{\sigma m}{P} L.$$

According to Apjohn's theory, this heat is supplied by the mixed air and vapour in cooling from θ_0 to θ_1.

If S is the specific heat of the air (which will not be sensibly different from that of dry air), this quantity of heat is

$$(\theta_0 - \theta_1) mS.$$

[*] Playfair's "Life of Hutton," *Edinburgh Transactions*, Vol. v. p. 67, note.

[†] *Encyc. Brit.*, 8th ed. Vol. I. "Dissertation Fifth," p. 764.

[‡] *Trans. Royal Irish Academy*, 1834.

Equating the two values we obtain

$$p_0 = p_1 - \frac{PS}{L\sigma}(\theta_0 - \theta_1).$$

Here p_0 is the pressure of the vapour in the atmosphere. The temperature—for which this is the maximum pressure—is the dew-point, and p_1 is the maximum pressure corresponding to the temperature θ_1 of the wet bulb. Hence this formula, combined with tables of the pressure of aqueous vapour, enables us to find the dew-point from observations of the wet and dry bulb thermometers.

We may call this the convection theory of the wet bulb, because we consider the temperature and humidity of a portion of air brought from a distance to be affected directly by the wet bulb without communication either of heat or of vapour with other portions of air.

Dr Everett has pointed out as a defect in this theory, that it does not explain how the air can either sink in temperature or increase in humidity unless it comes into absolute contact with the wet bulb. Let us, therefore, consider what we may call the conduction and diffusion theory in calm air, taking into account the effects of radiation.

The steady conduction of heat is determined by the conditions—

$\theta = \theta_0$ at a great distance from the bulb,

$\theta = \theta_1$ at the surface of the bulb,

$\nabla^2\theta = 0$ at any point of the medium.

The steady diffusion of vapour is determined by the conditions—

$p = p_0$ at a great distance from the bulb,

$p = p_1$ at the surface of the bulb,

$\nabla^2 p = 0$ at any point of the medium.

Now, if the bulb had been an electrified conductor, the conditions with respect to the potential would have been

$V = 0$ at a great distance,

$V = V_1$ at the surface,

$\nabla^2 V = 0$ at any point outside the bulb.

Hence the solution of the electrical problem leads to that of the other two. For if V is the potential at any point,

$$\theta = \theta_0 + (\theta_1 - \theta_0)\frac{V}{V_1}, \qquad p = p_0 + (p_1 - p_0)\frac{V}{V_1}.$$

If E is the electric charge of the conductor,

$$4\pi E = -\iint \frac{dV}{d\nu}\, dS,$$

where the double integral is extended over the surface of the bulb, and $d\nu$ is an element of a normal to the surface.

If H is the flow of heat in unit of time from the bulb,

$$H = -K \iint \frac{d\theta}{d\nu}\, dS,$$

and if Q is the flow of aqueous vapour from the bulb,

$$Q = -\frac{D}{k} \iint \frac{dp}{d\nu}\, dS,$$

where k is the ratio of the pressure of aqueous vapour to its density.

If C is the electrical capacity of the bulb, $E = CV_1$,

$$H = 4\pi CK\,(\theta_1 - \theta_0), \qquad Q = 4\pi C\frac{D}{k}\,(p_1 - p_0).$$

The heat which leaves the bulb by radiation to external objects at temperature θ_0 may be written

$$h = AR\,(\theta_1 - \theta_0),$$

where A is the surface of the bulb and R the coefficient of radiation of unit of surface.

When the temperature becomes constant

$$LQ + H + h = 0,$$

$$p_0 = -p_1 \frac{PS}{L\sigma}\left\{\frac{K}{D} + \frac{AR}{4\pi C\rho SD}\right\}(\theta_0 - \theta_1).$$

This formula gives the result of the theory of diffusion, conduction, and radiation in a still atmosphere. It differs from the formula of the convection theory only by the factor in the last term.

The first part of this factor $\dfrac{K}{D}$ is certainly less than unity, and probably about ·77.

If the bulb is spherical and of radius r, $A = 4\pi r^2$ and $C = r$, so that the second part is $\dfrac{Rr}{\rho SD}$.

Hence, the larger the wet bulb, the greater will be the ratio of the effect of radiation to that of conduction. If, on the other hand, the air is in motion, this will increase both conduction and diffusion, so as to increase the ratio of the first part to the second. By comparing actual observations of the dew-point with Apjohn's formula, it has been found that the factor should be somewhat greater than unity. According to our theory it ought to be greater if the bulb is larger, and smaller if there is much wind.

Relation between Diffusion and Electrolytic Conduction.

Electrolysis (see separate article) is a molecular movement of the constituents of a compound liquid in which, under the action of electromotive force, one of the components travels in the positive and the other in the negative direction, the flow of each component, when reckoned in electrochemical equivalents, being in all cases numerically equal to the flow of electricity.

Electrolysis resembles diffusion in being a molecular movement of two currents in opposite directions through the same liquid; but since the liquid is of the same composition throughout, we cannot ascribe the currents to the molecular agitation of a medium whose composition varies from one part to another as in ordinary diffusion, but we must ascribe it to the action of the electromotive force on particles having definite charges of electricity.

The force, therefore, urging an electro-chemical equivalent of either component, or *ion*, as it is called, in a given direction is numerically equal to the electromotive force at a given point of the electrolyte, and is therefore comparable with any ordinary force. The resistance which prevents the current from rising above a certain value is that arising from the encounters of the molecules of the ion with other molecules as they struggle forward through the liquid, and this depends on their relative velocity, and also on the nature of the ion, and of the liquid through which it has to flow.

The average velocity of the ions will therefore increase, till the resistance they meet with is equal to the force which urges them forward, and they will thus acquire a definite velocity proportional to the electric force at the point, but depending also on the nature of the liquid.

If the resistance of the liquid to the passage of the ion is the same for different strengths of solution, the velocity of the ion will be the same for different strengths, but the quantity of it, and therefore the quantity of electricity which passes in a given time, will be proportional to the strength of the solution.

Now, Kohlrausch has determined the conductivity of the solutions of many electrolytes in water, and he finds that for very weak solutions the conductivity is proportional to the strength. When the solution is strong the liquid through which the ions struggle can no longer be considered sensibly the same as pure water, and consequently this proportionality does not hold good for strong solutions.

Kohlrausch has determined the actual velocity in centimetres per second of various ions in weak solutions under an electro-motive force of unit value. From these velocities he has calculated the conductivities of weak solutions of electrolytes different from those of which he made use in calculating the velocity of the ions, and he finds the results consistent with direct experiments on those electrolytes.

It is manifest that we have here important information as to the resistance which the ion meets with in travelling through the liquid. It is not easy, however, to make a numerical comparison between this resistance and any results of ordinary diffusion, for, in the first place, we cannot make experiments on the diffusion of ions. Many electrolytes, indeed, are decomposed by the current into components, one or both of which are capable of diffusion, but these components, when once separated out of the electrolyte, are no longer ions—they are no longer acted on by electric force, or charged with definite quantities of electricity. Some of them, as the metals, are insoluble, and therefore incapable of diffusion; others, like the gases, though soluble in the liquid electrolyte, are not, when in solution, acted on by the current.

Besides this, if we accept the theory of electrolysis proposed by Clausius, the molecules acted on by the electro-motive force are not the whole of the molecules which form the constituents of the electrolyte, but only those which at a given instant are in a state of dissociation from molecules of the other kind, being forced away from them temporarily by the violence of the molecular agitation. If these dissociated molecules form a small proportion of the whole, the velocity of their passage through the medium must be much greater than the mean velocity of the whole, which is the quantity calculated by Kohlrausch.

On Processes by which the Mixture and Separation of Fluids can be effected in a Reversible Manner.

A physical process is said to be reversible when the material system can be made to return from the final state to the original state under conditions which at every stage of the reverse process differ only infinitesimally from the conditions at the corresponding stage of the direct process.

All other processes are called irreversible.

Thus the passage of heat from one body to another is a reversible process if the temperature of the first body exceeds that of the second only by an infinitesimal quantity, because by changing the temperature of either of the bodies by an infinitesimal quantity, the heat may be made to flow back again from the second body to the first.

But if the temperature of the first body is higher than that of the second by a finite quantity, the passage of heat from the first body to the second is not a reversible process, for the temperature of one or both of the bodies must be altered by a finite quantity before the heat can be made to flow back again.

In like manner the interdiffusion of two gases is in general an irreversible process, for in order to separate the two gases the conditions must be very considerably changed. For instance, if carbonic acid is one of the gases, we can separate it from the other by means of quicklime; but the absorption of carbonic acid by quicklime at ordinary temperatures and pressures is an irreversible process, for in order to separate the carbonic acid from the lime it must be raised to a high temperature.

In all reversible processes the substances which are in contact must be in complete equilibrium throughout the process; and Professor Gibbs has shewn the condition of equilibrium to be that not only the temperature and the pressure of the two substances must be the same, but also that the *potential* of each of the component substances must be the same in both compounds, and that there is an additional condition which we need not here specify.

Now, we may obtain complete equilibrium between quicklime and the mixture containing carbonic acid if we raise the whole to a temperature at which the pressure of dissociation of the carbonic acid in carbonate of lime is equal to the pressure of the carbonic acid in the mixed gases. By altering the temperature or the pressure very slowly we may cause carbonic acid to pass from the mixture to the lime, or from the lime to the mixture, in such a manner that the conditions of the system differ only by infinitesimal quantities at the corresponding stages of the direct and the inverse processes. The same thing may be done at lower temperatures by means of potash or soda.

If one of the gases can be condensed into a liquid, and if during the condensation the pressure is increased or the temperature diminished so slowly that the liquid and the mixed gases are always very nearly in equilibrium, the separation and mixture of the gases can be effected in a reversible manner.

The same thing can be done by means of a liquid which absorbs the gases in different proportions, provided that we can maintain such conditions as to temperature and pressure as shall keep the system in equilibrium during the whole process.

If the densities of the two gases are different, we can effect their partial separation by a reversible process which does not involve any of the actions commonly called chemical. We place the mixed gases in a very long horizontal tube, and we raise one end of the tube till the tube is vertical. If this is done so slowly that at every stage of the process the distribution of the two gases is sensibly the same as it would be at the same stage of the reverse process, the process will be reversible, and if the tube is long enough the separation of the gases may be carried to any extent.

In the *Philosophical Magazine* for 1876, Lord Rayleigh has investigated the thermodynamics of diffusion, and has shewn that if two portions of different gases are given at the same pressure and temperature, it is possible, by mixing them by a reversible process, to obtain a certain quantity of work. At the end of the process the two gases are uniformly mixed, and occupy a volume equal to the sum of the volumes they occupied when separate, but the temperature and pressure of the mixture are lower than before.

The work which can be gained during the mixture is equal to that which would be gained by allowing first one gas and then the other to expand from its original volume to the sum of the volumes ; and the fall of temperature

81—2

and pressure is equal to that which would be produced in the mixture by taking away a quantity of heat equivalent to this work.

If the diffusion takes place by an irreversible process, such as goes on when the gases are placed together in a vessel, no external work is done, and there is no fall of temperature or of pressure during the process.

We may arrive at this result by a method which, if not so instructive as that of Lord Rayleigh, is more general, by the use of a physical quantity called by Clausius the Entropy of the system.

The entropy of a body in equilibrium is a quantity such that it remains constant if no heat enters or leaves the body, and such that in general the quantity of heat which enters the body is

$$\int \theta d\phi,$$

where ϕ is the entropy, and θ the absolute temperature.

The entropy of a material system is the sum of the entropy of its parts.

In reversible processes the entropy of the system remains unchanged, but in all irreversible processes the entropy of the system increases.

The increase of entropy involves a diminution of the available energy of the system, that is to say, the total quantity of work which can be obtained from the system. This is expressed by Sir W. Thomson by saying that a certain amount of energy is *dissipated*.

The quantity of energy which is dissipated in a given process is equal to

$$\theta_0(\phi_2 - \phi_1),$$

where ϕ_1 is the entropy at the beginning, and ϕ_2 that at the end of the process, and θ_0 is the temperature of the system in its ultimate state, when no more work can be got out of it.

When we can determine the ultimate temperature we can calculate the amount of energy dissipated by any process; but it is sometimes difficult to do this, whereas the increase of entropy is determined by the known states of the system at the beginning and end of the process.

The entropy of a volume v_1 of a gas at pressure p_1 and temperature θ_1 exceeds its entropy where its volume is v_0 and its temperature θ_0 by the quantity

$$\frac{p_1 v_1}{\theta_1} \left\{ \frac{1}{\gamma - 1} \log \frac{\theta_1}{\theta_0} + \log \frac{v_1}{v_0} \right\}.$$

Hence if volumes v_1 and v_2 of two gases at the same temperature and pressure are mixed so as to occupy a volume $v_1 + v_2$ at the same temperature and pressure, the entropy of the system increases during the process by the quantity

$$\frac{p}{\theta} \left\{ v_1 \log \frac{v_1 + v_2}{v_1} + v_2 \log \frac{v_1 + v_2}{v_2} \right\}.$$

Since in this case the temperature does not change during the process, we may calculate the quantity of energy dissipated by multiplying the gain of entropy by the temperature, and we thus find for the dissipation

$$p v_1 \log \frac{v_1 + v_2}{v_1} + p v_2 \log \frac{v_1 + v_2}{v_2},$$

or the sum of the work which would be done by the two portions of gas if each expanded under constant temperature to the volume $v_1 + v_2$.

It is greatest when the two volumes are equal, in which case it is $1 \cdot 386 \, pv$, where p is the pressure and v the volume of one of the portions.

Let us now suppose that we have in a vessel two separate portions of gas of equal volume, and at the same pressure and temperature, with a movable partition between them. If we remove the partition the agitation of the molecules will carry them from one side of the partition to the other in an irregular manner, till ultimately the two portions of gas will be thoroughly and uniformly mixed together. This motion of the molecules will take place whether the two gases are the same or different, that is to say, whether we can distinguish between the properties of the two gases or not.

If the two gases are such that we can separate them by a reversible process, then, as we have just shewn, we might gain a definite amount of work by allowing them to mix under certain conditions; and if we allow them to mix by ordinary diffusion, this amount of work is no longer available, but is dissipated for ever. If, on the other hand, the two portions of gas are the same, then no work can be gained by mixing them. and no work is dissipated by allowing them to diffuse into each other.

It appears, therefore, that the process of diffusion does not involve dissipation of energy if the two gases are the same, but that it does if they can be separated from each other by a reversible process.

Now, when we say that two gases are the same, we mean that we cannot distinguish the one from the other by any known reaction. It is not probable, but it is possible, that two gases derived from different sources, but hitherto supposed to be the same, may hereafter be found to be different, and that a method may be discovered of separating them by a reversible process. If this should happen, the process of interdiffusion which we had formerly supposed not to be an instance of dissipation of energy would now be recognized as such an instance.

It follows from this that the idea of dissipation of energy depends on the extent of our knowledge. Available energy is energy which we can direct into any desired channel. Dissipated energy is energy which we cannot lay hold of and direct at pleasure, such as the energy of the confused agitation of molecules which we call heat. Now, confusion, like the correlative term order, is not a property of material things in themselves, but only in relation to the mind which perceives them. A memorandum-book does not, provided it is neatly written, appear confused to an illiterate person, or to the owner who understands it thoroughly, but to any other person able to read it appears to be inextricably confused. Similarly the notion of dissipated energy could not occur to a being who could not turn any of the energies of nature to his own account, or to one who could trace the motion of every molecule and seize it at the right moment. It is only to a being in the intermediate stage, who can lay hold of some forms of energy while others elude his grasp, that energy appears to be passing inevitably from the available to the dissipated state.

53. "Liquid Vapour Mixture"

Cambridge University Library.

Let a liquid be placed in a tall cylindrical vessel whose sides are impervious to heat but open to the air at the top. Let us suppose that in any stratum of the cylinder where the liquid the temperature is θ and the density and pressure ρ_1 and p_1, for the aqueous vapour ρ_2 and p_2, for dry air ρ and p for the mixture. Let u_1 be the velocity of vapour diffusing upwards u_2 the upward velocity of dry air and h the flux of heat in the same direction. Then when the system has reached a state of steady evaporation,

$$\frac{dp_1}{dx} = \kappa A_1 \rho_1 \rho_2 (u_2 - u_1) - g\rho_1$$

$$\frac{dp_2}{dx} = \kappa A_1 \rho_1 \rho_2 (u_1 - u_2) - g\rho_2$$

In this case u_2 is finally zero and the mass of vapour which ascends through unit area in unit time is

$$\rho_1 u_1 = -\frac{1}{\kappa A_1 \rho_2}\left(\frac{dp_1}{dx} + g\rho_1\right) = \text{a constant quantity}$$

We shall also have for the equation of conduction

$$h = -C\, d\theta/dx = \text{a constant}$$

where C is the coefficient of conduction which ⟨may depend on p_1 and θ⟩
If L is the latent heat of evaporation

$$L\rho_1 u_1 + h = 0$$

whence

$$\frac{L}{\kappa A_1 \rho_2}\left[\frac{dp_1}{dx} + g\rho_1\right] + C\frac{d\theta}{dx}$$

[End of sheet]

54. [Notes on the Evaporation of Liquids]

Cambridge University Library.

Let a liquid be placed at the bottom of a tall cylindrical vessel impervious to heat but open at the top and let evaporation go on until the state of the system no longer varies with time, the liquid being supposed to be constantly supplied at a constant temperature as it evaporates. Let Q be the quantity evaporated per unit area then, a quantity of heat LQ must descend through every stratum of the cylinder to evaporate it.

Let $\rho_1 \quad + \rho_2 \quad = \quad \rho \quad$ be the densities,

$ p_1 \quad + p_2 \quad = \quad p \quad$ the pressures

$ \underset{\text{vapour}}{\rho_1 u_1} + \underset{\text{air}}{\rho_2 u_2} = \underset{\substack{\text{mixed}\\\text{gas}}}{\rho u}$

$\rho_1 u_1$ is the mass of vapour which rises through unit of area in unit of time and $p_1 u_1$ is the volume of this unit of pressure.

The equation of diffusion is,

$$\rho_1 u_1 = \rho_1 u - D\left[\frac{dp_1}{dx} + g\rho_1\right]$$

where D is the coefficient of diffusion $= \dfrac{p_1 p_2}{\rho_1 \rho_2 k A_1} \dfrac{1}{p}$

When the motion is steady $u_2 = 0$ so that $u = (p_1/p)u_1$

Hence $Q = \rho_1 u_1 = \dfrac{-\rho_1 p}{p_1 p_2} D\left[\dfrac{dp_1}{dx} + g\rho_1\right]$

The quantity of heat which descends through unit of area of a surface rising with velocity u is,

$$h = -C\, d\theta/dx$$

where C is the coefficient of thermal conductivity and θ the temperature. But heat is also carried up by the motion of the mass with velocity u to the amount,

$$\cdot\ \frac{3}{2}\beta\rho u = \frac{3}{2}\beta p_1 u_1$$

Hence $LQ = -\dfrac{3}{2}\beta\dfrac{p}{p_2}D\left[\dfrac{dp_1}{dx} + g\rho_1\right] - C\dfrac{d\theta}{dx}$

Comparing

$$\cdot\ \left[L\frac{\rho_1}{p_1} + \frac{3}{2}\beta\right]\frac{p}{p_2}D\left[\frac{dp_1}{dx} + g\rho_1\right] - C\frac{d\theta}{dx} = 0$$

omitting the term in g,

$$\cdot\ d\theta = \left[L\frac{\rho_1}{p_1} + \frac{3}{2}\beta\right]\frac{p}{p - p_1}\frac{D}{C}[dp_1 + g\rho_1\, dx_1]$$

[End of sheet]

55. [Notes on Diffusion and Transpiration][a]

Cambridge University Library.

One of the Graham's most admirable discoveries was the fact that the law of the effusion of a gas through a small hole in a thin plate is quite different from that of the transpiration of a gas through a long narrow tube.[b] According to the ordinary theory of fluid motion there could be no great difference between these two phenomena, but Graham was led by his experiments to distinguish between them and to call one effusion and the other transpiration.

He found that the relative rates of effusion of different gases depended on their densities, the volume effused in a given time being inversely as the square root of the ⟨density⟩ specific gravity.

The relative rates of transpiration, however, had no regular relation to the specific gravity, and Graham therefore considered that they depended on a peculiar, and till then, unknown property to which we now give the name of Viscosity.

Sir William Thomson has pointed out, as the essential difference between effusion and transpiration, ⟨is⟩ that when a molecule ⟨comes up⟩ knocking about in a gas, comes up to a very small hole in a very thin plate, there is very little danger of its encountering another molecule before it has got fairly through. If, therefore, the hole is so small that the escape of molecules through it does not set up sensible currents in the gas, the quantity which effuses through the hole will depend on the product of the density into the molecular velocity, and it will be very little affected by the nature of the gas on the other side of the hole, so that in general there are intereffusion currents going ⟨in⟩ through the hole in opposite directions, but not sensibly interfering with each other.

[End of sheet]

But in the case of transpiration through a long tube, a molecule has almost no chance of getting through the tube without having many encounters with other molecules. The character of the motion is therefore mainly determined by the interval between these encounters, ⟨on⟩ the and this is the theoretical explanation of the viscosity of a gas.

The viscosities of gases are not even in the same order as their densities for ⟨that⟩ the viscosity of oxygen is greater than that of carbonic acid or hydrogen.

The experiments of Graham on the diffusion of gases through porous plates show that this phenomenon partakes much more of the character of effusion than that of transpiration, and the thermal effects obtained in other experiments of Professor Osborne Reynolds, as these also depend on the motion of gases through porous plates, ought to be called thermal effusion or diffusion rather than transpiration.

In fact Sir William Thomson has pointed out the portions of gas on opposite sides of the plate, being of different temperatures behave like gases of different specific gravities. The velocity of effusion is ⟨inversely⟩ as the square root of the absolute temperature and the quantity of gas which passes through is ⟨found by⟩ as the velocity into the density, that is as the density into the square root of the absolute temperature or as the pressure divided by the square root of the absolute temperature.

When the quantity which passes through the plate is the same in both directions, then, ⟨the pressures⟩ comparing the states of the gas on opposite sides, the pressure must be as the square root of the absolute temperature.

We have here supposed that the dimensions of the pores in the plate are exceedingly small compared with the length of the free path of the molecule. When this is not the case, the phenomenon becomes more and more of the nature of transpiration and depends more on the viscosity of the gas and less on its density, as the pores become larger or the gas denser.

a. These notes were probably written in 1879 as Maxwell refers to Reynolds's paper on transpiration, which he refereed for the Royal Society. See S. G. Brush and C. W. F. Everitt, "Maxwell, Osborne Reynolds, and the Radiometer," *Hist. Studies Phys. Sci.* 1 (1969), 105–125; Brush, *Kind of Motion we call Heat*, 221–227. Other documents on this subject will be presented in our next volume.

b. See the 1846 and 1849 papers by Graham cited in chapter I, note 57.

Index

Abdera, 138
Achilles, 177
Action and reaction, 474
Adams, John Couch, 130n
Adams Prize, 130n
Adams, W. G., 354n
Adiabatic compression or expansion, 258, 458–459
Aether. *See* Ether
Aetherhülle, 37
Agitation (local motion of molecules), 159, 159n, 160, 186, 187–188, 235–236
Air
 diffusion, 454–455, 532
 electrical conductivity, 515
 heat conduction, 318, 320, 471–472, 476
 specific heats, 231, 336, 396, 460
 viscosity, 320, 342, 351, 352, 370–371, 383, 467–468
Allen, B. M., 516n
Ammonia, 143
 diffusion, 144, 284, 455
Ampère, Andre-Marie, 2
Anaxagoras, 138, 138–155, 176
Andrews, Thomas, xvii, 134, 137n, 148, 157n, 158, 224, 247, 337, 338n, 413
Ångström, Anders Jonas, 494, 494n–495n
Apjohn, James, 537
Archives at Paris, 41, 152
Arcturus, 41, 152
Argon, 57n, 58n
Aristotle, 177

Arnold, D. H., 397n
Aronson, Samuel, 81n
Arrhenius, Svante, 157n
Atom, 176–181, 242, 246. *See also* Molecules; Vortex atom
 existence of, 38, 43, 60n, 85, 123–124, 138–139, 187
 hardness, 51n, 202–203
 as manufactured article, 41, 61n, 99–100, 152, 214–215, 241–242
 and molecule, relation to, 85, 138, 140, 211
Atomic models, 44
 structure of matter, 85, 96–98, 123–124, 138–140, 173–174, 182, 219
Avogadro, Amedeo, 238
Avogadro's hypothesis, 39, 85, 229, 279–280, 393, 483n. *See also* Equivalent volumes, law of
Avogadro's number, 39, 59n, 483–484

Bacon, Francis, 74, 105, 114
Bailey, Cyril, 68n, 83n, 108n
Baily, F., 107, 357, 361n, 371
Balmer series, 41
Bannawitz, E., 35
Barker, J. A., 58n
Barr, E. S., 49n
Barus, Carl, 30, 57n
Bayle, Pierre, 177
Bell, James F., 56n
Berger, Herbert, 48n
Berkeley, George Bishop of Cloyne, 87
Bernoulli, Daniel, 45, 49n, 82, 105, 109n, 141, 286, 336, 339, 416

Bernoulli (cont.)
  kinetic theory of gases, 4, 19
Bernoulli, Jean, 45
Bernoulli, Johann, 49n
Bernstein, Henry T., 62n, 83n, 108n
Bessel, Friedrich Wilhelm, 361n, 383
Billiard-ball model of gas, xx–xxii. *See
  also* Elastic sphere model of gas
Biology, 268
Blackwell, Richard J., 109n
Bobetic, M. V., 58n
Boltzmann, Ludwig, xvii, xxiii, 32, 46,
  52n, 57n, 61n, 62n, 143, 251, 274,
  275n, 493, 494n, 516n. *See also*
  Stefan-Boltzmann law; Maxwell, on
  Boltzmann
  on distribution of molecules acted on
    by force, 232–233, 515, 516n
  on *elastische Nachwirkung*, 251–253
  on electricity, 494n
  on gas molecules, 58n
  on heat conduction, 34
  on kinetic theory, xxi, xxiii–xxv, xxviin,
    9, 27, 32, 33, 34–35, 57n, 143, 227
  on Maxwell, 32, 33
  on molecular models, 43, 230–231
  on spectra, 41
Boltzmann's *H* Theorem, 476–477, 482n
Boole, George, 9, 11, 52n, 84, 104
Bosanquet, R. H. M., 43
Boscovich, R. J., 88, 179–180, 202, 211
Bowman, William, 350, 350n
Boyle and Mariotte's law, 298
Boyle, Robert, 5, 45, 105, 142
Boyle's law, 4, 126, 131–132, 142, 183–
  185, 191, 221–223, 263, 265, 298
  deviations from, 185, 222–223
Brace, D. B., 275n
Breger, Herbert, 48n
Brewster, Sir David, 24, 25, 56n
British Association for the Advancement
  of Science, 39, 43, 44, 46, 138
Brock, William H., 60n, 157n
Brodie, Sir Benjamin, 39, 60n, 84, 85n,
  108
Brompton Hospital for Consumption,
  338n
Brosche, Peter, 61n
Brown, Alexander Crum, 239n, 506,
  508n

letter to Maxwell, 506–508
Brush, Stephen G., 48n, 49n, 50n, 51n,
  52n, 53n, 54n, 56n, 57n, 58n, 59n,
  60n, 61n, 81n, 82n, 257n, 265n, 275n,
  477n, 482n, 550n
Bryan, G. H., 42, 62n
Buckle, Henry Thomas, 260, 262n
Bunsen, Robert Wilhelm Eberhard, 40,
  61n, 482n
Burr, A. C., 49n
Buys-Ballot, C. H. D., 6, 22, 51n, 61n

Cadmium, atomic volume, 507
Cadmium oxide, 507
Cagniard de la Tour, C., 247
Cajori, F., 265n
Calcium carbonate, 507
Caloric theory of heat, 2–3, 48n
*Cambridge and Dublin Mathematical
  Journal*, 5, 6
Cambridge Philosophical Transactions,
  358n
Cambridge University, 111
Campbell, Lewis, xviii, 9, 52n, 65, 319n
Canada balsam, 25
Canizzaro, Stanislao, 18, 39, 54n, 60n,
  483n
Capillarity, 525
Carbonate compounds, molecular vol-
  umes and angles, 506–507
Carbon dioxide
  critical point phenomena, 224
  deviation from Boyle's law, 185
  diffusion, 284, 454–455, 489, 514, 532,
    534
  heat conduction, 472
  molecular diameter, 479, 481, 490, 496,
    500, 504
  molecular mass, 488
  reversible separation, 543
  specific heats ratio, 472
  spherical particles, 341
  transpiration, 351, 467–468
  velocity of molecules, 488
  viscosity, 356, 467, 488, 494, 501
Carbonic acid gas. *See* Carbon dioxide
Carbonic oxide. *See* Carbon monoxide
Carbon monoxide, 479, 480, 481, 488–
  490, 494, 496, 500, 501, 504, 532

Cardwell, D. S. L., 47n, 48n
Carnot, Sadi, 48n
Carnot's function, 110n
Cartesian, 238
Cavendish, Hon. Henry, 62n, 105, 256, 262, 263–264, 266
Cavendish Laboratory, xxiii, 111, 116, 476n, 510n
Cay, Charles Hope, 349, 350n
Cay, Frances, xviii
Cayley, Arthur, 275n
  on dynamics, xxiii, xxviin
Cazelles, E., 158
Challis, James, 126, 127, 129, 130n, 131, 132
Chalmers, Dr. Thomas, 152, 214
Chandrasekhar, S., 495n
Chapman-Enskog gas theory, 58n
Chapman, Sydney, 31, 35, 60n
Charles, J. A. C., 3, 184
Charles' law, 3, 142, 184
Charlie, 349, 350n
Chemical Society of London, 39
Chemistry, 69–70, 187, 246, 271
  reduced to mechanics, 70, 217–218
  reduced to physics, 70
Chlorine, 284, 455
Chrystal, George, 510n
Cinematics (kinematics), 71, 72, 269
Clagett, M., 47n
Clarendon Press, 475
Clarke, Desmond N., 109n
Classification of physical quantities, 93–95
Clausius, Rudolf Julius Emmanuel, xvii, 45, 46, 53n, 61n, 62n, 82, 88, 97, 101, 106, 107, 136n, 137n, 142, 182, 183, 185, 189, 219–222, 225, 226, 227, 272, 279, 282n, 321, 338, 339, 346, 416, 422, 423, 428, 473, 474n, 481, 495, 496, 498, 508, 544. *See also* Maxwell, on Clausius
  on color of sky, 506n
  on diffusion, 301n
  on electrolysis, 133, 147, 541
  on entropy, 89
  on equipartition of energy, xx, 187, 258
  on heat conduction in gases, 12, 15–16, 21–23, 31–32, 301n, 314n, 348n, 471

on history of kinetic theory, 45
on kinetic theory of gases, xix–xxi, 4–6, 50n–51n, 280n, 281, 286
on Maxwell's kinetic theory, xxii, 15–16, 21–22
on mean free path, xix, 6, 55n, 146, 277, 295, 296n, 342
on molecular size, 38
on radiation exchange, 272
on rotation of molecules, 6, 258, 287
on specific heats ratio, 6, 230
on thermal conductivity, 32
on velocities of molecules, 5
vesicular theory, 506n
on virial theorem, 182–183, 220–221, 495
on viscosity, 12, 19
Clerk, Sir George, 352n
Clifford, William Kingdon, 61n, 156n, 157n, 158
Clifton, R. B., 41, 61n
Coefficient of viscosity, 281n
Collisions, 8, 80, 105, 106, 109n, 258, 287–289
  of nonspherical bodies, 314–318, 341
  number per second, 134, 279, 281, 283–284, 294, 488, 490n
Collocation of matter, 154, 214
Color blindness, 280
Colorimeter, 337, 338n
Colors seen by human eye, 282, 350n
Comte, Auguste, 337
Condensation coefficient, 484n–485n
Condensation of vapors, 133, 148, 156
Convection, 13, 14
Cornell, E. S., 49n
Coulomb, Charles, 262, 361n
Cowling, T. G., 60n
Critical point, 224, 247–248
Crombie, Alistair C., 52n
Crookes, William, xxii, 273, 516n
Crosland, M. P., 49n
Crystals, properties of, 173–174

Dallas, D. M., 85n
Dalton, John, 48n, 144
Dalton's law of partial pressures, 28, 233, 451
Dalton's law of thermal expansion, 3

Darwin, Charles, 11, 51n
Daub, E. E., 50n, 51n
Davy, Humphry, xix, 50n
de la Provostaye, F. H., 34
Democritus, 45, 138–139, 170, 176, 421
Density, 281n
  as a statistical quantity, 400–401
Desains, P., 34
Descartes, René, 46, 76, 106, 109n, 111, 130n, 178, 181, 337
Determinism, 260–261, 546
Devonshire Physical Laboratory. *See* Cavendish Laboratory
Dewar, Donald, 336n
Dewar, James, 257n
Dewar, Katherine Mary, xxi, 12, 239, 349
Dialysis, 545
Diffusion of gases, 16–17, 36–37, 60n, 88, 97, 141, 143–146, 161, 190, 279, 281, 301–312, 417, 451–456, 478–480, 491–492, 497–501, 511–514, 516–545
  of liquids, 133, 146–147, 525–526, 532–535
  table of numerical data, 155, 488–489
  thermal effects of, 461
  thermodynamics of, 543–544
di Luca, S., 54n
Dimensional analysis, 31, 368
Dissipation of energy, 101, 122, 132, 162, 544–546
  relation to our knowledge, 546
Döbereiner, J. W., on diffusion, 16
Domb, C., 79n
Donders, Francis Cornelius, 350, 350n
Doppler, J. C., 61n
Doppler shift, 40, 61n, 196
Dorling, J., xxviin
dp/dt (= J. C. Maxwell), 110n, 238
Droop, H. R., 336n, 353, 353–354
Du Bois Reymond, P., 45
Dubuat, M., 361n
Dulong-Petit cooling law, 13–14, 34
Dulong-Petit's specific heats law, 14, 107, 187–188, 231, 237
Dulong, P. L., 13–14, 34, 53n, 58n, 187
Duncan, David, 157n, 158, 160n, 162n

Dynamical theories, 420
Dynamics, 71, 72, 269

Earth, figure of, 413, 415n
Edinburgh Academy, xviii
Edinburgh Review, 9
Effusion, 16, 17, 502, 549–550
Eggen, Olin J., 130n
Elasticity, 23–26, 72, 131, 173, 202, 239–240, 249, 418, 425, 463
Elastic sphere model of gas, 7, 26, 29, 30, 31, 35, 36–37, 39–40, 43, 81, 189, 244, 282, 286–287, 319, 499, 528. *See also* Billiard-ball model of gas
  associated with "hard" or "rigid" spheres, 51n, 81, 286–287, 318
Electrical induction, xxv, 25
  resistance standards, 350
Electric conductivity of gases, 515
Electricity, 103, 218, 255–256, 539
  propagation of force, 473–474
Electrolytic conduction, electrolysis, 97, 133, 147–148, 540–541
Electron, 516n
Elkana, Yehuda, 47n
Ellicott, C. J., Bishop of Gloucester and Bristol, 241
  letter to Maxwell, 241–242
Elliptic functions, 410, 413, 414
Encounters between molecules, 26–27, 387–392, 398–400, 402–404, 429–444. *See also* Collisions
Encyclopedia Britannica, 47, 176, 271n
Energetics, 71, 72, 269
Energy, 71, 72. *See also* Dissipation of energy
  conservation, 129, 337
  location of, 473–474
Enskog D., 31, 35, 60n
Entropy, 88, 101, 544
  and information, 546
Epicurus, 45, 138–155
Equipartition theorem, xx, xxi, 18, 40, 42, 46, 132n, 186, 228, 281, 292–293, 318, 341, 393
Equity Bar, 336n
Equivalent volumes, law of, 143, 186–187, 229, 235, 299, 393, 417, 457–458
Ether, xxii, 2–3, 24, 30, 37, 44, 88, 126–

129, 131–132, 188, 236, 238, 243, 414n, 473, 505
created before matter, 243
Ethylene (olefiant gas), 7, 284, 285n, 318, 320, 455
Eucken, A., 35, 58n
Eucken factor, 34–35
Euclid, 105
Euler, Leonard, 49n, 50n
Evaporation, 133, 148, 535, 547–548
Everett, Joseph D., 421n, 538
Everitt, C. W. F., xviii, xxvin, 52n, 53n, 62n, 319n, 338n, 550n
Evolution, 153, 157n, 211
Ewing, James Alfred, 495, 496
Examination questions, 5, 352n, 354, 413, 415
Exchanges, theory of, 272–275
Exner, Franz, 513
Experiments of illustration, 112–113
of research, 113–114

Faraday lecture, 39
Faraday, Michael, 96, 118, 481, 483n, 503
Farrar, W. V., 60n
Feynmann, R. P., 475n
Fick, A., 155, 526, 533
Fischer, S. J., 49n
Fleck, G. M., 60n
Fluctuations, 261
Forbes, James David, xviii, 2, 52n, 56n, 157n, 419, 471, 494n
on statistical argument, 9
on visco-elastic processes and glaciers, 24
Forces between molecules, xxi, xxii, 30, 34, 36, 40, 60n, 179–180, 184–185, 434
inverse fifth power, xxi, 26–29, 31, 35, 36, 416, 435–437, 465
reduced to action of ether, 127–128, 131, 473–474
Fourier, Jean Baptiste Joseph, 88, 101, 272n, 275n
Fox, Robert, 4, 48n
Fraunhofer lines, 60n
Free will, 105, 150
French, A., 59n

Fresnel, Augustin, 2
Fusinieri, Ambrogio, 167, 168

Galilei, Galileo, 219–222
Galton, Francis, 192n
Garber, E. W., 50n, 51n, 52n, 53n, 54n, 55n, 59n, 280n, 506n
Garnett, James Clerk Maxwell, 516n
Garnett, William, 52n, 65, 515, 515n–516n
Gases. *See also* Kinetic theory of gases; Transport phenomena; Specific heats; Condensation
continuity with liquids, 225
electric properties, 235
Gassendi, Pierre, 45
Gauss, Karl Friedrich, 103, 115, 116, 218, 364
on theory of errors, 8
Gay-Lussac, J. L., 238, 458. *See also* Equivalent volumes, law of
free expansion experiment, 3
law of thermal expansion, 3, 5
Germ, 192
Gibbs, Josiah Willard, xvii, xxiii, 159n, 225, 274, 275n, 476n, 509, 510n, 525, 542. *See also* Maxwell, on Gibbs
Gillispie, Charles Coulston, 52n
Girault, C., 383
Glaciers, 24, 25, 56n, 157n, 354
Glenlair, xviii, 280n
Glennie, J. S. Stuart, 319n
Gore, Charles, 239
Göttingen Konigliche Akademie, 475n
Gough, J. B., 81n
Gower, Barry, 52n
Graham's law of diffusion, 309
Graham, Thomas, 53n, 55n, 107, 108, 136, 137n, 145, 152, 155, 188, 190, 234, 279, 280, 282n, 284–285, 307, 312, 320, 336, 351, 352n, 355, 361, 417, 418, 452–456, 465, 467–468, 477, 480, 482n, 497, 507, 513, 526, 533, 549, 550n
on diffusion, 7, 16–17, 18, 49n, 54n, 97
Gravity, 270
ether-wave theory, 127
kinetic theory, 81n, 105, 128, 170–173, 204–207

Green, George, 123, 256, 257n
Griffiths, G., 85n
Gulliver's Travels, 474
Guthrie, Francis, 46–47

Hacking, Ian, 53n
Hahnemann, Christian Friedrich
    Samuel, 508n
Hahn, Friedrich von, 61n
Haldane, J. S., 50n, 244n
Hall, Marie Boas, 109n
Hamilton, Sir William, xviii, xxv, 67n–
    68n, 338n
Hamilton, William Rowan, xix, 95, 103
    dynamical method, xxiii, 476
Hanley, H. J. M., 58n
Hansemann, Gustav, 108, 109n, 132,
    132n, 188n
Harman, Peter M., 48n, 52n
Hawthorne, R. M., Jr., 59n
Heat, 69, 72, 122, 235, 270
    conduction, xxi, 14–16, 31–35, 88, 101,
        133, 279, 312–314, 346–347, 404–
        405, 419, 449, 468–472, 476
    theories before 1850, 1–3
    transfer in gases, 13–14
Hegel, Georg Wilhelm Friedrich, 337
Heilbron, John, 48n
Heimann, Peter M. *See* Harman,
    Peter M.
Helmholtz, Hermann von, 44, 48n, 86,
    98, 129, 181, 197, 202, 269, 337, 361n,
    368
    on vortex atoms, 98
Helmholtz-Thomson vortex atom, 130n
Henry, Joseph, 48n
Herapath, John, 4, 17, 44–46, 49n–50n,
    51n, 54n, 62n, 82n, 106, 130n, 141,
    279, 281, 286, 312, 416, 422. *See also*
    Maxwell, on Herapath
    on Graham, on diffusion, 16–17
    on Huygens, 80
    on kinetic theory, 80–81
    on LeSage, 80
    letter to Maxwell, 80–81
    on Wren, 80
Heredity, 136, 192
Herschel, Sir John Frederick William, 9,
    52n, 61n, 153, 214, 242, 352n

on normal distribution of errors, 9–11
on statistics, 11
Herschel, Sir William, 2, 49n
Hirst, Thomas Arthur, 156, 157n
Hockin, C., xxviin
Homeopathic medicine, 136, 507, 508n
Homoiomereia, 139, 176
Hooke, Robert, 167
Hooykas, R., 49n
Hopkinson, John, 252n
Hopkins, William, xix, 273, 275n
Hoppe, R., 22
Hübner, Kurt, 109n
Humboldt, Alexander von, 114–115
Humidity, 537–538
Hutchinson, Keith, 47n
Hutton, James, 537
Huxley, Thomas Henry, 11, 51n, 157n
Huygens, Christian, 80
Hydrodynamics, 23–26, 28, 72, 126, 269,
    450, 463, 520, 522–523
    vortex motion, 197–201
Hydrogen
    diffusion, 16, 84, 284, 455, 514, 532,
        534
    heat conduction, 15, 32, 33, 472, 476,
        488–489
    mass of molecule, 60n, 136, 191, 481,
        488, 490, 496, 504
    mean free path, 190, 490, 502
    size of molecule, 351–352, 479, 481,
        488, 490, 496, 500, 504
    transpiration, 351
    velocity of molecule, 4, 106, 143, 488,
        490, 497, 499
    viscosity, 352, 356, 467–468, 480, 488,
        494, 501, 549
Hydrostatics, 180, 451

Ideal gas law, 5
Ikenberrry, E., 57n
Intemann, H., 58n
Interference between gases, coefficient of,
    466–468
Ions, 540–541
Iron, atomic volume, 507
Iron carbonate, 507
Iron oxide, 507
Irreversibility, 88, 100–101, 542

James, W. S., 49n
Jeans, James Hopwood, 3, 49n, 60n
Jochmann, E., 22
Johnson's dictionary, 138–155
Jones, R.V., 62n, 79n
Joule, James Prescott, 3, 4, 44, 46, 48n, 50n, 51n, 141, 185, 281, 282n, 286, 337, 416, 423. *See also* Maxwell, on Joule
Joule's equivalent, 110n
Joule-Thomson experiment, 30, 34, 185, 511
Journal de l'École Polytechnique, 397n
Junior Moderator, 83

Kant, Immanuel, 48n, 337
Kargon, Robert H., 48n, 52n, 319n, 516n
Karlsruhe Congress, 18, 483n
Keeler, James E., 61n
Kelvin (Lord). *See* Thomson, William
Kestin, J., 58n
Kinematics. *See* Cinematics
Kinetics, 71, 269
Kinetic theory of gases before 1859, xix, 1–6, 105–106, 141–142
Kings College, London, 62n
Kirchhoff, Gustav Robert, 40, 60n, 61n, 272, 275n
Knight, D. M., 60n
Knott, C. G., 47, 62n, 110n, 476n
Koch, R., 35
Kohlrausch, F. W. G., 251, 355, 358n, 360, 425, 541
Kohlrausch, Rudolph, 358n
Kopp, Hermann, 135, 137n, 481, 483n, 484n, 503, 507
Krönig, August Karl, 4, 51n, 106, 142, 286, 416
Kronstadt, Barbara, 49n
Kuhn, T. S., 47n, 48n
Kundt, Adolph, 33–34, 58n

Lagrange, Jean-Louis
dynamical method, xxiii–xxiv, 129, 219–222
hydrodynamics, 197
Laplace coefficients, 413, 414n, 415n
Laplace, Pierre Simon Marquis de, 52n, 60, 219–222, 260, 261n

on statistics, 9
on theory of errors, 8
on theory of sound, 3
Laplace's nebular hypothesis, 156n, 158
Larmor, Joseph, xxv, xxvin, 59n, 241n, 281n, 510n
Lavoisier, Antoine Laurent, 1
Lead, atomic volume, 507
Lead oxide, 507
Legendre, A.-M., tables of elliptical function, 413, 414, 435
Leibnitz, Gottfried Wilhelm, 74, 178, 337
Leidenfrost, W., 168
Leray, P., 204n
LeSage, George-Louis, 45, 46, 80, 81n, 82, 105, 106, 109n, 128–129, 130n, 132n, 141, 170, 171, 172, 204–205, 272, 416, 422. *See also* Maxwell, on LeSage
Leslie, Sir John, 49n
Light
aberration of, 24
effect of distortion of medium on, 24, 25
electromagnetic theory of, xxii, 350, 358n, 494n
nature of, 236
as radiation, 73
transmission through a gas, 234
velocity, 275
Lilley, S., 48n
Liouville, Joseph, 353n
Liouville's journal, 353n
Liquids
continuity with gases, 225
diffusion, 133, 146–147, 525–526, 532–535
dynamical/molecular theory, 146–149
molecular volumes, 480, 483n, 503, 507
surface tension, 165
Litchfield, R. B., 52n
Lockyer, J. N., 196
Lodge, Oliver, xxv, 42, 61n
Loeb, L. B., 60n
Lorentz, Hendrik Antoon, 3, 49n, 509, 510n
Lorenz, L. V., 103
on molecular size, 38, 59n

Loschmidt, Johann Josef, xx, 107, 108, 144, 147, 190, 191, 282n, 476, 477, 501
  on diffusion, 36–37, 39, 58n, 146, 155, 190, 476, 482n, 490n, 493, 526, 532
  on molecular diameter, xx–xxi, 38, 46, 59n, 135, 149, 191, 482n, 484n, 502
  number, 39, 60n, 149, 282n, 484, 499, 504
Louvre, 163
Lucrèce Newtonien, 82
Lucretius, 45, 46, 66, 67, 68n, 82–83, 105, 108n, 128, 138–155, 141, 150, 170, 171, 176, 177, 201, 202, 203–204, 258, 421–422
Lynx, 257n

Macadam (or McAdam), John, 238, 239n
McCarty, R. D., 58n
Macdonald, J. G., 59n
Macfarlane, Donald, 255
McGregor, James Gordon, 495, 496n
McGucken, William, 61n
MacLeod, Roy M., 157n
Macvicar, John G., 319n
Magnesium carbonate, 507
Magnetic Union, 116
Magnetism, 241n, 263
  terrestrial, 115–116
Magnus, Gustav, 31
  on thermal conductivity coefficients, 15, 18
Maier, C. L., 61n
Malthus, Thomas, 11
Manganese carbonate, 507
Mansel, Henry Longueville, 337, 338n
Manse of Corsock, 349
Marsden, Brian G., 275n
Marsh gas. *See* Methane
Mason, E. A., 49n, 53n, 59n
Mass, 87. *See also* Molecules
  of vortex atom, 203
Mathematical physics, 44
Mathematical Tripos, 119, 275n, 353n, 414n
Mathematics, 70, 77, 90–91, 104, 116–119
  design in, 256

Matter, properties of, 65–67, 87
Maxwell, Frances Cay. *See* Cay, Frances
Maxwell, James Clerk, xvii–xix
  on Ångström, 495
  on atoms, 39, 61n, 123, 170–175, 176–215
  on atoms and ether, 131–132
  on Avogadro's hypothesis, 18, 39
  on Avogadro's number, 483–487, 504
  on Boltzmann, 46–47, 186, 274, 493
  on equipartition, 47
  on Boltzmann's *H* theorem, 476–477
  on Boltzmann's theorem of energy distribution, 47, 232–233
  on Boltzmann's theory of elasticity, 252–253
  on Boole, 104
  on Boscovich's atomic theory, 179–180
  on Brodie, Sir Benjamin, 84–85
  on capillarity, 60n
  on Challis, James, 126–128
  on change of state, 247–248
  on Clausius, 22, 46, 220, 348n, 422, 473, 506n
  on electrolysis, 97, 133
  on kinetic theory, 39, 44, 46, 85, 106–107, 142, 185–186, 225–226, 279, 338, 416, 423
  on mean free path, 189, 502n, 508–509
  on thermal conductivity, 31
  on collision of elastic spheres, 287–289
  on color theory, 280, 283, 283n, 337, 338n
  on Comte, 337
  on constitution of bodies, 87–88, 217–237
  on continuity of gases and liquids, 148, 224–225
  on convertibility of energy, 337–338
  on Dalton's law of partial pressures, 26–29
  on determinism, 11
  on diffusion, xxv, 16–17, 36, 39–40, 58n–59n, 143–147, 279–280, 301–312, 348n, 351–352, 417, 451–456, 491–492, 494, 532, 510–514, 515, 516–523
  on diffusion data, 488–489
  on diffusion in liquids, 532–535

on dissipation of energy, 88
on dynamical theory of gases, 413, 415–419
on dynamics, xxiv, 102
on elasticity, 24–26, 56n, 424–426
on elasticity and viscosity, 239–241, 249–254
on electrical discharges in gases, 516n
on electrical induction, 473
on electrical lines of force, 62n
on electric force, 255–257
  propagation, 473–474
on electrolysis, 97, 133, 147
on electromagnetic fields, xxv, 44, 56n
on electromagnetism, xxii–xxv, 473–474
on encounter of two molecules, 8, 26–27, 387–392, 398–404, 429–431 (*see also* Maxwell, on forces between molecules)
on entropy, 88
on equilibrium of temperature, 392–395, 457–458, 469–470
on equipartition of energy, xx, xxi, 18, 39, 40, 46, 47, 142, 229–232, 314–318
on ether, xxii, 44, 236, 473
on evaporation, 246–247, 547–548
on experimental physics, 111–125
on Faraday's lines of force, 24
on forces between molecules, xxii, 26–27, 40, 44, 60, 107, 127, 354
on Gibbs, 159n, 274, 335, 509, 510n
on Graham, on diffusion, 26–29, 284–285
on Graham's law of diffusion, 16–17
on Hansemann, 108, 132
on heat conduction, 14–16, 31–32, 46, 107, 312–314, 339–347, 405–412, 468–469, 473
on Helmholtz, 125, 337
on Herapath, 44–46, 106, 279, 422
on history of kinetic theory, 44–46, 105–108, 141–142, 422–423
on history of science, 62n, 78–79, 114, 121, 272, 274
on irreversible processes, 100–101
on Joule, 46
on kinetic theory of gases, xix–xxiv, 7–18, 26–29, 51n, 56n, 83n, 85, 88–89, 185–187, 219–222, 225–233, 257–260, 262–264, 277–550
on Krönig, 142
on Lagrange, xxiv
on LeSage's theory of gravitation, 45, 105, 128, 132n, 170–173, 204–208, 422
letters or postcards to:
  Adams, W. G., 354n
  Campbell, L., 132n, 319, 337–338, 475–476
  Cay, C. H., 349–350
  Droop, H. R., 336, 353–354
  Ellicott, C. J., 242–243
  Garnett, W., 515
  Graham, Thomas, 351–352
  Hockin, C., xxviin
  Monro, C. J., 45, 506n
  Newcomb, Simon, 272–275
  Pattison, Mark, 44, 62n
  Rayleigh, J. W. S., Lord, 505, 506n
  Spencer, Herbert, 158–160
  Stokes, George Gabriel, 37–38, 44, 59n, 239–241, 277–280, 282
  Tait, 43, 47, 86–87, 87–88, 132n, 159n, 238–239, 255–257, 352n, 352–353, 413, 473–474, 493–494, 495–496, 508–509
  Thomson, 38, 59n, 89, 470
  unknown correspondent, 82–84
on liquid-vapor mixture, 547
on Lorentz, H. A., 509, 510n
on Loschmidt
  on diffusion, 36, 54n, 137n, 144, 477, 478, 479–481, 495–496, 497–498
  on molecular size, 59n, 107
on Loschmidt's number, 60n
on Lucretius, 82–84, 105, 170, 171, 176, 177, 201, 203, 204, 421
on mass of hydrogen molecule, 60n, 481
on mathematics and physics, 103–104, 118–119
on mathematics in physics, 92–93
on mean free path, xx, 7, 55n, 189–190, 226, 295–297, 299, 312, 342–344, 501–502, 508–509
on Meyer, O. E., 107, 357, 372, 386, 416, 465

Maxwell (cont.)
on molecular structure, 40–44, 59n, 182, 187–188
on molecules, 60n, 133–136, 136n universality of properties, 41, 99–100, 151–153
on natural philosophy, 68–79
on philosophy of science, xxiv, xxv, 74, 202
on physical sciences, 267–271
on Plateau, on soap bubbles, 163–169
on probability, 9
on properties of matter, 3, 65–68, 122–123, 246–255
on quaternions, 475–476
on radiant heat, 236, 272–275
on rarefied gas dynamics, xviii, xxii, 25–26
on Rayleigh, 506n
on relative velocity distribution function, 291–295, 321–335
on relaxation time for stress in gases, 23, 26, 465
on relaxation time of solids, 25
on Riemann, B., 473
on Royal Society of Edinburgh, 353–354
on Saturn's rings, xix, 12, 40, 130n, 475n
on scientific societies, 353–354
on size of molecules, 7, 38–40, 97–98, 99, 148–149, 190–193, 502–503, 505
on specific heats of gases, 40, 43, 47, 188, 231, 336, 395–396, 459–461
on specific heats ratio, 318, 321, 341, 396
on spectra, 41, 99, 193–197
on Spencer, Herbert, 158n–159n
on stability of molecules, 212–215
on statistical method, 150–151
on statistical regularity, 8, 260–261
on Stefan, Josef, 498
on Tait, P. G., 108, 475
on theories of electrodynamics, 103
on theories of light, 102–103
on thermal conductivity, 15, 31, 470–472
on thermodynamics, 122
on Thomson, 125, 413

on vortex atoms, 181
on transport phenomena, xx, xxii, 12, 22–23, 320–321, 418
on Tyndall, John, 354
on van der Waals, 40, 60n, 125, 202–203, 238
on velocity distribution function, xx–xxi, 7–8, 9–11, 12, 226–228, 283–284, 289–291, 339, 387–391, 437–441
on virial theorem, 182–184, 220–222
on viscosity experiments, 20, 24, 30, 55n, 107, 349–350, 351–353, 355–358, 360–386, 477, 497–498
on viscosity theory, xxi, 12–14, 23–26, 29–30, 53n, 145, 278–279, 299–300, 336, 342, 355–358, 396–397, 424–426, 461–468, 492–493
on vortex atom(s), 41–44, 86, 98, 125, 197–203, 238
on Watson's kinetic theory, 62n
Maxwell, sources for his velocity distribution function, 11–12
Maxwell's address to the British Association, 90–104
Maxwell's demon, xxi
Maxwell-Stefan theory of diffusion, 60n
Maxwellian molecules. *See* Forces between molecules, inverse fifth power
Maxwell, John Clerk, xviii
Maxwell, Katherine, 12, 239, 349
Mean free path, xix, xx, 6–7, 23, 26, 55n, 135, 146, 189–190, 277, 280n, 281, 286, 295–297, 342–344, 484n, 502
Mechanical world view, or mechaniam, xxv, 42, 43, 48n, 68, 126, 217, 219
Mechanics, 69
Melloni, Macedonio, 2, 49n
Mendoza, Eric, 47n, 48n, 49n, 50n, 82n
Mercury
specific heat, 34, 58n
thermal conductivity, 35
Metaphor, scientific, 102
Metaphysical Club, 157n
Metaphysics, 74, 125, 131
Meter, 41
Methane (marsh gas), 284, 455, 532
Method of least squares, 320
Meyer, Julius Lothar, 135, 137n, 191, 481, 483n, 503

Meyer, Lorenz. *See* Meyer, Julius
    Lothar
Meyer, Oskar Emil, 29, 53n, 54n, 57n,
    58n, 107, 109n, 137n, 145, 190, 251,
    355, 358n, 371, 386, 416, 423, 465,
    477, 480, 482n, 483n, 498. *See also*
    Maxwell, on Meyer
  on diffusion, 60n
  on *elastische Nachwirkung*, 251
  on forces between molecule, 30
  on heat conduction, 35
  on internal motions of molecules, 30
  on Joule-Thomson experiment, 30
  on kinetic theory, 482
  on thermal conductivity coefficients of
    gases, 18
  on viscosity, 19–20, 29–30, 53n, 55n,
    145, 155, 190, 355, 357, 358n, 361n,
    371–372, 383, 423, 480, 488, 498, 501
Michell, John, 9, 52n
Mixtures, separation of, xxv, 515, 542–
    545
Molecules, 134, 140, 187
  and atoms, relation of, xxviin, 85, 138,
    140, 211, 232, 238
  conductivity, 515
  data about, of different ranks, 134, 148,
    155
  diameter, xxi, 37–40, 107, 135, 148–
    149, 155, 191, 480
  internal energy of, 6, 32, 40–43, 187,
    230, 258, 341, 428
  mass of, 135, 148, 155
  motion of agitation, 159n
  properties, by creation, 124, 153, 210–
    214, 242
  species of, 211
  uniformity of, 100, 124, 151–154, 210–
    212, 234, 242
  velocity of, 4, 106, 141–143, 225, 278,
    281, 290
  vibrations of, 41, 194–196
  volumes of, 480–481, 502–503, 506–
    508
Monro, C. J., 242, 506n
Morris, R. J., Jr., 48n
Mott, Albert J., 130n, 131, 132
Motte, A., 265n
Muncaster, R. L., 52n

Munro, Hugh Andrew Johnstone, 45,
    68n, 83n, 108n
Mutual interference, coefficient of, 466–
    468

Nabla (∇), 476
Nablody, 476
Narr, Friedrich, 32–33, 57n
Natural philosophy, 68–70, 73–74, 126
Natural science, 268
Navier, C. L. M. H., 174
Navier-Stokes equation, 23
Nebular hypothesis, 156n–157n
Neumann, Franz, 103
Newcomb, Simon, 272, 275n
Newton, Isaac, 82n, 105, 111, 201, 203,
    218, 219–222, 257n, 262
  on atoms, 201
  gas pressure theory, xix, 4, 66, 68n,
    183–184, 221, 256, 263–264, 265–267
  law of cooling, 13
  on matter, 87
  scale of color, 167–168
Nitrogen
  heat conduction in, 472
  nonspherical particles of, 341
Nitrous oxide, 532
Niven, W. D., 55n, 280n–281n, 296n,
    314n, 348n
Nobert, F. A., 191
Normal distribution of errors, 8, 9–11

Obermayer, A. von, 57n
Observatory at Kew, 414n
Ohm's law, 509, 510n
Olefiant gas. *See* Ethylene
Olson, R. G., 49n
Osterbrock, Donald E., 61n
Oxygen
  diffusion of, 488–489, 532
  heat conduction, 472
  molecular diameter, 479, 481, 488, 490,
    500, 504
  nonspherical particles of, 341
  transpiration, 351
  viscosity, 494, 501, 549

Pacey, A. J., 49n, 50n
Pangenesis, 192

Parent, 45
Partington, J. R., 49n, 58n, 109n, 137n
Peacock, George, 59n
Pearson, Karl, 495n
Peclet, E., 313
Peterhouse, 5
Petit, A. L., 13–14, 34, 53n, 58n, 187. *See also* Dulong-Petit law
Phase space, xxiii
Philosophical Magazine, 277, 283, 320, 339, 416
Philosophical Transactions, 47
Physical science, 74, 267–268
Physical Society of London, 354n
Physics, 69
  experimental, 111–120
  reduction to mechanics, 70
Physiology, 192
Pietrowski, G. von, 361n, 368
Planck's quantum of action, 41
Plateau, J., 163–169
Plato, 319n
Playfair, Lyon, 537n
Plenum, 122, 125, 131, 361n
Poiseuille, J. L. M., 351, 361n
  on diffusion, 16
Poisson, Siméon Denis, 60n, 129, 174
  on elasticity, 24, 25, 56n, 397n, 417, 463
  on theory of sound, 3
Pompe, A., 58n
Pooh-bah, xx
Porter, Theodore M., 11, 53n
Potassium oxide, 481, 503
Pressure, spatial variation, 396
Preston, S. Tolver, 47, 243, 244n, 282n
  letter to Maxwell, 243–244
Prévost, Pierre, 45, 105, 109n, 204, 272, 275n, 416, 422
Price, Derek J. de Solla, 132n
Priestley, Joseph, 144
Prince Rupert's drops, 168
Probability theory, 9, 52n, 400. *See also* Statistics
Proceedings of the Royal Society, 358n, 361n
Proctor, Richard, 158
Prout's hypothesis, 211–212
Puluj, Johann, 30, 57n
Pythagoras, 274

Quantification, 38, 74–75, 92–93
  science reduced to, 114
Quarterly Journal of Science, 312
Quasisolidity, 25–26, 29
Quaternions, 95, 108, 239, 241n, 475–476
Queen Victoria, 508n
Quetelet, Adolphe, 9, 11
Quincke, G. H., 165

Radiation, heat, 2–3, 13–14
  relation to light, 2, 72–73, 236
  theory of exchanges, 272–274
Radiometer, 273
Raman, V. V., 49n
Randomness of atomic motion, 105, 150, 159–161, 170, 546
Random walk, 495n
Rankine, W. J. M., xix, xxvin, 44, 50n, 313, 341, 348n, 396, 460
Ratcliffe, J. A., 354n
Rayleigh (Lord), John William Strutt, 3rd Baron, 47, 63n, 196, 234, 241n, 272, 495n, 505, 506n, 544
  on viscosity, 31, 57n
Rayleigh (Lord), Robert John, 4th Baron, 62n
Refraction, double, 24, 25
  index of, for gases, 494n, 505
Regnault, H. V., 185, 222, 231, 396, 460, 489
Relaxation (of stresses), 23, 25, 26, 425, 463–465
Reversible processes, 542–545
Reynolds, Osborne, xvii, 549
Rhythmic motion, 160–161
Riemann, Georg Friedrich Bernard, 103, 157n, 169, 473–474, 475n
Rigidity of gas, 426, 464
Ritter, A., 157n
Ro, S. T., 58n
Ronge, Grete, 51n
Root mean square velocity, 281n
Roscoe, Henry E., 482n
Routh, E. J., xix, xxiii, xxviin
Rowlinson, J. S., 56n
Royal Society of Edinburgh, 239n, 353–354

Sabine, E., 13, 53n
Saint Claire-Deville, Henri Etienne, 513
Saturn's rings, xix, 12, 40, 130n, 156
Schleiermacher, A., 35, 58n
Schneider, Ivo, 51n, 506n
Schuster, Arthur, xxviin, 41, 61n, 509
Schwartze, W., 35
Sciences, definitions of and distinctions, 69–73, 267–271
Scientists, public image of, 120–121, 163
Scott, G. D., 59n
Scott, M., 86
Scott, Wilson L., 51n
Second law of thermodynamics, 32
Senate House, 44, 86, 116
Senior Wrangler, xix
Shakespeare, William, 259
Siegel, Daniel M., 275n
Silliman, Robert H., 130n
Simpson, James, 281n
Simpson, Thomas, 83n
Sirius, 41, 152
Size of molecules, xx–xxi, 4
  temperature variation, 30
Slip (of gas along solid), 356–357, 379
Smith, Crosbie, 48n
Smith, Edward, 337, 338n
Smith's Prize, xix
Smith, W. Robertson, 476n
Social sciences, 259
Socrates, 138
Solids
  atomic volumes, 506–507
  mechanics of, 23–25, 249–254, 421, 464
  molecular motion in, 147, 161, 253–254, 421
Soul, 337
Sound, propagation in air, 193, 235–236, 244, 245n, 282n
Specific gravity, 285n
Specific heat
  anomaly, xxi, 18
  of gases, 6, 41, 188, 230–232, 318, 341–342, 395–396, 459–460, 508
Spectra, 40–43, 99, 152, 193–197, 208–209, 232
  Doppler broadening of, 196–197
Spencer, Herbert, 156n, 157n
  on Laplace's nebular hypothesis, 156

letters to Maxwell, 156–158, 159n, 160–162, 162n
  on molecular motions, 160–162
Spice, 350, 350n
Spinoza, Baruch, 178
Spottiswoode, W., 90, 91
Statical theories, 420. *See also* Newton, gas pressure theory
Statics, 71, 72, 269
Statistics, 8–11, 123, 150–151, 226–227, 259, 260, 400
Steele, W. J., 5, 6
Stefan-Boltzmann law, 13–14
Stefan, Josef, 19, 30, 39, 54n, 55n, 146, 476, 477, 482n
  on diffusion, 36–37, 59n, 503
  on heat conduction, 32–33, 34, 57n, 58n, 155, 190, 476, 477, 489
Steffens, Henry John, 48n
Stephen, Leslie, 157n
Stewart, Balfour, 272, 275n, 413, 414n, 415
Stokes, George Gabriel, xvii, xix, 20, 53n, 55n, 56n, 107, 123, 130n, 241n, 275n, 277, 280n, 283, 320, 321, 398n,
  condition on coefficients of elasticity or viscosity, 24, 29, 463
  hydrodynamics, 23–24, 269, 375n, 397
  theory of elastic solids, 24, 174, 180
  on viscosity, 12, 13, 18, 19, 53n, 281, 300, 336, 342, 353, 355, 357, 361n, 463, 465, 498
Stoney, G. Johnstone, xxi, 61n, 107, 481, 482n
  on molecular diameter, xxi, 38, 46, 97, 135, 149, 191, 502
  on spectra, 41, 42
Strutt, John William. *See* Rayleigh (Lord), 3rd Baron
Strutt, Robert John. *See* Rayleigh (Lord), 4th Baron
Sulphurous acid gas, 284, 455, 496, 532
Sun
  created after light, 241, 242
  hollow sphere, 158n
  temperature of, 197
Surface tension, 165–167
Swift, Jonathan, 475n
Sylvester, James Joseph, 90, 91

T (Thomson, William), 110n
T' (Tait, Peter Guthrie), 110n, 238, 239n, 255
Tait, Peter Guthrie, xiv, xvii, xviii, xxiii, xxiv, xxviin, 6, 62n, 88, 88n, 103, 109n, 110n, 132n, 157n, 239n, 240, 257n, 275n, 352n, 413, 415, 473, 476n, 493, 494n, 496
  ether friction experiment, 414n
  letter to Maxwell, 414–415, 496
  on Maxwell, 1
  on thermal conductivity of metals, 494n
Talbot, G. R., 49n, 50n
Taylor, Richard, 475n
Taylor's theorem, 253
Temperature
  equilibrium between two gases, 392–394, 457, 515
  and molecular motion, 5, 80, 106, 235
  in vertical column of gas, 233, 419, 469–470, 516n
Temple of Karnac, 41, 152
Tennyson, Alfred Lord (1st Baron), 150
Tenth-meter, 483n
Terrestrial magnetism, 116
Thackray, Arnold, 48n
Thales, 125
The Athenaeum, 44
Thermal conductivity. *See* Heat conduction
Thermodynamics, 2–3, 122, 270. *See also* Dissipation of energy
  second law, 110n, 159n, 470
Thermometer, wet bulb, 536–537
Thiele, Joachim, 61n
Third Wrangler, 336n
Thomson, James, xvii
Thomson, J. J., xxv, 110n
Thomson, William (Lord Kelvin), xiv, xvii, xxi, xxiii, xxiv, xxv, xxviin, 3, 5, 26, 44, 46, 48n, 55n, 56n, 60n, 62n, 85n, 88n, 129, 130n, 185, 240, 269, 275n, 337, 349, 375, 413, 414n, 415
  on densities of strata in fluid, 526
  on diffusion and transpiration, 549, 550
  on dissipation of energy, 88–89, 101
  on heat, 3, 48n
  on heat conduction, 88
  on kinetic theory of gravity, 128, 171, 207
  on Maxwell's experiments on viscosity, 20–21
  on the mechanical theory of matter, 42, 88
  on molecular diameters, xxi, 38, 60n, 97, 109n, 135, 149, 191, 481, 483, 502
  on viscosity and elasticity, 20, 24, 251, 355, 358n, 361, 365
  on vortex atom, 98, 125, 181, 238
  on vortex theory, 86, 98, 129, 201–203
Time, divisibility of, 177
Tisza, L., 29, 56n
Todhunter, Isaac, 56n
Tolstoy, Ivan, xxvin
Torricelli, E., 87
Transaction of the Connecticut Academy of Sciences, 525
Transpiration, 16–17, 351, 356, 361, 465, 467–468, 549–550
Transport phenomena of gases, xx, xxii, 7, 12, 18–19, 28–29, 109n, 189–190, 445–451, 535–536. *See also* Diffusion; Viscosity; Heat conduction
Trinity College, Cambridge, 336n
Tripos day, 83
Troost, L., 513
Truesdell, Clifford, 5, 48n, 49n, 51n, 52n, 56n, 57n, 81n, 398n
  translation from Newton, 265n
Turner, Frank, 68n
Tyndall, John, 11, 39, 60n, 90–91, 150, 156, 157n, 160n, 354, 506n

Universal gravitation, 218
University of Cambridge, 111
University of Edinburgh, xviii

Vacuum, 65–66
van der Waals, J. D., 40, 165, 191, 225, 510n
van der Waals, Jr., J. D., 3, 49n
Vectors, 95, 108, 476
Velocity distribution law, xx, xxi, 7–8, 9–12, 15, 21, 106, 186, 226–228, 279, 283–284, 289–291, 339, 387–391, 437–441
Velocity (instead of speed), 50n

Virial theorem, 182–184, 220–222, 496n
Visco-elasticity, 23–26, 251, 354–355, 360–361
Viscosity, xx, xxi, 12–14, 19–20, 30–31, 34, 55n, 133, 145, 239–240, 248–249, 277–279, 299–300, 342, 349–350, 352–353, 354–386, 418, 464–465, 479, 492–494, 501, 549
Vis viva, 5, 6, 107, 287, 341, 427
Voit, W., 147, 155, 526, 533
Vortex atom, xix, 41–42, 44, 86, 98–99, 201–203

Wakeham, W., 58n
Warburg, Emil, 30, 33–34, 57n, 58n
Waterston, John James, 4, 47, 50n, 62n, on equipartition theorem, xx
on propagation of sound, 244, 282n
Waterston's kinetic theory, 243–244
Watson, Henry William, xvii, xxiii, 261n, 262n, 515, 516n
Wave theory of heat, 2, 3
Weber, Wilhelm, 103, 115, 116, 251, 358n, 425, 473–474, 475n
Weiner, Charles, 48n
Weinstein, Fred, 53n
Westfall, Richard S., 109n
Wheatstone, Charles, 354n
Wheeler, J. A., 475n
Wheeler, L. P., 510n
Whewell, William, xix
Wiedemann, Gustav Heinrich, 97, 252n, 494n
Wilberforce, Bishop Samuel, 51n
Williamson-Clausius hypothesis, 137n
Wilson, David B., 56n, 130n
Winkelmann, A., 34, 58n
Wolf, A., 49n
Wolstenholme, Joseph, 414n
Wren, Christopher, 80
Wright, P. G., 49n
Wroblewski, Sigmund von, 510–514

X-Club, 157n

Youmans, Edward Livingstone, 157n
Young, Charles Augustus, 158
Young, Thomas, 2, 59n
on molecular size, 38, 59n

Zaitseva, L. S., 35, 58n
Zeno of Elea, 177
Zinc carbonate, 507